Lecture Notes in Computer Science   6853

Commenced Publication in 1973
Founding and Former Series Editors:
Gerhard Goos, Juris Hartmanis, and Jan van Leeuwen

T0217693

Emmanuel Jeannot   Raymond Namyst
Jean Roman (Eds.)

Euro-Par 2011
Parallel Processing

17th International Conference, Euro-Par 2011
Bordeaux, France, August 29–September 2, 2011
Proceedings, Part II

 Springer

Volume Editors

Emmanuel Jeannot
INRIA
351, Cours de la Libération
33405 Talence Cedex, France
E-mail: emmanuel.jeannot@inria.fr

Raymond Namyst
Université de Bordeaux, INRIA
351, Cours de la Libération
33405 Talence Cedex, France
E-mail: raymond.namyst@labri.fr

Jean Roman
Université de Bordeaux, INRIA
351, Cours de la Libération
33405 Talence Cedex, France
E-mail: jean.roman@inria.fr

ISSN 0302-9743   e-ISSN 1611-3349
ISBN 978-3-642-23396-8   e-ISBN 978-3-642-23397-5
DOI 10.1007/978-3-642-23397-5

Springer Heidelberg Dordrecht London New York

Library of Congress Control Number: 2011934379

CR Subject Classification (1998): F.1.2, C.3, C.2.4, D.1, D.4, I.6, G.1, G.2, F.2, D.3

LNCS Sublibrary: SL 1 – Theoretical Computer Science and General Issues

Typesetting: Camera-ready by author, data conversion by Scientific Publishing Services, Chennai, India

Printed on acid-free paper

Springer is part of Springer Science+Business Media (www.springer.com)

Preface

Euro-Par is an annual series of international conferences dedicated to the promotion and advancement of all aspects of parallel and distributed computing.

Euro-Par covers a wide spectrum of topics from algorithms and theory to software technology and hardware-related issues, with application areas ranging from scientific to mobile and cloud computing.

Euro-Par provides a forum for the introduction, presentation and discussion of the latest scientific and technical advances, extending the frontier of both the state of the art and the state of the practice.

The main audience of Euro-Par are researchers in academic institutions, government laboratories and industrial organizations. Euro-Par's objective is to be the primary choice of such professionals for the presentation of new results in their specific areas. As a wide-spectrum conference, Euro-Par fosters the synergy of different topics in parallel and distributed computing. Of special interest are applications which demonstrate the effectiveness of the main Euro-Par topics.

In addition, Euro-Par conferences provide a platform for a number of accompanying, technical workshops. Thus, smaller and emerging communities can meet and develop more focussed topics or as-yet less established topics.

Euro-Par 2011 was the 17th conference in the Euro-Par series, and was organized by the INRIA (The French National Institute for Research in Computer Science and Control) Bordeaux Sud-Ouest center and LaBRI (Computer Science Laboratory of Bordeaux). Previous Euro-Par conferences took place in Stockholm, Lyon, Passau, Southampton, Toulouse, Munich, Manchester, Padderborn, Klagenfurt, Pisa, Lisbon, Dresden, Rennes, Las Palmas, Delft and Ischia. Next year the conference will take place in Rhodes, Greece. More information on the Euro-Par conference series and organization is available on the wesite http://www.europar.org.

The conference was organized in 16 topics. This year we introduced one new topic (16: GPU and Accelerators Computing) and re-introduced the application topic (15: High-Performance and Scientific Applications). The paper review process for each topic was managed and supervised by a committee of at least four persons: a Global Chair, a Local Chair, and two Members. Some specific topics with a high number of submissions were managed by a larger committee with more members. The final decisions on the acceptance or rejection of the submitted papers were made in a meeting of the Conference Co-chairs and Local Chairs of the topics.

The call for papers attracted a total of 271 submissions, representing 41 countries (based on the corresponding authors' countries). A total number of 1,065 review reports were collected, which makes an average of 3.93 review reports per paper. In total 81 papers were selected as regular papers to be presented at the conference and included in the conference proceedings, representing 27 countries from all continents, an yielding an acceptance rate of 29.9%. Three papers were selected as distinguished papers. These papers, which were presented in a separate session, are:

1. Lakshminarasimhan, Neil Shah, Stephane Ethier, Scott Klasky, Rob Latham, Rob Ross and Nagiza F. Samatova "Compressing the Incompressible with ISABELA: In-situ Reduction of Spatio-Temporal Data"
2. Aurelien Bouteiller, Thomas Herault, George Bosilca and Jack J. Dongarra "Correlated Set Coordination in Fault-Tolerant Message Logging Protocols"
3. Edgar Solomonik and James Demmel "Communication-Optimal Parallel 2.5D Matrix Multiplication and LU Factorization Algorithms".

Euro-Par 2011 was very happy to present three invited speakers of high international reputation, who discussed important developments in very interesting areas of parallel and distributed computing:

1. Pete Beckman (Argonne National Laboratory and the University of Chicago), "Facts and Speculations on Exascale: Revolution or Evolution?"
2. Toni Cortes Computer Architecture Department (DAC) in the Universitat Politècnica de Catalunya, Spain), "Why Trouble Humans? They Do Not Care"
3. Alessandro Curioni (IBM, Zurich Research Laboratory, Switzerland), "New Scalability Frontiers in Ab-Initio Molecular Dynamics"

In this edition, 12 workshops were held in conjunction with the main track of the conference. These workshops were:

1. CoreGRID/ERCIM Workshop on Grids, Clouds and P2P Computing (CGWS 2011)
2. Algorithms, Models and Tools for Parallel Computing on Heterogeneous Platforms (HeteroPar 2011)
3. High-Performance Bioinformatics and Biomedicine (HiBB)
4. System-Level Virtualization for High Performance Computing (HPCVirt 2011)
5. Algorithms and Programming Tools for Next-Generation High-Performance Scientific Software (HPSS 2011)
6. Managing and Delivering Grid Services (MDGS)
7. UnConventional High-Performance Computing 2011 (UCHPC 2011)
8. Cloud Computing Projects and Initiatives (CCPI)
9. Highly Parallel Processing on a Chip (HPPC 2011)
10. Productivity and Performance (PROPER 2011)
11. Resiliency in High-Performance Computing (Resilience) in Clusters, Clouds, and Grids
12. Virtualization in High-Performance Cloud Computing (VHPC 2011)

The 17th Euro-Par conference in Bordeaux was made possible thanks to the support of many individuals and organizations. Special thanks are due to the authors of all the submitted papers, the members of the topic committees, and all the reviewers in all topics, for their contributions to the success of the conference. We also thank the members of the Organizing Committee and people of the Sud Congrès Conseil. We are grateful to the members of the Euro-Par Steering Committee for their support. We acknowledge the help we had from Dick Epema of the organization of Euro-Par 2009 and Pasqua D'Ambra and Domenico Talia of the organization of Euro-Par 2010. A number of institutional and industrial sponsors contributed toward the organization of the conference. Their names and logos appear on the Euro-Par 2011 website http://europar2011.bordeaux.inria.fr/

It was our pleasure and honor to organize and host Euro-Par 2011 in Bordeaux. We hope all the participants enjoyed the technical program and the social events organized during the conference.

August 2011   Emmanuel Jeannot
                                                              Raymond Namyst
                                                                  Jean Roman

Organization

Euro-Par Steering Committee

Chair

Chris Lengauer	University of Passau, Germany

Vice-Chair

Luc Bougé	ENS Cachan, France

European Respresentatives

José Cunha	New University of Lisbon, Portugal
Marco Danelutto	University of Pisa, Italy
Emmanuel Jeannot	INRIA, France
Paul Kelly	Imperial College, UK
Harald Kosch	University of Passau, Germany
Thomas Ludwig	University of Heidelberg, Germany
Emilio Luque	Autonomous University of Barcelona, Spain
Tomàs Margalef	Autonomous University of Barcelona, Spain
Wolfgang Nagel	Dresden University of Technology, Germany
Rizos Sakellariou	University of Manchester, UK
Henk Sips	Delft University of Technology, The Netherlands
Domenico Talia	University of Calabria, Italy

Honorary Members

Ron Perrott	Queen's University Belfast, UK
Karl Dieter Reinartz	University of Erlangen-Nuremberg, Germany

Observers

Christos Kaklamanis	Computer Technology Institute, Greece

Euro-Par 2011 Organization

Conference Co-chairs

Emmanuel Jeannot   INRIA, France
Raymond Namyst   University of Bordeaux, France
Jean Roman   INRIA, University of Bordeaux, France

Local Organizing Committee

Olivier Aumage   INRIA, France
Emmanuel Agullo   INRIA, France
Alexandre Denis   INRIA, France
Nathalie Furmento   CNRS, France
Laetitia Grimaldi   INRIA, France
Nicole Lun   LaBRI, France
Guillaume Mercier   University of Bordeaux, France
Elia Meyre   LaBRI France

Euro-Par 2011 Program Committee

Topic 1: Support Tools and Environments
Global Chair
Rosa M. Badia   Barcelona Supercomputing Center and CSIC,
     Spain

Local Chair

Fabrice Huet   University of Nice Sophia Antipolis, France

Members
Rob van Nieuwpoort   VU University Amsterdam, The Netherlands
Rainer Keller   High-Performance Computing Center
     Stuttgart, Germany

Topic 2: Performance Prediction and Evaluation
Global Chair
Shirley Moore   University of Tennessee, USA

Local Chair

Derrick Kondo   INRIA, France

Members

Giuliano Casale	Imperial College London, UK
Brian Wylie	Jülich Supercomputing Centre, Germany

Topic 3: Scheduling and Load-Balancing

Global Chair

Leonel Sousa	INESC-ID/Technical University of Lisbon, Portugal

Local Chair

Frédéric Suter	IN2P3 Computing Center, CNRS, France

Members

Rizos Sakellariou	University of Manchester, UK
Oliver Sinnen	University of Auckland, New Zealand
Alfredo Goldman	University of São Paulo, Brazil

Topic 4: High Performance Architectures and Compilers

Global Chair

Mitsuhisa Sato	University of Tsukuba, Japan

Local Chair

Denis Barthou	University of Bordeaux, France

Members

Pedro Diniz	INESC-ID, Portugal
P. Saddayapan	Ohio State University, USA

Topic 5: Parallel and Distributed Data Management

Global Chair

Salvatore Orlando	Università Ca' Foscari Venezia, Italy

Local Chair

Gabriel Antoniu	INRIA, France

Members

Amol Ghoting	IBM T. J. Watson Research Center, USA
Maria S. Perez	Universidad Politecnica de Madrid, Spain

Topic 6: Grid, Cluster and Cloud Computing

Global Chair

Ramin Yahyapour   TU Dortmund University, Germany

Local Chair

Christian Pérez   INRIA, France

Members

Erik Elmroth   Umeå University, Sweden
Ignacio M. Llorente   Complutense University of Madrid, Spain
Francesc Guim   Intel, Portland, USA
Karsten Oberle   Alcatel-Lucent, Bell Labs, Germany

Topic 7: Peer to Peer Computing

Global Chair

Pascal Felber   University of Neuchâtel, Switzerland

Local Chair

Olivier Beaumont   INRIA, France

Members

Alberto Montresor   University of Trento, Italy
Amitabha Bagchi   Indian Institute of Technology Delhi, India

Topic 8: Distributed Systems and Algorithms

Global Chair

Dariusz Kowalski   University of Liverpool, UK

Local Chair

Pierre Sens   University Paris 6, France

Members

Antonio Fernandez Anta   IMDEA Networks, Spain
Guillaume Pierre   VU University Amsterdam, The Netherlands

Topic 9: Parallel and Distributed Programming

Global Chair

Pierre Manneback   University of Mons, Belgium

Local Chair

Thierry Gautier   INRIA, France

Members

Gudula Rünger   Technical University of Chemnitz, Germany
Manuel Prieto Matias   Universidad Complutense de Madrid, Spain

Topic 10: Parallel Numerical Algorithms

Global Chair

Daniela di Serafino   Second University of Naples and ICAR-CNR,
                                   Italy

Local Chair

Luc Giraud   INRIA, France

Members

Martin Berzins   University of Utah, USA
Martin Gander   University of Geneva, Switzerland

Topic 11: Multicore and Manycore Programming

Global Chair

Sabri Pllana   University of Vienna, Austria

Local Chair

Jean-François Méhaut   University of Grenoble, France

Members

Eduard Ayguade   Technical University of Catalunya and
                                   Barcelona Supercomputing Center, Spain
Herbert Cornelius   Intel, Germany
Jacob Barhen   Oak Ridge National Laboratory, USA

Topic 12: Theory and Algorithms for Parallel Computation

Global Chair

Arnold Rosenberg   Colorado State University, USA

Local Chair

Frédéric Vivien   INRIA, France

Members

Kunal Agrawal   Washington University in St Louis, USA
Panagiota Fatourou   University of Crete, Greece

Topic 13: High Performance Network and Communication

Global Chair

Jesper Träff   University of Vienna, Austria

Local Chair

Brice Goglin   INRIA, France

Members

Ulrich Bruening   University of Heidelberg, Germany
Fabrizio Petrini   IBM, USA

Topic 14: Mobile and Ubiquitous Computing

Global Chair

Pedro Marron   Universität Duisburg-Essen, Germany

Local Chair

Eric Fleury   INRIA, France

Members

Torben Weis   University of Duisburg-Essen, Germany
Qi Han   Colorado School of Mines, USA

Topic 15: High Performance and Scientific Applications

Global Chair

Esmond G. Ng   Lawrence Berkeley National Laboratory, USA

Local Chair

Olivier Coulaud   INRIA, France

Members

Kengo Nakajima   University of Tokyo, Japan
Mariano Vazquez   Barcelona Supercomputing Center, Spain

Topic 16: GPU and Accelerators Computing
Global Chair

Wolfgang Karl   University of Karlsruhe, Germany

Local Chair

Samuel Thibault   University of Bordeaux, France

Members

Stan Tomov   University of Tennessee, USA
Taisuke Boku   University of Tsukuba, Japan

Euro-Par 2011 Referees

Muresan Adrian	Jacob Barhen
Kunal Agrawal	Denis Barthou
Emmanuel Agullo	Rajeev Barua
Toufik Ahmed	Francoise Baude
Taner Akgun	Markus Bauer
Hasan Metin Aktulga	Ewnetu Bayuh Lakew
Sadaf Alam	Olivier Beaumont
George Almasi	Vicenç Beltran
Francisco Almeida	Joanna Berlinska
Jose Alvarez Bermejo	Martin Berzins
Brian Amedro	Xavier Besseron
Nazareno Andrade	Vartika Bhandari
Artur Andrzejak	Marina Biberstein
Luciana Arantes	Paolo Bientinesi
Mario Arioli	Aart Bik
Ernest Artiaga	David Boehme
Rafael Asenjo	Maria Cristina Boeres
Romain Aubry	Taisuke Boku
Cédric Augonnet	Matthias Bollhoefer
Olivier Aumage	Erik Boman
Eduard Ayguade	Michael Bond
Rosa M. Badia	Francesco Bongiovanni
Amitabha Bagchi	Rajesh Bordawekar
Michel Bagein	George Bosilca
Enes Bajrovic	François-Xavier Bouchez
Allison Baker	Marin Bougeret
Pavan Balaji	Aurelien Bouteiller
Sorav Bansal	Hinde Bouziane
Jorge Barbosa	Fabienne Boyer

Ivona Brandic
Francisco Brasileiro
David Breitgand
Andre Brinkman
François Broquedis
Ulrich Bruening
Rainer Buchty
j. Mark Bull
Aydin Buluc
Alfredo Buttari
Edson Caceres
Agustin Caminero
Yves Caniou
Louis-Claude Canon
Gabriele Capannini
Pablo Carazo
Alexandre Carissimi
Giuliano Casale
Henri Casanova
Simon Caton
José María Cela
Christophe Cerin
Ravikesh Chandra
Andres Charif-Rubial
Fabio Checconi
Yawen Chen
Gregory Chockler
Vincent Cholvi
Peter Chronz
IHsin Chung
Marcelo Cintra
Vladimir Ciric
Pierre-Nicolas Clauss
Sylvain Collange
Denis Conan
Arlindo Conceicao
Massimo Coppola
Julita Corbalan
Herbert Cornelius
Toni Cortes
Olivier Coulaud
Ludovic Courtès
Raphael Couturier
Tommaso Cucinotta
Angel Cuevas

Pasqua D'Ambra
Anthony Danalis
Vincent Danjean
Eric Darve
Sudipto Das
Ajoy Datta
Patrizio Dazzi
Pablo de Oliveira Castro
César De Rose
Ewa Deelman
Olivier Delgrange
Alexandre Denis
Yves Denneulin
Frederic Desprez
Gérard Dethier
Daniela di Serafino
François Diakhaté
James Dinan
Nicholas Dingle
Pedro Diniz
Alastair Donaldson
Antonio Dopico
Matthieu Dorier
Niels Drost
Maciej Drozdowski
Lúcia Drummond
Peng Du
Cedric du Mouza
Vitor Duarte
Philippe Duchon
Jörg Dümmler
Alejandro Duran
Pierre-François Dutot
Partha Dutta
Eiman Ebrahimi
Rudolf Eigenmann
Jorge Ejarque Artigas
Vincent Englebert
Dominic Eschweiler
Yoav Etsion
Lionel Eyraud-Dubois
Flavio Fabbri
Fabrizio Falchi
Catherine Faron Zucker
Montse Farreras

Panagiota Fatourou
Hugues Fauconnier
Mathieu Faverge
Gilles Fedak
Dror G. Feitelson
Pascal Felber
Florian Feldhaus
Marvin Ferber
Juan Fernández
Antonio Fernández Anta
Ilario Filippini
Salvatore Filippone
Eric Fleury
Aislan Foina
Pierre Fortin
Markos Fountoulakis
Rob Fowler
Vivi Fragopoulou
Felipe França
Emílio Francesquini
Sébastien Frémal
Davide Frey
Wolfgang Frings
Karl Fuerlinger
Akihiro Fujii
Nathalie Furmento
Edgar Gabriel
Martin Gaedke
Georgina Gallizo
Efstratios Gallopoulos
Ixent Galpin
Marta Garcia
Thierry Gautier
Stéphane Genaud
Chryssis Georgiou
Abdou Germouche
Michael Gerndt
Claudio Geyer
Amol Ghoting
Nasser Giacaman
Mathieu Giraud
Daniel Gmach
Brice Goglin
Spyridon Gogouvitis
Alfredo Goldman

Maria Gomes
Jose Gómez
Jose Gonzalez
José Luis González García
Rafael Gonzalez-Cabero
David Goudin
Madhusudhan Govindaraju
Maria Gradinariu
Vincent Gramoli
Fabíola Greve
Laura Grigori
Olivier Gruber
Serge Guelton
Gael Guennebaud
Stefan Guettel
Francesc Guim
Ronan Guivarch
Jens Gustedt
Antonio Guzman Sacristan
Daniel Hackenberg
Azzam Haidar
Mary Hall
Greg Hamerly
Qi Han
Toshihiro Hanawa
Mauricio Hanzich
Paul Hargrove
Masae Hayashi
Jiahua He
Eric Heien
Daniel Henriksson
Ludovic Henrio
Sylvain Henry
Francisco Hernandez
Enric Herrero
Pieter Hijma
Shoichi Hirasawa
Torsten Hoefler
Jeffrey Hollingsworth
Sebastian Holzapfel
Mitch Horton
Guillaume Houzeaux
Jonathan Hu
Ye Huang
Guillaume Huard

Fabrice Huet
Kévin Huguenin
Sascha Hunold
Costin Iancu
Aleksandar Ilic
Alexandru Iosup
Umer Iqbal
Kamil Iskra
Takeshi Iwashita
Julien Jaeger
Emmanuel Jeannot
Ali Jehangiri
Hideyuki Jitsumoto
Josep Jorba
Prabhanjan Kambadur
Yoshikazu Kamoshida
Mahmut Kandemir
Tejas Karkhanis
Wolfgang Karl
Takahiro Katagiri
Gregory Katsaros
Joerg Keller
Rainer Keller
Paul Kelly
Roelof Kemp
Michel Kern
Ronan Keryell
Christoph Kessler
Slava Kitaeff
Cristian Klein
Yannis Klonatos
Michael Knobloch
William Knottenbelt
Kleopatra Konstanteli
Miroslaw Korzeniowski
Dariusz Kowalski
Stephan Kraft
Sriram Krishnamoorthy
Diwakar Krishnamurthy
Rajasekar Krishnamurthy
Vinod Kulathumani
Raphael Kunis
Tilman Küstner
Felix Kwok
Dimosthenis Kyriazis

Renaud Lachaize
Ghislain Landry Tsafack
Julien Langou
Stefan Lankes
Lars Larsson
Alexey Lastovetsky
Guillaume Latu
Stevens Le Blond
Bertrand Le Cun
Erwan Le Merrer
Adrien Lèbre
Rich Lee
Erik Lefebvre
Arnaud Legrand
Christian Lengauer
Daniele Lezzi
Wubin Li
Charles Lively
Welf Loewe
Sebastien Loisel
João Lourenço
Kuan Lu
Jose Luis Lucas Simarro
Mikel Lujan
Ewing Lusk
Piotr Luszczek
Ignacio M. Llorente
Jason Maassen
Edmundo Madeira
Anirban Mahanti
Scott Mahlke
Sidi Mahmoudi
Nicolas Maillard
Constantinos Makassikis
Pierre Manneback
Loris Marchal
Ismael Marín
Mauricio Marin
Osni Marques
Erich Marth
Jonathan Martí
Xavier Martorell
Naoya Maruyama
Fabien Mathieu
Rafael Mayo

Abdelhafid Mazouz
Jean-François Méhaut
Wagner Meira
Alba Cristina Melo
Massimiliano Meneghin
Claudio Meneses
Andreas Menychtas
Jose Miguel-Alonso
Milan Mihajlovic
Alessia Milani
Cyriel Minkenberg
Neeraj Mittal
Flávio Miyazawa
Hashim Mohamed
Sébastien Monnet
Jesus Montes
Alberto Montresor
Shirley Moore
Matteo Mordacchini
Jose Moreira
Achour Mostefaoui
Miguel Mosteiro
Gregory Mounié
Xenia Mountrouidou
Hubert Naacke
Priya Nagpurkar
Kengo Nakajima
Jeff Napper
Akira Naruse
Bassem Nasser
Rajib Nath
Angeles Navarro
Philippe O. A. Navaux
Marco Netto
Marcelo Neves
Esmond Ng
Yanik Ngoko
Jean-Marc Nicod
Bogdan Nicolae
Dimitrios Nikolopoulos
Sébastien Noël
Ramon Nou
Alberto Nuñez
John O'Donnell
Satoshi Ohshima

Ariel Oleksiak
Stephen Olivier
Ana-Maria Oprescu
Anne-Cecile Orgerie
Salvatore Orlando
Per-Olov Ostberg
Herbert Owen
Sergio Pacheco Sanchez
Gianluca Palermo
George Pallis
Nicholas Palmer
Jairo Panetta
Alexander Papaspyrou
Michael Parkin
Davide Pasetto
George Pau
Christian Perez
Maria Perez-Hernandez
Francesca Perla
Jean-Jacques Pesqué
Franck Petit
Fabrizio Petrini
Frédéric Pétrot
Guillaume Pierre
Jean-Francois Pineau
Luis Piñuel
Jelena Pjesivac-Grbovic
Kassian Plankensteiner
Oscar Plata
Sabri Pllana
Leo Porter
Carlos Prada-Rojas
Manuel Prieto Matias
Radu Prodan
Christophe Prud'homme
Vivien Quema
Enrique Quintana-Ortí
Rajmohan Rajaraman
Lavanya Ramakrishnan
Pierre Ramet
Praveen Rao
Vinod Rebello
Pablo Reble
Sasank Reddy
Veronika Rehn-Sonigo

Table of Contents – Part II

Topic 11: Multicore and Manycore Programming

Topic 12: Theory and Algorithms for Parallel Computation

Topic 13: High Performance Networks and Communication

Topic 14: Mobile and Ubiquitous Computing

Topic 15: High-Performance and Scientific Applications

Topic 16: GPU and Accelerators Computing

Table of Contents – Part I

Topic 3: Scheduling and Load Balancing

Topic 4: High-Performance Architecture and Compilers

Topic 5: Parallel and Distributed Data Management

Topic 6: Grid Cluster and Cloud Computing

Topic 7: Peer to Peer Computing

Topic 8: Distributed Systems and Algorithms

Introduction

Pierre Manneback, Thierry Gautier, Gudula Rnger, and Manuel Prieto Matias

Topic chairs

Developing parallel or distributed applications is a hard task and it requires advanced algorithms, realistic modeling, efficient design tools, high-level programming abstractions, high-performance implementations, and experimental evaluation. Ongoing research in this field emphasizes the design and development of correct, high-performance, portable, and scalable parallel programs. Related to these central needs, important work addresses methods for reusability, performance prediction, large-scale deployment, self-adaptivity, and fault-tolerance. Given the rich history in this field, practical applicability of proposed methods, models, algorithms, or techniques is a key requirement for timely research. This topic is focusing on parallel and distributed programming in general, except for work specifically targeting multicore and manycore architectures, which has matured to becoming a Euro-Par topic of its own.

Each submission was reviewed by at least four reviewers and, finally, we were able to select five regular papers, spanning the topics scope, ranging from low-level issues like failure detectors, all the way up to parallelization of a parser.

In particular, Greve et al. in A Failure Detector for Wireless Networks with Unknown Membership propose a protocol for a new class of detector which tolerates mobility and message losses. In Correlated Set Coordination in Fault Tolerant Message Logging Protocols, Bouteiller et al. describe a hierarchical partitioning of a set of processes that takes benefit of a coordinated protocol on each many-core nodes, as well as a message logging protocol for scalability between nodes. Liu et al. contributed "Towards Systematic Parallel Programming over MapReduce", a framework based on list homomorphisms to derive MapReduce programs from sequential specification. In "HOMPI: A Hybrid Programming Framework for Expressing and Deploying Task-Based Parallelism", Dimakopoulos et al. present a framework to exploit cluster of multicores on task-based parallel programs. Last but not least, Cameron et al. present a original parallelization of the XML parser in their paper "Parallel Scanning with Bitstream Addition: An XML Case Study".

We are proud of the scientific program that we managed to assemble. Of course, this was only possible by combining the efforts of many. We would like to take the opportunity to thank the authors who submitted their contributions, and the external referees who have made the scientific selection process possible in the first place.

E. Jeannot, R. Namyst, and J. Roman (Eds.): Euro-Par 2011, LNCS 6853, Part II, p. 1, 2011.

Parallel Scanning with Bitstream Addition: An XML Case Study

Robert D. Cameron, Ehsan Amiri, Kenneth S. Herdy, Dan Lin,
Thomas C. Shermer, and Fred P. Popowich

Simon Fraser University, Surrey, BC, Canada
{cameron,eamiri,ksherdy,lindanl,shermer,popowich}@cs.sfu.ca

Abstract. A parallel scanning method using the concept of bitstream addition is introduced and studied in application to the problem of XML parsing and well-formedness checking. On processors supporting W-bit addition operations, the method can perform up to W finite state transitions per instruction. The method is based on the concept of parallel bitstream technology, in which parallel streams of bits are formed such that each stream comprises bits in one-to-one correspondence with the character code units of a source data stream. Parsing routines are initially prototyped in Python using its native support for unbounded integers to represent arbitrary-length bitstreams. A compiler then translates the Python code into low-level C-based implementations. These low-level implementations take advantage of the SIMD (single-instruction multiple-data) capabilities of commodity processors to yield a dramatic speed-up over traditional alternatives employing byte-at-a-time parsing.

Keywords: SIMD text processing, parallel bitstreams, XML, parsing.

1   Introduction

Although the finite state machine methods used in the scanning and parsing of text streams is considered to be the hardest of the "13 dwarves" to parallelize [1], parallel bitstream technology shows considerable promise for these types of applications [3,4]. In this approach, character streams are processed N positions at a time using the N-bit SIMD registers commonly found on commodity processors (e.g., 128-bit XMM registers on Intel/AMD chips). This is achieved by first slicing the byte streams into eight separate basis bitstreams, one for each bit position within the byte. These basis bitstreams are then combined with bitwise logic and shifting operations to compute further parallel bit streams of interest, such as the [<] bit stream marking the position of all opening angle brackets in an XML document.

Using these techniques as well as the *bit scan* instructions also available on commodity processors, the Parabix 1 XML parser was shown to considerably accelerate XML parsing in comparison with conventional byte-at-a-time parsers in applications such as statistics gathering [4] and as GML to SVG conversion [6]. Other efforts to accelerate XML parsing include the use of custom XML

E. Jeannot, R. Namyst, and J. Roman (Eds.): Euro-Par 2011, LNCS 6853, Part II, pp. 2–13, 2011.

chips [8], FPGAs [5], careful coding and schema-based processing[7] and multi-thread/multicore speedups based on data parallelism[9,10].

In this paper, we further increase the parallelism in our methods by introducing a new parallel scanning primitive using bitstream addition. In essence, this primitive replaces the sequential bit scan operations underlying Parabix 1 with a new approach that independently advances multiple marker bits in parallel using simple addition and logic operations. This paper documents the technique and evaluates it in application to the problem of XML parsing and well-formedness checking.

Section 2 reviews the basics of parallel bitstream technology and introduces our new parallel scanning primitive. Section 3 goes on to show how this primitive may be used in XML scanning and parsing, while Section 4 discusses the construction of a complete XML well-formedness checker based on these techniques. Section 5 then briefly describes the compiler technology used to generate the low level code for our approach. A performance study in Section 6 shows that the new Parabix 2 parser is dramatically faster than traditional byte-at-a-time parsers as well as the original Parabix 1 parser, particularly for dense XML markup. Section 7 concludes the paper.

2 The Parallel Bitstream Method

2.1 Fundamentals

A bitstream is simply a sequence of 0s and 1s, where there is one such bit in the bitstream for each character in a source data stream. For parsing, and other text processing tasks, we need to consider multiple properties of characters at different stages during the parsing process. A bitstream can be associated with each of these properties, and hence there will be multiple (parallel) bitstreams associated with a source data stream of characters.

The starting point for bitstream methods are *basis* bitstreams and their use in determining *character-class* bitstreams. The kth basis bitstream B_k consists of the kth bit (0-based, starting at the the least significant bit) of each character

```
source data ◁ ----173942---654----1----49731----321--
       B₇         ..................................
       B₆         ..................................
       B₅         111111111111111111111111111111111111111111
       B₄         ....111111...111....1....11111....111..
       B₃         1111...1..111...1111.1111.1...1111...11
       B₂         1111.1..1.1111111111.11111.1..1111...11
       B₁         .....11..1...1.............11.....11...
       B₀         11111111..111.1.111111111.111111111.111
      [0-9]       ....111111...111....1....11111....111..
```

Fig. 1. Basis and Character-Class Bitstreams

in the source data stream; thus each B_k is dependent on the encoding of the source characters (ASCII, UTF-8, UTF-16, etc.). Given these basis bitstreams, it is then possible to combine them using bitwise logic in order to compute character-class bitstreams, that is, streams that identify the positions at which characters belonging to a particular class occur. For example, the character class bitstream $D = $ [0-9] marks with 1s the positions at which decimal digits occur. These bitstreams are illustrated in Figure 1, for an example source data stream consisting of digits and hyphens. This figure also illustrates some of our conventions for figures: the left triangle $\lhd$ after "source data" indicates that all streams are read from right to left (i.e., they are in little-endian notation). We also use hyphens in the input stream represent any character that is not relevant to a character class under consideration, so that relevant characters stand out. Furthermore, the 0 bits in the bitstreams are represented by periods, so that the 1 bits stand out.

Transposition of source data to basis bitstreams and calculation of character-class streams in this way is an overhead on parallel bit stream applications, in general. However, using the SIMD capabilities of current commodity processors, these operations are fast, with an amortized overhead of about 1 CPU cycle per byte for transposition and less than 1 CPU cycle per byte for all the character classes needed for XML parsing [4].

Beyond the bitwise logic needed for character class determination, we also need *upshifting* to deal with sequential combination. The upshift $n(S)$ of a bit-stream S is obtained by shifting the bits in S one position forward, then placing a 0 bit in the starting position of the bitstream; n is meant to be mnemonic of "next". In $n(S)$, the last bit of S may be eliminated or retained for error-testing purposes.

2.2   A Parallel Scanning Primitive

In this section, we introduce the principal new feature of the paper, a parallel scanning method based on bitstream addition. Key to this method is the concept of *marker* bitstreams. Marker bitstreams are used to represent positions of interest in the scanning or parsing of a source data stream. The appearance of a 1 at a position in a marker bitstream could, for example, denote the starting position of an XML tag in the data stream. In general, the set of bit positions in a marker bitstream may be considered to be the current parsing positions of multiple parses taking place in parallel throughout the source data stream.

Figure 2 illustrates the basic concept underlying parallel parsing with bit-stream addition. All streams are shown in little-endian representation, with streams reading from right-to-left. The first row shows a source data stream that includes several spans of digits, together with other nondigit characters shown as hyphens. The second row specifies the parsing problem using a marker bitstream M_0 to mark four initial marker positions. In three instances, these markers are at the beginning (i.e., little end) of a span, while one is in the middle of a span. The parallel parsing task is to move each of the four markers forward (to the left) through the corresponding spans of digits to the immediately following positions.

```
    source data ◁        ----173942---654----1----49731----321--
         M₀               .........1.....1....1......1...........
      D = [0-9]           ....111111...111....1....11111....111..
       M₀ + D             ...1........1......1....1...11....111..
M₁ = (M₀ + D) ∧ ¬D       ...1........1......1....1.............
```

Fig. 2. Parallel Scan Using Bitstream Addition and Mask

The third row of Figure 2 shows the derived character-class bitstream D identifying positions of all digits in the source stream. The fourth row then illustrates the key concept: marker movement is achieved by binary addition of the marker and character class bitstreams. As a marker 1 bit is combined using binary addition to a span of 1s, each 1 in the span becomes 0, generating a carry to add to the next position to the left. For each such span, the process terminates at the left end of the span, generating a 1 bit in the immediately following position. These generated 1 bits represent the moved marker bits. However, the result of the addition also produces some additional bits that are not involved in the scan operation. These are easily removed as shown in the fifth row, by applying bitwise logic to mask off any bits from the digit bitstream; these can never be marker positions resulting from a scan. The addition and masking technique allows matching of the regular expression [0-9]* for any reasonable (conflict-free) set of initial markers specified in M_0.

In the remainder of this paper, the notation $s(M, C)$ denotes the operation to scan from an initial set of marker positions M through the spans of characters belonging to a character class C found at each position.

$$s(M, C) = (M + C) \wedge \neg C$$

3 XML Scanning and Parsing

We now consider how the parallel scanning primitive can be applied to the following problems in scanning and parsing of XML structures: (1) parallel scanning of XML decimal character references, and (2) parallel parsing of XML start tags. The grammar of these structures is shown in Figure 3.

```
   DecRef ::= '&#' Digit⁺ ';'
    Digit ::= [0-9]
     STag ::= '<' Name (W Attribute)* W? '>'
Attribute ::= Name W? '=' W? AttValue
 AttValue ::= ( '"' [^<"]* '"') | ( "'" [^<']* "'")
        W ::= (\x20 | \x9 | \xD | \xA)⁺
```

Fig. 3. XML Grammar: Decimal Character References and Start Tags

Figure 4 shows the parallel parsing of decimal references together with error checking. For clarity, the streams are now shown in left-to-right order as indicated by the $\triangleright$ symbol. The source data includes four instances of potential decimal references beginning with the & character. Of these, only the first one is legal according to the decimal reference syntax, the other three instances are in error. These references may be parsed in parallel as follows. The starting marker bitstream M_0 is formed from the [&] character-class bitstream as shown in the second row. The next row shows the result of the marker advance operation $n(M_0)$ to produce the new marker bitstream M_1. At this point, the grammar requires a hash mark, so the first error bitstream E_0 is formed using a bitwise "and" operation combined with negation, to indicate violations of this condition. Marker bitstream M_2 is then defined as those positions immediately following any M_1 positions not in error. In the following row, the condition that at least one digit is required is checked to produce error bitstream E_1. A parallel scan operation is then applied through the digit sequences as shown in the next row to produce marker bitstream M_3. The final error bitstream E_2 is produced to identify any references without a closing semicolon. In the penultimate row, the final marker bitstream M_4 marks the positions of all fully-checked decimal references, while the last row defines a unified error bitstream E indicating the positions of all detected errors.

Initialization of marker streams may be achieved in various ways, dependent on the task at hand. In the XML parsing context, we rely on an important property of well-formed XML: after an initial filtering pass to identify XML comments, processing instructions and CDATA sections, every remaining < in the file must be the initial character of a start, end or empty element tag, and every remaining & must be the initial character of a general entity or character reference. These assumptions permit easy creation of marker bitstreams for XML tags and XML references.

The parsing of XML start tags is a richer problem, involving sequential structure of attribute-value pairs as shown in Figure 3. Using the bitstream addition technique, our method is to start with the opening angle bracket of all tags as the initial marker bitstream for parsing the tags in parallel, advance through the element name and then use an iterative process to move through attribute-value pairs.

```
source data ▷         -&#978;-&9;--&#;--&#13!-
M₀                    .1......1....1....1.....
M₁= n(M₀)             ..1......1....1....1....
E₀ = M₁ ∧ ¬[#]        .........1..............
M₂= n(M₁ ∧ ¬E₀)       ...1...........1....1...
E₁ = M₂ ∧ ¬D          ..............1.........
M₃= s(M₂ ∧ ¬E₁, D)    ......1...............1.
E₂ = M₃ ∧ ¬[;]        ...................1.
M₄= M₃ ∧ ¬E₂          ......1.................
E  = E₀ | E₁ | E₂     .........1.....1......1.
```

Fig. 4. Parsing Decimal References

Figure 5 illustrates the parallel parsing of three XML start tags. The figure omits determination of error bitstreams, processing of single-quoted attribute values and handling of empty element tags, for simplicity. In this figure, the first four rows show the source data and three character class bitstreams: N for characters permitted in XML names, W for whitespace characters, and Q for characters permitted within a double-quoted attribute value string.

```
source data ▷              --<e a= "137">---<el2 a="17" a2="3379">---<x>--
N = name chars             11.1.1...111..111.111.1..11..11..1111..111.1.11
W = white space            ....1..1............1......1...............
Q = ¬["<]                  11.11111.111.1111.111111.11.1111.1111.1111.1111

M₀                         ..1............1....................1....
M₁ = n(M₀)                 ...1............1..................1...
M₀,₇ = s(M₁, N)            ....1............1...................1..
M₀,₈ = s(M₀,₇, W) ∧ ¬[>]  .....1............1...................

M₁,₁ = s(M₀,₈, N)          ......1............1................
M₁,₂ = s(M₁,₁, W)∧[=]      ......1............1................
M₁,₃ = n(M₁,₂)             .......1............1...............
M₁,₄ = s(M₁,₃, W)∧["]      ........1............1..............
M₁,₅ = n(M₁,₄)             .........1............1.............
M₁,₆ = s(M₁,₅, Q)∧["]      ...........1............1...........
M₁,₇ = n(M₁,₆)             ............1............1..........
M₁,₈ = s(M₁,₇, W) ∧ ¬[>]  ....................1............

M₂,₁ = s(M₁,₈, N)          .............................1......
M₂,₂ = s(M₂,₁, W)∧[=]      .............................1......
M₂,₃ = n(M₂,₂)             ..............................1.....
M₂,₄ = s(M₂,₃, W)∧["]      ..............................1.....
M₂,₅ = n(M₂,₄)             ...............................1....
M₂,₆ = s(M₂,₅, Q)∧["]      ....................................1......
M₂,₇ = n(M₂,₆)             .....................................1.....
M₂,₈ = s(M₂,₇, W) ∧ ¬[>]  ...........................
```

Fig. 5. Start Tag Parsing

The parsing process is illustrated in the remaining rows of the figure. Each successive row shows the set of parsing markers as they advance in parallel using bitwise logic and addition. Overall, the sets of marker transitions can be divided into three groups.

The first group M_0 through $M_{0,8}$ shows the initiation of parsing for each of the tags through the opening angle brackets and the element names, up to the first attribute name, if present. Note that there are no attribute names in the final tag shown, so the corresponding marker becomes zeroed out at the closing angle bracket. Since $M_{0,8}$ is not all 0s, the parsing continues.

The second group of marker transitions $M_{1,1}$ through $M_{1,8}$ deal with the parallel parsing of the first attribute-value pair of the remaining tags. After these operations, there are no more attributes in the first tag, so its corresponding marker becomes zeroed out. However, $M_{1,8}$ is not all 0s, as the second tag still has an unparsed attribute-value pair. Thus, the parsing continues.

The third group of marker transitions $M_{2,1}$ through $M_{2,8}$ deal with the parsing of the second attribute-value pair of this tag. The final transition to $M_{2,8}$ shows the zeroing out of all remaining markers once two iterations of attribute-value processing have taken place. Since $M_{2,8}$ is all 0s, start tag parsing stops.

The implementation of start tag processing uses a while loop that terminates when the set of active markers becomes zero, i.e. when some $M_{k,8} = 0$. Considered as an iteration over unbounded bitstreams, all start tags in the document are processed in parallel, using a number of iterations equal to the maximum number of attribute-value pairs in any one tag in the document. However, in block-by-block processing, the cost of iteration is considerably reduced; the iteration for each block only requires as many steps as there are attribute-value pairs overlapping the block.

Following the pattern shown here, the remaining syntactic features of XML markup can similarly be parsed with bitstream based methods. One complication is that the parsing of comments, CDATA sections and processing instructions must be performed first to determine those regions of text within which ordinary XML markups are not parsed (i.e., within each of these types of construct. This is handled by first parsing these structures and then forming a *mask bitstream*, that is, a stream that identifies spans of text to be excluded from parsing (comment and CDATA interiors, parameter text to processing instructions).

4   XML Well-Formedness

In this section, we consider the full application of the parsing techniques of the previous section to the problem of XML well-formedness checking [2]. We look not only at the question of well-formedness, but also at the identification of error positions in documents that are not well-formed.

Most of the requirements of XML well-formedness checking can be implemented using two particular types of computed bitstream: *error bitstreams*, introduced in the previous section, and *error-check bitstreams*. Recall that an error bitstream stream is a stream marking the location of definite errors in accordance with a particular requirement. For example, the E_0, E_1, and E_2 bitstreams as computed during parsing of decimal character references in Figure 4 are error bitstreams. One bits mark definite errors and zero bits mark the absence of an error. Thus the complete absence of errors according to the requirements listed may be determined by forming the bitwise logical "or" of these bitstreams and confirming that the resulting value is zero. An error check bitstream is one that marks potential errors to be further checked in some fashion during post-bitstream processing. An example is the bitstream marking the start positions of CDATA sections. This is a useful information stream computed during bit-stream processing to identify opening <![sequences, but also marks positions to subsequently check for the complete opening delimiter <![CDATA[at each position.

In typical documents, most of these error-check streams will be quite sparse or even zero. Many error conditions could actually be fully implemented using bitstream techniques, but at the cost of a number of additional logical and shift operations. In general, the conditions are easier and more efficient to check one-at-a-time using multibyte comparisons on the original source data stream. With very sparse streams, it is very unlikely that multiple instances occur within any given block, thus eliminating the benefit of parallel evaluation of the logic.

The requirement for name checking merits comment. XML names may use a wide range of Unicode character values. It is too expensive to check every instance of an XML name against the full range of possible values. However, it is possible and inexpensive to use parallel bitstream techniques to verify that any ASCII characters within a name are indeed legal name start characters or name characters. Furthermore, the characters that may legally follow a name in XML are confined to the ASCII range. This makes it useful to define a name scan character class to include all the legal ASCII characters for names as well as all non-ASCII characters. A namecheck character class bitstream will then be defined to identify non-ASCII characters found within namescans. In most documents this bitstream will be all 0s; even in documents with substantial internationalized content, the tag and attribute names used to define the document schema tend to be confined to the ASCII repertoire. In the case that this bitstream is nonempty, the positions of all 1 bits in this bitstream denote characters that need to be individually validated.

Attribute names within a single XML start tag or empty element tag must be unique. This requirement could be implemented using one of several different approaches. Standard approaches include: sequential search, symbol lookup, and Bloom filters [5].

Except for empty element tags, XML tags come in pairs with names that must be matched. To discharge this requirement, we form a bitstream consisting of the disjunction of three bitstreams formed during parsing: the bitstream marking the positions of start or empty tags (which have a common initial structure), the bitstream marking tags that end using the empty tag syntax ("/>"), and the bitstream marking the occurrences of end tags. In post-bitstream processing, we iterate through this computed bitstream and match tags using an iterative stack-based approach.

An XML document consists of a single root element within which all others contained; this constraint is also checked during post-bitstream processing. In addition, we define the necessary "miscellaneous" bitstreams for checking the prolog and epilog material before and after the root element.

Overall, parallel bitstream techniques are well-suited to verification problems such as XML well-formedness checking. Many of the character validation and syntax checking requirements can be conveniently and efficiently implemented using error streams. Other requirements are also supported by the computation of error-check streams for simple post-bitstream processing or composite stream over which iterative stack-based procedures can be defined for checking recursive syntax. To assess the completness of our analysis, we have confirmed that our implementations correctly handle all the well-formedness checks of the W3C XML Conformance Test Suite.

5   Compilation to Block-Based Processing

While our Python implementation of the techniques described in the previous section works on unbounded bitstreams, a corresponding C implementation needs to process an input stream in blocks of size equal to the SIMD register width of the processor it runs on. So, to convert Python code into C, the key question becomes how to transfer information from one block to the next.

The answer lies in the use of *carry bits*. The parallel scanning primitive uses only addition and bitwise logic. The logic operations do not require information flow accross block boundaries, so the information flow is entirely accounted by the carry bits for addition. Carry bits also capture the information flow associated with upshift operations, which move information forward one position in the file. In essence, an upshift by one position for a bitstream is equivalent to the addition of the stream to itself; the bit shifted out in an upshift is in this case equivalent to the carry generated by the additon.

Properly determining, initializing and inserting carry bits into a block-by-block implementation of parallel bitstream code is a task too tedious for manual implementation. We have thus developed compiler technology to automatically insert declarations, initializations and carry save/restore operations into appropriate locations when translating Python operations on unbounded bitstreams into the equivalent low-level C code implemented on a block-by-block bases. Our current compiler toolkit is capable of inserting carry logic using a variety of strategies, including both simulated carry bit processing with SIMD registers, as well as carry-flag processing using the processor general purpose registers and ALU. Details are beyond the scope of this paper, but are described in the on-line source code repository at parabix.costar.sfu.ca.

6   Performance Results

In this section, we compare the performance of our xmlwf implementation using the Parabix 2 technology described above with several other implementations. These include the original xmlwf distributed as an example application of the expat XML parser, implementations based on the widely used Xerces open source parser using both SAX and DOM interfaces, and an implementation using our prior Parabix 1 technology with bit scan operations.

Table 1 shows the document characteristics of the XML instances selected for this performance study, including both document-oriented and data-oriented XML files. The jawiki.xml and dewiki.xml XML files are document-oriented XML instances of Wikimedia books, written in Japanese and German, respectively. The remaining files are data-oriented. The roads.gml file is an instance of Geography Markup Language (GML), a modeling language for geographic information systems as well as an open interchange format for geographic transactions on the Internet. The po.xml file is an example of purchase order data, while the soap.xml file contains a large SOAP message. Markup density is defined as the ratio of the total markup contained within an XML file to the total XML document size. This metric is reported for each document.

Table 1. XML Document Characteristics

File Name	dewiki.xml	jawiki.xml	roads.gml	po.xml	soap.xml
File Type	document	document	data	data	data
File Size (kB)	66240	7343	11584	76450	2717
Markup Item Count	406792	74882	280724	4634110	18004
Attribute Count	18808	3529	160416	463397	30001
Avg. Attribute Size	8	8	6	5	9
Markup Density	0.07	0.13	0.57	0.76	0.87

Table 2 shows performance measurements for the various xmlwf implementations applied to the test suite. Measurements are made on a single core of an Intel Core 2 system running a stock 64-bit Ubuntu 10.10 operating system, with all applications compiled with llvm-gcc 4.4.5 optimization level 3. Measurements are reported in CPU cycles per input byte of the XML data files in each case. The first row shows the performance of the Xerces C parser using the tree-building DOM interface. Note that the performance varies considerably depending on markup density. Note also that the DOM tree construction overhead is substantial and unnecessary for XML well-formedness checking. Using the event-based SAX interface to Xerces gives much better results as shown in the second row. The third row shows the best performance of our byte-at-a-time parsers, using the original xmlwf based on expat.

The remaining rows of Table 2 show performance of parallel bitstream implementations, including post-bitstream processing. The first row shows the performance of our Parabix 1 implementation using bit scan instructions. While showing a substantial speed-up over the byte-at-a-time parsers in every case, note also that the performance advantage increases with increasing markup density, as expected. The last two rows show Parabix 2 implementations using different carry-handling strategies, with the "simd" row referring to carry computations performed with simulated calculation of propagated and generated carries using SIMD operations, while the "adc64" row referring to an implementation directly employing the processor carry flags and add-with-carry instructions on 64-bit

Table 2. Parser Performance (Cycles Per Byte)

Parser Class	Parser	dewiki.xml	jawiki.xml	roads.gml	po.xml	soap.xml
Byte	Xerces (DOM)	37.921	40.559	72.78	105.497	125.929
at-a	Xerces (SAX)	19.829	24.883	33.435	46.891	57.119
Time	expat	12.639	16.535	32.717	42.982	51.468
Parallel	Parabix1	8.313	9.335	13.345	16.136	19.047
Bit	Parabix2 (simd)	6.103	6.445	8.034	8.685	9.53
Stream	Parabix2 (adc64)	5.123	5.996	6.852	7.648	8.275

general registers. In both cases, the overall performance is impressive, with the increased parallelism of parallel bit scans clearly paying off in improved performance for dense markup.

7 Conclusion

In application to the problem of XML parsing and well-formedness checking, the method of parallel parsing with bitstream addition is effective and efficient. Using only bitstream addition and bitwise logic, it is possible to handle all of the character validation, lexical recognition and parsing problems except for the recursive aspects of start and end tag matching. Error checking is elegantly supported through the use of error streams that eliminate separate if-statements to check for errors with each byte. The techniques are generally very efficient particularly when markup density is high. However, for some conditions that occur rarely and/or require complex combinations of upshifting and logic, it may be better to define simpler error-check streams that require limited postprocessing using byte matching techniques.

The techniques have been implemented and assessed for present-day commodity processors employing current SIMD technology. As processor advances see improved instruction sets and increases in width of SIMD registers, the relative advantages of the techniques over traditional byte-at-a-time sequential parsing methods is likely to increase substantially. Of particular benefit to this method, instruction set modifications that provide for more convenient carry propagation for long bitstream arithmetic would be most welcome.

A significant challenge to the application of these techniques is the difficulty of programming. The method of prototyping on unbounded bitstreams has proven to be of significant value in our work. Using the prototyping language as input to a bitstream compiler has also proven effective in generating high-performance code. Nevertheless, direct programming with bitstreams is still a specialized skill; our future research includes developing yet higher level tools to generate efficient bitstream implementations from grammars, regular expressions and other text processing formalisms.

References

1. Asanovic, K., Bodik, R., Catanzaro, B.C., Gebis, J.J., Husbands, P., Keutzer, K., Patterson, D.A., Plishker, W.L., Shalf, J., Williams, S.W., Yelick, K.A.: The landscape of parallel computing research: A view from Berkeley. Technical Report UCB/EECS-2006-183, EECS Department, University of California, Berkeley (December 2006)
2. Bray, T., Paoli, J., Sperberg-McQueen, C.M., Maler, E., Yergeau, F.: Extensible markup language (XML) 1.0, 5th edn. W3C Recommendation (2008)
3. Cameron, R.D.: A Case Study in SIMD Text Processing with Parallel Bit Streams. In: ACM Symposium on Principles and Practice of Parallel Programming (PPoPP), Salt Lake City, Utah (2008)
4. Cameron, R.D., Herdy, K.S., Lin, D.: High performance XML parsing using parallel bit stream technology. In: CASCON 2008: Proceedings of the 2008 Conference of the Center for Advanced Studies on Collaborative Research, pp. 222–235. ACM Press, New York (2008)
5. Dai, Z., Ni, N., Zhu, J.: A 1 cycle-per-byte XML parsing accelerator. In: FPGA 2010: Proceedings of the 18th Annual ACM/SIGDA International Symposium on Field Programmable Gate Arrays, pp. 199–208. ACM Press, New York (2010)
6. Herdy, K.S., Burggraf, D.S., Cameron, R.D.: High performance GML to SVG transformation for the visual presentation of geographic data in web-based mapping systems. In: Proceedings of SVG Open 2008 (August 2008)
7. Kostoulas, M.G., Matsa, M., Mendelsohn, N., Perkins, E., Heifets, A., Mercaldi, M.: XML Screamer: An Integrated Approach to High Performance XML Parsing, Validation and Deserialization. In: Proceedings of the 15th International Conference on World Wide Web (WWW 2006), pp. 93–102 (2006)
8. Leventhal, M., Lemoine, E.: The XML chip at 6 years. In: International Symposium on Processing XML Efficiently: Overcoming Limits on Space, Time, or Bandwidth (August 2009)
9. Shah, B., Rao, P.R., Moon, B., Rajagopalan, M.: A data parallel algorithm for XML DOM parsing. In: Bellahsène, Z., Hunt, E., Rys, M., Unland, R. (eds.) XSym 2009. LNCS, vol. 5679, pp. 75–90. Springer, Heidelberg (2009)
10. Zhang, Y., Pan, Y., Chiu, K.: Speculative p-DFAs for parallel XML parsing. In: 2009 International Conference on High Performance Computing (HiPC), pp. 388–397 (December 2009)

HOMPI: A Hybrid Programming Framework for Expressing and Deploying Task-Based Parallelism*

Vassilios V. Dimakopoulos and Panagiotis E. Hadjidoukas

Department of Computer Science
University of Ioannina, Ioannina, Greece, GR-45110
{dimako,phadjido}@cs.uoi.gr

Abstract. This paper presents HOMPI, a framework for programming and executing task-based parallel applications on clusters of multiprocessors and multi-cores, while providing interoperability with existing programming systems such as MPI and OpenMP. HOMPI facilitates expressing irregular and adaptive master-worker and divide-and-conquer applications avoiding explicit MPI calls. It also allows hybrid shared-memory / message-passing programming, exploiting fully the availability of multiprocessor and multi-core nodes, as it integrates by design with OpenMP; the runtime infrastructure presents a unified substrate that handles local threads and remote tasks seamlessly, allowing both programming flexibility and increased performance opportunities.

Keywords: cluster programming, task-based parallelism, load balancing, MPI

1 Introduction

The pool-of-tasks (or master-worker) paradigm is one of the most widely used paradigms for programming a multitude of applications on a variety of parallel computing platforms. According to this model, the master assigns tasks to a set of workers, providing them with any required input data, and waits for the results. The number of tasks usually exceeds the number of workers and the master may generate new tasks dynamically, depending on the received results. In the simple case, a few primary message passing (MPI) calls are enough to implement the model on a distributed-memory platform with a self scheduling mechanism where inactive workers dynamically probe the master for work. On the other hand, limitations and difficulties arise if advanced functionality is needed. First, because of the bottleneck at the master, the model may suffer from low scalability. Hierarchical task parallelism and techniques like distributed task queues require additional and non-trivial programming effort. Finally, a pure MPI-based implementation cannot easily adapt to take advantage of a multi-core node's physically shared memory.

---

* This work is supported in part by the Artemisia SMECY project (grant 100230).

E. Jeannot, R. Namyst, and J. Roman (Eds.): Euro-Par 2011, LNCS 6853, Part II, pp. 14–26, 2011.

Although exploring new languages and programming models is currently a major issue in the parallel processing research community (and possibly the ultimate solution to leveraging current and emerging parallel hardware), other pragmatic approaches seem more promising for wide adoption in the short- to medium-term. Programming constructs that extend without changing a popular language have proven quite successful, OpenMP [2] being the most prominent example. Along the same lines, interoperability with popular programming models is another important requirement, easing the utilization of existing codebases.

In this work we present HOMPI, an infrastructure for programming and executing task-based applications on clusters of multi-cores. It consists of a source-to-source compiler that provides for simple directive-based definition and execution of tasks and a runtime library that orchestrates the execution over a variety of platforms, including pure shared-memory systems and clusters of such nodes. HOMPI targets message passing, shared address space and hybrid programs. In the standard master-worker case, the programmer does not have to use low-level message passing primitives at all, hiding away the communication details while providing load balancing transparently. For more advanced functionality, HOMPI integrates the tasking model into traditional MPI programs, allowing one or more MPI processes to independently spawn tasks. Each task may also spawn OpenMP-based parallelism, allowing seamless hybrid programming possibilities. The compiler supports OpenMP by design while the runtime system provides unified support for OpenMP threads as well as remotely executed tasks.

A number of programming tools and languages for task parallelism have been proposed recently for contemporary and emerging architectures. On shared-memory platforms, OpenMP has been extended in V3.0 with support for a tasking model [2], similar to Cilk [1]. For the Cell BE, runtime libraries include MPI microtask [3], ALF [4] and StarPU [5]. HOMPI's programming model borrows the #pragma-based annotation style of OpenMP and is reminiscent of other proposals such as HMPP [6] and StarSs [7], which combine runtime and compiler support to provide a (limited) task-based environment. These proposals target mainly accelerator-equipped systems and, in contrast to HOMPI, they do not support the divide-and-conquer model, since they do not allow recursive parallelism.

The contribution of this work is twofold. First, we introduce an easy-to-use programming framework which preserves the base language, while providing convenient code annotation for expressing task-based master/slave and divide-and-conquer parallel algorithms. While the annotation style is in the spirit of other proposals, to the best of our knowledge, HOMPI is the first of its kind targeting (and fully exploiting) clusters of SMPs/multi-cores. Second, albeit self-contained, our framework is fully interoperable with standard programming systems like MPI and OpenMP, allowing legacy or already parallelized code to be trivially integrated in an application. In our opinion this is a crucial attribute for the viability of any programming model proposal.

```
#pragma hompi taskdef in(n) out(res)
void fib (int n, unsigned long *res) {
    unsigned long res1, res2;

    if (n <= 1) {
        *res = n;
    } else {
        #pragma hompi task
        fib(n-1, &res1);
        #pragma hompi task
        fib(n-2, &res2);
        #pragma hompi tasksync
        *res = res1+res2;
    }
}

void main(int argc, char *argv[]) {
    unsigned long res;
    fib(50, &res);
}
```

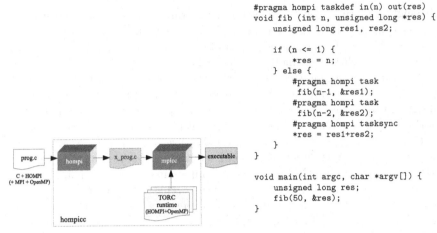

Fig. 1. HOMPI compilation procedure   **Fig. 2.** Recursive Fibonacci in HOMPI

2 Programming Environment

HOMPI is based on a source-to-source translator that can handle #pragma-based directives within the user code, similar to OpenMP. Fig. 1 shows the compilation steps: from the annotated source code, the source-to-source compiler (hompi) produces an intermediate, transformed C file (x_prog.c) augmented by run-time calls. This file is then compiled by the system's native mpicc compiler and linked with HOMPI's runtime libraries to produce the final executable. The whole process is automated by the hompicc script.

HOMPI's execution model assumes that an application consists of multiple MPI processes with private memory, running on cluster nodes. Furthermore, multi-threading is used to exploit the multi-processor/core configuration of a node; each process consists of one or more kernel threads sharing the process memory.

A task in HOMPI corresponds to the remote execution of a function on a set of data that are passed as arguments to this function, in the spirit of remote procedure calls. Tasks are executed asynchronously and in any order, without any data dependencies or point-to-point communication between them. Tasks have a parent-child relationship and can be arbitrarily nested, allowing multiple levels of parallelism and straightforward coding of divide-and-conquer algorithms.

The HOMPI programming model in essence requires that the programmer only designates which of the program functions can be used as tasks and be executed on (possibly) remote nodes. In many cases, just two directives are enough for the application to take advantage of the infrastructure, resulting in minimal programmer effort. The directive for designating a function as an independent task is taskdef and is placed right before the definition of a C function. A taskdef directive may contain *intent* clauses, similar to intent attributes of Fortan 90, which specify the intended usage of the function arguments: in(*variable-list*), for variables that are to be passed to the function by value, out(*variable-list*), for results returned by the function and inout(*variable-list*) for variables whose values are passed to the function but will also be used to return a result.

An example is given in Fig. 2, which presents a complete HOMPI application that uses a recursive function (fib) to compute the 50th Fibonacci number. The taskdef directive designates the fib function as a task that accepts an argument by value (n) and computes a result (res). If any of the arguments is an array, the number of elements must be known, and this is either determined by hompi from the function prototype (if a size expression exists) or must be specified explicitly in the intent clauses of the taskdef directive.

The actual execution of a function as a task occurs with the task directive, which must be placed right before the function call. Finally, task joining (blocking until all child tasks finish their work) is possible anywhere in the code through the tasksync directive. In Fig. 2, the fib function generates two new tasks that are distributed across the available workers and waits for their results. Notice (i) the complete absence of explicit messaging and (ii) that if the directives are ignored by the compiler, the program's semantics remain the same to pure sequential execution.

2.1 Callbacks, Reductions and Detached Tasks

Normally, a parent task creates an arbitrary number of tasks and uses the tasksync directive to suspend itself until all child tasks have finished and their results have been returned. HOMPI supports *callback* functions, which allow for asynchronous execution of post-processing code on the process where the parent task runs on, even if the parent task is suspended. The callback function is defined immediately following the task definition through a callback directive; the callback function specifier is generated by the compiler and assumes the exact same arguments as the corresponding task function, providing thus access to the input parameters and the result of the task. An example where the callback just prints the results of each generated task, is depicted below.

```
#pragma hompi taskdef in(a) inout(b[2])
void taskfunc(int a,int *b) {
    b[1] = b[0] + a + 1;
}
#pragma hompi callback
{
    printf("result = %d\n", b[1]);
}
```

HOMPI also supports *reduction operations* (which can be actually seen as special cases of callbacks) for the common scenario where each child task computes a partial result which is collected by the parent to produce the final result. These operations (summation, product, etc.) replace the out intent clause and are specified exactly in OpenMP style, as seen below:

```
#pragma hompi taskdef in(a) reduction(+:b)
void taskfunc(int a, int *b) {
    *b = a;
}
```

Reduction operations are supported for *both* scalar variables and arrays.

Finally HOMPI supports *detached* tasks, that is tasks that execute without the parent being able to wait on them. In such cases task management is left up to the programmer. Detached tasks are executed as such by including the `detached` clause in the `task` directive. They can be combined with callbacks which actually provide the only way for them to synchronize with their parents; for example, within a callback, a detached task can modify a condition on which the parent task is explicitly waiting. Moreover, new tasks can be created within the callback routine. Detached tasks combined with callbacks offer a powerful mechanism; for example, they can be used for implementing dependencies among arbitrary subsets of tasks.

2.2   Task Distribution and Scheduling

Although not always necessary, in many cases one needs to control how tasks are distributed across workers or cluster nodes (e.g. due to particular load balancing needs). HOMPI offers two ways for achieving this. First, it provides a standard cyclic distribution scheme with tunable parameters. This is especially useful when tasks are created within a iterative control structure (e.g. while, for). The parameters of this scheme include the *scope* (whether tasks are distributed per node or per worker), the *starting* point (node id or worker id) and the *stride* (increment). These parameters are specified using a `taskschedule` directive:

```
#pragma hompi taskschedule scope(workers) start(0) stride(1)
for (t = 0; t < 8; t++) {
    #pragma hompi task
    func();
}
```

If the stride is zero then all tasks are submitted to the target node or worker. The default scheduling policy is represented with the tuple (nodes, -1, 1), i.e. distribution across nodes with stride 1 starting from the current node.

The second mechanism allows the user to explicitly specify the node or worker where a task will be submitted for execution. This is achieved with the `atnode(x)` and `atworker(y)` clauses in the `task` directive, where x and y are the identities of the intended node and worker respectively, e.g.

```
for (t = 0; t < K; t++) {
    #pragma hompi task atworker(t % hompi_total_workers())
    func();
}
```

It must be noted that the runtime system of HOMPI, which is presented next, includes a work-stealing mechanism whereby tasks may be stolen from a node and executed at another. If this mechanism is activated then all the above refer to the *initial* placement of a task; the actual node/worker that will ultimately execute it may be different. To explicitly control this, tasks can also be classified as *tied* and *untied* (using homonymous clauses in the `task` directive), similarly to OpenMP 3.0; a tied task can never be stolen, and will run on the process it was initially submitted for execution.

3 TORC: The Runtime System

In this section we give a short overview of TORC, the runtime environment of HOMPI. More details can be found in [10]. TORC uses exclusively POSIX and MPI calls for portability and performance, and integrates seamlessly hardware shared-memory and message passing. It provides application adaptability to the same application code, or even binary, on both shared-memory multiprocessors/multi-cores and clusters of them. TORC views an application as a collection of MPI processes. Each process consists of one or more POSIX kernel threads that execute tasks and a *server* thread that is responsible for the remote queue management and the asynchronous data movement. There exist private and public worker-specific and node-specific ready queues where tasks can be submitted for execution. A two-level threading model is implemented, where each kernel thread is a worker that continuously dispatches and executes ready-to-run tasks.

Tasks are associated with the process (home node) they were created on and can be executed either locally or remotely. In the latter case, explicit but transparent to the user data movement takes place. A worker thread executes a task by calling the task function with the locally stored arguments. When it finishes, it sends a notification message back to the home node, along with any arguments that represent results (out/inout). These are received asynchronously by the server thread and copied on their actual memory locations in the address space of the home process. A running task that spawns parallelism can suspend its execution, waiting for the termination of all its child tasks. The execution state of the current task is saved, releasing the underlying kernel thread, which runs the scheduling loop for selecting the next-to-run task. When all child tasks have completed (and all callbacks, if any, have finished), the suspended task becomes ready for execution and eventually resumes. A callback is implemented as a tied task, submitted for local execution when the corresponding user task finishes and notifies its parent task. The submission is performed by either the worker thread that executes the user task (if this is executed locally) or the server thread of the same process.

Data transfer mechanisms. The low level communication subsystem of TORC is based on MPI. However, other data transfer mechanisms have also been considered. In particular, we have successfully integrated two more mechanisms: MPI-2's one-side communications and software distributed shared memory (SDSM). MPI-2's remote memory access (RMA) [8] supports data transfer through one-sided operations. In TORC, the MPI_Get routine is used for fetching input data and MPI_Put for writing the results back to the home node. On the other hand, SDSM implements the notion of global memory on distributed computing environments and provides implicit data movement through the memory consistency protocol [9]. One-sided operations and SDSM provide receiver-initiated data movement for remotely executed tasks, performed by the worker thread just before or during the execution of the task function. This on-demand data movement does not allow data pre-fetching opportunities for server threads but may avoid unnecessary data transfers if task stealing is enabled.

Dynamic load balancing. Spawning a large number of tasks can be an effective approach to distribute the work evenly among the available workers. The user can specify the node or worker where each task will be submitted for execution and then employ the internal stealing mechanism for untied tasks that TORC provides. A idle worker extracts and executes a task from its local ready queue. If this is empty and task stealing is enabled, the worker first searches for work in the rest of the ready queues of the same node and then visits randomly the remote nodes. The worker waits synchronously for a response from the server thread of the target node. The answer is either a message that denotes unavailability of work at the target node, or an untied task descriptor that will be immediately executed upon receipt. Remote task stealing includes the corresponding data movement, unless the task returns to its home node.

4   Mixed-Mode and Hybrid Programming

Although the default execution model of HOMPI is that of master-worker, mixing it with SPMD execution and dynamically switching between them may be beneficial or required. For instance, a task parallel program may take advantage of common scientific SPMD libraries built on top of MPI.

The atnode(*) clause is a special case in the **task** creation directive that provides the above functionality; at runtime, the application creates as many tasks as the number of available nodes. These tasks are marked as tied and are distributed across the cluster nodes. The specified task function is executed by a single worker on every node. This approach matches the execution model of MPI and at the same time allows for hybrid MPI + OpenMP programming. When all the tasks have finished, the execution model switches back to master-worker. In Fig. 3, we demonstrate the flexibility of the atnode(*) clause; the master broadcasts the global array (ga) to the other nodes by issuing a collective call to the native MPI_Bcast function (with all workers participating). In this way, the application does not need to send ga with every task, avoiding thus unnecessary data transfers.

The atnode(*) clause improves the programmability of our system by facilitating the insertion of legacy MPI codes into the supported task-based execution environment. Following a similar approach, native MPI applications can be seamlessly enriched with the tasking model that HOMPI provides. Specifically, by setting a particular environmental variable (HOMPI_MODE), the TORC library is initialized for SPMD execution and thus the primary thread of all MPI processes executes the main routine. Switching the MPI application's execution model to master-worker is possible with a special directive (spmd_barrier). In Fig. 4, one of the MPI processes (e.g. the one with rank 0) becomes the master task that spawns work, while the rest of the processes block at the spmd_barrier directive which converts them to workers, activating the scheduling loop in TORC. After task completion, the master process reaches spmd_barrier and all MPI processes resume their execution, while the execution model switches back to SPMD.

```
int ga[16];

#pragma hompi taskdef in(root)
void spmdfunc(int root) {
  /* legacy MPI code can run here */
  MPI_Bcast(ga, 16, MPI_INT, root, MPI_COMM_WORLD);
}


#pragma hompi taskdef out(b[16])
void func(int *b) {
  for (int i = 0; i < 16; i++) b[i] = ga[i];
}

main() {
  int res[8][16], root = hompi_node();
  for (int i = 0; i < 16; i++) ga[i] = i;

  #pragma hompi atnode(*)
  spmdfunc(root);

  for (int t = 0; t < 8; t++) {
    #pragma hompi task tied
    func(res[t]);
  }
  #pragma hompi tasksync
}
```

Fig. 3. Example of `atnode(*)`

```
#pragma hompi taskdef
void func() { ... }

main(int argc, char *argv[]) {
  /* legacy MPI code */
  MPI_Init(&argc, &argv);
  MPI_Comm_rank(MPI_COMM_WORLD, &rank);
  ...

  if (rank==0) {    /* tasking */
    for (int i = 0; i < N; i++) {
      #pragma hompi task
      func();
    }
    #pragma hompi tasksync
  }
  #pragma hompi spmd_barrier

  /* legacy MPI code continues */
}
```

Fig. 4. Tasking in MPI code

Implementation of hybrid programming. HOMPI allows expressing the intra-node parallelism of a task function with OpenMP directives, in accordance to the hybrid MPI + OpenMP programming model. However, because task functions are executed by TORC's underlying worker threads, the utilized OpenMP compiler must support interoperability between OpenMP and independent POSIX threads. Moreover, caution is needed because the combination of TORC threads and OpenMP threads can easily oversubscribe the system, a situation resulting in performance degradation [11].

To cope with the above problem, we have constructed a *unified library* that handles both levels of parallelism within the same compilation and runtime environment. In particular, we have introduced a threading layer that is implemented on top of TORC into the OMPi OpenMP compiler [12]. Thanks to the layered architecture of OMPi, TORC was attached as an opaque OpenMP thread provider, thus letting OMPi control OpenMP execution through TORC-provided threads while at the same time TORC handles tasks independently.

The HOMPI translator was implemented by extending OMPi's translator, in order to have it parse and transform the new directives. Both HOMPI tasks and OpenMP threads are executed within the same runtime infrastructure. Internally, as worker threads first access the private and then the public ready queues of their node, OpenMP parallelism has a higher priority with respect to internode parallelism expressed with HOMPI tasks.

5   Experimental Evaluation

In this section we present preliminary experimental evaluation of our HOMPI prototype. We report both benchmarking results and results from full applications executed on a Sun Fire x4100 cluster of 16 nodes interconnected with Gigabit Ethernet. Each node has 2 dual core AMD Opteron-275 processors running at 2.2GHz giving a total of 64 cores. The cluster nodes are running Linux 2.6, while HOMPI was built with GNU GCC 4.3 as the system's native C compiler and the MPICH2 implementation of MPI. Thanks to the design of TORC, the same application binary can exploit the 4 processor cores of a single node with several combinations in the number of processes and workers. Therefore, our performance results refer to both distributed and shared-memory organizations.

Data transfer overheads. To evaluate the three different data transfer methods we discussed in Section 3 (MPI, RMA and SDSM) implemented in TORC, we measure the time required for the remote execution of a single task with input argument an array of double-precision floating point numbers that has been initialized by the parent before task creation. The task computes the sum of the array elements. For a fair comparison, we spawn exactly one task and thus preclude any data prefetching through the server thread when MPI calls are used. Besides the data movement for the argument of the task, the measured time includes the overhead for the allocation of the descriptor and its insertion in the queue of the remote process, the execution of the task function and the notification of the parent task at the first process.

Figs. 5 and 6 illustrate the overhead of the three methods with respect to the argument size, for `in` and `inout` argument types respectively. Regarding SDSM, we provide results for two libraries: Mome [13] and Mocha [14]. To enforce the `inout` semantics for the SDSM case, the parent task on process 0 accesses the array after task completion. Due to the relaxed consistency model of Mocha, SDSM barrier calls were introduced in the benchmark code. We observe that the overhead of the MPI and RMA methods, which both involve explicit communication, is almost identical. SDSM exhibits significantly higher overhead, due to the page-based consistency protocol and the multiple invocations of the page-fault handler. The performance difference between Mocha and Mome is because the

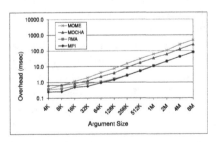

Fig. 5. Task execution overhead for the three data transfer methods (`in`)

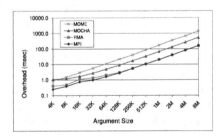

Fig. 6. Task execution overhead for the three data transfer methods (`inout`)

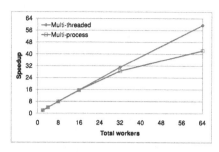
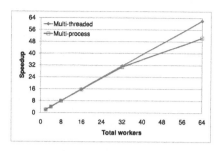

Fig. 7. EP performance   **Fig. 8.** PMCMC performance

former uses an 8KB (instead of 4KB) page size, resulting in a 50% reduction in the number of page faults.

Applications. For evaluating the performance of HOMPI we used two applications: EP and PMCMC. EP is an Embarrassingly Parallel benchmark that involves minimal inter-processor communication. The number of spawned tasks is equal to the number of workers, while the results of the tasks are accumulated through a reduction (+) operation. PMCMC implements an embarrassingly parallel Markov Chain Monte Carlo algorithm of the hard-disk problem. Each task is assigned a seed and performs a large number of Markov Chain computations. The code of this application was adapted from the ADLB library [15].

Fig. 7 depicts the performance of EP for 2^{28} random numbers and specifically the best observed speedup for a particular number of workers, using the multi-threaded and multiprocess configurations. When a single process per node and multiple workers per process are used, we observe that EP scales almost linearly; the slight performance degradation on 32 and 64 processors is mostly attributed to load imbalance effects due to the small number of generated tasks. The performance of the same application is significantly lower when multiple processes, of a single worker thread each, are deployed at each node of the cluster. The drawbacks of this configuration are the increased number of explicit messages and the oversubscription of processor cores due to the the multiple server threads on each node. Similarly, Fig. 8 presents the performance results for PMCMC when 128 independent tasks are used. We observe that the application exhibits almost perfect scalability for the multi-threaded approach. The performance of the multi-process approach is similar for all but the 64-process case, where its efficiency is significantly reduced to 78.87%.

Load balancing. We demonstrate the load balancing mechanism of HOMPI using the Mandelbrot application included in the LAM/MPI software package, rewritten to follow the tasking model of HOMPI. In our case, the main routine of the application creates a single task for each image block. The task receives as input arguments the coordinates of the block and as output argument an array for the image block. Each task is also associated with a callback routine which copies the processed block to the image region. Tasks are either distributed cyclically across the available workers or inserted in the queue of the master process.

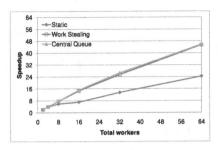

Fig. 9. Performance of task scheduling schemes on Mandelbrot

Table 1. Speedup of Mandelbrot on the (16,1,4) configuration for various task numbers

Nodes	Processes per node	Workers per process	Total workers	Total tasks	Static Scheduling	Work Stealing	Central Queue
16	1	4	64	256	24.30	45.47	45.79
16	1	4	64	512	26.70	50.74	50.32
16	1	4	64	1024	24.78	56.83	55.12

Table 2. Speedup of Mandelbrot for the hybrid programming model (256 tasks)

Nodes	Processes per node	Workers per process	OpenMP threads	Total workers	Static Scheduling	Work Stealing	Central Queue
1	1	1	4	4	3.79	3.79	3.79
2	1	1	4	8	7.33	7.52	7.54
4	1	1	4	16	13.30	14.94	14.98
8	1	1	4	32	22.14	29.02	29.26
16	1	1	4	64	24.30	54.99	56.62

Fig. 9 presents the speedup of the Mandelbrot application on the Sun cluster for an image of 2048x2048 pixels, 50000 maximum iterations for each pixel, and blocks of 128x128 pixels (256 tasks). We spawn a single process per node and provide results for the cyclic task distribution scheme, having the inter-node stealing mechanism disabled (termed 'Static Scheduling') or enabled (termed 'Work Stealing'). In addition, we evaluate the central-queue approach, where workers access the queue of the master process to get a task to execute. We observe that as the number of cores increases, the application manages to scale efficiently only if task stealing has been activated. For instance, the speedup of the application on 64 cores is approximately 24 and 45 for the static and work stealing approach respectively. The attained performance of the cyclic distribution and central queue are almost identical because, for this particular experiment, the latter does not suffer from bottlenecks as the server thread manages to handle the stealing requests efficiently. The scalability of the application declines with the number of processors, mostly because of the overhead for storing the results through the callback routine.

Table 1 studies the behavior of Mandelbrot for the (16 nodes, 1 process per node, 4 workers per node) configuration for smaller block sizes and thus a larger number of spawned tasks. It is apparent that better load balancing is achieved if the work stealing mechanism is enabled and thus the scalability of Mandelbrot is further improved. As the number of tasks of finer granularity increases, the cyclic distribution scheme with work stealing begins to outperform the central queue approach.

Our last experiment demonstrates the effectiveness of hybrid programming for the Mandelbrot application. Specifically, a single process is spawned on each cluster node and the loop-level intra-task parallelism is expressed with OpenMP. The performance results are depicted in table 2. We observe that the attained performance of the hybrid programming approach is higher than that of the corresponding configurations in the previously presented experiments. This is attributed to better load balancing, as the 256 tasks are distributed to a smaller number of processes, and the full utilization of OpenMP threads as the serial fraction of the task function in this particular application is negligible.

6   Conclusion

This paper presents HOMPI, a directive-based programming and runtime environment for task-parallel applications on clusters of multiprocessor/multi-core nodes. The framework consists of a source-to-source C compiler that understands a small number of `#pragma`-based directives which allow for rather straightforward task creation and scheduling across the cluster. The output of the compiler contains calls to TORC, a sophisticated runtime library that handles all the task execution details, providing transparent load balancing and resulting in significant performance figures. The HOMPI infrastructure integrates features of several parallel programming models, from threads and OpenMP to MPI and remote procedure calls and in addition it is fully interoperable with them. As such, we believe its applicability will be quite general. We are currently extending our infrastructure on heterogeneous platforms and computational grids and introducing fault tolerance mechanisms.

References

1. Blumofe, R.D., Joerg, C.F., et al.: Cilk: An efficient multithreaded runtime system. J. Parallel Distrib. Comput. 37(1), 55–69 (1996)
2. OpenMP Architecture Review Board: OpenMP Specifications, http://www.openmp.org
3. Ohara, M., Inoue, H., Sohda, Y., Komatsu, H., Nakatani, T.: MPI microtask for programming the cell broadband engine processor. IBM Syst. Journal 45(1) (2006)
4. IBM Corporation: Accelerated Library Framework (ALF) for Cell Broadband Engine programmer's guide and API reference. SDK for Multicore Acceleration V3.0
5. Augonnet, C., Thibault, S., Namyst, R., Wacrenier, P.-A.: STARPU: A unified platform for task scheduling on heterogeneous multicore architectures. In: Sips, H., Epema, D., Lin, H.-X. (eds.) Euro-Par 2009. LNCS, vol. 5704, pp. 863–874. Springer, Heidelberg (2009)
6. Dolbeau, R., et al.: HMPP: A hybrid multi-core parallel programming environment. In: 1st Wrkshp on General Purpose Processing on GPUs, Boston, MA (2007)

7. Planas, J., Badia, R.M., et al.: Hierarchical task-based programming with StarSs. Int'l. J. of High Perf. Comput. Applic. 23(3), 284–299 (2009)
8. Geist, A., Gropp, W., Lusk, E., et al.: MPI-2: Extending the message-passing interface. In: Fraigniaud, P., Mignotte, A., Robert, Y., Bougé, L. (eds.) Euro-Par 1996. LNCS, vol. 1124, Springer, Heidelberg (1996)
9. Li, K., Hudak, P.: Memory coherence in shared virtual memory systems. ACM Trans. on Computer Systems 7(4), 321–359 (1989)
10. Hadjidoukas, P.E., Dimakopoulos, V.V.: TORC: a tasking library for multicore clusters. Tech. Report TR-2011-6, CS Dept., University of Ioannina, Greece (2011)
11. Hadjidoukas, P.E., Dimakopoulos, V.V.: Nested parallelism in the OMPi OpenMP C compiler. In: Kermarrec, A.-M., Bougé, L., Priol, T. (eds.) Euro-Par 2007. LNCS, vol. 4641, Springer, Heidelberg (2007)
12. Philos, G.C., Dimakopoulos, V.V., Hadjidoukas, P.E.: A runtime architecture for ubiquitous support of OpenMP. In: 7th Int'l. Symposium on Parallel and Distrib. Comput., Krakow, Poland (2008)
13. Jegou, Y.: Implementation of page management in Mome, a user-level DSM. In: 3rd IEEE Int'l. Symposium on Cluster Comput. and the Grid, Tokyo, Japan (2003)
14. Kise, K., et al.: Evaluation of the acknowledgment reduction in a sDSM system. In: 6th Int'l. Conf. on Parallel Processing and Applied Math., Poznan, Poland (2005)
15. ADLB library, http://www.cs.mtsu.edu/~rbutler/adlb/

A Failure Detector for Wireless Networks with Unknown Membership

Fabíola Greve[1,*], Pierre Sens[2], Luciana Arantes[2], and Véronique Simon[2]

[1] Department of Computer Science,
Federal University of Bahia (UFBA), Bahia - Brazil
[2] LIP6, University of Paris 6, CNRS,
INRIA, 4 - Place Jussieu, 75005, Paris, France

Abstract. The distributed computing scenario is rapidly evolving for integrating self-organizing and dynamic wireless networks. Unreliable failure detectors are classical mechanisms which provide information about process failures and can help systems to cope with the high dynamism of these networks. A number of failure detection algorithms has been proposed so far; nonetheless, most of them assume a global knowledge about the membership as well as a fully communication connectivity; additionally, they are timer-based, requiring that eventually some bound on the message transmission will hold. These assumptions are no longer appropriate to the new scenario. This paper presents a new failure detector protocol which implements a new class of detectors, namely $\diamond S^{\mathcal{M}}$, which adapts the properties of the $\diamond S$ class to a dynamic network with an unknown membership. It has the interesting feature to be time-free, so that it does not rely on timers to detect failures; moreover, it tolerates mobility of nodes and message losses.

Keywords: Unreliable failure detector, dynamic distributed systems, wireless mobile networks, asynchronous systems.

1 Introduction

The distributed computing scenario is rapidly evolving for integrating unstructured, self-organizing and dynamic systems, like MANETs (mobile ad-hoc networks) [1]. Nonetheless, the issue of designing reliable services which can cope with the high dynamism of these systems is a challenge. Failure detector is a fundamental service, able to help in the development of fault-tolerant distributed systems. *Unreliable failure detectors*, namely FD, can informally be seen as a per process oracle, which periodically provides a list of processes suspected of having crashed [2]. In this paper, we are interested in the class of eventually strong FDs, denoted $\diamond S$. Those FDs can make an arbitrary number of mistakes; yet, there is a time after which some correct process is never suspected (*eventual weak accuracy* property). Moreover, eventually, every process that crashes is permanently suspected by every correct process (*strong completeness* property). $\diamond S$ is the weakest class allowing to solve consensus in an asynchronous system (with the additional assumption that a majority of processes are correct) and

---

* The work of F. Greve is supported by grants from CAPES-Brazil and Paris City Hall, France.

E. Jeannot, R. Namyst, and J. Roman (Eds.): Euro-Par 2011, LNCS 6853, Part II, pp. 27–38, 2011.

consensus is as the heart of important middleware, e.g., group communication services, transactions and replication servers.

The nature of wireless mobile networks creates important challenges for the development of failure detection protocols. The inherent dynamism of these environments prevents processes from gathering a global knowledge of the system's properties. The network topology is constantly changing and the best that a process can have is a local perception of these changes. Global assumptions, such as the knowledge about the whole membership, the maximum number of crashes, full connectivity or reliable communication, are no more realistic.

This paper proposes a FD algorithm that implements the class $\Diamond S^M$ of failure detectors. This class adapts the properties of the $\Diamond S$ class to a dynamic system with an unknown membership. It is suitable for wireless mobile networks and has the following innovative features that allow for scalability and adaptability: (i) it is conceived for a network whose membership is unknown and whose communication graph is not complete; (ii) it tolerates node mobility, beyond arbitrary joins and leaves; (iii) the failure detection uses local information (for the membership of the neighborhood), instead of traditional global information, such as n (the total number of nodes) and f (the maximum number of faults); (iv) the failure detection is time-free, thus the satisfaction of the properties of the FD does not rely on traditional synchrony assumptions, but on a message exchange pattern followed by the nodes; (v) the message exchange pattern is based on local exchanged information among neighbors and not on global exchanges among nodes in the system. As far as we are aware of, this is the first time-free FD algorithm for networks with unknown membership that tolerates mobility of nodes.

1.1 Related Work

A number of failure detection algorithms has been proposed so far. Nonetheless, most of current implementations of FDs are based on an all-to-all communication approach where each process periodically sends "I am alive" messages to all processes [3]. As they usually consider a fully connected set of known nodes, these implementations are not adequate for dynamic environments. Furthermore, they are usually timer-based, assuming that eventually some bound of the transmission will permanently hold. Such an assumption is not suitable for dynamic environments where communication delays between two nodes can vary due to mobility of nodes. In [4], Mostefaoui *et al.* have proposed an asynchronous implementation of FDs which is time-free. It is based on an exchange of messages which just uses the values of f and n. However, their computation model consists of a set of fully connected initially known nodes. Some works [5–7] focus on the heartbeat FD for sparsely connected networks with unknown membership. The heartbeat FD is a special class of FD which is time-free and is able to implement quiescent reliable communication. But, instead of lists of suspects, it outputs a vector of unbounded counters; if a process crashes, its counter eventually stops increasing. It is worth remarking that none of these works tolerate mobility of nodes.

Few implementations of unreliable FDs focus on wireless mobile networks [8–10]. The fundamental difference between these works and ours is the fact that all of them are timer-based. Friedman and Tcharny [8] propose a simple gossiping protocol which exploits the natural broadcast range of wireless networks to delimit the local membership of a node in a mobile network. Contrarily to our approach, this work assumes a known number of nodes and provides probabilistic guarantees for the FD properties. Tai *et al.* [9] exploit a cluster-based communication architecture to propose a hierarchical gossiping FD protocol for a network of non-mobile nodes. The FD is implemented both via intra-cluster heartbeat diffusion and failure report diffusion across clusters, i.e., if a failure is detected in a local cluster, it will be further forwarded across the clusters. Unlike our solution, this work considers a cluster-based communication architecture and provides probabilistic guarantees for the accuracy and completeness properties; moreover, it does not consider mobility. Sridhar [10] adopts a hierarchical design to propose a deterministic local FD. He introduces the notion of *local failure detection* and restraints the scope of detection to the neighborhood of a node and not to the whole system. While our approach allows the implementation of a $\Diamond S^M$ FD, this work implements an eventually perfect *local* failure detector of the class $\Diamond P$, i.e., it provides perfect failure detection, but with regard to a node's neighborhood. As soon as we are aware of, the only work to follow a time-free detection strategy has been proposed by [11] in order to implement a *leader* FD of the Ω class. This class ensures that, each process will be provided by a unique leader, elected among the set of correct processes, in spite of crashes. Differently from ours, this work is for a specific infra-structured network composed of mobile and static nodes. We believe that our FD of class $\Diamond S^M$ may be successful adopted to implement coordination protocols in a dynamic set, such as the one proposed by Greve *et al.*[12], who present a solution for the fault-tolerant consensus in a network of unknown participants with minimal synchrony assumptions.

The rest of the paper is organized as follows. Section 2 defines the model and specifies the $\Diamond S^M$ FD class. Section 3 identifies assumptions to implement those FDs. Section 4 presents a time-free FD of the $\Diamond S^M$ class. Section 5 concludes the paper. In an extended report [13], one can find complete correctness proofs, a thorough related work section and performance experiments showing that the proposed FD exhibits a good reactivity to detect failures and revoke false suspicions, even in presence of mobility.

2   Model and Problem Definition

The wireless mobile network is a dynamic system composed of infinitely many processes; but each run consists of a finite set Π of $n > 1$ mobile nodes, namely, $\Pi = \{p_1, \ldots, p_n\}$. Contrarily to a static network, the membership is unknown, thus processes are not aware about Π and n, because, moreover, these values can vary from run to run; this coincides with the *finite arrival model* [14]. This model is suitable for long-lived or unmanaged applications, as for example, sensor networks deployed to support crises management or help on dealing with natural disasters. There is one process per node; each process knows its own identity, but it does not necessarily knows the identities of the others. Nonetheless, nodes communicate by sending and receiving messages via a packet radio network and may make use of the broadcast facility of this communication medium to know one another. There are no assumptions on the relative speed of processes or on message transfer delays, thus the system is *asynchronous*; there is no global clock, but to simplify the presentation, we take the range $\mathcal{T}$ of the clock's tick to be the set of natural numbers. A process may fail by *crashing*, i.e., by prematurely or by deliberately halting (switched off); a crashed process does not recover.

The network is represented by a communication graph $G = (V, E)$ in which $V = \Pi$ represents the set of mobile nodes and E represents the set of logical links. The topology of G is dynamic due to arbitrary joins, leaves, crashes and moves. A bidirectional link between nodes p_i and p_j means that p_i is within the wireless transmission range of p_j and vice-versa. If this assumption appears to be inappropriate for a mobile environment, one can use the strategy proposed in [15] for allowing a protocol originally designed for bidirectional links to work with unidirectional links. Let R_i be the transmission range of p_i, then all the nodes that are at distance at most R_i from p_i in the network are considered 1-hop *neighbors*, belonging to the same *neighborhood*. We denote N_i to be the set of 1-hop *neighbors* from p_i; thus, $(p_i, p_j) \in E$ iff $(p_i, p_j) \in N_i$. Local broadcast between 1-hop neighbors is *fair-lossy*. This means that messages may be lost, but, if p_i broadcasts m to processes in its neighborhood an infinite number of times, then every p_j in the neighborhood receives m from p_i an infinite number of times, or p_j is faulty. This condition is attained if the MAC layer of the underlying wireless network provides a protocol that reliably delivers broadcast data, even in presence of unpredictable behaviors, such as fading, collisions, and interference; solutions in this sense have been proposed in [16–18]. Nodes in Π may be mobile and they can keep continuously *moving* and *pausing* in the system. When a node p_m moves, its neighborhood may change. We consider a *passive mobility* model, i.e., the node that is moving does not know that it is moving. Hence, the mobile node p_m cannot notify its neighbors about its moving. Then, for the viewpoint of a neighbor, it is not possible to distinguish between a moving, a leave or a crash of p_m. During the neighborhood changing, p_m keeps its state, that is, the values of its variables.

2.1 Stability Assumptions

In order to implement unreliable failure detectors with an unknown membership, processes should interact with some others to be known. If there is some process in the system such that the rest of processes have no knowledge whatsoever of its identity, there is no algorithm that implements a failure detector with weak completeness, even if links are reliable and the system is synchronous [19]. In this sense, the characterization of the *actual membership* of the system, that is, the set of processes which might be considered for the computation is of utmost importance for our study. We consider then that after have joined the system for some point in time, a mobile process p_i must communicate somehow with the others in order to be known. Afterwards, if p_i leaves, it can re-enter the system with a new identity, thus, it is considered as a new process. Processes may join and leave the system as they wish, but the number of re-entries is bounded, due to the finite arrival assumption. One important aspect concerns the time period and conditions in which processes are connected to the system. During unstable periods, certain situations, as for example, connections for very short periods, the rapid movement of nodes, or numerous joins or leaves along the execution (characterizing a churn) could block the application and prevent any useful computation. Thus, the system should present some stability conditions that when satisfied for longtime enough will be sufficient for the computation to progress and terminate.

Definition 1. *Membership Let $t, t' \in \mathcal{T}$. Let $UP(t) \subset \Pi$ be the set of mobile processes that are in the system at time t, that is, after have joined the system before t, they neither* *leave it nor crash before t. Let p_i, p_j be mobile nodes. Let the $known_j$ set denotes the partial knowledge of p_j about the system's membership. The membership of the system is the* KNOWN *set.*

STABLE $\overset{def}{=} \{p_i : \exists t, t', s.t. \forall t' \geq t,\ p_i \in UP(t')\}$.

FAULTY $\overset{def}{=} \{p_i : \exists t, t',\ t < t',\ p_i \in UP(t) \land p_i \notin UP(t')\}$.

KNOWN $\overset{def}{=} \{p_i : (p_i \in$ STABLE $\cup$ FAULTY$) \land (p_i \in known_j, p_j \in$ STABLE$)\}$.

The actual membership is in fact defined by the KNOWN set. A process is *known* if, after have joined the system, it has been identified by some stable process. A *stable* process is thus a mobile process that, after had entered the system for some point in time, never departs (due to a crash or a leave); otherwise, it is *faulty*. A process is faulty after time t, when, after had entered the system at t, it departs at $t' > t$. The STABLE set corresponds to the set of *correct* processes in the classical model of static systems.

Assumption 1. *Connectivity Let* $G($KNOWN $\cap$ STABLE$) = G(S) \subseteq G$ *be the graph obtained from the stable known processes. Then,* $\exists t \in \mathcal{T}$, *s.t., in* $G(S)$ *there is a path between every pair of processes* $p_i, p_j \in G(S)$.

This connectivity assumption states that, in spite of changes in the topology of G, from some point in time t, the set of known stables forms a *strongly connected component* in G. This condition is frequently present in the classical model of static networks and is indeed mandatory to ensure dissemination of messages to all stable processes and thus to ensure the global properties of the failure detector [2, 19–21].

2.2 A Failure Detector of Class $\Diamond S^M$

Unreliable failure detectors provide information about the liveness of processes in the system [2]. Each process has access to a local failure detector which outputs a list of processes that it currently suspects of being faulty. The failure detector is *unreliable* in the sense that it may erroneously add to its list a process which is actually correct. But if the detector later believes that suspecting this process is a mistake, it then removes the process from its list. Failure detectors are formally characterized by two properties: (i) *Completeness* characterizes its capability of suspecting every faulty process permanently; (ii) *Accuracy* characterizes its capability of not suspecting correct processes. Our work is focused on the class of *Eventually Strong* detectors, also known as $\Diamond S$. Nonetheless, we adapt the properties of this class in order to implement a FD in a dynamic set. Then, we define the class of *Eventually Strong Failure Detectors* with *Unknown Membership*, namely $\Diamond S^M$. This class keeps the same properties of $\Diamond S$, except that they are now valid to known processes, that are stable and faulty.

Definition 2. *Eventually Strong FD with Unknown Membership* $(\Diamond S^M)$ *Let* $t, t' \in \mathcal{T}$. *Let* p_i, p_j *be mobile nodes. Let* $susp_j$ *be the list of processes that* p_j *currently suspects of being faulty. The* $\Diamond S^M$ *class contains all the failure detectors that satisfy:*

Strong completeness $\overset{def}{=} \{\exists t, t', s.t. \forall t' \geq t,\ \forall p_i \in$ KNOWN $\cap$ FAULTY $\Rightarrow p_i \in susp_j,\ \forall p_j \in$ KNOWN $\cap$ STABLE$\}$.

Eventual weak accuracy $\overset{def}{=} \{\exists t, t', s.t. \forall t' \geq t,\ \exists p_i \in$ KNOWN $\cap$ STABLE $\Rightarrow p_i \notin susp_j,\ \forall p_j \in$ KNOWN $\cap$ STABLE$\}$.

3   Towards a Time-Free Failure Detector for the $\Diamond S^M$ Class

None of the failure detector classes can be implemented in a purely asynchronous system [2]. Indeed, while completeness can be realized by using "I am alive" messages and timeouts, accuracy cannot be safely implemented for all system executions. Thus, some additional assumptions on the underlying system should be made in order to implement them. With this aim, two orthogonal approaches can be distinguished: the timer-based and the time-free failure detection [22]. The timer-based model is the traditional approach and supposes that channels in the system are eventually timely; this means that, for every execution, there are bounds on process speeds and on message transmission delays. However, these bounds are not known and they hold only after some unknown time [2]. An alternative approach suggested by [4] and developed so far by [11, 20] considers that the system satisfies a message exchange pattern on the execution of a *query-based* communication and is time-free. While the timer-based approach imposes a constraint on the physical time (to satisfy message transfer delays), the time-free approach imposes a constraint on the logical time (to satisfy a message delivery order). These approaches are orthogonal and cannot be compared, but, they can be combined at the link level in order to implement hybrid protocols with combined assumptions [22].

3.1   Stable Query-Response Communication Mechanism

Our failure detector is time-free and based on a local QUERY-RESPONSE communication mechanism [20] adapted to a network with unknown membership. At each *query-response* round, a node systematically broadcasts a QUERY message to the nodes in its neighborhood until it possibly crashes or leaves the system. The time between two consecutive queries is finite but arbitrary. Each couple of QUERY-RESPONSE messages are uniquely identified in the system. A process p_i launches the primitive by sending a QUERY(m) with a message m. When a process p_j delivers this query, it updates its local state and systematically answers by sending back a RESPONSE() to p_i. Then, when p_i has received at least α_i responses from different processes, including a stable one, the current QUERY-RESPONSE *terminates*. Without loss of generality, the response for p_i itself is among the α_i responses. An implementation of a QUERY-RESPONSE communication over fair-lossy local channels can be done by the repeated broadcast of the query by the sender p_i until it has received at least α_i responses from its neighbors. Formally, the stable QUERY–RESPONSE primitive has the following properties:

(i) QR-Validity: If a QUERY(m) is delivered by process p_j, it has been sent by p_i;
(ii) QR-Uniformity: A QUERY(m) is delivered at most once by a process;
(iii) QR-Stable-Termination: If a process p_i is not faulty (it does not crash nor leave the system) while it is issuing a query, that query generates at least α_i responses.

The value associated to α_i should correspond to the expected number of processes with whom p_i can communicate, in spite of moves and faults. Since communication is local, α_i is a local parameter and can be defined as the value of the neighborhood density of p_i (i.e., $|N_i|$) minus the maximum number of faulty processes in its neighborhood; let f_i be this number; that is, $\alpha_i = |N_i| - f_i$. This local choice for α_i changes from previous works which consider a global value either proportional to the number of correct processes [4] or the number of stable processes [20] or the global number of faults [11]. Moreover, it follows recent works on fault tolerant communication in radio networks which propose a "local" fault model, instead of a "global" fault model, as an adequate strategy to deal with the dynamism and unreliability of wireless channels in spite of failures [17]. To reliably delivery data in spite of crashes, the maximum number of local failures should be $f_i < |N_i|/2$ [23]. From Assumption 1 about the network connectivity over time, at least one stable known node p_j will receive the QUERY and send a RESPONSE to p_i, since moreover channels are fair-lossy. Thus, the following property holds:

Property 1. **Stable Termination Property** ($\mathcal{S}at\mathcal{P}$). Let p_i be a node which issues a QUERY. Let X_i be the set of processes that issued a RESPONSE to that query. Thus, $\exists p_j \in X_i, p_j \in$ KNOWN $\cap$ STABLE, $p_j \neq p_i$.

For the failure detection problem, the *stable termination* is important for the diffusion of the information to the whole network and consequent satisfaction of the accuracy and completeness properties. Moreover, it ensures that the first QUERY issued by p_i, when it joins the network, will be delivered by at least one stable process in such a way that p_i may take part to the membership of the system.

3.2   Behavioral Properties

Node p_i can keep continuously moving and pausing, but, infinitely often, p_i should stay within its neighborhood for a sufficient period of time in order to be able to update its state with recent information regarding suspicions and mistakes; otherwise, it would not update its state properly and thus completeness and accuracy properties would not be ensured. Recent information is gathered by p_i from its neighbors via the delivery of a QUERY message. Hence, the following *mobility property*, namely $\mathcal{M}obi\mathcal{P}$, has been defined and should be satisfied by all nodes. It ensures that, after reaching a new neighborhood at t', there will be a time $t > t'$ at which p_i should have received QUERY messages from at least one stable neighbor p_j, beyond itself. Since channels are fair-lossy, the QUERY sent by p_j will be received by p_i, except if p_i is faulty.

Property 2. **Mobility Property** ($\mathcal{M}obi\mathcal{P}$). Let $t', t \in \mathcal{T}, t' < t$. Let p_i be a node. Let t' be the time after which p_i has changed of neighborhood. Let SQ_i^t be the set of processes from which p_i has received a QUERY message after t' and before or at t. Process p_i satisfies $\mathcal{M}obi\mathcal{P}$ at time t if:

$$\mathcal{M}obi\mathcal{P}^t(p_i) \stackrel{def}{=} \exists p_{j,j\neq i} \in SQ_i^t, t > t' : p_j \in \text{KNOWN} \cap \text{STABLE} \ \lor \ p_i \ is faulty$$
after t'.

Instead of synchrony assumptions, to ensure the accuracy of the detection, the time-free model establishes conditions on the logical time the messages are delivered by processes. These are unified in the *stabilized responsiveness property*, namely $\mathcal{S}\mathcal{R}\mathcal{P}$. Thus, $\mathcal{S}\mathcal{R}\mathcal{P}(p_i)$ states that eventually, for any process p_j (which had received a response from p_i in the past), the set of responses received by p_j to its last QUERY always includes a response from p_i, that is, the response of p_i is always a winning response [22].

Property 3. **Stabilized Responsiveness Property** ($\mathcal{SRP}$). Let $t'', t', t \in \mathcal{T}$. Let p_i be a stable known node. Let $rec\_from_j^{t'}$ ($rec\_from_j^{t''}$) be the set of processes from which p_j has received responses to its last QUERY that terminated at or before $t'(t'')$. Process p_i satisfies $\mathcal{SRP}$ at time t if:

$$\mathcal{SRP}^t(p_i) \stackrel{def}{=} \forall t' \geq t, \forall t'' > t', p_i \in rec\_from_j^{t'} \Rightarrow p_i \in rec\_from_j^{t''} \vee$$
p_j *is faulty* after t.

This property denotes the ability of a stable known node p_i to reply, among the first α_i nodes, to a QUERY sent by a node p_j, who had received responses from p_i before. It should hold for at least one stable known node p_i; thus preventing p_i to be permanently suspected. As a matter of comparison, in the timer-based model, this property would approximate the following: there is a time t after which the output channels from a stable process p_i to every other process p_j that knows p_i are eventually timely.

In order to implement a $\Diamond S^M$ FD, the following behaviors should be satisfied:
1) $\forall p_i \in$ KNOWN : $\mathcal{MobiP}^t(p_i)$ holds after p_i moves and changes of neighborhood;
2) $\exists p_i \in$ KNOWN $\cap$ STABLE : $\mathcal{SRP}^t(p_i)$ eventually holds.

A discussion about how to satisfy in practice the properties and assumptions of the model is done in Section 4.2 after the protocol's explanation.

4   A Failure Detector Algorithm for the $\Diamond S^M$ Class

4.1   Algorithm Description

Algorithm 1 describes our protocol for implementing a FD of class $\Diamond S^M$ for a network of KNOWN mobile nodes that satisfies the model stated in Sections 2 and 3.

Notations. We use the following notations:
- $susp_i$: denotes the current set of processes suspected of being faulty by p_i. Each element of this set is a tuple of the form $\langle id, ct \rangle$, where id is the identifier of the suspected node and ct is the tag associated to this information.
- $mist_i$: denotes the set of nodes which were previously suspected of being faulty but such suspicions are currently considered to be a mistake. Similar to the $susp_i$ set, the $mist_i$ is composed of tuples of the form $\langle id, ct \rangle$.
- rec_from_i: denotes the set of nodes from which p_i has received responses to its last QUERY message.
- $known_i$: denotes the partial knowledge of p_i about the system's membership, i.e., it denotes the current knowledge of p_i about its neighborhood.
- $Add(set, \langle id, ct \rangle)$: is a function that includes $\langle id, ct \rangle$ in set. If an $\langle id, - \rangle$ already exists in set, it is replaced by $\langle id, ct \rangle$.

Description. The algorithm is composed of two tasks $T1$ and $T2$.

Task $T1$: Generating suspicions. This task is made up of an infinite loop. At each round, a QUERY($susp_i$, $mist_i$) message is sent to all nodes of p_i's neighborhood (line 5). Node p_i waits for at least α_i responses, which includes p_i's own response (line 6). Then, p_i detects new suspicions (lines 8-13). It starts suspecting each node p_j, not previously suspected ($p_j \notin susp_i$), which it knows ($p_j \in known_i$), but from which it does not receive a RESPONSE to its last QUERY. If a previous mistake information related to this new suspected node exists in the mistake set $mist_i$, it is removed from it (line 11) and the suspicion information is then included in $susp_i$ with a tag which is greater than the previous mistake tag (line 10). If p_j is not in the $mist$ set (i.e., it is the first time p_j is suspected), p_i suspected information is tagged with 0 (line 13).

Algorithm 1. Time-Free Implementation of a $\Diamond S^M$ Failure Detector

```
1    init:
2        suspᵢ ← ∅; mistᵢ ← ∅  ;  knownᵢ ← ∅
3    Task T1:
4    Repeat forever
5        broadcast QUERY(suspᵢ, mistᵢ)
6        wait until RESPONSE received from ≥ αᵢ processes
7        rec_fromᵢ ← all pⱼ, a RESPONSE is received in line 6
8        For all pⱼ ∈ knownᵢ \ rec_fromᵢ | ⟨pⱼ,−⟩ ∉ suspᵢ do
9            If  ⟨pⱼ,ct⟩ ∈ mistᵢ
10               Add(suspᵢ,⟨pⱼ,ct+1⟩)
11               mistᵢ = mistᵢ \ {⟨pⱼ,−⟩}
12           Else
13               Add(suspᵢ,⟨pⱼ,0⟩)
14   End repeat
15
16   Task T2:
17   Upon reception of QUERY (suspⱼ,mistⱼ) from pⱼ do
18   knownᵢ ← knownᵢ ∪ {pⱼ}
19   For  all ⟨pₓ,ctₓ⟩ ∈ suspⱼ do
20   If ⟨pₓ,−⟩ ∉ suspᵢ ∪ mistᵢ or (⟨pₓ,ct⟩ ∈ suspᵢ ∪ mistᵢ and ct < ctₓ)
21       If pₓ = pᵢ
22           Add(mistᵢ,⟨pᵢ,ctₓ+1⟩)
23       Else
24           Add(suspᵢ,⟨pₓ,ctₓ⟩)
25           mistᵢ = mistᵢ \ {⟨pₓ,−⟩}
26   For  all ⟨pₓ,ctₓ⟩ ∈ mistⱼ do
27   If ⟨pₓ,−⟩ ∉ suspᵢ ∪ mistᵢ or (⟨pₓ,ct⟩ ∈ suspᵢ ∪ mistᵢ and ct < ctₓ)
28       Add(mistᵢ,⟨pₓ,ctₓ⟩)
29       suspᵢ = suspᵢ \ {⟨pₓ,−⟩}
30       If (pₓ ≠ pⱼ)
31           knownᵢ ← knownᵢ \ {pₓ}
32   send RESPONSE to pⱼ
```

Task T2: Propagating suspicions and mistakes. This task allows a node to handle the reception of a QUERY message. A QUERY message contains the information about suspected nodes and mistakes kept by the sending node. However, based on the tag associated to each piece of information, the receiving node only takes into account the ones that are more recent than those it already knows or the ones that it does not know at all. The two loops of task $T2$ respectively handle the information received about suspected nodes (lines 19–25) and about mistaken nodes (lines 26–31). Thus, for each node p_x included in the suspected (respectively, mistake) set of the QUERY message, p_i includes the node p_x in its $susp_i$ (respectively, $mist_i$) set only if the following condition is satisfied: p_i received a more recent information about p_x status (failed or mistaken) than the one it has in its $susp_i$ and $mist_i$ sets. Furthermore, in the first loop of task $T2$, a new mistake is detected if the receiving node p_i is included in the suspected set of the QUERY message (line 21) with a greater tag. At the end of the task (line 32), p_i sends to the querying node a RESPONSE message.

Dealing with mobility and generating mistakes. When a node p_m moves to another destination, the nodes of its old destination will start suspecting it, since p_m is in their *known* set and it cannot reply to QUERY messages from the latter anymore. Hence, QUERY messages that include p_m as a suspected node will be propagated to nodes of the network. Eventually, when p_m reaches its new neighborhood, it will receive such suspicion messages. Upon receiving them, p_m will correct such a mistake by including itself (p_m) in the mistake set of its corresponding QUERY messages with a greater tag (lines 21–22). Such information will be propagated over the network. On the other hand, p_m will start suspecting the nodes of its old neighborhood since they are in its $known_m$ set. It then will broadcast this suspicion in its next QUERY message. Eventually, this information will be corrected by the nodes of its old neighborhood and the corresponding generated mistakes will spread over the network, following the same principle.

In order to avoid a "ping-pong" effect between information about suspicions and mistakes, lines 30–31 allow the updating of the *known* sets of both the node p_m and of those nodes that belong to the original destination of p_m. Then, for each mistake $\langle p_x, ct_x \rangle$ received from p_j, such that p_i keeps an old information about p_x, p_i verifies whether p_x is the sending node p_j (line 30). If they are different, p_x should belong to a remote neighborhood, because otherwise, p_i would have received the mistake by p_x itself. Notice that only the node can generate a new mistake about itself (line 21). Thus, p_x is removed from the $known_i$ set (line 31). Notice, however, that this condition is not sufficient to detect the mobility, because p_x can be a neighbor of p_i and due to an asynchronous race, the QUERY sent by p_x with the mistake has not yet arrived at p_i. In fact, the propagated mistake sent by p_j has arrived at p_i firstly. If that is the case, p_x has been unduly removed from $known_i$. Fortunately, since local broadcast is fair-lossy, the QUERY from p_x is going to eventually arrive at p_i, if p_i is stable, and, as soon as the QUERY arrives, p_i will once again add p_x to $know_i$ (lines 17–18).

4.2   Practical Issues

The *stable termination* of the QUERY-RESPONSE primitive and the $\mathcal{M}obi\mathcal{P}$ property may be satisfied if the time of pause, between changes in direction and/or speed, is defined to be greater than the time to transmit the QUERY and receive the RESPONSE messages. This condition is attained when for example, the most widely used Random Waypoint Mobility Model [24] is considered. In practice, the value of α_i (the number of responses that a process p_i should wait in order to implement a QUERY-RESPONSE) relates not only with the application density and the expected number of local faults, but also with the type of network considered (either WMN, WSN, etc.) and the current topology of the network during execution. Thus, it can be defined on the fly, based on the current behavior of the network. Wireless Mesh Network (WMN), Wireless Sensor

Network (WSN), and infra-structured mobile networks [11, 25] are a good examples of platforms who would satisfy the assumptions of our model, specially the $\mathcal{SRP}$. In a WMN, the nodes move around a fixed set of nodes (the core of the network) and each mobile node eventually connects to a fix node. A WSN is composed of stationary nodes and can be organized in clusters, so that communication overhead can be reduced; one node in each cluster is designated the cluster head (CH) and the other nodes, cluster members (CMs). Communication inter-clusters is always routed through the respective CHs which act as gateway nodes and are responsible for maintaining the connectivity among neighboring CHs. An infra-structured mobile network is composed of mobile hosts (MH) and mobile support stations (MSS). A MH is connected to a MSS if it is located in its transmission range and two MHs can only communicate through MSSs, but, due to mobility, an MH can leave and enter the area covered by other MSSs. The system is composed of N MSSs but infinitely many MHs. However, in each run the protocol has only finitely many MHs. There are some works to implement a leader oracle [11] and to solve consensus in this type of network [25].

For all these platforms, special nodes (the fixed node for WMN, CHs for WSN or MSSs for infra-structured networks) eventually form a strongly connected component of stable nodes; additionally, they can be regarded as fast, so that they will always answer to a QUERY faster than the other nodes, considered as slow nodes (the mobile node for WMN, CMs for WSN or MHs for infra-structured networks). Thus, one of these fast nodes may satisfy the $\mathcal{SRP}$ property. The $\mathcal{SRP}$ may seem strong, but in practice it should just hold during the time the application needs the strong completeness and eventual weak accuracy properties of FDs of class $\diamond S^M$, as for instance, the time to execute a consensus algorithm.

5   Conclusion

This paper has presented a new algorithm for an unreliable failure detector suitable for mobile wireless networks, such as WMNs or WSNs. It implements failure detectors of class $\diamond S^M$ (eventually strong with unknown membership) when the exchanged pattern of messages satisfies some behavioral properties. As a future work, we plan to adapt the algorithm and properties to implement other classes of failure detectors.

References

1. Conti, M., Giordano, S.: Multihop ad hoc networking: The theory. IEEE Communications Magazine 45(4), 78–86 (2007)
2. Chandra, T., Toueg, S.: Unreliable failure detectors for reliable distributed systems. Journal of the ACM 43(2), 225–267 (1996)
3. Devianov, B., Toueg, S.: Failure detector service for dependable computing. In: Proc. of the 1st Int. Conf. on Dependable Systems and Networks, pp. 14–15 (2000)
4. Mostefaoui, A., Mourgaya, E., Raynal, M.: Asynchronous implementation of failure detectors. In: Proc. of Int. Conf. on Dependable Systems and Networks (2003)
5. Aguilera, M.K., Chen, W., Toueg, S.: Heartbeat: A timeout-free failure detector for quiescent reliable communication. In: Proc. of the 11th International Workshop on Distributed Algorithms, pp. 126–140 (1997)

6. Hutle, M.: An efficient failure detector for sparsely connected networks. In: Proc. of the IASTED International Conference on Parallel and Distributed Computing and Networks, pp. 369–374 (2004)

7. Tucci-Piergiovanni, S., Baldoni, R.: Eventual leader election in infinite arrival message-passing system model with bounded concurrency. In: Dependable Computing Conference (EDCC), pp. 127–134 (2010)

8. Friedman, R., Tcharny, G.: Evaluating failure detection in mobile ad-hoc networks. Int. Journal of Wireless and Mobile Computing 1(8) (2005)

9. Tai, A., Tso, K., Sanders, W.: Cluster-based failure detection service for large-scale ad hoc wireless network applications. In: Int. Conf. on Dependable Systems and Networks, pp. 805–814 (2004)

10. Sridhar, N.: Decentralized local failure detection in dynamic distributed systems. In: The 25th IEEE Symp. on Reliable Distributed Systems, pp. 143–154 (2006)

11. Cao, J., Raynal, M., Travers, C., Wu, W.: The eventual leadership in dynamic mobile networking environments. In: 13th Pacific Rim Intern. Symp. on Dependable Computing, pp. 123–130 (2007)

12. Greve, F., Tixeuil, S.: Knowledge conectivity vs. synchrony requirements for fault-tolerant agreement in unknown networks. In: Int. Conf. on Dependable Systems and Networks, pp. 82–91 (2007)

13. Sens, P., Arantes, L., Bouillaguet, M., Simon, V., Greve, F.: Asynchronous implementation of failure detectors with partial connectivity and unknown participants. Technical Report, RR6088, INRIA - France, http://hal.inria.fr/inria-00122517/fr/

14. Aguilera, M.K.: A pleasant stroll through the land of infinitely many creatures. SIGACT News 35(2), 36–59 (2004)

15. Ramasubramanian, V., Chandra, R., Mossé, D.: Providing a bidirectional abstraction for unidirectional adhoc networks. In: Proc. of the 21st IEEE International Conference on Computer Communications (2002)

16. Min-Te, S., Lifei, H., Arora, A.A., Ten-Hwang, L.: Reliable mac layer multicast in ieee 802.11 wireless networks. In: Proc. of the Intern, August 2002, pp. 527–536 (2002)

17. Koo, C.Y.: Broadcast in radio networks tolerating byzantine adversarial behavior. In: 23th Symp. on Principles of Distributed Computing, pp. 275–282 (2004)

18. Bhandari, V., Vaidya, N.H.: Reliable local broadcast in a wireless network prone to byzantine failures. In: The 4th Int. Work. on Foundations of Mobile Computing (2007)

19. Jiménez, E., Arévalo, S., Fernández, A.: Implementing unreliable failure detectors with unknown membership. Inf. Process. Lett. 100(2), 60–63 (2006)

20. Mostefaoui, A., Raynal, M., Travers, C., Patterson, S., Agrawal, D., Abbadi, A.: From static distributed systems to dynamic systems. In: Proc. of the 24th IEEE Symposium on Reliable Distributed Systems, pp. 109–118 (2005)

21. Bhandari, V., Vaidya, N.H.: Reliable broadcast in radio networks with locally bounded failures. IEEE Trans. on Parallel and Distributed Systems 21, 801–811 (2010)

22. Mostefaoui, A., Raynal, M., Travers, C.: Time-free and timer-based assumptions can be combined to obtain eventual leadership. IEEE Trans. Parallel Distrib. Syst. 17(7), 656–666 (2006)

23. Bhandari, V., Vaidya, N.H.: On reliable broadcast in a radio network. In: 24th Symp. on Principles of Distributed Computing, pp. 138–147. ACM, New York (2005)

24. Camp, T., Boleng, J., Davies, V.: A survey of mobility models for ad hoc network research. Wireless Communications & Mobile Computing: Special issue on Mobile Ad Hoc Networking: Research, Trends and Applications 2, 483–502 (2002)

25. Wu, W., Cao, J., Yang, J., Raynal, M.: Design and performance evaluation of efficient consensus protocols for mobile ad hoc networks. IEEE Trans. Comput. 56(8), 1055–1070 (2007)

Towards Systematic Parallel Programming over MapReduce

Yu Liu[1], Zhenjiang Hu[2], and Kiminori Matsuzaki[3]

[1] The Graduate University for Advanced Studies, Japan
yuliu@nii.ac.jp
[2] National Institute of Informatics, Japan
hu@nii.ac.jp
[3] School of Information, Kochi University of Technology, Japan
matsuzaki.kiminori@kochi-tech.ac.jp

Abstract. MapReduce is a useful and popular programming model for data-intensive distributed parallel computing. But it is still a challenge to develop parallel programs with MapReduce systematically, since it is usually not easy to derive a proper divide-and-conquer algorithm that matches MapReduce. In this paper, we propose a homomorphism-based framework named Screwdriver for systematic parallel programming with MapReduce, making use of the program calculation theory of list homomorphisms. Screwdriver is implemented as a Java library on top of Hadoop. For any problem which can be resolved by two sequential functions that satisfy the requirements of the third homomorphism theorem, Screwdriver can automatically derive a parallel algorithm as a list homomorphism and transform the initial sequential programs to an efficient MapReduce program. Users need neither to care about parallelism nor to have deep knowledge of MapReduce. In addition to the simplicity of the programming model of our framework, such a calculational approach enables us to resolve many problems that it would be nontrivial to resolve directly with MapReduce.

1 Introduction

Google's MapReduce [5] is a programming model for data-intensive distributed parallel computing. It is the de facto standard for large scale data analysis, and has emerged as one of the most widely used platforms for data-intensive distributed parallel computing.

Despite the simplicity of MapReduce, it is still a challenge for a programmer to systematically solve his or her (nontrivial) problems. Consider the *maximum prefix sum* problem of a sequence. For instance, if the input sequence is

$$\underline{3, -1, 4, 1, -5}, 9, 2, -6, 5$$

the maximum of the prefix sums should be 13 to which the underlined prefix corresponds. It is not obvious how to solve this problem efficiently with MapReduce (and we encourage the reader to pause to think how to solve this). Such problems widely exist in the real world, e.g, financial time series analysis.

E. Jeannot, R. Namyst, and J. Roman (Eds.): Euro-Par 2011, LNCS 6853, Part II, pp. 39–50, 2011.
© Springer-Verlag Berlin Heidelberg 2011

Our basic idea to resolve such problems is to wrap MapReduce with *list homomorphisms* (or homomorphisms for short) [2]. We propose a simpler programming model based on the theory of list homomorphisms, and implement an associated framework called *Screwdriver*[1] for systematic parallel programming over MapReduce. Screwdriver provides users with an easy-to-use programming interface, where users just need to write two sequential programs: one for solving the problem itself and the other for solving the inverse problem. By building an algorithmic parallelization layer upon MapReduce, Screwdriver automatically generates homomorphisms from user-specific programs and efficiently executes them with MapReduce. We implemented this homomorphism-based framework efficiently in Java on top of an open source MapReduce framework Hadoop[2].

The rest of the paper is organized as follows. In Section 2 we review the concept of MapReduce and the theory of homomorphisms. The design and implementation of the homomorphism-based algorithmic layer on top of MapReduce are illustrated in Section 3. Then, we demonstrate the usefulness of our system with the maximum prefix sum problem above in Section 4, and report some experiment results in Section 5. The conclusion and highlight of future work are summarized in Section 6.

2   MapReduce and List Homomorphisms

The notations are mainly based on the functional language Haskell [1]. Function application is denoted with a space with its argument without parentheses, i.e., $f\ a$ equals to $f(a)$. Functions are curried and bound to the left, and thus $f\ a\ b$ equals to $(f\ a)\ b$. Function application has higher precedence than using operators, so $f\ a \oplus b = (f\ a) \oplus b$. We use two operators $\circ$ and $\vartriangle$ over functions: by definition, $(f \circ g)\ x = f\ (g\ x)$ and $(f \vartriangle g)\ x = (f\ x, g\ x)$. Function id is the identity function.

Tuples are written like (a, b) or (a, b, c). Function fst (snd) extracts the first (the second) element of the input tuple.

We denote lists with square brackets. We use $[\ ]$ to denote an empty list, and $+\!\!+$ to denote the list concatenation: $[3, 1, 4] +\!\!+ [1, 5] = [3, 1, 4, 1, 5]$. A list that has only one element is called a *singleton*. Operator $[\cdot]$ takes a value and returns a singleton list with it.

2.1   MapReduce and MapReduce Programming Model

Figure 1 depicts the MapReduce computation model. The input and output data of MapReduce are managed as a *set* of key-value pairs and stored in distributed file system over a cluster. MapReduce computation mainly consists of three phases: the MAP phase, SHUFFLE & SORT phase, and the REDUCE phase[3].

---

[1] It is available online http://code.google.com/p/screwdriver/

[2] http://hadoop.apache.org/

[3] For readability, we use MAP and REDUCE to denote the phases in MapReduce, and f_{MAP} and f_{REDUCE} for the parameter functions used in the MAP and REDUCE phases. When unqualified, *map* and *reduce* refer to the functions of Haskell.

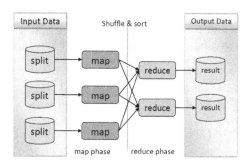

Fig. 1. MapReduce Computation

In the MAP phase, each input key-value pair is processed independently and a list of key-value pairs will be produced. Then in the SHUFFLE & SORT phase, the key-value pairs are grouped based on the key. Finally, in the REDUCE phase, the key-value pairs of the same key are processed to generate a result.

To make the discussion precise, we introduce a specification of the MapReduce programming model in a functional programming manner. Note that the specification in this paper is based on that in [11] but is more detailed. In this model, users need to provide four functions to develop a MapReduce application. Among them, the f_{MAP} and f_{REDUCE} functions performs main computation.

- Function f_{MAP}.

$$f_{\text{MAP}} :: (k_1, v_1) \rightarrow [(k_2, v_2)]$$

  This function is invoked during the MAP phase, and it takes a key-value pair and returns a list of intermediate key-value pairs.

- Function f_{SHUFFLE}.

$$f_{\text{SHUFFLE}} :: k_2 \rightarrow k_3$$

  This function is invoked during the SHUFFLE&SORT phase, takes a key of an intermediate key-value pair, and generate a key with which the key-value pairs are grouped.

- Function f_{SORT}.

$$f_{\text{SORT}} :: k_2 \rightarrow k_2 \rightarrow \{-1, 0, 1\}$$

  This function is invoked during the SHUFFLE&SORT phase, and compares two keys in sorting the values.

- Function f_{REDUCE}.

$$f_{\text{REDUCE}} :: (k_3, [v_2]) \rightarrow (k_3, v_3)$$

  This function is invoked during the REDUCE phase, and it takes a key and a list of values associated to the key and merges the values.

Now a functional specification of the MapReduce framework can be given as follows, which accepts four functions f_{MAP}, f_{SHUFFLE}, f_{SORT}, and f_{REDUCE} and transforms a set of key-value pairs to another set of key-value pairs.

$$MapReduce :: ((k_1, v_1) \to [(k_2, v_2)]) \to (k_2 \to k_3) \to (k_2 \to k_2 \to \{-1, 0, 1\})$$
$$\to ((k_3, [v_2]) \to (k_3, v_3)) \to \{(k_1, v_1)\} \to \{(k_3, v_3)\}$$

$$MapReduce\ f_{\text{MAP}}\ f_{\text{SHUFFLE}}\ f_{\text{SORT}}\ f_{\text{REDUCE}}\ input$$
$$= \textbf{let}\ sub1 = map_{\textbf{S}}\ f_{\text{MAP}}\ input$$
$$sub2 = map_{\textbf{S}}\ (\lambda(k', kvs).\ (k', map\ snd\ (sortByKey\ f_{\text{SORT}}\ kvs)))$$
$$(shuffleByKey\ f_{\text{SHUFFLE}}\ sub1)$$
$$\textbf{in}\ map_{\textbf{S}}\ f_{\text{REDUCE}}\ sub2$$

Function $map_{\textbf{S}}$ is a set version of the map function: i.e., it applies the input function to each element in the set. Function $shuffleByKey$ takes a function f_{SHUFFLE} and a set of a list of key-value pairs, flattens the set, and groups the key-value pairs based on the new keys computed by f_{SHUFFLE}. The result type after $shuffleByKey$ is $\{(k_3, \{k_2, v_2\})\}$. Function $sortByKey$ takes a function f_{SORT} and a set of key-value pairs, and sorts the set into a list based on the relation computed by f_{SORT}.

2.2   List Homomorphism and Homomorphism Theorems

List homomorphisms are a class of recursive functions on lists, which match very well the divide-and-conquer paradigm [2, 4, 5, 12, 15, 16]. They are attractive in parallel programming, not only because they are suitable for parallel implementation, but also because they enjoy many nice algebraic properties, among which, the three well-known homomorphism theorems form the basis of *systematic development of parallel programs* [4, 6, 7, 9, 10]. Recently, it has been shown [8, 14] that homomorphisms can be automatically developed for solving various kinds of problems. All these have indeed motivated us to see how to apply these results to parallel programming with MapReduce.

Definition 1 (List homomorphism). *Function h is said to be a list homomorphism, if and only if there is a function f and an associative operator $\odot$ such that the function h is defined as follows.*

$$h\ [a]\qquad = f\ a$$
$$h\ (x + y) = h\ x \odot h\ y$$

Since h is uniquely determined by f and $\odot$, we write $h = (\!|f, \odot|\!)$.

For instance, the function that sums up the elements in a list can be described as a list homomorphism $(\!|id, +|\!)$:

$$sum\ [a]\qquad = a$$
$$sum\ (x + y) = sum\ x + sum\ y.$$

Below is the well-known theorem for homomorphisms [6]. It provides a necessary and sufficient condition for the existence of a list homomorphism.

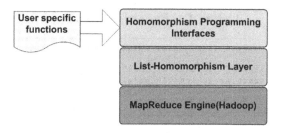

Fig. 2. System Overview

Theorem 1 (The third homomorphism theorem). *Let h be a given function and $\ominus$ and $\oslash$ be binary operators. If and only if the following two equations hold for any element a and list y*

$$h\ ([a] + y) = a \ominus h\ y$$
$$h\ (y + [a]) = h\ y \oslash a$$

then the function h is a homomorphism.

In order to show how to automatically derive a list homomorphism, firstly, we introduce the concept of function's right inverse.

Definition 2 (Right inverse). *For a function h, its right inverse h° is a function that satisfies $h \circ h^\circ \circ h = h$.*

By taking use of right inverse, we can obtain the list-homomorphic definition as follows.
$$h = (\!| f, \odot |\!) \quad \textbf{where } f\ a = h\ [a]$$
$$l \odot r = h\ (h^\circ\ l + h^\circ\ r)$$

With this property, the third homomorphism theorem is also an important and useful theorem for automatic derivation of list homomorphisms [14]. Our parallelization algorithm is mainly based on the third homomorphism theorem.

3   A Homomorphism-Based Framework for Parallel Programming with MapReduce

The main contribution of our work is a novel programming model and its framework for systematic programming over MapReduce, based on theorems of list homomorphisms [8,14]. Our framework *Screwdriver* is built on top of Hadoop, purely in Java.

As shown in Fig. 2, *Screwdriver* consists of three layers: the interface layer for easy parallel programming, the homomorphism layer for implementing homomorphism, and the base layer of the MapReduce engine (Hadoop).

Listing 1.1. Programming Interface

```
1  public abstract class ThirdHomomorphismTheorem<T1,T2> {
2  ...
3      public abstract T2 fold(ArrayList<T1> values);
4      public abstract ArrayList<T1> unfold(T2 value);
5  ...
6  }
```

3.1 Programming Interface and Homomorphism Derivation

The first layer of Screwdriver provides a simple programming interface and generates a homomorphism based on the third homomorphism theorem.

Users specify a pair of sequential functions instead of specifying a homomorphism directly: one for solving the problem itself and the other for a right inverse of the problem. Consider the summing-up example again. A right inverse $sum°$ of the function sum takes a value (the result of sum) and yields a singleton list whose element is the input value itself. The functional definition of $sum°$ is: $sum°\ s = [s]$.

Listing 1.1 shows the programming interface provided in Screwdriver, where users should write a program by inheriting the `ThirdHomomorphismTheorem` class. The function `fold` corresponds to the sequential function that solves the problem, and the function `unfold` corresponds to the sequential function that computes a right inverse. In a functional specification, the types of the two functions are $fold :: [t_1] \rightarrow t_2$ and $unfold :: t_2 \rightarrow [t_1]$. The concrete Java source code with Screwdriver for the summing-up example can be found on our project's site.

To utilize the third homomorphism theorem, users are requested to confirm that the two functions satisfy the following conditions. Firstly, the function $unfold$ should be a right inverse of the function $fold$. In other words, the equation $fold \circ unfold \circ fold = fold$ should hold. Secondly, for the $fold$ function there should exist two operators $\ominus$ and $\oplus$ as stated in Theorem 1. A sufficient condition for this second requirement is that the following two equations hold respectively for any a and x.

$$fold([a] ++ x) = fold([a] ++ unfold(fold(x))) \tag{1}$$

$$fold(x ++ [a]) = fold(unfold(fold\ x) ++ [a]) \tag{2}$$

Note that we can use some tools (such as QuickCheck [3]) in practice to verify whether Equations (1) and (2) hold or not.

Under these conditions, Screwdriver automatically derives a list homomorphism from the pair of $fold$ and $unfold$ functions. A list homomorphism $([f, \oplus])$ that computes $fold$ can be obtained by composing user's input programs, where the parameter functions f and $\oplus$ are defined as follows.

$$
\begin{aligned}
f\ a &= fold([a]) \\
x \oplus y &= fold(unfold\ x ++ unfold\ y).
\end{aligned}
$$

3.2   Homomorphism Implementation on MapReduce

In the second layer, Screwdriver provides an efficient implementation of list homomorphisms over MapReduce. In particular, the implementation consists of two passes of MapReduce.

Manipulation of Ordered Data

The computation of a list homomorphism obeys the order of elements in the input list, while the input data of MapReduce is given as a set stored on the distributed file system. This means we need to represent a list as a set.

On Screwdriver, we represent each element of a list as an *(index, value)* pair where *index* is an integer indicating the position of the element. For example, a list $[a, b, c, d, e]$ may be represented as a set $\{(3, d), (1, b), (2, c), (0, a), (4, e)\}$. Note that the list can be restored from this set representation by sorting the elements in terms of their indices. Such indexed pairs permit storing data in arbitrary order on the distributed file systems

Implementing Homomorphism by two Passes of MapReduce

For the input data stored as a set on the distributed file system, Screwdriver computes a list homomorphism in parallel by two passes of MapReduce computation. Here, the key idea of the implementation is that we group the elements consecutive in the list into some number of sublists and then apply the list homomorphism in parallel to those sublists.

In the following, we summarize our two-pass implementation of homomorphism $([f, \oplus])$. Here, *hom* f $(\oplus)$ denotes a sequential version of $([f, \oplus])$, *comp* is a comparing function defined over the *Int* type, and *const* is a constant value defined by the framework.

$$hom_{\mathbf{MR}} :: (\alpha \to \beta) \to (\beta \to \beta \to \beta) \to \{(Int, \alpha)\} \to \beta$$
$$hom_{\mathbf{MR}} \; f \; (\oplus) = getValue \circ MapReduce \; ([\cdot]) \; g_{\text{SHUFFLE}} \; comp \; g_{\text{REDUCE}}$$
$$\circ \; MapReduce \; ([\cdot]) \; f_{\text{SHUFFLE}} \; comp \; f_{\text{REDUCE}}$$

     where

        $f_{\text{SHUFFLE}} :: Int \to Int$
        $f_{\text{SHUFFLE}} \; k = k/const$

        $f_{\text{REDUCE}} :: (Int, [\alpha]) \to (Int, \beta)$
        $f_{\text{REDUCE}} \; (k, as) = (k, hom \; f \; (\oplus) \; as)$

        $g_{\text{SHUFFLE}} :: Int \to Int$
        $g_{\text{SHUFFLE}} \; k = 1$

        $g_{\text{REDUCE}} :: (Int, [\beta]) \to (Int, \beta)$
        $g_{\text{REDUCE}} \; (1, bs) = (1, hom \; id \; (\oplus) \; bs)$

        $getValue :: \{(Int, \beta)\} \to \beta$
        $getValue \; \{(1, b)\} = b$

First pass of MapReduce: The first pass of MapReduce divides the list into some sublists, and computes the result of the homomorphism for each sublist. Firstly in the MAP phase, we do no computation (except for wrapping the key-value pair into a singleton list). Then in the SHUFFLE&SORT phase, we group the pairs so that the set-represented list is partitioned into some number of sublists and sort each grouped elements by their indices. Finally, we apply the homomorphism to each sublist in the REDUCE phase.

Second pass of MapReduce: The second pass of MapReduce computes the result of the whole list from the results of sublists given by the first pass of MapReduce. Firstly in the MAP phase, we do no computation as in the first pass. Then in the SHUFFLE&SORT phase, we collect the subresults into a single set and sort them by the their indices. Finally, we reduce the subresults using the associative operator of the homomorphism.

Finally, by the *getValue* function, we picked the result of the homomorphism out from the set (of single value).

Implementation Issues

In terms of the parallelism, the number of the MAP tasks in the first pass is decided by the data splitting mechanism of Hadoop. For one split data of the input, Hadoop spawns one MAP task which applies f_{MAP} to each record. The number of the REDUCE tasks in the first pass of MapReduce should be chosen properly with respect to the total number of the task-trackers inside the cluster. By this number of REDUCE task, the parameter *const* in the program above is decided. In the REDUCE phase in the second pass of MapReduce, only one REDUCE task is invoked because all the subresults are grouped into a single set.

4   A Programming Example

In this section we demonstrate how to develop parallel programs with our framework, by using the maximum prefix sum problem in the introduction as our example. As discussed in Section 3, users need to define a Java class that inherits the Java class shown in Listing 1.1 and implement the two abstract methods *fold* and *unfold*.

Recall the maximum prefix sum problem in the introduction. It is not difficult to develop a sequential program for computing the maximum prefix sum:

$$mps\ [1, -2, 3, ...] = 0 \uparrow 1 \uparrow (1 + (-2)) \uparrow (1 + (-2) + 3) \uparrow (1 + (-2) + 3 + ...)$$

where $a \uparrow b$ returns a if $a > b$ otherwise returns b.

Although the *mps* function cannot be represented by a homomorphism in the sense that it cannot be described at the same time, it is not difficult to see, as discussed in [14], that the tupled function *mps* △ *sum* can be described leftwards and rightwards.

Listing 1.2. Our Parallel Program for Solving MPS Problem

```
1   import ...
2
3   public class Mps extends ThirdHomomorphismTheorem <LongWritable , LongPair > {
4
5       // Computing mps and sum at a time.
6       public LongPair fold(ArrayList <LongWritable> values) {
7           long mps = 0;
8           long sum = 0;
9           for (LongWritable v : values) {
10              sum += v.get();
11              if (sum > mps) mps = sum;
12          }
13
14          return new LongPair(mps, sum);
15      }
16
17      // A right inverse of fold.
18      public ArrayList <LongWritable> unfold(LongPair value) {
19          long m = value.getFirst();
20          long s = value.getSecond();
21
22          ArrayList <LongWritable> rst = new ArrayList <LongWritable >();
23          rst.add(new LongWritable(m));
24          rst.add(new LongWritable(s-m));
25          return rst;
26      }
27  }
```

What we need to do now is to develop an efficient sequential program for computing $mps \vartriangle sum$ and an efficient sequential program for computing a right inverse of $mps \vartriangle sum$. These two sequential programs are not difficult to obtain. A simple sequential program for computing the tupled function $(mps \vartriangle sum)$ can be defined by

$$
\begin{aligned}
(mps \vartriangle sum)\,[a] &= (a \uparrow 0, a) \\
(mps \vartriangle sum)\,(x + [a]) &= \mathbf{let}\ (m, s) = (mps \vartriangle sum)\ x\ \mathbf{in}\ (m \uparrow (s + a), s + a)
\end{aligned}
$$

and a right inverse of $(mps \vartriangle sum)$ can be defined as follows.

$$
(mps \vartriangle sum)^\circ\ (m, s) = [m, s - m].
$$

That is all for our development. We can now use $(mps \vartriangle sum)$ as the *fold* function and $(mps \vartriangle sum)^\circ$ as *unfold* function. Listing 1.2 gives the concrete Java program for solving the maximum prefix sum problem using Screwdriver.

5   Experiments

In this section, we report experiment results that evaluate the performance of our framework on PC clusters. We evaluated the scalability of programs on our framework, the overhead of our framework compared with the direct Hadoop program, and the overhead of the non-trivial parallel program compared with sequential program.

We configured clusters with 2, 4, 8, 16, and 32 virtual machines (VM) inside the *EdubaseCloud* system in National Institute of Informatics. Each VM has one CPU (Xeon E5530@2.4GHz, 1 core), 3 GB memory, and 5 GB disk space. We installed Hadoop (version 0.20.2.203) on each VM. Three sets of programs are used for the evaluation: *SUM* computes the sum of 64-bit integers; *VAR* computes the variance of 32-bit floating-point numbers; *MPS* solves the maximum-prefix-sum problem for a list of 64bit-integers. We both implemented the programs with the Hadoop APIs directly (*SUM-MR, VAR-MR, MPS-MR*), and with our Screwdriver (*SUM-LH, VAR-LH, MPS-LH*). Also a sequential program is implemented (*MPS-Seq*). The input for SUM and MPS was a list of 10^8 64bit-integer elements (593 MB), and the input for VAR is a list of 10^8 32bit-floating-point numbers (800 MB). Note that the elements of lists are indexed as in Section 3.2 with the type information (each element has a 64bit-integer index), stored in the Avro data format, and put in the HDFS.

The experiment results are summarized in Fig. 3 and Table 1. Note that the relative speedup is calculated with respect to the result of 2 nodes. The execution of the parallel programs on our framework and on Hadoop failed on 1 node, due to the limitation of disk space for the intermediate data.

All the programs achieved good scalability with respect to the number of nodes: the speedup ratios for 32 nodes against 2 nodes are more than 10 times. This shows that our framework does not spoil the strong advantage of MapReduce framework, namely *scalable data processing*. For the summation problem, the SUM-LH program on our framework cannot use combiner due to the limitation of Hadoop's implementation, so SUM-MR which uses combiner doing local reductionism can run almost twice faster. for almost all MapReduce programs combiners usually can increase performance very much. So we will work on to let our framework taking full use of data-locality. And we think it will bring notable performance improvement. Besides this, two-passes MapReduce processing and sorting with respect to the keys, which are unnecessary for the summation problem. In other words, with these overheads we can extend the MapReduce framework to support computations on *ordered* lists.

Finally we discuss the execution times for the maximum-prefix-sum problem. Although the parallel program on our framework MPS-LH shows good scalability (as well as MPS-MR), it ran slower on 32 nodes than the sequential program MPS-Seq. We consider this is a reasonable result: first, in this case of the maximum-prefix-sum problem, the parallel algorithm becomes more complex than that of sequential one, and in particular we produced (large) intermediate data when doing parallel processing on our framework but it is not the case for sequential processing. Second, the test data is not big enough, so the sequential program can still handle it. Because the limitation of our cloud, we cannot test big enough data. An important work to improve the performance in future is to make use of data locality to optimize the parallel execution.

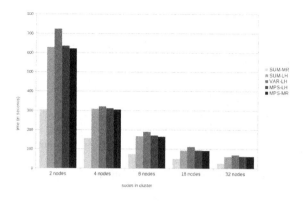

Fig. 3. Time Consuming

Table 1. Execution Time (second) and Relative Speedup w.r.t. 2 Nodes

Program	1 node	2 nodes	4 nodes	8 nodes	16 nodes	32 nodes
SUM-MR	NA (NA)	304 (1.00)	156 (1.95)	75 (4.05)	50 (6.08)	26 (11.69)
SUM-LH	NA (NA)	628 (1.00)	309 (2.03)	166 (3.78)	93 (6.75)	61 (10.30)
VAR-LH	NA (NA)	723 (1.00)	321 (2.25)	189 (3.82)	111 (6.50)	69 (10.45)
MPS-LH	NA (NA)	635 (1.00)	311 (2.04)	169 (3.76)	93 (6.78)	62 (10.24)
MPS-MR	NA (NA)	621 (1.00)	304 (2.04)	163 (3.81)	91 (6.82)	61 (10.22)
MPS-Seq	37	NA (NA)	NA (NA)	NA (NA)	NA (NA)	NA (NA)

6   Concluding Remarks

The research on list homomorphisms and the third homomorphism theorem indicate a systematic and constructive way to parallelization, by which this work was inspired. List homomorphisms and the theory related to them are very suitable for providing a method of developing MapReduce programs systematically.

In this paper, we presented a new approach to systematic parallel programming over MapReduce based on program calculation, and gave a concrete implementation to verify the approach. We believe that the calculation theorems for list homomorphisms can provide a higher abstraction over MapReduce, and such abstraction brings good algebraic properties of list homomorphisms to parallel programming with MapReduce.

We introduced a novel parallel programming approach based on list homomorphism to wrapping MapReduce. And we believe such approach is not limited to wrapping MapReduce, but also can be adopted to other parallel programming environments to provide higher-level programming interfaces.

As a future work, we plan to extend this framework for resolving parallel programming problems on trees and graphs. It will enlarge the computation capability of Screwdriver.

References

1. Bird, R.S.: Introduction to Functional Programming using Haskell. Prentice-Hall, Englewood Cliffs (1998)
2. Bird, R.S.: An introduction to the theory of lists. In: Broy, M. (ed.) Logic of Programming and Calculi of Discrete Design. NATO ASI Series F, vol. 36, pp. 5–42. Springer, Heidelberg (1987)
3. Claessen, K., Hughes, J.: QuickCheck: a lightweight tool for random testing of Haskell programs. In: Odersky, M., Wadler, P. (eds.) ICFP 2000: Proceedings of the Fifth ACM SIGPLAN International Conference on Functional Programming, pp. 268–279. ACM Press, New York (2000)
4. Cole, M.: Parallel programming with list homomorphisms. Parallel Processing Letters 5(2), 191–203 (1995)
5. Dean, J., Ghemawat, S.: MapReduce: Simplified data processing on large clusters. In: 6th Symposium on Operating System Design and Implementation (OSDI 2004), pp. 137–150 (2004)
6. Gibbons, J.: The third homomorphism theorem. Journal of Functional Programming 6(4), 657–665 (1996)
7. Gorlatch, S.: Systematic extraction and implementation of divide-and-conquer parallelism. In: Kuchen, H., Swierstra, S.D. (eds.) PLILP 1996. LNCS, vol. 1140, pp. 274–288. Springer, Heidelberg (1996)
8. Hu, Z.: Calculational parallel programming. In: HLPP 2010: Proceedings of the Fourth International Workshop on High-level Parallel Programming and Applications, p. 1. ACM Press, New York (2010)
9. Hu, Z., Iwasaki, H., Takeichi, M.: Formal derivation of efficient parallel programs by construction of list homomorphisms. ACM Transactions on Programming Languages and Systems 19(3), 444–461 (1997)
10. Hu, Z., Takeichi, M., Chin, W.N.: Parallelization in calculational forms. In: POPL 1998, Proceedings of the 25th ACM SIGPLAN-SIGACT Symposium on Principles of Programming Languages, pp. 316–328. ACM Press, New York (1998)
11. Lämmel, R.: Google's MapReduce programming model — Revisited. Science of Computer Programming 70(1), 1–30 (2008)
12. Matsuzaki, K., Iwasaki, H., Emoto, K., Hu, Z.: A library of constructive skeletons for sequential style of parallel programming. In: InfoScale 2006: Proceedings of the 1st International Conference on Scalable Information Systems. ACM International Conference Proceeding Series, vol. 152. ACM Press, New York (2006)
13. Morihata, A., Matsuzaki, K., Hu, Z., Takeichi, M.: The third homomorphism theorem on trees: downward & upward lead to divide-and-conquer. In: Shao, Z., Pierce, B.C. (eds.) POPL 2009: Proceedings of the 36th Annual ACM SIGPLAN-SIGACT Symposium on Principles of Programming Languages, pp. 177–185. ACM Press, New York (2009)
14. Morita, K., Morihata, A., Matsuzaki, K., Hu, Z., Takeichi, M.: Automatic inversion generates divide-and-conquer parallel programs. In: ACM SIGPLAN 2007 Conference on Programming Language Design and Implementation (PLDI 2007), pp. 146–155. ACM Press, New York (2007)
15. Rabhi, F.A., Gorlatch, S. (eds.): Patterns and Skeletons for Parallel and Distributed Computing. Springer, Heidelberg (2002)
16. Steele Jr. G.L.: Parallel programming and parallel abstractions in fortress. In: Hagiya, M. (ed.) FLOPS 2006. LNCS, vol. 3945, pp. 1–1. Springer, Heidelberg (2006)

Correlated Set Coordination in Fault Tolerant Message Logging Protocols

Aurelien Bouteiller[1], Thomas Herault[1], George Bosilca[1],
and Jack J. Dongarra[1,2]

[1] Innovative Computing Laboratory,
The University of Tennessee
[2] Oak Ridge National Laboratory
{bouteill,herault,bosilca,dongarra}@eecs.utk.edu

Abstract. Based on our current expectation for the exascale systems, composed of hundred of thousands of many-core nodes, the mean time between failures will become small, even under the most optimistic assumptions. One of the most scalable checkpoint restart techniques, the message logging approach, is the most challenged when the number of cores per node increases, due to the high overhead of saving the message payload. Fortunately, for two processes on the same node, the failure probability is correlated, meaning that coordinated recovery is free. In this paper, we propose an intermediate approach that uses coordination between correlated processes, but retains the scalability advantage of message logging between independent ones. The algorithm still belongs to the family of event logging protocols, but eliminates the need for costly payload logging between coordinated processes.

1 Introduction

High Performance Computing, as observed by the Top 500 ranking[1], has exhibited a constant progression of the computing power by a factor of two every 18 months for the last 15 years. Following this trend, the exaflops milestone should be reached as soon as 2019. The International Exascale Software Project (IESP) [7] proposes an outline of the characteristics of an exascale machine, based on the foreseeable limits of the hardware and maintenance costs. A machine in this performance range is expected to be built from gigahertz processing cores, with thousands of cores per computing node (up to 10^{12} flops per node), thus requiring millions of computing nodes to reach the exascale. Software will face the challenges of complex hierarchies and unprecedented levels of parallelism.

One of the major concerns is reliability. If we consider that failures of computing nodes are independent, the reliability probability of the whole system (i.e. the probability that all components will be up and running during the next time unit) is the product of the reliability probability of each of the components. A conservative assumption of a ten years mean time to failure translates into a

---

[1] http://www.top500.org/

E. Jeannot, R. Namyst, and J. Roman (Eds.): Euro-Par 2011, LNCS 6853, Part II, pp. 51–64, 2011.
© Springer-Verlag Berlin Heidelberg 2011

probability of 0.99998 that a node will still be running in the next hour. If the system consists of a million of nodes, the probability that at least one unit will be subject to a failure during the next hour jumps to $1 - 0.99998^{10^6} > 0.99998$. This probability being disruptively close to 1, one can conclude that many computing nodes will inevitably fail during the execution of an exascale application.

Automatic fault tolerant algorithms, which can be provided either by the operating system or the middleware, remove some of the complexity in the development of applications by masking failures and the ensuing recovery process. The most common approaches to automatic fault tolerance are replication, which consumes a high number of computing resources, and rollback recovery. Rollback recovery stores system-level checkpoints of the processes, enabling rollback to a saved state when failures happen. Consistent sets of checkpoints must be computed, using either coordinated checkpointing or some variant of uncoordinated checkpointing with message logging (for brevity, in this article, we use indifferently message logging or uncoordinated checkpointing). Coordinated checkpointing minimizes the overhead of failure-free operations, at the expense of a costly recovery procedure involving the rollback of all processes. Conversely, message logging requires every communication to be tracked to ensure consistency, but its uncoordinated recovery procedure demonstrates unparalleled efficiency in failure prone environments.

Although the low mean time to failure of exascale machines calls for preferring an uncoordinated checkpoint approach, the overhead on communication of message logging is bound to increase with the advent of many-core nodes. Uncoordinated checkpointing has been designed with the idea that failures are mostly independent, which is not the case in many-core systems where multiple cores crash when the node is struck by a failure. Not only do simultaneous failures negate the advantage of uncoordinated recovery, but the logging of messages between cores is also a major performance issue. All interactions between two uncoordinated processes have to be logged, and a copy of the transaction must be kept for future replay. Since making a copy has the same cost as doing the transaction itself (as the processes are on the same node we consider the cost of communications equal to the cost of memory copies), the overhead is unacceptable. It is disconcerting that the most resilient fault tolerant method is also the most bound to suffer, in terms of performance, on expected future systems.

In this paper, we consider the case of *correlated failures*: we say that two processes are correlated or co-dependent if they are likely to be subject to a simultaneous failure. We propose a hybrid approach between coordinated and non coordinated checkpointing, that prevents the overhead of keeping message copies for communications between correlated processes, but retains the more scalable uncoordinated recovery of message logging for processes whose failure probability is independent. The coordination protocol we present is a split protocol, which takes into account the fragmentation of messages, to avoid long waiting cycles, while still implementing a transactional semantic for whole messages.

2   Rollback Recovery Background

2.1   Execution Model

Events and States: Each computational or communication step of a process is an event. An execution is an alternate sequence of events and process states, with the effect of an event on the preceding state leading the process to the new state. As the system is basically asynchronous, there is no direct time relationship between events occurring on different processes. However, Lamport defines a causal partial ordering between events with the *happened before* relationship [14].

Events can be classified into two categories. An event is *deterministic* when, from the current state, all executions lead to the same outcome state. On the contrary, if in different executions, the same event happening on a particular state can result in several different outcome states, then it is *nondeterministic*. Examples of nondeterministic events are message receptions, which depend on external influences like network jitter.

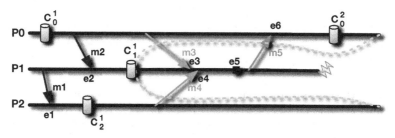

Fig. 1. Recovery line based on rollback recovery of a failed process

Recovery Line: Rollback recovery addresses mostly fail-stop errors: a failure is the loss of the complete state and actions of a process. A checkpoint is a copy of a past state of a particular process stored on some persistent memory (remote node, disk, ...), and used to restore the process in case of failure. The recovery line is the configuration of the entire application after some processes have been reloaded from checkpoints. If the checkpoints can happen at arbitrary dates, some messages can cross the recovery line. Consider the example execution of Figure 1. When the process P_1 fails, it rolls back to checkpoint C_1^1. If no other process rolls back, messages m_3, m_4, m_5 are crossing the recovery line. A recovery set is the union of the saved states (checkpoint, messages, events) and a recovery line.

In-transit Messages: Messages m_3 and m_4 are crossing the recovery line from the past, they are called *in-transit* messages. The *in-transit* messages are necessary for the progression of the recovered processes, but are not available anymore, as the corresponding send operation is in the past of the recovery line. For a recovery line to form a complete recovery set, every *in-transit* message must be added to the recovery line.

Orphan Messages: Message m_5 is crossing the recovery line from the future to the past; such messages are referred to as *orphan* messages. By following the happened-before relationship, the current state of P_0 depends on the reception of m_5; by transitivity, it also depends on events e_3, e_4, e_5 that occurred on P_1 since C_1^1. Since the channels are asynchronous, the reception of m_3 and m_4, from different senders, can occur in any order during re-execution, leading to a recovered state of P_1 that can diverge from the initial execution. As a result, the current state of P_0 depends on a state that P_1 might never reach after recovery. Checkpoints leading to such inconsistent states are useless and must be discarded; in the worst case, a domino effect can force all checkpoints to be discarded.

2.2   Building a Consistent Recovery Set

Two different strategies can be used to create consistent recovery sets. The first one is to create checkpoints at a moment in the history of the application where no *orphan* messages exist, usually through coordination of checkpoints. The second approach avoids coordination, but instead saves all *in-transit* messages to be able to replay those without rollback, and keep track of nondeterministic events, so that *orphan* messages can be regenerated identically. We focus our work on this second approach, deemed more scalable.

Coordinated Checkpoint: Checkpoint coordination aims at eliminating *in-transit* and *orphan* messages from the recovery set. Several algorithms have been proposed to coordinate checkpoints, the most usual being the Chandy-Lamport algorithm [6] and the blocking coordinated checkpointing, [5,17], which silences the network. In these algorithms, waves of tokens are exchanged to form a recovery line that eliminates *orphan* messages and detects *in-transit* messages. Coordinated algorithms have the advantage of having almost no overhead outside of checkpointing periods, but require that every process, even if unaffected by failures, rolls back to its last checkpoint, as this is the only recovery line that is guaranteed to be consistent.

Message Logging: Message Logging is a family of algorithms that attempt to provide a consistent recovery set from checkpoints taken at independent dates. As the recovery line is arbitrary, every message is potentially *in-transit* or *orphan*. Event Logging is the mechanism used to correct the inconsistencies induced by *orphan* messages, and nondeterministic events, while Payload Copy is the mechanism used to keep the history of *in-transit* messages. While introducing some overhead on every exchanged message, this scheme can sustain a much more adverse failure pattern, which translates to better efficiency on systems where failures are frequent [15].

Event Logging: In event logging, processes are considered *Piecewise deterministic*: only sparse nondeterministic events occur, separating large parts of deterministic computation. Event logging suppresses future nondeterministic events by adding the outcome of nondeterministic events to the recovery set, so that it can be forced to a deterministic outcome (identical to the initial execution) during recovery. The network, more precisely the order of reception, is considered the unique source of nondeterminism. The relative ordering of messages from different senders (e_3, e_4 in fig. 1), is the only information necessary to be logged. For a recovery set to be consistent, no unlogged nondeterministic event can precede an *orphan* message.

Payload Copy: When a process is recovering, it needs to replay any reception that happened between the last checkpoint and the failure. Consequently, it requires the payload of *in-transit* messages (m_3, m_4 in fig.1). Several approaches have been investigated for payload copy, the most efficient one being the sender-based copy [18]. During normal operation, every outgoing message is saved in the sender's volatile memory. The surviving processes can serve past messages to recovering processes on demand, without rolling back. Unlike events, sender-based data do not require stable or synchronous storage (although this data is also part of the checkpoint). Should a process holding useful sender-based data crash, the recovery procedure of this process replays every outgoing send and thus rebuilds the missing messages.

3   Group-Coordinated Message Logging

3.1   Shared Memory and Message Logging

Problem Statement: In uncoordinated checkpoint schemes, the ordering between checkpoint and message events is arbitrary. As a consequence, every message is potentially *in-transit*, and must be copied. Although the cost of the sender-based mechanism involved to perform this necessary copy is not negligible, the cost of a memory copy is often one order of magnitude lower than the cost of the network transfer. Furthermore, the copy and the network operation can overlap. As a result, proper optimization greatly mitigates the performance penalty suffered by network communications (typically to less than 10%, [2,3]). One can hope that future engineering advances will further reduce this overhead.

Unlike a network communication, a shared memory communication is a strongly memory-bound operation. In the worst case, memory copy induced by message logging doubles the volume of memory transfers. Because it competes for the same scarce resource - memory bandwidth - the cost of this extra copy cannot be overlapped, hence the time to send a message is irremediably doubled.

A message is *in-transit* (and needs to be copied) if it crosses the recovery line from the past to the future. The emission and reception dates of messages are beyond the control of the fault tolerant algorithm: one could delay the emission or reception dates to match some arbitrary ordering with checkpoint events, but these delays would obviously defeat the goal of improving communication performance. The only events that the fault tolerant algorithm can alter, to enforce an ordering between message events and checkpoint events, are checkpoint dates. Said otherwise, the only way to suppress *in-transit* messages is to synchronize checkpoints.

Correlated Failures: Fortunately, although many-core machines put a strain on message logging performance, a new opportunity opens, thanks to the side effect that failures do not have an independent probability on such an environment. All the processes hosted by a single many-core node are prone to fail simultaneously: they are located on the same piece of silicon, share the same memory bus, network interface, cooling fans, power supplies, operating system, and are subject to the same physical interferences (rays, heat, vibrations, ...). One of the motivating properties of message logging is that it tolerates a large number of independent failures very well. If failures are correlated, the fault tolerant algorithm can be more synchronous without decreasing its effective efficiency.

The leading idea of our approach is to propose a partially coordinated fault tolerant algorithm, that retains message logging between sets of processes experiencing independent failure probability, but synchronize the checkpoints of processes that have a strong probability of simultaneous failures, what we call a *correlated set*. It leverages the correlated failures property to avoid message copies that have a high chance of being useless.

3.2   Correlated Set Coordinated Message Logging

Whenever a process of a correlated set needs to take a checkpoint, it forces a synchronization with all other processes of the set. If a failure hits a process, all processes of that set have to roll back to their last checkpoint (see the recovery line in example execution depicted in figure 2). Considering a particular correlated set, every message can be categorized as either *ingoing* (m_1, m_2), *outgoing* (m_5), or *internal* (m_3, m_4). Between sets, no coordination is enforced. A process failing in another correlated set does not trigger a rollback, but messages between sets have no guaranteed properties with respect to the recovery line, and can still be *orphan* or *in-transit*. Therefore, regular message logging, including payload copy and event logging must continue for outgoing and ingoing messages.

As checkpoints are coordinated, all *orphan* and *in-transit* messages are eliminated between processes of the correlated set. However, as the total recovery

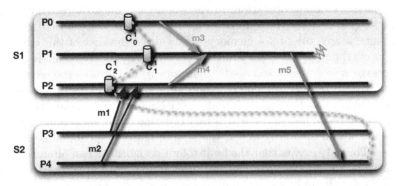

Fig. 2. An execution of the Correlated Set Coordinated Message Logging Algorithm

set does contain *in-transit* and *orphan* messages, the consistency proof of coordinated checkpoint does not hold for the recovery set formed by the union of the coordinated sets. In an uncoordinated protocol, a recovery set is consistent if all *in-transit* messages are available, and no *orphan* message depends on the outcome of a non-deterministic event. In the next paragraphs, we demonstrate that payload copy can be disabled for internal messages, but that event logging must apply to all types of messages.

Intra-set Payload Copy: By the direct application of the coordination algorithm, no message is *in-transit* between any pair of synchronized processes at the time of checkpoint (in the case of the Chandy/Lamport algorithm, occasional *in-transit* messages are integrated inside the checkpoint, hence they are not *in-transit* anymore). Because an internal message cannot be *in-transit*, it is never sent before the recovery line and received after. Therefore, the payload copy mechanism, used to recover past sent messages during the recovery phase, is unnecessary for internal messages.

Intra-set Event Logging:

Theorem 1. *In a fault tolerant protocol creating recovery sets with at least two distinct correlated sets, if the nondeterministic outcome of any internal messages preceding an outgoing message is omitted from the recovery set, there exists an execution that reaches an inconsistent state.*

Outgoing messages are crossing a non-coordinated portion of the recovery line, hence the execution follows an arbitrary ordering between checkpoint events and message events. Therefore, for any outgoing message there is an execution in which it is *orphan*. Consider the case of the execution depicted in figure 2. In this execution, the message m_5, between the sets S_1 and S_2 is *orphan* in the recovery line produced by a rollback of the processes of S_1.

Let's suppose that Event logging of internal messages is unnecessary for building a consistent recovery set. The order between the internal receptions and any other reception of the same process on another channel is nondeterministic. By transitivity of the Lamport relationship, this nondeterminism is propagated to the dependent outgoing message. Because an execution in which this outgoing message is *orphan* exists, the recovery line in this execution is inconsistent. The receptions of messages m_3, m_4 are an example: the nondeterministic outcome created by the unknown ordering of messages in asynchronous channels is propagated to P_4 through m_5. The state of the correlated set S_2 depends on future nondeterministic events of the correlated set S_1, therefore the recovery set is inconsistent. One can also remark that the same proof holds for ingoing messages (as illustrated by m_1 and m_2).

As a consequence of this theorem, it is necessary to log all message receptions, even if the emitter is located in the same correlated set as the receiver. Only the payload of this message can be spared.

3.3 Implementation

We have implemented the correlated set coordinated message logging algorithm inside the Open MPI library. Open MPI [9] is one of the leading Message Passing Interface standard implementations [19]. In Open MPI, the PML-V framework enables researchers to express their fault tolerant policies. The Vprotocol Pessimist is such an implementation of a pessimistic message logging protocol [3]. In order to evaluate the performance of our new approach, we have extended this fault tolerant component with the capabilities listed below.

Construction of the Correlated Set, Based on Hardware Proximity: Open MPI enables the end user to select a very precise mapping of his application on the physical resources, up to pinning a particular MPI rank to a particular core. As a consequence, the Open MPI's runtime instantiates a process map detailing node hierarchies and ranks allocations. The detection of correlated sets parses this map and extracts the groups of processes hosted on the same node.

Internal Messages Detection: In Open MPI, the couple formed by the rank and the communicator is translated into a list of endpoints, each one representing a channel to the destination (eth0, ib0, shared memory, ...). During the construction of the correlated set, all endpoints pertaining to a correlated process are marked. When the fault tolerant protocol considers making a sender-based copy, the endpoint is checked to determine if the message payload has to be copied.

Checkpoint Coordination in a Correlated Set: The general idea of a network-silence based coordination is simple: processes send a marker in their communication channels to notify other processes that no other message will be sent before the end of the phase. When all output channels and input channels have been notified, the network is silenced, and the processes can start communicating again. However, MPI communications do not exactly match the theoretical model, which assumes message emissions or receptions are atomic events. In practice, an MPI message is split into several distinct events. The most important include the emission of the first fragment (also called eager fragment), the matching of an incoming fragment with a receive request, and the delivery of the last fragment. Most of those events are unordered, in particular, a fragment can overtake another fragment, even from the same message (especially with channel bonding). Fortunately, because the MPI matching has to be FIFO, in Open MPI, eager fragments are FIFO, an advantageous property that our algorithm leverages. Our coordination algorithm has three phases: it silences eager fragments so that all posted sends are matched; it completes any matched receives; it checkpoints processes in the correlated set.

Eager silence: When a process enters the checkpoint synchronization, it sends a token to all correlated opened endpoints. Any send targeting a correlated endpoint, if posted afterwards, is stalled upon completion of the algorithm. When a process not yet synchronizing receives a token, it enters the synchronization immediately. The eager silence phase is complete for a process when it has received a token from every opened endpoint. Because no new message can inject an eager fragment after the token, and eager fragments are FIFO, at the end of this phase, all posted sends of processes in the correlated set have been matched.

Rendez-vous Silence: Unlike eager fragments, the remainder fragments of a message can come in any order. Instead of a complex non-FIFO token algorithm, the property that any fragment left in the channel belongs to an already matched message can be leveraged to drain remaining fragments. In the rendez-vous silence phase, every receive request is considered in turn. If a request has matched an eager fragment from a process of the correlated set, the progress engine of Open MPI is called repeatedly until it is detected that this particular request completed. When all such requests have completed, all fragments of internal messages to this process have been drained.

Checkpoint phase: When a process has locally silenced its internal inbound channels, it enters a local barrier. After the barrier, all channels are guaranteed to be empty. Each process then takes a checkpoint. A second barrier denotes that all processes finished checkpointing and that subsequent sends can be resumed.

4   Experimental Evaluation

4.1   Experimental Conditions

The Pluto platform features 48 cores, and is our main testbed for large shared memory performance evaluations. Pluto is based on four 12-core AMD opteron 6172 processors with 128GB of memory. The operating system is Red Hat 4.1.2 with the Linux 2.6.35.7 kernel. Despite the NUMA hierarchies, in this machine, the bandwidth is almost equal between all pairs of cores. The Dancer cluster is an 8 node cluster, where each node has two quad-core Intel Xeon E5520 CPUs, with 4GB of memory. The operating system is Caos NSA with the 2.6.32.6 Linux kernel. Nodes are connected through an Infiniband 20G network.

All protocols are implemented in Open MPI devel r20284. Vanilla Open MPI means that no fault tolerant protocol is enabled, regular message logging means that the pessimistic algorithm is used, and coordinated message logging denotes that cores of the same node belong to a correlated set. The evaluation includes synthetic benchmarks, such as NetPIPE 3.7 and IMB 3.3, and application benchmarks, such as the NAS 3.3 and HPL (with MKL BLAS10.2). The different benchmarks of the NAS suite accept a constrained number of processes (some expect a square number of processes, others a power of two). In all cases, we ran the largest possible experiment, for a given benchmark and a given parallel machine.

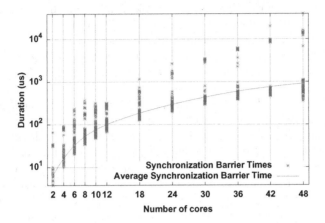

Fig. 3. Time to synchronize a correlated set (Pluto platform, log/log scale)

4.2 Shared Memory Performance

Coordination Cost: The cost of coordinating a growing number of cores is presented in the figure 3. The first token exchange is a complete all-to-all, that cannot rely on a spanning tree algorithm. Although, all other synchronizations are simple barriers, the token exchange dominates the execution time, which grows quadratically with the number of processes. Note, however, that this synchronization happens only during a checkpoint, and that its average cost is comparable to sending a 10KB message. Clearly, the cost of transmitting a checkpoint to the I/O nodes overshadows the cost of this synchronization.

Ping Pong: Figure 4 presents the results of the NetPIPE benchmark on shared memory with a logarithmic scale. Processes are pinned to two cores sharing an L2 cache, a worst case scenario for regular message logging. The maximum bandwidth reaches 53Gb/s, because communication cost is mostly related to accessing the L2 cache. The sender-based algorithm decreases the bandwidth to 11Gb/s, because it copies data to a buffer that is never in the cache. When the coordination algorithm allows for disabling the sender-based mechanism, event logging obtains the same bandwidth as the non fault tolerant execution.

NAS Benchmarks: Figure 5 presents the performance of the NAS benchmarks on the shared memory Pluto platform. BT and SP run on 36 cores, all others run on 32. One can see that avoiding payload copy enables the coordinated message logging algorithm to experience at most a 7% slowdown, and often no overhead, while the regular message logging suffers from up to 17% slowdown.

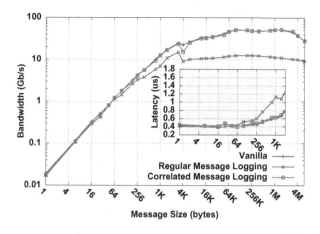

Fig. 4. Ping pong performance (Dancer node, shared memory, log/log scale)

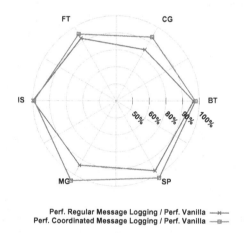

Fig. 5. NAS performance (Pluto platform, shared memory, 32/36 cores)

4.3   Cluster of Multicore Performance

Figure 6 presents the performance of the HPL benchmark on the Dancer cluster, with a one process per core deployment. For small matrix sizes, the behavior is similar between the three MPI versions. However, for slightly larger matrix sizes, the performance of regular message logging suffers. Conversely the coordinated message logging algorithm performs better, and only slightly slower than the non fault tolerant MPI, regardless of the problem size.

On the Dancer cluster, the available 500MB of memory per core is a strong limitation. In this memory envelope, the maximum computable problem size on this cluster is N=28260. The extra memory consumed by payload copy limits the maximum problem size to only N=12420 for regular message logging, while the

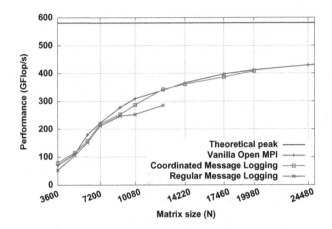

Fig. 6. HPL cluster performance (Dancer cluster, IB20G, 8 nodes, 64 cores)

reduction on the amount of logged messages enables the coordinated message logging approach to compute problems as large as N=19980. Not only does partial coordination of the message logging algorithm increase communication performance, it also decreases memory consumption.

5 Related Works

Recent advances in message logging have decreased the cost of event logging [3]. As a consequence, more than the logging scheme adopted, the prominent source of overhead in message logging is the copy of message payload caused by *in-transit* messages [4]. While attempts at decreasing the cost of payload copy have been successful to some extent [2], these optimizations are hopeless at improving shared memory communication speed. Our approach circumvents this limitation by completely eliminating the need for copies inside many-core processors.

Communication Induced Checkpoint (CIC) [12] is another approach that aims at constructing a consistent recovery set without coordination. The CIC algorithm maintains the dependency graph of events and checkpoints to compute *Z-paths* as the execution progresses. Forced checkpoints are taken whenever a Z-path would become a consistency breaking *Z-cycle*. This approach has several drawbacks: it adds piggyback to messages, and is notably not scalable because the number of forced checkpoints grows uncontrollably [1].

Group coordinated checkpoint have been proposed in MVAPICH2 [10] to solve I/O storming issues in coordinated checkpointing. In this paper, the group coordination refers to a particular scheduling of the checkpoint traffic, intended to avoid overwhelming the I/O network. Unlike our approach, which is partially uncoordinated, this algorithm builds a completely coordinated recovery set.

In [11], Ho, Wang and Lau propose a group-based approach that combines coordinated and uncoordinated checkpointing, similar to the technique we use in this paper, to reduce the cost of message logging in uncoordinated checkpointing. Their work, however, focuses on communication patterns of the application, to reduce the amount of message logging. Similarly, in the context of Charm++ [13], and AMPI[16], Meneses, Mendes and Kalé have proposed in [8] a team-based approach to reduce the overhead of message logging. The Charm++ model advocates a high level of oversubscription, with a ratio of user-level thread per core much larger than one. In their work, teams are of fixed, predetermined sizes. The paper does not explicitly explain how teams are built, but an emphasis on communication patterns seems preferred. In contrast, our work takes advantage of hardware properties of the computing resources, proposing to build correlated groups based on likeliness of failures, and relative efficiency of the communication medium.

6 Concluding Remarks

In this paper, we proposed a novel approach combining the best features of coordinated and uncoordinated checkpointing. The resulting fault tolerant protocol, belonging to the event logging protocol family, spares the payload logging for messages belonging to a correlated set, but retains uncoordinated recovery scalability. The benefit on shared memory point-to-point performance is significant, which translates into an observable improvement of many application types. Even though inter-node communications are not modified by this approach, the shared memory speedup translates into a reduced overhead on cluster of multicore type platforms. Last, the memory required to hold message payload is greatly reduced; our algorithm provides a flexible control of the tradeoff between synchronization and memory consumption. Overall, this work greatly improves the applicability of message logging in the context of distributed systems based on a large number of many-core nodes.

Acknowledgement. This work was partially supported by the DOE Cooperative Agreement DE-FC02-06ER25748, and the INRIA-Illinois Joint Laboratory for Petascale Computing and the ANR RESCUE project.

References

1. Alvisi, L., Elnozahy, E., Rao, S., Husain, S.A., Mel, A.D.: An analysis of communication induced checkpointing. In: 29th Symposium on Fault-Tolerant Computing (FTCS 1999). IEEE CS Press, Los Alamitos (1999)
2. Bosilca, G., Bouteiller, A., Herault, T., Lemarinier, P., Dongarra, J.J.: Dodging the cost of unavoidable memory copies in message logging protocols. In: Keller, R., Gabriel, E., Resch, M., Dongarra, J. (eds.) EuroMPI 2010. LNCS, vol. 6305, pp. 189–197. Springer, Heidelberg (2010)
3. Bouteiller, A., Bosilca, G., Dongarra, J.: Redesigning the message logging model for high performance. In: ISC 2008, Wiley, Dresden (June 2008) (p. to appear)
4. Bouteiller, A., Ropars, T., Bosilca, G., Morin, C., Dongarra, J.: Reasons to be pessimist or optimist for failure recovery in high performance clusters. In: IEEE (ed.) Proceedings of the 2009 IEEE Cluster Conference (September 2009)

5. Buntinas, D., Coti, C., Herault, T., Lemarinier, P., Pilard, L., Rezmerita, A., Rodriguez, E., Cappello, F.: Blocking vs. non-blocking coordinated checkpointing for large-scale fault tolerant MPI protocols. Future Generation Computer Systems 24(1), 73–84 (2008),
   http://www.sciencedirect.com/science/article/B6V06-4N2KT6H-1/2/00e790651475028977cc3031d9ea3980
6. Chandy, K.M., Lamport, L.: Distributed snapshots: Determining global states of distributed systems. Transactions on Computer Systems 3(1), 63–75 (1985)
7. Dongarra, J., Beckman, P., et al.: The international exascale software roadmap. Intl. Journal of High Performance Computer Applications 25(11) (to appear) (2011)
8. Esteban Meneses, C.L.M., Kalé, L.V.: Team-based message logging: Preliminary results. In: 3rd Workshop on Resiliency in High Performance Computing (Resilience) in Clusters, Clouds, and Grids (CCGRID 2010) (May 2010)
9. Gabriel, E., Fagg, G.E., Bosilca, G., Angskun, T., Dongarra, J.J., Squyres, J.M., Sahay, V., Kambadur, P., Barrett, B., Lumsdaine, A., Castain, R.H., Daniel, D.J., Graham, R.L., Woodall, T.S.: Open MPI: Goals, concept, and design of a next generation MPI implementation. In: Proceedings, 11th European PVM/MPI Users' Group Meeting, Budapest, Hungary, pp. 97–104 (September 2004)
10. Gao, Q., Huang, W., Koop, M.J., Panda, D.K.: Group-based coordinated checkpointing for mpi: A case study on infiniband. In: International Conference on Parallel Processing, ICPP 2007 (2007)
11. Ho, J.C.Y., Wang, C.L., Lau, F.C.M.: Scalable Group-based Checkpoint/Restart for Large-Scale Message-Passing Systems. In: Proceedings of the 22nd IEEE International Symposium on Parallel and Distributed Processing (IPDPS), pp. 1–12. IEEE, Los Alamitos (2008)
12. Hlary, J.M., Mostefaoui, A., Raynal, M.: Communication-induced determination of consistent snapshots. IEEE Transactions on Parallel and Distributed Systems 10(9), 865–877 (1999)
13. Kale, L.: Charm++. In: Padua, D. (ed.) Encyclopedia of Parallel Computing, Springer, Heidelberg (to appear) (2011)
14. Lamport, L.: Time, clocks, and the ordering of events in a distributed system. Communications of the ACM 21(7), 558–565 (1978)
15. Lemarinier, P., Bouteiller, A., Herault, T., Krawezik, G., Cappello, F.: Improved message logging versus improved coordinated checkpointing for fault tolerant MPI. In: IEEE International Conference on Cluster Computing. IEEE CS Press, Los Alamitos (2004)
16. Negara, S., Pan, K.C., Zheng, G., Negara, N., Johnson, R.E., Kale, L.V., Ricker, P.M.: Automatic MPI to AMPI Program Transformation. Tech. Rep. 10-09, Parallel Programming Laboratory (March 2010)
17. Plank, J.S.: Efficient Checkpointing on MIMD Architectures. Ph.D. thesis, Princeton University (June 1993),
   http://www.cs.utk.edu/~plank/plank/papers/thesis.html
18. Rao, S., Alvisi, L., Vin, H.M.: The cost of recovery in message logging protocols. In: 17th Symposium on Reliable Distributed Systems (SRDS), October 1998, pp. 10–18. IEEE CS Press, Los Alamitos (1998)
19. The MPI Forum: MPI: a message passing interface. In: Supercomputing 1993: Proceedings of the 1993 ACM/IEEE Conference on Supercomputing, pp. 878–883. ACM Press, New York (1993)

Introduction

Martin Berzins, Daniela di Serafino, Martin Gander, and Luc Giraud

Topic chairs

The solution of Computational Science problems relies on the availability of accurate and efficient numerical algorithms and software capable of harnessing the processing power of modern parallel and distributed computers. Such algorithms and software allow to prototype and develop new large-scale applications, as well as to improve existing ones, by including up-to-date numerical methods, or well-assessed ones re-designed in the light of the new architectures.

This conference topic is aimed at discussing new developments in the design and implementation of numerical algorithms for modern parallel architectures, including multi-core systems, multi-GPU based computers, clusters and the Grid. Different aspects, ranging from fundamental algorithmic concepts to software design techniques and performance analysis, are considered.

The papers submitted to this topic came from Austria, Australia, the Czech Republic, France, India, Italy, Japan, the Netherlands, Russia, Spain, and the USA. Each paper received at least three reviews and, finally, we selected three regular papers, all related to numerical linear algebra. E. Solomonik and J. Demmel describe and analyze a class of 2.5D linear algebra algorithms for matrix-matrix multiplication and LU factorization, that use extra memory to reduce bandwidth and latency costs. X. Dong and G. Cooperman present a parallel version of the ILU(k) preconditioner, that preserves stability properties. L. Bergamaschi and A. Martinez describe a parallel implementation of an inexact constraint preconditioner, based on sparse approximate inverse, for generalized saddle-point linear systems. We think that these papers provide a significant contribution to the scientific programme of Euro-Par 2011 and will contribute to the success of the conference.

Finally, we would like to thank all the authors for their submissions, the referees for helping us to select high-quality papers, and the Euro-Par Organizing Committee for the coordination of all the conference topics.

E. Jeannot, R. Namyst, and J. Roman (Eds.): Euro-Par 2011, LNCS 6853, Part II, p. 65, 2011.
© Springer-Verlag Berlin Heidelberg 2011

A Bit-Compatible Parallelization for ILU(k) Preconditioning

Xin Dong[*] and Gene Cooperman[*]

College of Computer Science, Northeastern University
Boston, MA 02115, USA
{xindong,gene}@ccs.neu.edu

Abstract. ILU(k) is a commonly used preconditioner for iterative linear solvers for sparse, non-symmetric systems. It is often preferred for the sake of its stability. We present TPILU(k), the first efficiently parallelized ILU(k) preconditioner that maintains this important stability property. Even better, TPILU(k) preconditioning produces an answer that is bit-compatible with the sequential ILU(k) preconditioning. In terms of performance, the TPILU(k) preconditioning is shown to run faster whenever more cores are made available to it — while continuing to be as stable as sequential ILU(k). This is in contrast to some competing methods that may become unstable if the degree of thread parallelism is raised too far. Where Block Jacobi ILU(k) fails in an application, it can be replaced by TPILU(k) in order to maintain good performance, while also achieving full stability. As a further optimization, TPILU(k) offers an optional *level-based incomplete inverse method* as a fast approximation for the original ILU(k) preconditioned matrix. Although this enhancement is not bit-compatible with classical ILU(k), it is bit-compatible with the output from the single-threaded version of the same algorithm. In experiments on a 16-core computer, the enhanced TPILU(k)-based iterative linear solver performed up to 9 times faster. As we approach an era of many-core computing, the ability to efficiently take advantage of many cores will become ever more important.

1 Introduction

This work introduces a parallel preconditioner, TPILU(k), with good stability and performance across a range of sparse, non-symmetric linear systems. For a large sparse linear system $Ax = b$, parallel iterative solvers based on ILU(k) [1,2] often suffer from instability or performance degradation. In particular, most of today's commonly used algorithms are domain decomposition preconditioners, which become slow or unstable with greater parallelism. This happens as they attempt to approximate a linear system by more and smaller subdomains to provide the parallel work for an increasing number of threads. The restriction to subdomains of ever smaller dimension must either ignore more of the off-diagonal

---

[*] This work was partially supported by the National Science Foundation under Grant CCF 09-16133.

E. Jeannot, R. Namyst, and J. Roman (Eds.): Euro-Par 2011, LNCS 6853, Part II, pp. 66–77, 2011.

matrix elements, or must raise the complexity by including off-diagonals into the computation for an optimal decomposition. The former tends to create instability for large numbers of threads (i.e., for small subdomains), and the latter is slow.

Consider the parallel preconditioner PILU [3,4] as an example. PILU would experience performance degradation unless the matrix A is *well-partitionable* into subdomains. This condition is violated by linear systems generating many fill-ins (as occurs with higher initial density or higher level k) or by linear solvers employing many threads. Another parallel preconditioner BJILU [5] (Block Jacobi ILU(k)), would fail to converge as the number of threads w grows. This is especially true for linear systems that are not diagonally dominant, in which the solver might become invalid by ignoring significant off-diagonal entries. This kind of performance degradation or instability is inconsistent with the widespread acceptance of parallel ILU(k) for varying k to provide efficient preconditioners.

In contrast, TPILU(k) is as stable as sequential ILU(k) and its performance increases with the number of cores. TPILU(k) can capture both properties simultaneously — precisely because it is not based on domain decomposition. In the rest of this paper, we will simply write that *TPILU(k) is stable* as a shortened version of the statement that TPILU(k) is stable for any number of threads whenever sequential ILU(k) is stable.

TPILU(k) uses a task-oriented parallel ILU(k) preconditioner for the base algorithm. However, it optionally first tries a different, level-based incomplete inverse submethod (*TPIILU(k)*). The term *level-based incomplete inverse* is used to distinguish it from previous methods such as "threshold-based" incomplete inverses [6]. The level-based submethod either succeeds or else it fails to converge. If it doesn't converge fast, TPILU(k) quickly reverts to the stable, base task-oriented parallel ILU(k) algorithm.

A central point of novelty of this work concerns bit-compatibility. The base task-oriented parallel component of TPILU(k) is bit-compatible with classical sequential ILU(k), and the level-based optimization produces a new algorithm that is also bit-compatible with the single-threaded version of that same algorithm. Few numerical parallel implementations can guarantee this stringent standard. The order of operations is precisely maintained so that the low order bits due to round-off do not change under parallelization. Further, the output remains bit-compatible as the number of threads increases — thus eliminating worries whether scaling a computation will bring increased round-off error.

In practice, bit-compatible algorithms are well-received in the workplace. A new bit-compatible version of code may be substituted with little discussion. In contrast, new versions of code that result in output with modified low-order bits must be validated by a numerical analyst. New versions of code that claim to produce more accurate output must be validated by a domain expert.

A prerequisite for an efficient implementation in this work was the use of thread-private memory allocation arenas. The implementation derives from [7], where we first noted the issue. The essence of the issue is that any implementation of POSIX-standard "malloc" libraries must be prepared for the case that a second thread frees memory originally allocated by a first thread. This requires a centralized data structure, which is slow in many-core architectures. Where it is known that memory allocated by a thread will be freed by that same thread, one can use a thread-private (per-thread) memory allocation arena. The issue arises in the memory allocations for "fill-ins" for symbolic factorization. In LU-factorization based algorithms, the issue is still more serious than incomplete LU, since symbolic factorization is a relatively larger part of the overall algorithm.

The rest of this paper is organized as follows. Section 2 reviews LU factorization and sequential ILU(k) algorithm. Section 3 presents task-oriented parallel TPILU(k), including the base algorithm (Sections 3.1 through 3.2) and the level-based incomplete inverse submethod (Section 3.3). Section 4 analyzes the experimental results. We review related work in Section 5.

2   Review of the Sequential ILU(k) Algorithm

A brief sketch is provided. See [8] for a detailed review of ILU(k). LU factorization decomposes a matrix A into the product of a lower triangular matrix L and an upper triangular matrix U. From L and U, one efficiently computes A^{-1} as $U^{-1}L^{-1}$. While computation of L and U requires $O(n^3)$ steps, once done, the computation of the inverse of the triangular matrices proceeds in $O(n^2)$ steps.

For sparse matrices, one contents oneself with solving x in $Ax = b$ for vectors x and b, since A^{-1}, L and U would all be hopelessly dense. Iterative solvers are often used for this purpose. An ILU(k) algorithm finds sparse approximations, $\widetilde{L} \approx L$ and $\widetilde{U} \approx U$. The preconditioned iterative solver then implicitly solves $A\widetilde{U}^{-1}\widetilde{L}^{-1}$, which is close to the identity. For this purpose, triangular solve operations are integrated into each iteration to obtain a solution y such that

$$\widetilde{L}\widetilde{U}y = p \tag{1}$$

where p varies for each iteration. This has faster convergence and better numerical stability. Here, the *level limit* k controls how many elements should be computed in the process of incomplete LU factorization. A level limit of $k = \infty$ yields full LU-factorization.

Similarly to LU factorization, ILU(k) factorization can be implemented by the same procedure as Gaussian elimination. Moreover, it also records the elements of a lower triangular matrix $\widetilde{L}$. Because the diagonal elements of $\widetilde{L}$ are defined to be 1, we do not need to store them. Therefore, a single *filled matrix* F is sufficient to store both $\widetilde{L}$ and $\widetilde{U}$.

2.1   Terminology for ILU(k)

For a huge sparse matrix, a standard dense format would be wasteful. Instead, we just store the position and the value of non-zero elements. Similarly, incomplete LU factorization does not insert all elements that are generated in the process of factorization. Instead, it employs some mechanisms to control how many elements are stored. ILU(k) [1,2] uses the level limit k as the parameter to implement a more flexible mechanism. We next review some definitions.

Definition 2.1. A fill entry, *or* entry *for short, is an element stored in memory.* *(Elements that are not stored are called zero elements.)*

Definition 2.2. Fill-in: *Consider Figure 1a. If there exists h such that $i, j > h$ and both f_{ih} and f_{hj} are fill entries, then the ILU(k) factorization algorithm may fill in a non-zero value when considering rows i and j. Hence, this element f_{ij} is called a* fill-in; *i.e., an entry candidate. We say the fill-in f_{ij} is caused by the existence of the two entries f_{ih} and f_{hj}. The entries f_{ih} and f_{hj} are the* causative entries *of f_{ij}. The causality will be made clearer in the next subsection.*

Definition 2.3. Level: *Each entry f_{ij} is associated with a level, denoted as* level (i, j) *and defined recursively by*

$$level\ (i,j) = \begin{cases} 0, & if\ a_{ij} \neq 0 \\ \min_{1 \leq h < \min(i,j)} level\ (i,h) + level\ (h,j) + 1, & otherwise \end{cases}$$

The *level limit k* is used to control how many fill-ins should be inserted into the filled matrix during ILU(k) factorization. Those fill-ins with a level smaller than or equal to k are inserted into the filled matrix F. Other fill-ins are ignored. By limiting fill-ins to level k or less, ILU(k) maintains a sparse filled matrix.

2.2 ILU(k) Algorithm and Its Parallelization

For LU factorization, the defining equation $A = LU$ is expanded into $a_{ij} = \sum_{h=1}^{min(i,j)} l_{ih} u_{hj}$, since $l_{ih} = 0$ for $i > j$ and $u_{hj} = 0$ for $i < j$. When $i > j$, $f_{ij} = l_{ij}$ and we can write $a_{ij} = \sum_{h=1}^{j-1} l_{ih} u_{hj} + f_{ij} u_{jj}$. When $i \leq j$, $f_{ij} = u_{ij}$ and we can write $a_{ij} = \left(\sum_{h=1}^{i-1} l_{ih} u_{hj} \right) + l_{ii} f_{ij} = \left(\sum_{h=1}^{i-1} l_{ih} u_{hj} \right) + f_{ij}$. Rewriting them yields the equations for LU factorization.

$$f_{ij} = \begin{cases} \left(a_{ij} - \sum_{h=1}^{j-1} l_{ih} u_{hj} \right) / u_{jj}, & i > j \\ a_{ij} - \sum_{h=1}^{i-1} l_{ih} u_{hj}, & i \leq j \end{cases} \tag{2}$$

The equations for ILU(k) factorization are similar except that an entry f_{ij} is computed only if $level(i, j) \leq k$. Hence, ILU(k) factorization is separated into two passes: *symbolic factorization* and *numeric factorization*. Symbolic factorization computes the levels of all entries less than or equal to k. Numeric factorization computes the numerical values in the filled matrix of all fill entries with level less than or equal to k. While the remaining description considers numeric factorization, the algorithm applies equally to symbolic factorization.

The ILU(k) algorithm reorganizes the above Equations (2) for efficient use of memory. The filled matrix F is initialized to A. As the algorithm proceeds, additional terms of the form $-l_{ih} u_{hj}$ are added to f_{ij}. Figure 1a illustrates f_{ij} accumulating an incremental value based on the previously computed values of f_{ih} (i.e., l_{ih}) and f_{hj} (i.e., u_{hj}).

The algorithmic flow of control is to factor the rows in order from first to last. In the factorization of row i, h varies from 1 to i in an outer loop, while j varies

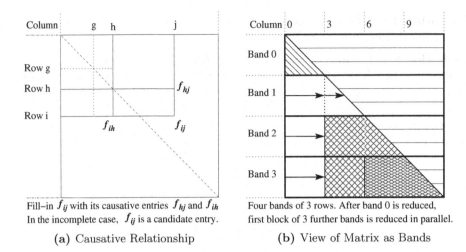

Fill–in f_{ij} with its causative entries f_{hj} and f_{ih}. In the incomplete case, f_{ij} is a candidate entry.

(a) Causative Relationship

Four bands of 3 rows. After band 0 is reduced, first block of 3 further bands is reduced in parallel.

(b) View of Matrix as Bands

Fig. 1. Parallel Incomplete LU Factorization

from h to n in an inner loop. In the example of Figure 1a, f_{hj} has clearly already been fully completed. Before the inner loop, f_{ih} is divided by u_{hh} following the case $i > j$ of Equations (2) since $i > h$. This is valid because f_{ih} depends on terms of the form $l_{ig}u_{gh}$ only for the case $g < h$, and those terms have already been accumulated into f_{ih} by previous inner loops. Inside the inner loop, we just subtract $l_{ih}u_{hj}$ from f_{ij} as indicated by Equations (2).

The algorithm has some of the same spirit as Gaussian elimination if one thinks of ILU(k) as using the earlier row h to *reduce* the later row i. This is the crucial insight in the parallel ILU(k) algorithm of this paper. One splits the rows of F into bands, and reduces the rows of a later band by the rows of an earlier band. Distinct threads can reduce distinct bands simultaneously, as illustrated in Figure 1b.

3   TPILU(k): Task-Oriented Parallel ILU(k) Algorithm

3.1   Parallel Tasks and Static Load Balancing

To describe a general parallel model valid for Gaussian elimination as well as ILU(k) and ILUT, we introduce the definition *frontier*: the maximum number of rows that are currently factored completely. The frontier i is the limit up to which the remaining rows can be partially factored except for the $(i+1)^{th}$ row. The $(i+1)^{th}$ row can be factored completely. That changes the frontier to $i+1$.

Threads synchronize on the frontier. To balance and overlap computation and synchronization, the matrix is organized as bands to make the granularity of the computation adjustable, as demonstrated in Figure 1b. A task is associated to a band and is defined as the computation to partially factor the band to the current frontier.

For each band, the program must remember up to what column this band has been partially factored. We call this column the *current position*, which is the start point of factorization for the next task attached to this band. In addition, it is important to use a variable to remember the first band that has not been factored completely. After the first unfinished band is completely factored, the frontier global value is increased by the number of rows in the band.

The smaller the band size, the larger the number of synchronization points. However, TPILU(k) prefers a smaller band size, that leads to more parallel tasks. Moreover, the lower bound of the factorization time is the time to factor the last band, which should not be very large. Luckily, shared memory allows for a smaller band size because the synchronization here is to read/write the frontier, which has a small cost.

While the strategy of bands is well known to be efficient for dense matrices (e.g., see [9]), researchers hesitate to use this strategy for sparse matrices because they may find only a small number of relatively dense bands, while all other bands are close to trivial. The TPILU(k) algorithm works well on sparse matrices because successive factoring of bands produces many somewhat dense bands (with more fill-ins) near the end of the matrix. TPILU(k) uses static load balancing whereby each worker is assigned a fixed group of bands chosen round robin so that each thread will also be responsible for some of the denser bands.

3.2   Optimized Symbolic Factorization

Static Load Balancing and TPMalloc. Simultaneous memory allocation for fill-ins is a performance bottleneck for shared-memory parallel computing. TPILU(k) takes advantage of a thread-private malloc library to solve this issue as discussed in [7]. TPMalloc is a non-standard extension to a standard allocator implementation, which associates a thread-private memory allocation arena to each thread. A thread-local global variable is also provided, so that the modified behavior can be turned on or off on a per-thread basis. By default, threads use thread-private memory allocation arenas. The static load balancing strategy guarantees that if a thread allocates memory, then the same thread will free it, which is consistent with the use of a thread-private allocation arena.

Optimization for the Case $k = 1$. When $k = 1$, it is possible to symbolically factor the bands and the rows within each band in any desired order. This is because if either f_{ih} or f_{hj} is an entry of level 1, the resulting fill-in f_{ij} must be an element of level 2 or level 3. So f_{ij} is not inserted into the filled matrix F. As a first observation, the symbolic factorization now becomes pleasingly parallel since the processing of each band is independent of that of any other.

Second, since the order can be arbitrary, even the purely sequential processing within one band by a single thread can be made more efficient. Processing rows in reverse order from last to first is the most efficient, while the more natural first-to-last order is the least efficient. First-to-last is inefficient, because we add level 1 fill-ins to the sparse representation of earlier rows, and we must then skip over those earlier level 1 fill-ins in determining level 1 fill-ins of later rows. Processing from last to first avoids this inefficiency.

3.3   Optional Level-Based Incomplete Inverse Method

The goal of this section is to describe the level-based incomplete inverse method for solving $\widetilde{L}x = p$ by matrix-vector multiplication: $x = \widetilde{L}^{-1}p$. This avoids the sequential bottleneck of using forward substitution on $\widetilde{L}x = p$. We produce incomplete inverses $\widetilde{L}^{-1}$ and $\widetilde{U}^{-1}$ so that the triangular solve stage of the linear solver (i.e., solving for y in $\widetilde{L}\widetilde{U}y = p$ as described in Equation (1) of Section 2) can be trivially parallelized ($y = \widetilde{U}^{-1}\widetilde{L}^{-1}p$) while also enforcing bit compatibility. Although details are omitted here, the same ideas are then used in a second stage: using the solution x to solve for y in $\widetilde{U}y = x$.

   Below, denote the matrix $(-\beta_{it})_{t \leq i}$ to be the lower triangular matrix $\widetilde{L}^{-1}$. Recall that $\beta_{ii} = 1$, just as for $\widetilde{L}$. First, we have Equation (3a), i.e., $x = \widetilde{L}^{-1}p$. Second, we have Equation (3b), i.e., the equation for solving $\widetilde{L}x = p$ by forward substitution. Obviously, Equation (3a) and Equation (3b) define the same x.

$$x_i = \sum_{t<i}(-\beta_{it})p_t + p_i \qquad \text{(3a)} \qquad\qquad x_i = p_i - \sum_{h<i} f_{ih}x_h \qquad \text{(3b)}$$

Substituting Equation (3a) into Equation (3b), one has Equation (4).

$$x_i = p_i - \sum_{h<i} f_{ih}\left(\sum_{t<h}(-\beta_{ht})p_t + p_h\right) = \sum_{t<i}\left(-\left(f_{it} - \sum_{t<h<i} f_{ih}\beta_{ht}\right)\right)p_t + p_i$$

$$\text{(4)}$$

Combining the right hand sides of equations (3a) and (4) yields Equation (5), the defining equation for β_{it}.

$$\beta_{it} = f_{it} - \sum_{t<h<i} f_{ih}\beta_{ht} \qquad\qquad \text{(5)}$$

Equation (5) is the basis for computing $\widetilde{L}^{-1}$ (a.k.a. $(-\beta_{it})_{t \leq i}$). Recall that f_{ij} was initialized to the matrix A. In algorithm steps (6a) and (6b) below, row i is factored using ILU(k) factorization, which computes $\widetilde{L}$ and $\widetilde{U}$ as part of a single matrix. These steps are reminiscent of Gaussian elimination using pivoting element f_{hh}. Steps (6a) and (6b) are used in steps (6c) and (6d) to compute $\widetilde{L}^{-1}$.

$$f_{ih} \leftarrow f_{ih}f_{hh}^{-1} \qquad \text{(6a)} \qquad \forall j > h, f_{ij} \leftarrow f_{ij} - f_{ih}f_{hj} \qquad \text{(6b)}$$
$$\forall t < h, f_{it} \leftarrow f_{it} - f_{ih}f_{ht} \qquad \text{(6c)} \qquad \forall t < i, f_{it} \leftarrow -f_{it} \qquad \text{(6d)}$$

The matrix $\widetilde{L}^{-1}$ is in danger of becoming dense. To maintain the sparsity, we compute the level-based incomplete inverse matrix $\widetilde{L}^{-1}$ following the same non-zero pattern as $\widetilde{L}^{-1}$. The computation for $\widetilde{L}^{-1}$ can be combined with the original numeric factorization phase. A further factorization phase is added to compute $\widetilde{U}^{-1}$ by computing matrix entries in reverse order from last row to first and from right to left within a given row.

Given the above algorithm for $\widetilde{L}^{-1}$ and a similar algorithm for $\widetilde{U}^{-1}$, the triangular solve stage is reduced to matrix-vector multiplication, which can be trivially parallelized. Inner product operations are not parallelized for two reasons: first, even when sequential, they are fast; second, parallelization of inner products would violate bit-compatibility by changing the order of operations.

4   Experimental Results

We evaluate the performance of the bit-compatible parallel ILU(k) algorithm, TPILU(k), by comparing with two commonly used parallel preconditioners, PILU [3] and BJILU [5] (Block Jacobi ILU(k)). Both PILU and BJILU are based on *domain decomposition*. Under the framework of Euclid [10, Section 6.12], both preconditioners appear in Hypre [10], a popular linear solver package under development at Lawrence Livermore National Laboratory since 2001.

The primary test platform is a computer with four Intel Xeon E5520 quad-core CPUs (16 cores total). Figure 3 demonstrates the scalability of TPILU(k) both on this primary platform and a cluster including two nodes connected by Infiniband. Each node has a single Quad-Core AMD Opteron 2378 CPU. The operating system is CentOS 5.3 (Linux 2.6.18) and the compiler is gcc-4.1.2 with the "-O2" option. The MPI library is OpenMPI 1.4. Within Hypre, the same choice of iterative solver is used to test both Euclid (PILU and BJILU) and TPILU(k). The chosen iterative solver is preconditioned stabilized bi-conjugate gradients with the default tolerance $rtol = 10^{-8}$. Note that the Euclid framework employs multiple MPI processes communicating via MPI's shared-memory architecture, instead of directly implementing a single multi-threaded process.

Driven Cavity Problem. This set of test cases [11] consists of some difficult problems from the modeling of the incompressible Navier-Stokes equations. These test cases are considered here for the sake of comparability. They had previously been chosen to demonstrate the features of PILU by [4]. Here, we test on three representatives: $e20r3000$, $e30r3000$ and $e40r3000$. Figure 2 shows that both Euclid PILU and Euclid BJILU are influenced by the number of processes and the level k when solving driven cavity problems. With more processes or larger k, both the PILU and BJILU preconditioners tend to slow down, break down or diverge.

Euclid registers its best solution time for $e20r3000$ by using PILU(2) with 1 process, for $e30r3000$ by using BJILU with 2 processes, and for $e40r3000$ by using PILU(1) with 2 processes. The reason that Euclid PILU obtains only a small speedup for these problems is that PILU requires the matrix to be *well-partitionable*, which is violated when using a larger level k or when employing more processes. Similarly, Euclid BJILU must approximate the original matrix by a number of subdomains equal to the number of processes. Therefore, higher parallelism forces BJILU to ignore even more off-diagonal matrix entries with

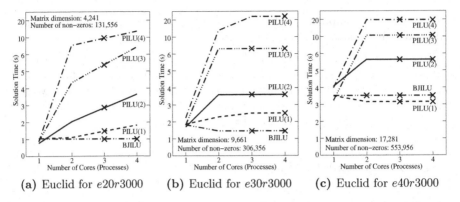

Fig. 2. Euclid PILU and BJILU for Driven Cavity Problem using a Single AMD Opteron (4 Cores). "X" means fail, and the time is arbitrarily shown to be an interpolated value or the same as for the preceding number of threads. Note that in Figure 2(a), PILU(k) actually breaks down for 3 threads, while then succeeding for 4 threads.

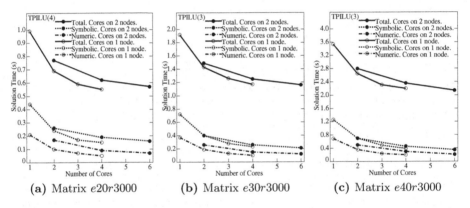

Fig. 3. TPILU(k) for the Driven Cavity Problem Using 2 AMD Opteron (2×4 Cores). The experimental runs for 1,2,3,4 threads are all for a 4-core shared memory CPU. The experimental runs for 2,4,6 threads are all for two nodes with 4-cores per node, while an additional thread per node is reserved for communication between nodes in order to replicate bands.

more blocks of smaller block dimension, and eventually the BJILU computation just breaks down.

In contrast, TPILU(k) is bit-compatible. Greater parallelization only accelerates the computation, while also never introducing instabilities or other negative side effects. Figure 3a illustrates that for the $e20r3000$ case, TPILU with level $k = 4$ and 4 threads leads to a better performance (0.55 s) than Euclid's 0.78 s (Figure 2a). For the $e30r3000$ case, TPILU(k) finishes in 1.16 s (Figure 3b), as compared to 1.47 s for BJILU and 1.64 s for PILU (Figure 2b). For the $e40r3000$ case, TPILU(k) with $k = 3$ finishes in 2.14 s (Figure 3c), as compared to 3.15 s for PILU and 3.52 s for BJILU (Figure 2c). Figure 3c demonstrates the potential of TPILU(k) for further performance improvements when a hybrid architecture is used to provide additional cores: the hybrid architecture with 6 CPU cores over two nodes connected by Infiniband is even better (2.14 s) than the shared-memory model with a single quad-core CPU (2.20 s).

3D 27-point Central Differencing. As pointed out in [4], ILU(k) preconditioning is amenable to performance analysis since the non-zero patterns of the resulting ILU(k) preconditioned matrices are identical for any partial differential equation (PDE) that has been discretized on a grid with a given stencil. However, a parallelization based on domain decomposition may eradicate this feature since it generally relies on re-ordering to maximize the independence among subdomains. The re-ordering is required for domain decomposition since it would otherwise face a higher cost dominated by the resulting denser matrix. As Figure 4a shows, Euclid PILU degrades with more processes when solving a linear system generated by 3D 27-point central differencing for Poisson's equation. The performance degradation also increases rapidly as the level k grows.

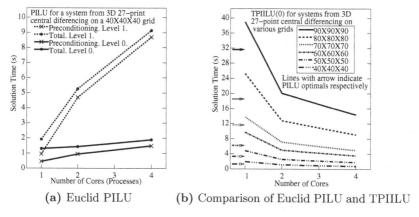

(a) Euclid PILU   (b) Comparison of Euclid PILU and TPIILU

Fig. 4. Solving Linear System from 3D 27-point Central Differencing on Grid using a Single AMD Quad-Core Opteron. Focusing on the algorithm only, the comparison ignores reusing the domain decomposition over multiple linear system solutions.

This performance degradation is not an accident. The domain-decomposition computation dominates when the number of non-zeros per row is larger (about 27 in this case). Therefore, the sequential algorithm with the level $k = 0$ wins over the parallelized PILU in the contest for the best solution time. This observation holds true for all grid sizes tested: from $50 \times 50 \times 50$ to $90 \times 90 \times 90$. In contrast, for all of these test cases, TPIILU (the level-based incomplete inverse submethod of TPILU(k)) leads to improved performance using 4 cores, as seen in Figure 4b.

Model for DNA Electrophoresis: cage15. The cage model of DNA electrophoresis [12] describes the drift, induced by a constant electric field, of homogeneously charged polymers through a gel. We test on the largest case in this problem set: *cage*15. For *cage*15, TPIILU(0) obtains a speedup of 2.93 using 8 threads (Figure 5a). The ratio of the number of FLoating point arithmetic OPerations (FLOPs) to the number of non-zero entries is less than 5. This implies that ILU(k) preconditioning just passes through matrices with few FLOPs. In other words, the computation is too "easy" to be further sped up.

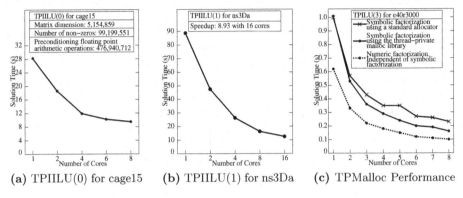

(a) TPIILU(0) for cage15   (b) TPIILU(1) for ns3Da   (c) TPMalloc Performance

Fig. 5. TPIILU(k)/TPILU(k) using 4 Intel Xeon E5520 (4 × 4 Cores)

Computational Fluid Dynamics Problem: ns3Da. The problem $ns3Da$ [12] is used as a test case in FEMLAB, developed by Comsol, Inc. Because there are zero diagonal elements in the matrix, we use TPIILU with level $k = 1$ as the preconditioner. Figure 5b shows a speedup of 8.93 with 16 threads since the preconditioning is floating-point intensive.

TPMalloc Performance. For a large level k, the symbolic factorization time will dominate. To squeeze greater performance from this first phase, glibc's standard malloc is replaced with a thread-private malloc (TPMalloc). Figure 5c demonstrates that the improvement provided by TPMalloc is significant whenever the number of cores is greater than 2.

4.1   Experimental Analysis

Given a denser matrix, or a higher level k or more CPU cores, the time for domain-decomposition based parallel preconditioning using Euclid's PILU(k) can dominate over the time for the iterative solving phase. This degrades the overall performance, as seen both in Figure 4a and in Figures 2(a,b,c). A second domain-decomposition based parallel preconditioner, Euclid's BJILU, generally produces a preconditioned matrix of lower quality than ILU(k) in Figure 2(a,b,c). This happens because it ignores off-diagonal non-zero elements. Therefore, where Euclid PILU(k) degrades the performance, it is not reasonable to resort to Euclid BJILU. Figures 2a and 2c show that the lower quality of BJILU-based solvers often performed worse than PILU(k). Figure 3 shows TPILU(k) to perform better than either while maintaining the good scalability expected of a bit-compatible algorithm. TPILU(k) is also robust enough to perform reasonably even in a configuration with two quad-core nodes. Additionally, Figures 4b and 5 demonstrate very good scalability on a variety of applications when using the optional level-based incomplete inverse optimization.

5   Related Work

ILU(k) [1] was formalized to solve the system of linear equations arising from finite difference discretizations in 1978. In 1981, ILU(k) was extended to apply to more general problems [2]. Some previous parallel ILU(k) preconditioners include [3,13,14]. The latter two methods, whose parallelism comes from level/backward scheduling, are stable and were studied in the 1980's and achieved a speedup of about 4 or 5 on an Alliant FX-8 [5, 1st edition, page 351] and a speedup of 2 or 3 on a Cray Y-MP. The more recent work [3] is directly compared with in the current work, and is not stable.

References

1. Gustafsson, I.: A Class of First Order Factorization Methods. BIT Numerical Mathematics, Springer Netherlands 18(2), 142–156 (1978)
2. Watts III, J.W.: A Conjugate Gradient-Truncated Direct Method for the Iterative Solution of the Reservoir Simulation Pressure Equation. SPE Journal 21(3), 345–353 (1981)
3. Hysom, D., Pothen, A.: A Scalable Parallel Algorithm for Incomplete Factor Preconditioning. SIAM J. Sci. Comput 22, 2194–2215 (2000)
4. Hysom, D., Pothen, A.: Efficient Parallel Computation of ILU(k) Preconditioners. In: Supercomputing 1999 (1999)
5. Saad, Y.: Iterative Methods for Sparse Linear Systems, 2nd edn. SIAM, Philadelphia (2003)
6. Bollhöfer, M., Saad, Y.: On the Relations between ILUs and Factored Approximate Inverses. SIAM J. Matrix Anal. Appl. 24(1), 219–237 (2002)
7. Dong, X., Cooperman, G., Apostolakis, J.: Multithreaded Geant4: Semi-Automatic Transformation into Scalable Thread-Parallel Software. In: Euro-Par 2010 (2010)
8. Saad, Y., van der Vorst, H.A.: Iterative Solution of Linear Systems in the 20th Century. J. Comput. Appl. Math. 123(1-2), 1–33 (2000)
9. Cooperman, G.: Practical Task-Oriented Parallelism for Gaussian Elimination in Distributed Memory. Linear Algebra and Its Applications 275-276, 107–120 (1998)
10. hypre: High Performance Preconditioners. User's Manual, version 2.6.0b, https://computation.llnl.gov/casc/hypre/download/hypre-2.6.0b_usr_manual.pdf
11. Matrix Market.: Driven Cavity from the SPARSKIT Collection, http://math.nist.gov/MatrixMarket/data/SPARSKIT/drivcav/drivcav.html
12. UF Sparse Matrix Collection, http://www.cise.ufl.edu/research/sparse/matrices/
13. Anderson, E.: Parallel Implementation of Preconditioned Conjugate Gradient Methods for Solving Sparse Systems of Linear Equations. Master's Thesis, Center for Supercomputing Research and Development, University of Illinois (1988)
14. Heroux, M.A., Vu, P., Yang, C.: A Parallel Preconditioned Conjugate Gradient Package for Solving Sparse Linear Systems on a Cray Y-MP. Appl. Num. Math. 8, 93–115 (1991)

Parallel Inexact Constraint Preconditioners for Saddle Point Problems

Luca Bergamaschi and Angeles Martinez

Department of Mathematical Methods and Models for Scientific Applications
University of Padua, via Trieste 63, 35121 Padova, Italy
berga@dmsa.unipd.it, acalomar@dmsa.unipd.it

Abstract. In this paper we propose a parallel implementation of the FSAI preconditioner to accelerate the PCG method in the solution of symmetric positive definite linear systems of very large size. This preconditioner is used as *building block* for the construction of an indefinite Inexact Constraint Preconditioner (ICP) for saddle point-type linear systems arising from Finite Element (FE) discretization of 3D coupled consolidation problems. The FSAI-ICP preconditioner, based on an efficient approximation of the inverse of the $(1,1)$ block proves very effective in the acceleration of the BiCGSTAB iterative solver in parallel environments. Numerical results on a number of realistic test cases of size up to 6×10^6 unknowns and 3×10^8 nonzeros show the almost perfect scalability of the overall code up to 512 processors.

Keywords: Parallel computing Preconditioning Krylov subspace methods coupled consolidation.

1 Introduction

The time-dependent displacements and fluid pore pressure in porous media are controlled by the consolidation theory. This was first mathematically described by Biot [8], who coupled the elastic equilibrium equations with a continuity or mass balance equation to be solved under appropriate boundary and initial flow and loading conditions.

The coupled consolidation equations are typically solved numerically using FE in space, thus giving rise to a system of first-order differential equations whose solution is addressed by an appropriate time marching scheme. A major computational issue is the repeated solution in time of the resulting discretized indefinite equations, which can be generally written as

$$\mathcal{A}x = b, \qquad \text{where} \qquad \mathcal{A} = \begin{bmatrix} K & B^T \\ B & -C \end{bmatrix}. \tag{1}$$

The sub-matrices K and C are both symmetric and positive definite (SPD). Denoting with m the number of FE nodes, $C \in \mathbb{R}^{m \times m}$, $B \in \mathbb{R}^{m \times n}$, and $K \in \mathbb{R}^{n \times n}$, where n is equal to $2m$ or $3m$ according to the spatial dimension of the problem.

E. Jeannot, R. Namyst, and J. Roman (Eds.): Euro-Par 2011, LNCS 6853, Part II, pp. 78–89, 2011.

Matrix $\mathcal{A}$ in (1) is a classical example of saddle point problem, which is encountered in other fields as well including constrained optimization, least squares, and Navier-Stokes equations. Because of the large size of realistic three-dimensional (3D) consolidation models (and particularly so in problems related to fluid withdrawal/injection from/into geological formations) the use of iterative solvers is strongly recommended against direct factorization methods. However, well established iterative methods such as Krylov subspace methods are very slow or even fail to converge if not conveniently preconditioned. The constraint preconditioners for Krylov solvers in the solution of saddle point problems have been studied by a number of authors [1,4,5,11,14]. In this work we propose a fully explicit parallel ICP based on the FSAI preconditioner [16] of the matrices K and S where S is an approximate Schur complement of a block matrix $\mathcal{M}$ resembling $\mathcal{A}$. The FSAI preconditioner is based on *prefiltration* and *postfiltration* techniques and allows to choose nonzeros in the preconditioner factors in the same position as those of $\widetilde{A}^{d_K}$, where $\widetilde{A}$ is an sparse approximation of A obtained by eliminating the small entries below a given threshold and $d_K = 1, 2, 4$.

We have developed parallel codes which implement both the FSAI-PCG solver for solution of $K\boldsymbol{x} = \boldsymbol{b}$ and the BiCGSTAB solver preconditioned with the parallel FSAI-ICP preconditioner described above. We show numerical results obtained in the solution of a number of problems of large size arising from 3D FE discretization of realistic engineering problems.

The paper is organized as follows. Section 2 gives a brief description of the consolidation equations. In Section 3 we describe the Inexact Constraint Preconditioner and recall the main spectral properties of the block preconditioned matrices. Section 4 describes the parallel preconditioner used in this work and explains in detail how it is implemented and applied during the BiCGSTAB iteration. Section 5 contains the numerical results obtained with PCG accelerated with FSAI preconditioner on seven test cases arising from realistic engineering applications as well as the results of the FSAI-ICP code on a difficult problem arising from a Coupled Consolidation model. We include also a scalability study of the parallel solution of system (1). Finally, some conclusions are stated in Section 6.

2   Finite Element Coupled Consolidation Equations

The system of partial differential equations governing the 3D coupled consolidation process in fully saturated porous media is derived from the classical Biot's formulation [8] and successive modifications as:

$$(\lambda + \mu)\frac{\partial \epsilon}{\partial i} + \mu \nabla^2 u_i = \alpha \frac{\partial p}{\partial i} \qquad i = x,\, y,\, z \tag{2}$$

$$\frac{1}{\gamma}\nabla(k\nabla p) = [\phi\beta + c_{br}(\alpha - \phi)]\frac{\partial p}{\partial t} + \alpha\frac{\partial \epsilon}{\partial t} \tag{3}$$

where c_{br} and β are the volumetric compressibility of solid grains and water, respectively, ϕ is the porosity, k the medium hydraulic conductivity, ϵ the medium volumetric dilatation, α the Biot coefficient, λ and μ are the Lamé constant and the shear modulus of the porous medium, respectively, γ is the specific weight of water, t is time, and p and u_i are the incremental pore pressure and the components of incremental displacement along the $i-$direction, respectively.

Use of standard linear Galerkin FE in space yields a system of first order differential equations which can be integrated by the Crank-Nicolson scheme. The resulting linear system has to be repeatedly solved to obtain the transient displacements and pore pressures. The nonsymmetric matrix controlling the solution scheme reads:

$$A = \begin{bmatrix} K/2 & -Q/2 \\ \dfrac{Q^T}{\Delta t} & H/2 + \dfrac{P}{\Delta t} \end{bmatrix} \tag{4}$$

where K, H, P and Q are the elastic stiffness, flow stiffness, flow capacity and flow-stress coupling matrices, respectively. Matrix A can be readily symmetrized by multiplying the upper set of equations by 2 and the lower set by $-\Delta t$, thus obtaining the sparse 2×2 block symmetric indefinite matrix (1) where $B = -Q^T$ and $C = \Delta t H/2 + P$.

3   Inexact Constraint Preconditioners

To solve system (1) we look for a preconditioner $\mathcal{M}^{-1}$ where

$$\mathcal{M} = \begin{bmatrix} G_1 & B^T \\ B & -C \end{bmatrix},$$

with G_1 an SPD approximation of the 1×1 block K. Its inverse, G_1^{-1}, which can be viewed as a preconditioner for K, is assumed to be explicitly known. To fulfill such a requirement we compute G_1^{-1} using FSAI [15,16] which is readily available in the factorized form $K^{-1} \simeq G_1^{-1} = W_1^T W_1$. The Inexact Constraint Preconditioner (ICP) is written as $\mathcal{M}_I^{-1}$ where:

$$\mathcal{M}_I^{-1} = \begin{bmatrix} I_n & -G_1^{-1}B^T \\ 0 & I_m \end{bmatrix} \begin{bmatrix} G_1^{-1} & 0 \\ 0 & -G_S^{-1} \end{bmatrix} \begin{bmatrix} I_n & 0 \\ -BG_1^{-1} & I_m \end{bmatrix}. \tag{5}$$

I_i begin $i \times i$ identity matrix and G_S^{-1} an approximation of the inverse of the Schur complement matrix S relative to $\mathcal{M}$: $S = BG_1^{-1}B^T + C$.

A further approximation can be used by simply neglecting the right matrix in the above expression thus obtaining a Triangular ICP preconditioner:

$$\mathcal{M}_T^{-1} = \begin{bmatrix} I_n & -G_1^{-1}B^T \\ 0 & I_m \end{bmatrix} \begin{bmatrix} G_1^{-1} & 0 \\ 0 & -G_S^{-1} \end{bmatrix}. \tag{6}$$

Following the approach in [3], we construct an approximate Schur complement $\widehat{S} = BG_2^{-1}B^T + C$, with the aim of reducing its fill-in. G_2^{-1} is computed as a further (sparser) FSAI approximation for the inverse of the structural block. A third FSAI preconditioner is used to approximate the inverse of $\widehat{S}$, $G_S^{-1} \approx \widehat{S}^{-1}$.

3.1 Eigenvalue Distribution of the Preconditioned Matrices

Let G_1 and G_S be SPD approximations of K and $S = C + BG_1^{-1}B^\top$, respectively. G_1^{-1} and G_S^{-1} can also be viewed as preconditioners for the corresponding matrices, so that we can define the following SPD preconditioned matrices:

$$K_P = G_1^{-1/2}KG_1^{-1/2} \quad \text{and} \quad S_P = G_S^{-1/2}SG_S^{-1/2}$$

Let us assume that

$$0 < \alpha_K = \lambda_{\min}(K_P) < 1 < \lambda_{\max}(K_P) = \beta_K,$$
$$0 < \alpha_S = \lambda_{\min}(S_P) < 1 < \lambda_{\max}(S_P) = \beta_S. \tag{7}$$

The conditions $1 \in [\alpha_K, \beta_K]$ and $1 \in [\alpha_S, \beta_S]$ are very often fulfilled in practice since preconditioners G_1 and G_S are expected to cluster eigenvalues around unit.

The following two theorems give bounds on the eigenvalues of the preconditioned matrix using ICP and TICP. They show that the eigenvalues of the preconditioned matrix are clustered around one if those of the preconditioned K and the preconditioned Schur complement are so. An exhaustive spectral analysis can be found in [2]. We denote a generally complex eigenvalue λ as $\lambda_R + i\lambda_I$.

Theorem 1

If $\beta_K < 2$ then the real eigenvalues of the ICP preconditioned matrix satisfy:

$$\min\left\{\alpha_K, \frac{\alpha_S}{\beta_K}\right\} \leq \lambda \leq \max\{(2 - \alpha_K)\beta_S, \beta_K\}.$$

If $\lambda_I \neq 0$ then

$$\frac{\alpha_K + \alpha_S(2 - \beta_K)}{2} \leq \lambda_R \leq \frac{\beta_K + \beta_S(2 - \alpha_K)}{2} \qquad |\lambda_I| \leq \sqrt{\beta_S}\max\{1 - \alpha_K, \beta_K - 1\}.$$

Proof. See proof of Theorem 3 in [2].

Theorem 2

The eigenvalues of $\mathcal{M}_T^{-1}\mathcal{A}$ satisfy the following bounds. If $\lambda_I \neq 0$ then

$$|\lambda - 1| \leq \sqrt{1 - \alpha_K}, \qquad \text{and} \qquad \frac{\alpha_K}{2} \leq \lambda_R \leq \min\left\{\frac{1 + \beta_S}{2}, 2\right\}.$$

The real eigenvalues satisfy:

$$\min\left\{\alpha_K, \frac{\alpha_S}{\beta_K + \alpha_S}\right\} \leq \lambda_R \leq \beta_S + \beta_K.$$

Proof. See proof of Theorem 5 in [2].

4   FSAI-Based ICP

The FSAI preconditioner, initially proposed in [15] and [16], has been later developed and implemented in parallel by Bergamaschi et al. in [6]. Here, we only shortly recall the main features of this preconditioner. Given and SPD matrix K the FSAI preconditioner approximately factorize its inverse as a product of two sparse triangular matrices as

$$K^{-1} \approx G^{-1} = W^T W.$$

The choice of nonzeros in W are based on a sparsity pattern which in our work may be the same as $\widetilde{K}^k$ where $\widetilde{K}$ is the result of *prefiltration* [7] of K i.e. dropping of all elements below of a threshold parameter δ. In the present paper we allow the power k to be equal to $1, 2$ or 4. The entries of W are computed by minimizing the Frobenius norm of $I - WL$ where L is the exact Cholesky factor of K. The computed W is then sparsified by dropping all the elements which are below a second tolerance parameter (ε). The final FSAI preconditioner is therefore related to the following three parameters: δ, prefiltration threshold; $d_K = 1, 2, 4$, power of K generating the sparsity pattern; ε, postfiltration threshold.

Recalling equation (5), the full ICP can be written as:

$$
\begin{aligned}
\mathcal{M}_I^{-1} &= \begin{bmatrix} I_n & -W_1^T W_1 B^T \\ 0 & I_m \end{bmatrix} \begin{bmatrix} W_1^T W_1 & 0 \\ 0 & -W_S^T W_S \end{bmatrix} \begin{bmatrix} I_n & 0 \\ -BW_1^T W_1 & I_m \end{bmatrix} \\
&= \begin{bmatrix} W_1^T & -W_1^T W_1 B^T W_S^T \\ 0 & W_S^T \end{bmatrix} \begin{bmatrix} W_1 & 0 \\ W_S B W_1^T W_1 & -W_S \end{bmatrix}
\end{aligned}
\tag{8}
$$

where $G_1^{-1} = W_1^T W_1$ and W_S is the FSAI factor of the approximate Schur complement matrix $\widetilde{S}$, $\widetilde{S}^{-1} = W_S^T W_S$. The Schur complement matrix S is evaluated as $S = BW_1^T W_2 B^T + C = S_0 + C$, W_2 being the triangular factor of a sparser FSAI approximation of K^{-1}, obtained from W_1 by a further postfiltration.

Analogously the Triangular ICP can be written as

$$
\mathcal{M}_T^{-1} = \begin{bmatrix} W_1^T & -W_1^T W_1 B^T W_S^T \\ 0 & W_S^T \end{bmatrix} \begin{bmatrix} W_1 & 0 \\ 0 & -W_S \end{bmatrix}.
\tag{9}
$$

The application of $\mathcal{M}^{-1}$ requires the explicit computation of the Schur complement matrix S whose construction may be time and memory consuming, However, it should be noted that the evaluation of $S_0 = BW_2^T W_2 B^T$, which involves the main computational burden in building S, is independent of the time step Δt, and therefore can be done just once at the beginning of the simulation. The construction of the preconditioner is therefore based on the following parameters:

1. δ_1, d_K and ε_1, for the 1st FSAI preconditioner (W_1).
2. ε_2, postfiltration threshold for W_2
3. δ_S, d_S and ε_S, for the FSAI preconditioner applied to the Schur complement matrix (W_S).

4.1 Parallel Implementation

Our code is written in FORTRAN 90 and exploits the MPI library for exchanging data among the processors. We used a block row distribution of all matrices, that is, with complete rows assigned to different processors. All these matrices are stored in static data structures in CSR format.

Any row i of matrix W of FSAI preconditioner is computed independently of each other, by solving a small SPD dense linear system of size n_i equal to the number of nonzeros allowed in row i of W. Some of the rows which contribute to form this linear system may be non local to processor i and should be received from other processors. To this aim we implemented a routine called *get_extra_rows* which carries out all the row exchanges among the processors, before starting the computation of W, which proceed afterwards entirely in parallel. Since the number of non local rows needed by each processor is relatively small we chose to temporarily replicate these rows on auxiliary data structures. Once W is obtained a parallel transposition routine provides every processor with its part of W^T.

The FSAI and the FSAI-ICP preconditioners will be used to accelerate the PCG and the BiCGSTAB Krylov subspace methods. These iterative solvers are essentially based on matrix-vector products. We made use of an optimized parallel matrix-vector product which has been developed in [17] showing its effectiveness up to 1024 processors.

5 Numerical Results

5.1 Solution of $Kx = b$.

Since the key of the success of ICP is related to the goodness of the preconditioner for matrix K (numerical experience shows that the Schur complement matrix is instead well-conditioned), we analyze the performance of our FSAI preconditioner when used within the PCG method to solve a linear system $Kx = b$.

The test cases are all realistic examples of large size arising from 2D and 3D FE discretization of geomechanical problems. In detail:

1. FAULT-639: arises from the numerical solution by a linear FE of the inequality-constrained minimization problem governing the mechanical equilibrium of a 3D body with contact surfaces [12]. The contact is solved with the aid of a penalty formulation that gives rise to an SPD ill-conditioned linear system.
2. STOCF-729: arises from the FE integration of the diffusion partial differential equation governing the 3D transient flow of groundwater in saturated porous media. The problem is solved assuming a stochastic distribution of the hydraulic conductivity tensor with a large permeability contrast in adjacent elements.
3. GEO-1438: arises from a regional geomechanical model of the sedimentary basin underlying the Venice lagoon discretized by a linear FE with randomly heterogeneous properties [18].

4. FLAN-1565: arises from the mechanical equilibrium of a steel flange discretized by a 3D 8-node brick FE [13].
5. HOOK-1498: arises from the mechanical equilibrium of a steel hook discretized by 3D 4-node tetrahedral FE [13].
6. PO-878: arises in the simulation of the consolidation of a real gas reservoir of the Po Valley, Italy, used for underground gas storage purposes (for details, see [9]).
7. CUBE-6536: simulates the compaction of a shallow confined aquifer due to groundwater withdrawal in a representative 3D sedimentary basin at a regional scale. The discretization employs 1 171140 grid nodes, giving raise to a very large problem of more than 6 million unknowns.

The size and number of nonzero terms for each matrix is provided in Table 1. The linear system is solved by PCG using the exact solution as a vector of all ones. The exit test for the iterative solver is $\frac{\|r_k\|}{\|b\|} \leq 10^{-10}$, r_k being the relative residual at iteration k. Each matrix has been preliminarily reordered by a Reverse Cuthill McKee (RCM) algorithm [10].

Table 1. Size n and number of nonzeros **nnz** of the test matrices

name	n	nnz
FAULT-639	638 812	14 626 683
STOCF-729	729 400	10 765 586
GEO-1438	1 437 960	63 156 690
FLAN-1565	1 564 794	117 406 044
HOOK-1498	1 498 023	60 917 445
PO-878	878 355	38 896 749
CUBE-6353	6 353 100	282 438 234

All tests are performed on the IBM SP6/5376 cluster at the CINECA Centre for HCP, equipped with IBM Power6 processors at 4.7 GHz with 168 nodes, 5376 computing cores, and 21 Tbytes of internal network RAM. The code is written in Fortran 90 and compiled with `-O4 -q64 -qarch=pwr6 -qtune=pwr6 -qnoipa -qstrict -bmaxdata:0x70000000` options.

In Table 2 we report the results of the PCG runs for the seven test cases and a number of combination of the FSAI parameters. In particular we provide the number of iteration (iter) the density of the FSAI preconditioner computed as $\rho = \frac{\text{nnz}(G_1^{-1})}{\text{nnz}(K)}$ as well as three CPU times referring to the cost of FSAI computation (T_P), the cost of iterative solver (T_{sol}) and the total time ($T_{tot} = T_P + T_{sol}$.) For a fixed test case all the runs have been performed using a fixed number of processors.

Inspection of Table 2 reveals that the choice of $d_K = 4$ produces in all tests the smallest number of iterations and (with the only exception of Problem FLAN-1565) the smallest T_{sol} CPU time. However, in some instances the large cost to compute the FSAI preconditioner may greatly influence the total CPU time.

5.2 Parallel Results and Scalability

We will use a strong scaling measure to see how the CPU times vary with the number of processors for a fixed total problem size. We will denote with T_p

Table 2. Iteration number, iter, density ρ of the preconditioner, CPU times obtained using a fixed number of processors for each combination of parameters. Best iteration number, smallest T_{sol} and T_{tot} for each test are printed in boldface.

name	p	d_K	δ	ϵ	iter	ρ	T_P	T_{sol}	T_{tot}
FLAN-1565	64	4	0.1	0.1	4546	0.12	12.60	67.62	80.22
		4	0.1	0.01	**2785**	1.17	11.79	82.06	93.85
		4	0.1	0.05	3909	0.29	12.47	63.44	75.91
		2	0.1	0.1	5414	0.10	0.81	**62.49**	**63.30**
		1	0.01	0.1	6064	0.09	0.72	75.55	76.27
FAULT-639	16	4	0.1	0.01	**674**	1.32	5.90	21.92	27.82
		4	0.2	0.01	986	0.18	0.35	**13.54**	**13.89**
		2	0.2	0.01	1667	0.10	0.23	26.35	26.58
		2	0	0.01	938	1.41	8.03	29.64	37.67
		1	0	0.01	1745	0.56	0.83	38.25	39.08
HOOK-1498	16	4	0.1	0.1	3511	0.28	49.29	**142.05**	191.34
		4	0.1	0.01	**2362**	2.76	46.38	267.64	314.02
		2	0.2	0.01	5195	0.10	0.49	215.56	216.05
		1	0.01	0.1	4164	0.18	1.12	149.00	**150.12**
		1	0.01	0.01	3416	0.66	0.96	168.83	169.79
GEO-1438	16	4	0.1	0.1	585	0.34	20.12	**20.34**	40.46
		4	0.1	0.01	**405**	2.13	26.77	42.93	69.70
		2	0.1	0.1	766	0.21	1.24	34.06	**35.30**
		2	0.1	0.01	671	0.58	1.42	38.65	40.07
		1	0.0	0.01	818	0.65	1.13	45.03	46.16
STOCF-729	16	4	0.1	0.05	**755**	1.61	1.96	17.06	19.02
		4	0.1	0.1	881	0.95	1.51	**9.96**	11.47
		2	0.1	0.01	1230	1.11	0.30	11.75	12.05
		2	0.2	0.1	2030	0.24	0.17	11.00	**11.17**
		1	0.01	0.01	1699	0.77	0.20	15.67	15.87
PO-878	64	4	0.2	0.1	844	0.14	0.27	4.47	**4.74**
		4	0.1	0.1	728	0.26	2.99	**3.55**	6.54
		4	0.1	0.01	**698**	1.42	2.75	7.30	10.05
		2	0.1	0.1	1414	0.17	0.34	6.27	6.61
		1	0.01	0.1	2297	0.13	0.22	8.31	8.53
CUBE-6353	256	4	0.1	0.01	**459**	1.13	5.24	12.29	17.53
		4	0.1	0.1	649	0.20	5.76	**8.56**	**14.32**
		2	0.01	0.01	511	1.09	3.68	16.02	19.70

the total CPU elapsed times expressed in seconds on p processors. As relative measures of the parallel efficiency achieved by the code we denote as $S_p^{(\bar{p})}$ the pseudo speedup computed with respect to the smallest number of processors $(\bar{p})$ used to solve a given problem and $E_p^{(\bar{p})}$ the corresponding efficiency:

$$S_p^{(\bar{p})} = \frac{T_{\bar{p}}\bar{p}}{T_p}, \qquad E_p^{(\bar{p})} = \frac{S_p^{(\bar{p})}}{p} = \frac{T_{\bar{p}}\bar{p}}{T_p p}.$$

Scalability of FSAI-PCG. In Table 3 we report number of iterations and timings in solving problems GEO-1438 and CUBE-6536 by FSAI-PCG with varying number of processors. The parameters used are: $d_K = 4, \delta = 0.1$ and $\varepsilon = 0.1$ for both cases. We also report the scaled speedups and efficiencies for the total CPU time. Speedups larger than p and efficiencies larger than 1 are printed in boldface. They can be put in connection both with cache effects and with the not optimal use of the memory for small number of processors which slow down the performance the code. We note from the table that our code scales almost perfectly up to 128 processors for problem GEO-1438 and up to $p = 512$ for problem CUBE-6536 which is roughly 4 times larger. This is also accounted by the results of Figure 1 where pseudo-speedups vs processor number are displayed in a log-log plot.

Table 3. Number of iterations and timings of FSAI-PCG in the solution of problems GEO-1438 (left) and CUBE-6536 (right)

p	iter	T_P	T_{sol}	T_{tot}	$S_p^{(2)}$	$E_p^{(2)}$
2	585	195.0	175.4	370.4		
4	585	83.5	95.5	179.0	**4.1**	**1.03**
8	585	45.1	40.7	85.8	**8.6**	**1.08**
16	585	20.1	20.3	41.4	**17.9**	**1.12**
32	585	11.0	10.4	21.4	**34.6**	**1.08**
64	585	5.9	5.2	11.1	**66.7**	**1.04**
128	585	3.1	2.7	5.9	125.6	0.98
256	585	2.0	1.8	3.8	195.0	0.76
512	585	1.0	1.4	2.4	308.7	0.60

p	iter	T_P	T_{sol}	T_{tot}	$S_p^{(16)}$	$E_p^{(16)}$
16	459	76.9	198.4	275.3		
32	459	43.6	88.2	131.8	**33.6**	**1.05**
64	459	22.3	45.4	67.7	**65.3**	**1.02**
128	459	10.0	24.1	34.1	**129.8**	**1.01**
256	459	5.2	12.3	17.5	252.2	0.99
512	459	3.2	6.7	9.9	444.8	0.87

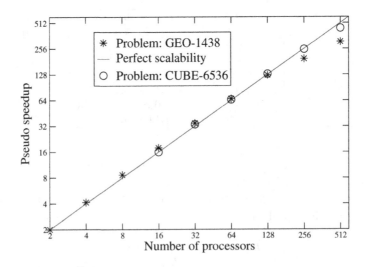

Fig. 1. Speedups vs number of processors. Problems GEO-1438 and CUBE-6536

Scalability of ICP preconditioner. We report in this Section the results obtained in the solution of our saddle point problem with PO-878 as the test example whose main features are summarized as follows.

m	n	N	nnz(K)	nnz(B)	nnz(C)	nnz($\mathcal{A}$)
292785	878355	1 171 140	38 896 749	12 965 583	4 321 861	69 039 776

We choose this problem among the seven presented in Section 5.1 since it is the most challenging one due to the complexity and the heterogeneity of the geological domain which give raise to a large number of distorted tetrahedra. This produces a very ill-conditioned matrix $\mathcal{A}$, especially for small timesteps. Moreover, the bandwidth is very large and this forces a large amount of interprocessor communication. We solved symmetrized system (4) using $\Delta t = 1$ after an intensive testing to tune the parameters. We choose BiCGSTAB as the iterative solver with the same exit test of Section 5.1.

Table 4. Combinations of parameters and results for problem PO-878 on 128 processors

Run	δ_1	d_K	ϵ_1	ϵ_2	δ_S	d_S	ϵ_S	ρ	iter	T_{P1}	T_{P2}	T_{sol}	T_{tot}
ICP 1	0	1	0	0.01	0.01	1	0	1.23	> 10000	1.4	0.2	> 200.0	> 200.0
ICP 2	0.01	2	0.01	0.1	0.01	1	10^{-3}	1.36	4945	2.9	1.4	127.8	129.2
ICP 3	0.1	4	0.1	0.1	0.01	1	10^{-3}	0.72	1254	2.3	2.2	24.3	26.6
TICP	0.1	4	0.1	0.1	0.01	1	10^{-3}	0.72	3669	3.6	2.3	66.1	68.4

In Table 4 we report for each run the parameters related to the three FSAI approximations as described in the previous sections. We also provide a measure ρ of the density of the preconditioner matrices as:

$$\rho = \rho_1 + \rho_2 = \frac{\text{nnz}(G_1^{-1})}{\text{nnz}(\mathcal{A})} + \frac{\text{nnz}(G_S^{-1})}{\text{nnz}(\mathcal{A})}$$

Parameter ρ gives an indication of the additional core memory needed for computing and storing the preconditioner. We present the following timings, all given in seconds: T_{P1} is the preprocessing time needed to construct G_1^{-1}, G_2^{-1} and S_0, T_{P2} refers to the construction of G_S^{-1} and T_{sol} to the CPU time required by the iterative solver. Finally, $T_{tot} = T_{P2} + T_{sol}$ is the total CPU time.

We report in Table 4 the results of three ICP and one TICP runs employing the three different patterns for the FSAI preconditioner in the approximation of K (with $p = 128$). Using $d_K = 1$ no convergence is attained within 10000 iterations, $d_K = 2$ yields 4945 iterations while with $d_K = 4$ the iterative method obtains convergence after 1254 iterations. From the table we see that only a sparsity pattern for the block K which uses nonzeros far away from the diagonal ($d_K = 4$) allows for a (relatively) fast convergence. We note on passing that the TICP with the same parameters as the third ICP run yields more than twice the ICP iterations and roughly twice CPU time. This is again a consequence of the ill conditioning of this problem.

Table 5. Parallel performance of FSAI-ICP (TICP) code for problem PO-878

run	p	T_{P1}	$S_P^{(2)}$	iter	T_{P2}	T_{sol}	T_{tot}	$S_p^{(2)}$	$E_p^{(2)}$
	2	99.7		1409	83.1	1667.4	1750.5		
	4	42.8	**4.7**	1521	32.3	693.0	725.3	**4.8**	**1.21**
	8	23.6	**8.5**	1518	17.6	350.5	368.1	**9.5**	**1.19**
ICP 3	16	13.6	14.7	1407	10.1	171.8	181.9	**19.3**	**1.20**
	32	7.9	25.2	1397	5.7	92.9	98.6	**35.5**	**1.11**
	64	4.3	46.7	1521	3.4	55.3	58.7	59.7	0.93
	128	2.3	86.4	1254	2.2	24.3	26.6	**131.8**	**1.03**
	2	86.5		3726	67.9	2998.5	3066.4		
	4	42.3	**4.1**	3916	32.2	1523.7	1556.0	3.9	0.99
	8	23.6	7.3	3754	17.5	767.6	783.1	7.8	0.98
TICP	16	13.6	12.7	3842	10.0	397.1	407.1	15.1	0.94
	32	7.9	21.9	3737	5.7	206.8	212.5	28.9	0.90
	64	4.2	41.2	3834	3.4	115.5	118.9	51.6	0.81
	128	2.3	75.2	3669	2.2	57.5	59.7	102.7	0.80

We present in the sequel the results of the scalability study carried out with the FSAI–ICP code when used to solve the PO-878 test problem. We show in Table 5 the results obtained running our FSAI-ICP code using $p = 2$ to $p = 128$, regarding the two preconditioners ICP3 and TICP of Table 4. These results show that our code exhibits almost perfect scalability both on the preprocessing stage and the iterative part. As before, superspeedups can occur due to cache effects and also to the variable number of iterations with different processor number p.

Acknowledgments. We acknowledge the CINECA Iscra Award PARPSEA (2010) for the availability of HPC resources and support. We also thank the four anonymous reviewers who helped improve the overall quality of the paper.

6   Conclusions

This paper describes a parallel block preconditioner for saddle point type linear systems based on an FSAI preconditioner with variable sparsity pattern. We first show that our FSAI-PCG code is efficient and scalable for the solution of $Kx = b$. Then the FSAI preconditioner is used to develop a parallel fully explicit ICP within the BiCGSTAB Krylov subspace solver. We have presented a portable parallel code implemented in Fortran 90 using MPI for interprocessor communications. This ensures portability on a whole range of supercomputers. The efficiency of our code is evaluated on realistic engineering applications arising from 3D FE discretization of a coupled consolidation problem exhibiting almost perfect scalability both on the preprocessing stage and the iterative part as well as satisfactory computational efficiency.

References

1. Benzi, M., Golub, G.H., Liesen, J.: Numerical solution of saddle point problems. Acta Numer. 14, 1–137 (2005)
2. Bergamaschi, L.: Eigenvalue distribution of constraint preconditioned saddle point matrices. Numer. Lin. Alg. Appl (submitted) (2011)
3. Bergamaschi, L., Ferronato, M., Gambolati, G.: Mixed constraint preconditioners for the solution to FE coupled consolidation equations. J. Comp. Phys. 227(23), 9885–9897 (2008)
4. Bergamaschi, L., Gondzio, J., Venturin, M., Zilli, G.: Inexact constraint preconditioners for linear systems arising in interior point methods. Comput. Optim. Appl. 36(2–3), 136–147 (2007)
5. Bergamaschi, L., Gondzio, J., Zilli, G.: Preconditioning indefinite systems in interior point methods for optimization. Comput. Optim. Appl. 28(2), 149–171 (2004)
6. Bergamaschi, L., Martínez, Á.: Parallel acceleration of Krylov solvers by factorized approximate inverse preconditioners. In: Daydé, M., Dongarra, J., Hernández, V., Palma, J.M.L.M. (eds.) VECPAR 2004. LNCS, vol. 3402, pp. 623–636. Springer, Heidelberg (2005)
7. Bergamaschi, L., Martínez, A., Pini, G.: An efficient parallel MLPG method for poroelastic models. CMES: Computer and Modeling in Engineering & Sciences 49(3), 191–216 (2009)
8. Biot, M.A.: General theory of three-dimensional consolidation. J. Appl. Phys. 12(2), 155–164 (1941)
9. Castelletto, N., Ferronato, M., Gambolati, G., Janna, C., Teatini, P., Marzorati, D., Cairo, E., Colombo, D., Ferretti, A., Bagliani, A., Mantica, S.: 3D geomechanics in UGS projects: a comprehensive study in northern Italy. In: Proceedings of the 44th US Rock Mechanics Symposium, Salt Lake City, UT (2010)
10. Cuthill, E., McKee, J.: Reducing the bandwidth of sparse symmetric matrices. In: Proceedings of the 1969 24th National Conference, pp. 157–172. ACM, New York (1969)
11. D'Apuzzo, M., De Simone, V., di Serafino, D.: On mutual impact of numerical linear algebra and large-scale optimization with focus on interior point methods. Comput. Optim. Appl. 45(2), 283–310 (2010)
12. Ferronato, M., Janna, C., Gambolati, G.: Mixed constraint preconditioning in computational contact mechanics. Comp. Methods App. Mech. Engrg. 197(45-48), 3922–3931 (2008)
13. Janna, C., Comerlati, A., Gambolati, G.: A comparison of projective and direct solvers for finite elements in elastostatics. Adv. Engrg. Soft. 40(8), 675–685 (2009)
14. Keller, C., Gould, N.I.M., Wathen, A.J.: Constraint preconditioning for indefinite linear systems. SIAM J. Matrix Anal. Appl. 21, 1300–1317 (2000)
15. Kolotilina, L.Yu., Nikishin, A.A., Yeremin, A.Yu.: Factorized sparse approximate inverse preconditionings IV. Simple approaches to rising efficiency. Numer. Lin. Alg. Appl. 6, 515–531 (1999)
16. Kolotilina, L.Yu., Yeremin, A.Yu.: Factorized sparse approximate inverse preconditionings I. Theory. SIAM J. Matrix Anal. Appl. 14, 45–58 (1993)
17. Martínez, A., Bergamaschi, L., Caliari, M., Vianello, M.: A massively parallel exponential integrator for advection-diffusion models. J. Comput. Appl. Math. 231(1), 82–91 (2009)
18. Teatini, P., Ferronato, M., Gambolati, G., Bau, D., Putti, M.: Anthropogenic Venice uplift by seawater pumping into a heterogeneous aquifer system. Water Resour. Res. 46 (2010)

Communication-Optimal Parallel 2.5D Matrix Multiplication and LU Factorization Algorithms

Edgar Solomonik and James Demmel

Department of Computer Science
University of California at Berkeley, Berkeley, CA, USA
solomon@eecs.berkeley.edu, demmel@eecs.berkeley.edu

Abstract. Extra memory allows parallel matrix multiplication to be done with asymptotically less communication than Cannon's algorithm and be faster in practice. "3D" algorithms arrange the p processors in a 3D array, and store redundant copies of the matrices on each of $p^{1/3}$ layers. '2D' algorithms such as Cannon's algorithm store a single copy of the matrices on a 2D array of processors. We generalize these 2D and 3D algorithms by introducing a new class of "2.5D algorithms". For matrix multiplication, we can take advantage of any amount of extra memory to store c copies of the data, for any $c \in \{1, 2, ..., \lfloor p^{1/3} \rfloor\}$, to reduce the bandwidth cost of Cannon's algorithm by a factor of $c^{1/2}$ and the latency cost by a factor of $c^{3/2}$. We also show that these costs reach the lower bounds, modulo polylog(p) factors. We introduce a novel algorithm for 2.5D LU decomposition. To the best of our knowledge, this LU algorithm is the first to minimize communication along the critical path of execution in the 3D case. Our 2.5D LU algorithm uses communication-avoiding pivoting, a stable alternative to partial-pivoting. We prove a novel lower bound on the latency cost of 2.5D and 3D LU factorization, showing that while c copies of the data can also reduce the bandwidth by a factor of $c^{1/2}$, the latency must *increase* by a factor of $c^{1/2}$, so that the 2D LU algorithm ($c = 1$) in fact minimizes latency. We provide implementations and performance results for 2D and 2.5D versions of all the new algorithms. Our results demonstrate that 2.5D matrix multiplication and LU algorithms strongly scale more efficiently than 2D algorithms. Each of our 2.5D algorithms performs over 2X faster than the corresponding 2D algorithm for certain problem sizes on 65,536 cores of a BG/P supercomputer.

1 Introduction

Goals of parallelization include minimizing communication, balancing the work load, and reducing the memory footprint. In practice, there are tradeoffs among these goals. For example, some problems can be made embarrassingly parallel by replicating the entire input on each processor. However, this approach may use much more memory than necessary and require significant redundant computation. At the other extreme, one stores exactly one copy of the data spread

E. Jeannot, R. Namyst, and J. Roman (Eds.): Euro-Par 2011, LNCS 6853, Part II, pp. 90–109, 2011.
© Springer-Verlag Berlin Heidelberg 2011

evenly across the processors, tries to balance the load, and minimize communication subject to this constraint.

However, some parallel algorithms do successfully take advantage of limited extra memory to increase parallelism or decrease communication. In this paper, we examine the trade-off between memory usage and communication cost in linear algebra algorithms. We introduce 2.5D algorithms (the name is explained below), which have the property that they can utilize any available amount of extra memory beyond the memory needed to store one distributed copy of the input and output. 2.5D algorithms use this extra memory to provably reduce the amount of communication they perform to a theoretical minimum.

We measure costs along the critical path to make sure our algorithms are well load balanced as well as communication efficient. In particular, we measure the following quantities along the critical path of our algorithms (which determines the running time):

- F, the computational cost, is the number of flops done along the critical path.
- W, the bandwidth cost, is the number of words sent/received along the critical path.
- S, the latency cost, is the number of messages sent/received along the critical path.
- M, the memory footprint, is the maximum amount of memory, in words, utilized by any processor at any point during algorithm execution.

Our communication model does not account for network topology. However, it does assume that all communication has to be synchronous. So, a processor cannot send multiple messages at the cost of a single message. Under this model a reduction or broadcast among p processors costs $O(\log p)$ messages but a one-to-one permutation requires only $O(1)$ messages. This model aims to capture the behavior of low-dimensional mesh or torus network topologies. Our LU communication lower-bound is independent of the above collective communication assumptions, however, it does leverage the idea of the critical path.

Our starting point is n-by-n dense matrix multiplication, for which there are known algorithms that minimize both bandwidth and latency costs in two special cases:

1. Most algorithms assume that the amount of available memory, M, is enough for one copy of the input/output matrices to be evenly spread across all p processors (so $M \approx 3n^2/p$). If this is the case, it is known that Cannon's Algorithm [7] simultaneously balances the load (so $F = \Theta(n^3/p)$), minimizes the bandwidth cost (so $W = \Theta(n^2/p^{1/2})$), and minimizes the latency cost (so $S = \Theta(p^{1/2})$) [15,5]. We call Cannon's algorithm a "2D algorithm" because it is naturally expressed by laying out the matrices across a $p^{1/2}$-by-$p^{1/2}$ grid of processors.

2. "3D algorithms" assume the amount of available memory, M, is enough for $p^{1/3}$ copies of the input/output matrices to be evenly spread across all p processors (so $M \approx 3n^2/p^{2/3}$). Given this much memory, it is known

that algorithms presented in [8,1,2,16] simultaneously balance the load (so $F = \Theta(n^3/p)$), minimize the bandwidth cost (so $W = \Theta(n^2/p^{2/3})$), and minimize the latency cost (so $S = \Theta(\log p)$) [15,5]. These algorithms are called "3D" because they are naturally expressed by laying out the matrices across a $p^{1/3}$-by-$p^{1/3}$-by-$p^{1/3}$ grid of processors.

The contributions of this paper are as follows.

1. We present a new matrix multiplication algorithm that uses $M \approx 3cn^2/p$ memory for $c \in \{1, 2, ..., \lfloor p^{1/3} \rfloor\}$, sends $c^{1/2}$ times fewer words than the 2D (Cannon's) algorithm, and sends $c^{3/2}$ times fewer messages than Cannon's algorithm. We call the new algorithm *2.5D matrix multiplication*, because it has the 2D and 3D algorithms as special cases, and effectively interpolates between them, by using a processor grid of shape $(p/c)^{1/2}$-by-$(p/c)^{1/2}$-by-c. Our 2.5D matrix multiplication algorithm attains lower bounds (modulo polylog(p) factors) on the number of words and messages communicated.

   Our implementation of 2.5D matrix multiplication achieves better strong scaling and efficiency than Cannon's algorithm and ScaLAPACK's PDGEMM [6]. On 2048 nodes of BG/P, our 2.5D algorithm multiplies square matrices of size $n = 65,536$ 5.3X faster than PDGEMM and 1.2X faster than Cannon's algorithm. On 16,384 nodes of BG/P, our 2.5D algorithm multiplies a small square matrix ($n = 8192$), 2.6X faster than Cannon's algorithm.

2. We present a 2.5D LU algorithm that also reduces the number of words moved by a factor of $c^{1/2}$ in comparison with standard 2D LU algorithms. 2.5D LU attains the same lower bound on the number of words moved as 2.5D matrix multiplication Our 2.5D LU algorithm uses *tournament pivoting* as opposed to partial pivoting [9,12]. Tournament pivoting is a stable alternative to partial pivoting that was used to minimize communication (both number of words and messages) in the case of 2D LU. We will refer to tournament pivoting as communication-avoiding pivoting (CA-pivoting) to emphasize the fact that this type of pivoting attains the communication lower-bounds.

   We present 2.5D LU implementations without pivoting and with CA-pivoting. Our results demonstrate that 2.5D LU reduces communication and runs more efficiently than 2D LU or ScaLAPACK's PDGETRF [6]. For an LU factorization of a square matrix of size $n = 65,536$, on 2048 nodes of BG/P, 2.5D LU with CA-pivoting is 3.4X faster than PDGETRF with partial pivoting. Further, on 16384 nodes of BG/P, 2.5D LU without pivoting and with CA-pivoting are over 2X faster than their 2D counterparts.

3. 2.5D LU does not, however, send fewer messages than 2D LU; instead it sends a factor of $c^{1/2}$ *more* messages. Under minor assumptions on the algorithm, we demonstrate an inverse relationship among the latency and bandwidth costs of any LU algorithms. This relation yields a lower bound on the latency cost of an LU algorithm with a given bandwidth cost. We show that 2.5D LU attains this new lower bound. Further, we show that using extra memory cannot reduce the latency cost of LU below the 2D algorithm, which sends $\Omega(p^{1/2})$ messages. These results hold for LU with CA-pivoting and without pivoting.

2 Previous Work

In this section, we detail the motivating work for our algorithms. First, we recall linear algebra communication lower bounds that are parameterized by memory size. We also detail the main motivating algorithm for this work, 3D matrix multiplication, which uses extra memory but performs less communication. The communication complexity of this algorithm serves as a matching upper-bound for our general lower bound.

2.1 Communication Lower Bounds for Linear Algebra

Recently, a generalized communication lower bound for linear algebra has been shown to apply for a large class of matrix-multiplication-like problems [5]. The lower bound applies to either sequential or parallel distributed memory, and either dense or sparse algorithms. The distributed memory lower bound is formulated under a communication model identical to that which we use in this paper. This lower bound states that for a fast memory of size M (e.g. cache size or size of memory space local to processor) the lower bound on communication bandwidth is

$$W = \Omega \left(\frac{\#arithmetic\ operations}{\sqrt{M}} \right)$$

words, and the lower bound on latency is

$$S = \Omega \left(\frac{\#arithmetic\ operations}{M^{3/2}} \right)$$

messages. On a parallel machine with p processors and a local processor memory of size M, this yields the following lower bounds for communication costs of matrix multiplication of two dense n-by-n matrices as well as LU factorization of a dense n-by-n matrix:

$$W = \Omega \left(\frac{n^3/p}{\sqrt{M}} \right), \quad S = \Omega \left(\frac{n^3/p}{M^{3/2}} \right)$$

These lower bounds are valid for $\frac{n^2}{p} < M < \frac{n^2}{p^{2/3}}$ and suggest that algorithms can reduce their communication cost by utilizing more memory. If $M < \frac{n^2}{p}$, the entire matrix won't fit in memory. As explained in [5], conventional algorithms, for example those in ScaLAPACK [6], mostly do not attain both these lower bounds, so it is of interest to find new algorithms that do.

2.2 3D Linear Algebra Algorithms

Consider we have p processors arranged into a 3D grid as in Figure 1(a), with each individual processor indexed as $P_{i,j,k}$. We replicate input matrices on 2D layers of this 3D grid so that each processor uses $M = \Omega\left(\frac{n^2}{p^{2/3}}\right)$ words of memory. In this decomposition, the lower bound on bandwidth is

$$W_{3d} = \Omega\left(n^2/p^{2/3}\right).$$

According to the general lower bound the lower bound on latency is trivial: $\Omega(1)$ messages. However, for any blocked 2D or 3D layout,

$$S_{3d} = \Omega(\log p).$$

This cost arises from the row and column dependencies of dense matrix-multiplication-like problems. Information from a block row or block column of can only be propagated to one processor with $\Omega(\log p)$ messages.

Algorithm 1. $[C]$ = 3D-matrix-multiply(A,B,n,p)

Input: n-by-n matrix A distributed so that P_{ij0} owns $\frac{n}{p^{1/3}}$-by-$\frac{n}{p^{1/3}}$ block A_{ij} for each i, j
Input: n-by-n matrix B distributed so that P_{0jk} owns $\frac{n}{p^{1/3}}$-by-$\frac{n}{p^{1/3}}$ block B_{jk} for each j, k
Output: n-by-n matrix $C = A \cdot B$ distributed so that P_{i0k} owns $\frac{n}{p^{1/3}}$-by-$\frac{n}{p^{1/3}}$ block C_{ik} for each i, k

```
// do in parallel with all processors
forall i, j, k ∈ {0, 1, ..., p^{1/3} − 1} do
    P_{ij0} broadcasts A_{ij} to all P_{ijk}        /* replicate A on each ij layer */
    P_{0jk} broadcasts B_{jk} to all P_{ijk}        /* replicate B on each jk layer */
    C_{ijk} := A_{ij} · B_{jk}
    P_{ijk} contributes C_{ijk} to a sum-reduction to P_{i0k}
end
```

3D matrix multiplication. For matrix multiplication, Algorithm 1 [8,1,2,16] achieves the 3D bandwidth and latency lower bounds. The amount of memory used in this 3D matrix multiplication algorithm is $M = \Theta\left(\frac{n^2}{p^{2/3}}\right)$ so the 3D communication lower bounds apply. The only communication performed is the reduction of C and, if necessary, a broadcast to spread the input. So the bandwidth cost is $W = O\left(\frac{n^2}{p^{2/3}}\right)$, which is optimal, and the latency cost is $S = O(\log p)$, which is a optimal for a blocked layout.

Memory efficient matrix multiplication. McColl and Tiskin [18] present a memory efficient variation on the 3D matrix multiplication algorithm for a PRAM-style model. They partition the 3D computation graph to pipeline the work and therefore reduce memory in a tunable fashion. However, their theoretical model is not reflective of modern supercomputer architectures, and we see no clear way to reformulate their algorithm to be communication optimal. Nevertheless, their research is in very similar spirit to and serves as a motivating work for the new 2.5D algorithms we present in later sections.

Previous work on 3D LU factorization. Irony and Toledo [14] introduced a 3D LU factorization algorithm that minimizes total communication volume (sum of the number of words moved over all processors), but does not minimize either bandwidth or latency along the critical path. This algorithm distributes A and B cyclically on each processor layer and recursively calls 3D LU and 3D TRSM routines on sub-matrices.

Neither the 3D TRSM nor the 3D LU base-case algorithms given by Irony and Toledo minimize communication along the critical path which, in practice, is the bounding cost. We define a different 2.5D LU factorization algorithm that does minimize communication along its critical path.

Ashcraft [4,3] suggested that total communication volume can be reduced for LU and Cholesky via the use of *aggregate data*. Aggregate data is a partial sum of updates, rather than simply the matrix entries. Our 2.5D LU algorithm uses aggregate data to reduce communication by the amount Ashcraft predicted.

3   2.5D Lower and Upper Bounds

The general communication lower bounds are valid for a range of M in which 2D and 3D algorithms hit the extremes. 2.5D algorithms are parameterized to be able to achieve the communication lower bounds for any valid M. Let $c \in \{1, 2, \ldots, \lfloor p^{1/3} \rfloor\}$ be the number of replicated copies of the input matrix. Consider the processor grid in Figure 1(b) (indexed as $P_{i,j,k}$) where each processor has local memory size $M = \Omega\left(\frac{cn^2}{p}\right)$. The lower bounds on communication are

$$W_{2.5d} = \Omega\left(\frac{n^2}{\sqrt{cp}}\right) \qquad S_{2.5d} = \Omega\left(\frac{p^{1/2}}{c^{3/2}}\right).$$

The lower bound in Section 6 of [5] is valid while $c <= p^{1/3}$. When $c = p^{1/3}$, the latency lower bound is trivial, $\Omega(1)$ messages, and the bandwidth lower bound is $\Omega(n^2/p^{2/3})$ words. If the initial data is not replicated, we claim the $\Omega(n^2/p^{2/3})$ bandwidth lower bound also holds for $c > p^{1/3}$. A total of $\Omega(cn^2 - n^2) = \Omega(cn^2)$ words must be communicated to produce the replicated copies without local entry duplicates. Therefore, some processor must communicate $\Omega(cn^2/p)$ words. When $c > p^{1/3}$, this replication bandwidth cost is bound from below by $cn^2/p = \Omega(n^2/p^{2/3})$ words.

From a performance-tuning perspective, by formulating 2.5D linear algebra algorithms, we are essentially adding an extra tuning parameter to the algorithm. Also, as a sanity check for our 2.5D algorithms, we made sure they reduced to practical 2D algorithms when $c = 1$ and to practical 3D algorithms when $c = p^{1/3}$.

3.1   2.5D Matrix Multiplication

For matrix multiplication, Algorithm 2 achieves the 2.5D bandwidth lower bound and gets within a factor of $O(\log p)$ of the 2.5D latency lower bound (likely

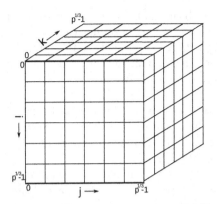

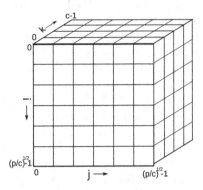

(a) 3D processor grid of dimension $p^{1/3}$-by-$p^{1/3}$-by-$p^{1/3}$.

(b) 2.5D processor grid of dimension $(p/c)^{1/2}$-by-$(p/c)^{1/2}$-by-c (replication factor c).

Fig. 1.

optimal). Algorithm 2 generalizes Cannon's algorithm (set $c = 1$). At a high level, our 2.5D algorithm does a portion of Cannon's algorithm on each set of copies of matrices A and B, then combines the results. To make this possible, we adjust the initial shift done by Cannon's algorithm to be different for each set of copies of matrices A and B.

Our 2.5D algorithm doesn't quite generalize Algorithm 1 since C is reduced in a different dimension and shifted initially. However, in terms of complexity, only two extra matrix shift operations are required by the 3D version of our 2.5D algorithm. Further, the 2.5D algorithm has the nice property that C ends up spread over the same processor layer that both A and B started on. The algorithm moves $W = O\left(\frac{n^2}{\sqrt{cp}}\right)$ words and sends $S = O\left(\sqrt{p/c^3} + \log c\right)$ messages. This cost is optimal according to the general communication lower bound. The derivations of these costs are in Appendix A in [19].

We also note that if the latency cost is dominated by the intra-layer communication $S = O(\sqrt{p/c^3})$, our 2.5D matrix multiplication algorithm can achieve perfect strong scaling in certain regimes. Suppose we want to multiply $n \times n$ matrices, and the maximum memory available per processor is $M_{\max}$. Then we need to use at least $p_{\min} = \Theta(n^2/M_{\max})$ processors to store one copy of the matrices. The 2D algorithm uses only one copy of the matrix and has a bandwidth cost of $W_{p_{\min}} = O(n^2/\sqrt{p_{\min}})$ words and latency cost of $S_{p_{\min}} = O(\sqrt{p_{\min}})$ messages. If we use $p = c \cdot p_{\min}$ processors, with a total available memory of $p \cdot M_{\max} = c \cdot p_{\min} \cdot M_{\max}$, we can afford to have c copies of the matrices. The 2.5D algorithm can store a matrix copy on each of c layers of the p processors. Utilizing c copies reduces the bandwidth cost to $W_p = O(n^2/\sqrt{cp}) = O(n^2/(c\sqrt{p_{\min}})) = O(W_{p_{\min}}/c)$ words, and the latency cost to $S_p = O(\sqrt{p/c^3}) = O(\sqrt{p_{\min}}/c) = O(S_{p_{\min}}/c)$ messages. This strong scaling is

---

Algorithm 2. $[C] = $ 2.5D-matrix-multiply(A,B,n,p,c)

Input: square n-by-n matrices A, B distributed so that P_{ij0} owns $\frac{n}{\sqrt{p/c}}$-by-$\frac{n}{\sqrt{p/c}}$ blocks A_{ij} and B_{ij}
   for each i, j

Output: square n-by-n matrix $C = A \cdot B$ distributed so that P_{ij0} owns $\frac{n}{\sqrt{p/c}}$-by-$\frac{n}{\sqrt{p/c}}$ block C_{ij}
   for each i, j

```
/* do in parallel with all processors                                          */
```
forall $i, j \in \{0, 1, ..., \sqrt{p/c} - 1\}$, $k \in \{0, 1, ..., c - 1\}$ **do**

   P_{ij0} broadcasts A_{ij} and B_{ij} to all P_{ijk} `/* replicate input matrices */`

   $s := \mod (j - i + k\sqrt{p/c^3}, \sqrt{p/c})$ `/* initial circular shift on A */`

   P_{ijk} sends A_{ij} to A_{local} on P_{isk}

   $s' := \mod (i - j + k\sqrt{p/c^3}, \sqrt{p/c})$ `/* initial circular shift on B */`

   P_{ijk} sends B_{ij} to B_{local} on $P_{s'jk}$

   $C_{ijk} := A_{\text{local}} \cdot B_{\text{local}}$

   $s := \mod (j + 1, \sqrt{p/c})$

   $s' := \mod (i + 1, \sqrt{p/c})$

   for $t = 1$ to $\sqrt{p/c^3} - 1$ **do**

      P_{ijk} sends A_{local} to P_{isk} `/* rightwards circular shift on A */`

      P_{ijk} sends B_{local} to $P_{s'jk}$ `/* downwards circular shift on B */`

      $C_{ijk} := C_{ijk} + A_{\text{local}} \cdot B_{\text{local}}$

   end

   P_{ijk} contributes C_{ijk} to a sum-reduction to P_{ij0}

end

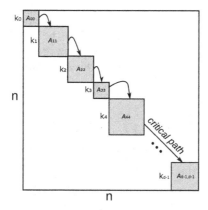

Fig. 2. LU diagonal block dependency path. These blocks must be factorized in order and communication is required between each block factorization.

perfect because all three costs (flops, bandwidth and latency) fall by a factor of c. (up to a factor of $c = p^{1/3}$, and ignoring the $\log(c)$ latency term).

4   2.5D LU Communication Lower Bound

We argue that for Gaussian-elimination style LU algorithms that achieve the bandwidth lower bound, the latency lower bound is actually much higher, namely $S_{lu} = \Omega\left(\sqrt{cp}\right)$.

Given a parallel LU factorization algorithm, we assume the algorithm must uphold the following properties

1. Consider the largest k-by-k matrix A_{00} factorized sequentially such that $A = \begin{bmatrix} A_{00} & A_{01} \\ A_{10} & A_{11} \end{bmatrix}$ (we can always pick some A_{00} since at least the top left element of A is factorized sequentially), the following conditions must hold,
   (a) $\Omega(k^3)$ flops must be done before A_{11} can be factorized (it can be updated but Gaussian elimination cannot start).
   (b) $\Omega(k^2)$ words must be communication before A_{11} can be factorized.
   (c) $\Omega(1)$ messages must be sent before A_{11} can be factorized.
2. The above condition holds recursively (for factorization of A_{11} in place of A).

We now lower bound the communication cost for any algorithm that follows the above restrictions. Any such algorithm must compute a sequence of diagonal blocks $\{A_{00}, A_{11}, \ldots, A_{d-1,d-1}\}$. Let the dimensions of the blocks be $\{k_0, k_1, \ldots, k_{d-1}\}$. As done in Gaussian Elimination and as required by our conditions, the factorizations of these blocks are on the critical path and must be done in strict sequence.

Given this dependency path (shown in Figure 2), we can lower bound the complexity of the algorithm by counting the complexity along this path. The latency cost is $\Omega(d)$ messages, the bandwidth cost is $\sum_{i=0}^{d-1} \Omega(k_i^2)$ words and the computational cost is $\sum_{i=0}^{d-1} \Omega(k_i^3)$ flops. Due to the constraint, $\sum_{i=0}^{d-1} k_i = n$, it is best to pick all $k_i = k$, for some k (we now get $d = n/k$), to minimize bandwidth and flop costs. Now we see that the algorithmic costs are

$$F_{lu} = \Omega(nk^2) \quad S_{lu} = \Omega(n/k) \quad W_{lu} = \Omega(nk).$$

Evidently, if we want to do $O(n^3/p)$ flops we need $k = O\left(\frac{n}{\sqrt{p}}\right)$, which would necessitate $S = \Omega(\sqrt{p})$. Further, the cost of sacrificing flops for latency is large. Namely, if $S = O\left(\frac{\sqrt{p}}{r}\right)$, the computational cost is $F = \Omega\left(\frac{r^2 n^3}{p}\right)$, a factor of r^2 worse than optimal. Since we are very unlikely to want to sacrifice so much computational cost to lower the latency cost, we will not attempt to design algorithms that achieve a latency smaller than $\Omega(\sqrt{p})$.

If we want to achieve the bandwidth lower bound we need,

$$W_{lu} = O\left(n^2/\sqrt{cp}\right) \quad k = O\left(n/\sqrt{cp}\right) \quad S_{lu} = \Omega(\sqrt{cp}).$$

A latency cost of $O(\sqrt{cp}/r)$, would necessitate a factor of r larger bandwidth cost. So, an LU algorithm can do minimal flops, bandwidth, and latency as defined in the general lower bound, only when $c = 1$. For $c > 1$, we can achieve optimal bandwidth and flops but not latency.

It is also worth noting that the larger c is, the higher the latency cost for LU will be (assuming bandwidth is prioritized). This insight is the opposite of that of the general lower bound, which lower bounds the latency as $\Omega(1)$ messages for

3D ($c = p^{1/3}$). However, if a 3D LU algorithm minimizes the number of words communicated, it must send $\Omega(p^{2/3})$ messages. This tradeoff suggests that c should be tuned to balance the bandwidth cost and the latency cost.

5   2.5D Communication Optimal LU

In order to write down a 2.5D LU algorithm, it is necessary to find a way to meaningfully exploit extra memory. A 2D parallelization of LU typically factorizes a vertical and a top panel of the matrix and updates the remainder (the Schur complement). The dominant cost in a typical parallel LU algorithm is the update to the Schur complement. Our 2.5D algorithm exploits this by accumulating the update over layers. However, in order to factorize each next panel we must reduce the contributions to the Schur complement. We note that only the panel we are working on needs to be reduced and the remainder can be further accumulated. Even so, to do the reductions efficiently, a block-cyclic layout is required. This layout allows more processors to participate in the reductions and pushes the bandwidth cost down to the lower bound.

Algorithm 3. $[L, U] = $ 2.5D-LU-factorization(A,n,p,c)

Input: n-by-n matrix A distributed so that for each l, m, (n/c)-by-(n/c) block A_{lm} is spread over P_{ij0} in $(n/\sqrt{pc})$-by-$(n/\sqrt{pc})$ blocks.

Output: triangular n-by-n matrices L, U such that $A = L \cdot U$ and for each l, m, (n/c)-by-(n/c) blocks L_{lm}, U_{lm} are spread over P_{ij0}.

P_{ij0} broadcasts its portion of A to each P_{ijk}

for $t = 0$ **to** $c - 1$ **do**

    $[L_{tt}, U_{tt}] = $ 2D-LU(A_{tt})   /* *redundantly factorize top right (n/c)-by-(n/c) block* */

    $[L_{t+k+1,t}^T] = $ 2D-TRSM($U_{tt}^T, A_{t+k+1,t}^T$)   /* *perform TRSMs on (n/c)-by-(n/c) blocks* */

    $[U_{t,t+k+1}] = $ 2D-TRSM($L_{tt}, A_{t,t+k+1}$)

    P_{ijk} broadcasts its portions of $L_{t+k+1,t}$ and $U_{t,t+k+1}$ to $P_{ijk'}$ for all k' /* *all-gather panels* */

    if $\lfloor k\sqrt{p/c^3} \rfloor \le j < \lfloor (k+1)\sqrt{p/c^3} \rfloor$ **then**   /* *broadcast sub-panels of L* */

        P_{ijk} broadcasts its portion of $L_{t+1:c-1,t}$ to each $P_{ij'k}$ for all j'

    end

    if $\lfloor k\sqrt{p/c^3} \rfloor \le i < \lfloor (k+1)\sqrt{p/c^3} \rfloor$ **then**   /* *broadcast sub-panels of U* */

        P_{ijk} broadcasts its portion of $U_{t,t+1:c-1}$ to each $P_{i'jk}$ for all i'

    end

    P_{ijk} computes and accumulates its portion of the Schur complement

    update S /* *multiply sub-panels* */

    All-reduce (sum and subtract from A) $S_{t+1:c-1,t+1}, S_{t+1,t+2:c-1}$ /* *reduce next big block panels* */

end

Algorithm 3 (work-flow diagram in Figure 3) is a communication optimal LU factorization algorithm for the entire range of $c \in \{1, 2, \ldots, \lfloor p^{1/3} \rfloor\}$. The algorithm replicates the matrix A on each layer and partitions it block cyclically across processors with block size $(n/\sqrt{pc})$-by-$(n/\sqrt{pc})$. Note that this block dimension corresponds to the lower bound derivations in the previous section. Every processor owns one such block within each bigger block of size n/c-by-n/c. We will sometimes refer to big blocks (block dimension n/c) and small blocks (block dimension $n/\sqrt{pc}$) for brevity.

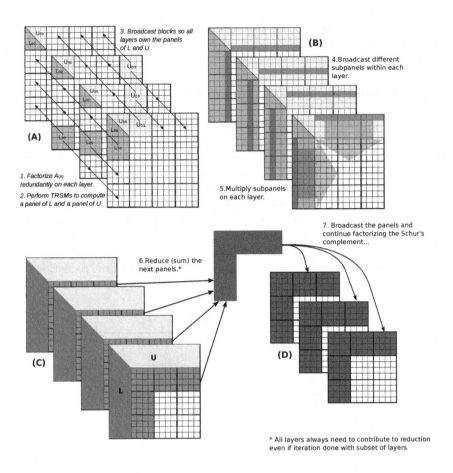

Fig. 3. 2.5D LU algorithm work-flow

Algorithm 3 has a bandwidth cost of $W = O\left(\frac{n^2}{\sqrt{cp}}\right)$ words and a latency cost of $S = O\left(\sqrt{cp}\log(p)\right)$ messages. Therefore, it is asymptotically communication optimal for any choice of c (modulo a $\log(p)$ factor for latency). Further, it is also always asymptotically computationally optimal (the redundant work is a low order cost). These costs are derived in Appendix B in [19].

6   2.5D Communication Optimal LU with Pivoting

Regular partial pivoting is not latency optimal because it requires $\Omega(n)$ messages if the matrix is in a 2D blocked layout. $\Omega(n)$ messages are required by partial pivoting since a pivot needs to be determined for each matrix column

---

Algorithm 4. $[V, L, U] = 2.5\text{D-TSLU-pivot-factorization}(A,n,m,p,c)$

Let $[V]$ = CA-Pivot$_l$ (A_l,n,b) be a function that performs CA-pivoting with block size b on A of size n-by-b
and outputs the pivot matrix V to all processors.

Input: n-by-m matrix A distributed so that for each i, j, P_{ijk} owns $\frac{m}{\sqrt{p/c}}$-by-$\frac{m}{\sqrt{p/c}}$ blocks A_{l_ij} for

$$l_i \in \{i, i + \sqrt{p/c}, i + 2\sqrt{p/c}, \ldots, i + (n/m - 1)\sqrt{p/c}\}.$$

Output: n-by-n permutation matrix V and triangular matrices L, U such that $V \cdot A = L \cdot U$ and for each

i, j, P_{ijk} owns $\frac{n}{\sqrt{pc}}$-by-$\frac{n}{\sqrt{pc}}$ blocks L_{l_ij} and U_{ij} for each i, l_i, and j.

for $s = 0$ **to** $\sqrt{p/c} - 1$ **do**

  P_{isk} compute $[V_s]$ = CA-Pivot$_{k\sqrt{p/c}+i}(A_{k\sqrt{p/c}+i,s}, n, m)$

  P_{ijk} pivots rows between $A_{k\sqrt{p/c}+i,j}$ and each copy of A_{sj} stored on each P_{sjk} according to V_s

  $A_{ss} := V_s'^T L_{ss} U_{ss}$ /* factorize top left small block redundantly using GEPP */

  $U_{sj} := L_{ss}^{-1} V_s' A_{sj}$ for $j > s$ /* do TRSMs on top small block row redundantly */

  $L_{is}^T := U_{ss}^{-T} A_{is}^T$ for $i > s$ /* do TRSMs on the top part of a small block column redundantly */

  $L_{k\sqrt{p/c}+i,s}^T := U_{ss}^{-T} A_{k\sqrt{p/c}+i,s}^T$ /* do TRSMs on rest of small block column */

  P_{isk} broadcasts L_{is} and $L_{k\sqrt{p/c}+i,s}$ to all P_{ijk}

  P_{sjk} broadcasts U_{sj} to all P_{ijk}

  $A_{ij} := A_{ij} - L_{is} \cdot U_{sj}$ for $i, j > s$ /* update top big block redundantly */

  $A_{k\sqrt{p/c}+i,j} := A_{k\sqrt{p/c}+i,j} - L_{k\sqrt{p/c}+i,s} \cdot U_{sj}$ for $j > s$ /* update remaining big blocks */

  Update V with V_s

end

P_{ijk} broadcasts $L_{k\sqrt{p/c}+i,j}$ to $P_{ijk'}$ for all k'

---

which always requires communication unless the entire column is owned by one processor. However, tournament pivoting (CA-pivoting) [9], is a new LU pivoting strategy that can satisfy the general communication lower bound. We will incorporate this strategy into our 2.5D LU algorithm.

CA-pivoting simultaneously determines b pivots by forming a tree of factorizations as follows,

1. Factorize each $2b$-by-b block $[A_{0,2k}, A_{0,2k+1}]^T = P_k^T L_k U_k$ for $k \in [0, \frac{n}{2b} - 1]$ using GEPP.
2. Write $B_k = P_k[A_{0,2k}, A_{0,2k+1}]^T$, and $B_k = [B_k', B_k'']^T$. Each B_k' represents the 'best rows' of each sub-panel of A.
3. Now recursively perform steps 1-3 on $[B_0', B_1', \ldots, B_{n/(2b)-1}']^T$ until the number of total best pivot rows is b.

For a more detailed and precise description of the algorithm and stability analysis see [9,12].

To incorporate CA-pivoting into our LU algorithm, we would like to do pivoting with block size $b = n/\sqrt{pc}$. The following modifications need to be made to accomplish this,

1. Previously, we did the big-block side panel Tall-Skinny LU (TSLU) via a redundant top block LU-factorization and TRSMs on lower blocks. To do pivoting, the TSLU factorization needs to be done as a whole rather

---

Algorithm 5. $[V, L, U] = $ 2.5D-LU-pivot-factorization(A,n,p,c)

Input: n-by-n matrix A distributed so that for each l, m, (n/c)-by-(n/c) block A_{lm} is spread over P_{ij0} in
    $(n/\sqrt{pc})$-by-$(n/\sqrt{pc})$ blocks.

Output: n-by-n matrices V and triangular L, U such that $V \cdot A = L \cdot U$ and for each l, m, (n/c)-by-(n/c)
    blocks L_{lm}, U_{lm} are spread over P_{ij0}.

$S_{1:n,1:n} := 0$ /* S will hold the accumulated Schur complemented updates to A */

P_{ij0} broadcasts its portion of A to each P_{ijk}

for $t = 0$ **to** $c - 1$ **do**
    $[V_t, L_{t:n/c-1,t}, U_{tt}] = $ 2.5D-TSLU-pivot-factorization$(A_{t:c-1,t}, n - tn/c, n/c, p, c)$
    Update V with V_t
    Swaps any rows as required by V_t to $(A, S)_{t,1:c-1}$ /* pivot remainder of matrix redundantly */
    All-reduce (sum and subtract from A) $S_{t,t+1:c-1}$ /* reduce big block top panel */
    $[U_{t,t+k+1}] = $ 2D-TRSM$(L_{tt}, A_{t,t+k+1})$ /* perform TRSMs on (n/c)-by-(n/c) blocks */
    P_{ijk} broadcasts its portion of $U_{t,t+k+1}$ to each $P_{ijk'}$ for all k' /* all-gather top panel */
    if $\lfloor k\sqrt{p/c^3} \rfloor \leq j < \lfloor (k+1)\sqrt{p/c^3} \rfloor$ **then**   /* broadcast sub-panels of L */
       P_{ijk} broadcasts its portion of $L_{t+1:c-1,t}$ to each $P_{ij'k}$ for all j'
    end
    if $\lfloor k\sqrt{p/c^3} \rfloor \leq i < \lfloor (k+1)\sqrt{p/c^3} \rfloor$ **then**   /* broadcast sub-panels of U */
       P_{ijk} broadcasts its portion of $U_{t,t+1:c-1}$ to each $P_{i'jk}$ for all i'
    end
    P_{ijk} computes and accumulates its portion the Schur complement update S /* multiply sub-panels */
    All-reduce (sum and subtract from A) $S_{t+1:c-1,t+1}$ /* reduce next big block vertical panel */
end

---

than in blocks. We can still have each processor layer compute a different 'TRSM block' but we need to interleave this computation with the top block LU factorization and communicate between layers to determine each set of pivots as follows (Algorithm 4 gives the full TSLU algorithm),

(a) For every small block column, we perform CA-pivoting over all layers to determine the best rows.

(b) We pivot the rows within the panel on each layer. Interlayer communication is required, since the best rows are spread over the layers (each layer updates a subset of the rows).

(c) Each ij processor layer redundantly performs small TRSMs and the Schur complement updates in the top big block.

(d) Each ij processor layer performs TRSMs and updates on a unique big-block of the panel.

2. After the TSLU, we need to pivot rows in the rest of the matrix. We do this redundantly on each layer, since each layer will have to contribute to the update of the entire Schur complement.

3. We still reduce the side panel (the one we do TSLU on) at the beginning of each step but we postpone the reduction of the top panel until pivoting is complete. Basically, we need to reduce the 'correct' rows which we know only after the TSLU.

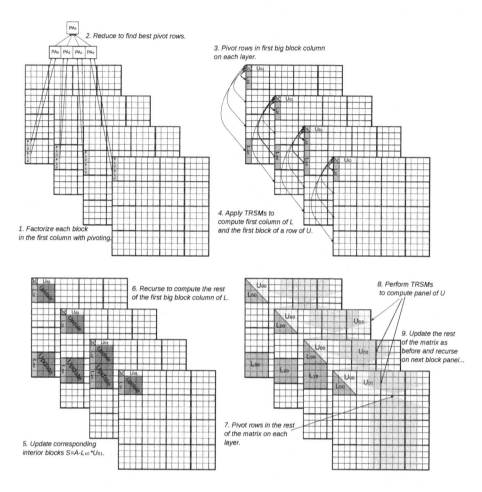

Fig. 4. 2.5D LU with pivoting panel factorization (step A in Figure 3)

Algorithm 5 details the entire 2.5D LU with CA-pivoting algorithm and Figure 4 demonstrates the workflow of the new TSLU with CA-pivoting. Asymptotically, 2.5D LU with CA-pivoting has almost the same communication and computational cost as the original algorithm. Both the flops and bandwidth costs gain an extra asymptotic $\log p$ factor (which can be remedied by using a smaller block size and sacrificing some latency). Also, the bandwidth cost derivation requires a probabilistic argument about the locations of the pivot rows, however, the argument should hold up very well in practice. For the full cost derivations of this algorithm see Appendix C in [19].

7   Performance Results

We implemented 2.5D matrix multiplication and LU factorization using MPI [13] for inter-processor communication. We perform most of the sequential work using BLAS routines: DGEMM for matrix multiplication, DGETRF, DTRSM, DGEMM, for LU. We found it was fastest to use provided multi-threaded BLAS libraries rather than our own threading. All the results presented in this paper use threaded ESSL routines.

We benchmarked our implementations on a Blue Gene/P (BG/P) machine located at Argonne National Laboratory (Intrepid). We chose BG/P as our target platform because it uses few cores per node (four 850 MHz PowerPC processors) and relies heavily on its interconnect (a bidirectional 3D torus with 375 MB/sec of achievable bandwidth per link). On this platform, reducing inter-node communication is vital for performance.

BG/P also provides topology-aware partitions, which 2.5D algorithms are able to exploit. For node counts larger than 16, BG/P allocates 3D cuboid partitions. Since 2.5D algorithms have a parameterized 3D virtual topology, a careful choice of c allows them to map precisely to the allocated partitions (provided enough memory).

Topology-aware mapping can be very beneficial since all communication happens along the three dimensions of the 2.5D virtual topology. Therefore, network contention is minimized or, in certain scenarios, completely eliminated. Topology-aware mapping also allows 2.5D algorithms to utilize optimized line multicast and line reduction collectives provided by the DCMF communication layer [11,17].

We study the strong scaling performance of 2.5D algorithms on a 2048 node partition (Figures 5(a), 6(a), 7(a)). The 2048 node partition is arranged in a 8-by-8-by-32 torus. In order to form square layers, our implementation uses 4 processes per node (1 process per core) and folds these processes into the X dimension. Now, each XZ virtual plane is 32-by-32. We strongly scale 2.5D algorithms from 256 nodes $c = Y = 1$ to 2048 nodes $c = Y = 8$. For ScaLAPACK we use smp or dual mode on these partitions, since it is not topology-aware.

We also compare performance of 2.5D and 2D algorithms on 16,384 nodes (65,536 cores) of BG/P. The 16,384 node partition is a 16-by-32-by-32 torus. We run both 2D and 2.5D algorithms in SMP mode. For 2.5D algorithms, we use $c = 16$ YZ processor layers.

7.1   2.5D Matrix Multiplication Performance

Our 2.5D matrix multiplication implementation is a straight-forward adjustment of Cannon's algorithm. We assume square and correctly padded matrices, as does Cannon's algorithm. A more general 2.5D matrix multiplication algorithm ought to be built on top of a more general 2D algorithm (e.g. the SUMMA algorithm [20]). However, our algorithm and implementation provide an idealistic and easily reproducible proof of concept.

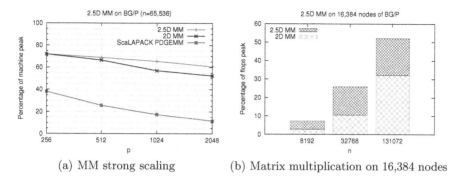

(a) MM strong scaling   (b) Matrix multiplication on 16,384 nodes

Fig. 5. Performance of 2.5D MM on BG/P

Figure 5(a) demonstrates that 2.5D matrix multiplication achieves better strong scaling than its 2D counter-part. However, both run at high efficiency (over 50%) for this problem size, so the benefit is minimal. The performance of the more general ScaLAPACK implementation lags behind the performance of our code by a large factor.

Figure 5(b) shows that 2.5D matrix multiplication outperforms 2D matrix multiplication significantly for small matrices on large partitions. The network latency and bandwidth costs are reduced, allowing small problems to execute much faster (up to 2.6X for the smallest problem size).

7.2   2.5D LU Performance

We implemented a version of 2.5D LU without pivoting. While this algorithm is not stable for general dense matrices, it provides a good upper-bound on the performance of 2.5D LU with pivoting. The performance of 2.5D LU is also indicative of how well a 2.5D Cholesky implementation might perform.

Our 2.5D LU implementation has a structure closer to that of Algorithm 5 rather than Algorithm 3. Processor layers perform updates on different big-block subpanels of the matrix as each corner small block gets factorized.

Our 2.5D LU implementation made heavy use of subset broadcasts (multicasts). All communication is done in the form of broadcasts or reductions along axis of the 3D virtual topology. This design allowed our code to utilize efficient line broadcasts on the BG/P supercomputer.

Figure 6(a) shows that 2.5D LU achieves more efficient strong scaling than 2D LU. 2D LU maps well to the 2D processor grid on 256 nodes. However, the efficiency of 2D LU suffers when we use more nodes, since the network partition becomes 3D. On 3D partitions, the broadcasts within 2D LU are done via topology-oblivious binomial trees and suffer from contention. For this problem configuration, 2.5D LU achieves a 2.2X speed-up over the 2D algorithm on 2048 nodes.

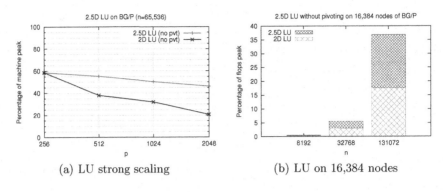

(a) LU strong scaling   (b) LU on 16,384 nodes

Fig. 6. Performance of 2.5D LU without pivoting on BG/P

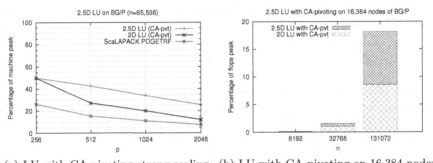

(a) LU with CA-pivoting strong scaling   (b) LU with CA-pivoting on 16,384 nodes

Fig. 7. Performance of 2.5D LU with pivoting on BG/P

Figure 6(b) demonstrates that 2.5D LU is also efficient and beneficial at a larger scale. However, the efficiency of both 2D and 2.5D LU falls off for small problem sizes. The best efficiency and relative benefits are seen for the largest problem size ($n = 131,072$). We did not test larger problem sizes, since execution becomes too time consuming. However, we expect better performance and speedups for larger matrices.

7.3   2.5D LU with CA-Pivoting Performance

2.5D LU performs pivoting in two stages. First, pivoting is performed only in the big-block panel. Then the rest of the matrix is pivoted according to a larger, accumulated pivot matrix. We found it most efficient to perform the subpanel pivoting via a broadcast and a reduction, which minimize latency. For the rest of the matrix, we performed scatter and gather operations to pivot, which minimize bandwidth. We found that this optimization can also be used to improve the performance of 2D LU and used it accordingly.

Figure 7(a) shows that 2.5D LU with CA-pivoting strongly scales with higher efficiency than its 2D counter-part. It also outperforms the ScaLAPACK PDGETRF implementation. Though, we note that ScaLAPACK uses partial pivoting rather than CA-pivoting and therefore computes a different answer.

On 16,384 nodes, 2D and 2.5D LU run efficiently only for larger problem sizes (see Figure 7(b)). The latency costs of pivoting heavily deteriorate the performance of the algorithms when the matrices are small. Since, 2.5D LU does not reduce latency cost, not much improvement is achieved for very small matrix sizes. However, for medium sized matrices ($n = 131,072$) over 2X gains in efficiency are achieved. We expect similar trends and better efficiency for even larger matrices.

The absolute efficiency achieved by our 2.5D LU with CA-pivoting algorithm is better than ScaLAPACK and can be improved even further. Our implementation does not exploit overlap between communication and computation and does not use prioritized scheduling. We observed that, especially at larger scales, processors spent most of their time idle (waiting to synchronize). Communication time, on the other hand, was heavily reduced by our techniques and was no longer a major bottleneck.

8   Future Work

Preliminary analysis suggests that a 2.5D algorithm for TRSM can be written using a very similar parallel decomposition to what we present in this paper for LU. We will formalize this analysis.

Our 2.5D LU algorithm can also be modified to do Cholesky. Thus, using Cholesky-QR we plan to formulate many other numerical linear algebra operations with minimal communication. As an alternative, we are also looking into adjusting the algorithms for computing QR, eigenvalue decompositions, and the SVD which use Strassen's algorithm [10] to using our 2.5D matrix multiplication algorithm instead. Further, we plan to look for the most efficient and stable 2.5D QR factorization algorithms. In particular, the 2D parallel Householder algorithm for QR has a very similar structure to LU, however, we have not found a way to accumulate Householder updates across layers. The Schur complement updates are subtractions and therefore commute, however, each step of Householder QR orthogonalizes the remainder of the matrix with the newly computed panel of Q. This orthogonalization is dependent on the matrix remainder and is a multiplication, which means the updates do not commute. Therefore it seems to be difficult to accumulate Housholder updates onto multiple buffers.

We plan to implement a more general 2.5D MM algorithm based on SUMMA [20]. We also plan to further tune our 2.5D LU algorithms. Incorporating better scheduling and overlap should improve the absolute efficiency of our implementation. We hope to apply these implementations to accelerate scientific simulations that solve distributed dense linear algebra problems. Our motivating scientific domain has been quantum chemistry applications, which spend a significant fraction of execution time performing small distributed dense matrix multiplications and factorizations.

Acknowledgements. The first author was supported by a Krell Department of Energy Computational Science Graduate Fellowship, grant number DE-FG02-97ER25308. Research was also supported by Microsoft (Award #024263) and Intel (Award #024894) funding and by matching funding by U.C. Discovery (Award #DIG07-10227). This material is supported by U.S. Department of Energy grants numbered DE-SC0003959, DE-SC0004938, and DE-FC02-06-ER25786. This research used resources of the Argonne Leadership Computing Facility at Argonne National Laboratory, which is supported by the Office of Science of the U.S. Department of Energy under contract DE-AC02-06CH11357. We acknowledge the support of NSF grant OCI-1032639.

References

1. Agarwal, R.C., Balle, S.M., Gustavson, F.G., Joshi, M., Palkar, P.: A three-dimensional approach to parallel matrix multiplication. IBM J. Res. Dev. 39, 575–582 (1995)
2. Aggarwal, A., Chandra, A.K., Snir, M.: Communication complexity of PRAMs. Theoretical Computer Science 71(1), 3–28 (1990)
3. Ashcraft, C.: A taxonomy of distributed dense LU factorization methods. Boeing Computer Services Technical Report ECA-TR-161 (March 1991)
4. Ashcraft, C.: The fan-both family of column-based distributed Cholesky factorization algorithms. In: Alan George, J.R.G., Liu, J.W.H. (eds.) Graph Theory and Sparse Matrix Computation. IMA Volumes in Mathematics and its Applications, vol. 56, pp. 159–190. Springer, Heidelberg (1993)
5. Ballard, G., Demmel, J., Holtz, O., Schwartz, O.: Minimizing communication in numerical linear algebra. To appear in SIAM J. Mat. Anal. Appl., UCB Technical Report EECS-2009-62 (2010)
6. Blackford, L.S., Choi, J., Cleary, A., D'Azeuedo, E., Demmel, J., Dhillon, I., Hammarling, S., Henry, G., Petitet, A., Stanley, K., Walker, D., Whaley, R.C.: ScaLA-PACK User's Guide, Society for Industrial and Applied Mathematics, Philadelphia, PA, USA (1997)
7. Cannon, L.E.: A cellular computer to implement the Kalman filter algorithm. Ph.D. thesis, Bozeman, MT, USA (1969)
8. Dekel, E., Nassimi, D., Sahni, S.: Parallel matrix and graph algorithms. SIAM Journal on Computing 10(4), 657–675 (1981)
9. Demmel, J., Grigori, L., Xiang, H.: A Communication Optimal LU Factorization Algorithm. EECS Technical Report EECS-2010-29, UC Berkeley (March 2010)
10. Demmel, J., Dumitriu, I., Holtz, O.: Fast linear algebra is stable. Numerische Mathematik 108, 59–91 (2007)
11. Faraj, A., Kumar, S., Smith, B., Mamidala, A., Gunnels, J.: MPI collective communications on the Blue Gene/P supercomputer: Algorithms and optimizations. In: 17th IEEE Symposium on High Performance Interconnects HOTI 2009, pp. 63–72 (2009)
12. Grigori, L., Demmel, J.W., Xiang, H.: Communication avoiding Gaussian elimination. In: Proceedings of the 2008 ACM/IEEE Conference on Supercomputing SC 2008, pp. 29:1–29:12. IEEE Press, Piscataway (2008)
13. Gropp, W., Lusk, E., Skjellum, A.: Using MPI: portable parallel programming with the message-passing interface. MIT Press, Cambridge (1994)

14. Irony, D., Toledo, S.: Trading replication for communication in parallel distributed-memory dense solvers. Parallel Processing Letters 71, 3–28 (2002)
15. Irony, D., Toledo, S., Tiskin, A.: Communication lower bounds for distributed-memory matrix multiplication. Journal of Parallel and Distributed Computing 64(9), 1017–1026 (2004)
16. Johnsson, S.L.: Minimizing the communication time for matrix multiplication on multiprocessors. Parallel Comput. 19, 1235–1257 (1993)
17. Kumar, S., Dozsa, G., Almasi, G., Heidelberger, P., Chen, D., Giampapa, M.E., Michael, B., Faraj, A., Parker, J., Ratterman, J., Smith, B., Archer, C.J.: The deep computing messaging framework: generalized scalable message passing on the Blue Gene/P supercomputer. In: Proceedings of the 22nd Annual International Conference on Supercomputing ICS 2008, pp. 94–103. ACM, New York (2008)
18. McColl, W.F., Tiskin, A.: Memory-efficient matrix multiplication in the BSP model. Algorithmica 24, 287–297 (1999)
19. Solomonik, E., Demmel, J.: Communication-optimal parallel 2.5D matrix multiplication and LU factorization algorithms. Tech. Rep. UCB/EECS-2011-10, EECS Department, University of California, Berkeley (February 2011), http://www.eecs.berkeley.edu/Pubs/TechRpts/2011/EECS-2011-10.html
20. Van De Geijn, R.A., Watts, J.: SUMMA: scalable universal matrix multiplication algorithm. Concurrency: Practice and Experience 9(4), 255–274 (1997)

Introduction

Sabri Pllana, Jean-François Méhaut, Eduard Ayguade,
Herbert Cornelius, and Jacob Barhen

Topic chairs

Modern multicore and manycore systems offer impressive performance for various applications. However, achieving this performance is a challenging task. While multicore and manycore processors alleviate several problems that are related to single-core processors – known as memory wall, power wall, or instruction-level parallelism wall – they raise the issue of the programmability wall. The multicore and manycore programmability wall calls for new parallel programming methods and tools. Therefore, this topic focuses on novel solutions for efficient programming of multicore and manycore processors in the context of general-purpose and embedded systems.

The quality of submissions was very high. Papers have been selected based on the recommendations of at least four reviewers. The nine accepted papers address a representative set of issues related to the multicore and manycore programming.

The paper – "Hardware and Software Tradeoffs for Task Synchronization on Manycore Architectures" by Yonghong Yan, Sanjay Chatterjee, Daniel Orozco, Elkin Garcia, Zoran Budimlic, Jun Shirako, Robert Pavel, Guang R. Gao, and Vivek Sarkar – describes an implementation of the "phasers" synchronization construct on the IBM Cyclops64 manycore processor.

In the paper – "OpenMPspy: Leveraging Quality Assurance for Parallel Software" by Victor Pankratius, Fabian Knittel, Leonard Masing, and Martin Walser – authors describe OpenMPspy. This tool may be used for detecting mistakes that occur while the code is typed in Eclipse and for collecting statistics on the use of OpenMP language constructs.

The paper – "A Generic Parallel Collection Framework" by Aleksandar Prokopec, Phil Bagwell, Tiark Rompf, and Martin Odersky – describes an approach for development of parallel containers such as parallel arrays or hash maps.

In the paper – "Progress Guarantees when Composing Lock-free Objects" by Nhan Nguyen Dang and Philippas Tsigas – authors describe a novel synchronization mechanism for composing lock-free data objects that guarantees lock-free progress.

The paper – "Engineering a multicore Radix Sort" by Jan Wassenberg and Peter Sanders – describes a novel variant of radix sorting algorithm that is based on a micro-architecture-aware variant of counting sort.

E. Jeannot, R. Namyst, and J. Roman (Eds.): Euro-Par 2011, LNCS 6853, Part II, pp. 110–111, 2011.
© Springer-Verlag Berlin Heidelberg 2011

In the paper – "Accelerating code on multicores with FastFlow" by Marco Aldinucci, Marco Danelutto, Peter Kilpatrick, Massimiliano Meneghin, and Massimo Torquati – authors describe an approach for parallelization of sequential codes via thread-offloading. Basically, a thread uses other threads as software accelerators.

The paper – "A Novel Shared-Memory Thread-Pool Implementation for Hybrid Parallel CFD Solvers" by Jens Jgerskpper and Christian Simmendinger – describes an approach for shared-memory parallelization of grid-based CFD solvers.

In the paper – "A Fully Empirical Autotuned Dense QR Factorization for Multicore Architectures" by Emmanuel Agullo, Jack Dongarra, Rajib Nath, and Stanimire Tomov – authors describe an empirical approach for tuning dense linear algebra libraries on multicore architectures.

The paper – "Parallelizing a Real-Time Physics Engine Using Transactional Memory" by Jaswanth Sreeram and Santosh Pande – describes experiences of authors in parallelizing the ODE physics engine that is used in computer games.

We are grateful to all authors for submitting their high-quality papers to this topic and to reviewers for their efforts to evaluate submitted papers. Furthermore, we would like to acknowledge the encouragement and support of conference chairs Emmanuel Jeannot, Raymond Namyst, and Jean Roman.

Hardware and Software Tradeoffs for Task Synchronization on Manycore Architectures

Yonghong Yan[1], Sanjay Chatterjee[1], Daniel A. Orozco[2], Elkin Garcia[2],
Zoran Budimlić[1], Jun Shirako[1], Robert S. Pavel[2],
Guang R. Gao[2], and Vivek Sarkar[1]

[1] Department of Computer Science, Rice University
{yanyh,sanjay.chatterjee,zoran,shirako,vsarkar}@rice.edu
[2] Department of Electrical Engineering, University of Delaware
{egarcia@,orozco@eecis.,rspavel@,ggao@capsl.}udel.edu

Abstract. Manycore architectures – hundreds to thousands of cores per processor – are seen by many as a natural evolution of multicore processors. To take advantage of this massive parallelism in practice requires a productive parallel programming model, and an efficient runtime for the scheduling and coordination of concurrent tasks. A critical prerequisite for an efficient runtime is a scalable synchronization mechanism to support task coordination at different levels of granularity.

This paper describes the implementation of a high-level synchronization construct called *phasers* on the IBM Cyclops64 manycore processor, and compares phasers to lower-level synchronization primitives currently available to Cyclops64 programmers. Phasers support synchronization of dynamic tasks by allowing tasks to register and deregister with a phaser object. It provides a general unification of point-to-point and collective synchronizations with easy-to-use interfaces, thereby offering productivity advantages over hardware primitives when used on manycores. We have experimented with several approaches to phaser implementation using software, hardware and a combination of both to explore their portability and performance. The results show that a highly-optimized phaser implementation delivered comparable performance to that obtained with lower-level synchronization primitives. We also demonstrate the success of the hardware optimizations proposed for phasers.

1 Introduction

Manycore architectures, with hundreds to thousands of cores per processor, are seen by many as a natural evolution of multicore processors. In practice, a productive parallel programming model, and an efficient runtime for thread execution and coordination, are essential to take advantage of this massive parallelism. Programming models using dynamic task parallelism, such as the ones introduced in the programming languages of the DARPA HPCS program (X10 [1] and Chapel [2]), present a promising approach to productive parallel programming on manycore processors. However, the overhead of communication and synchronization between concurrent tasks typically presents one of the greatest obstacles

E. Jeannot, R. Namyst, and J. Roman (Eds.): Euro-Par 2011, LNCS 6853, Part II, pp. 112–123, 2011.

to achieving high performance and scalability on parallel systems. To support diverse workloads on manycore architectures, synchronization mechanisms that provide high-level operations such as barrier using different granularity levels, would be highly desirable.

Phasers, first introduced in the Habanero-Java multicore programming system [3], are synchronization constructs for task parallel programs. Phasers unify barrier operation and point-to-point synchronization in a single interface, and feature deadlock-freedom and phase-ordering. The current Habanero-Java phaser implemented on a Java virtual machine does not leverage hardware support for synchronization and only works on top of a work-sharing runtime, a much less scalable choice for task parallel runtime than workstealing [4]. In this paper, we present the evaluations of phaser implementations in a workstealing runtime using a C-based Habanero-C parallel programming language. Using the IBM Cyclops64 (C64) manycore architecture [5], we have experimented with several approaches to phaser implementations using software, hardware, and a combination of both to explore their portability and performance. The results show that a highly-optimized phaser implementation delivered comparable performance to that obtained with lower-level synchronization primitives. We also demonstrate the success of the hardware optimizations proposed for phasers.

The contributions of this work includes the following. First, we have provided a highly-optimized spin-based implementation of phasers. It is software-based and portable across POSIX-compliant systems. Secondly, we have optimized a phaser implementation that leverages hardware support for synchronization to deliver superior performance over the software approach while maintaining the same interfaces and features. Finally, we have provided a runtime that is able to switch between software and hardware based implementations to better leverage hardware support, if available.

In the rest of the paper, Section 2 presents the Habanero-C task parallel programming language, and the portable software implementation of phasers. Section 3 describes the phaser implementations on Cyclops64, taking advantage of its hardware features. Section 4 presents the experimental results. Finally, Section 5 discusses related work and Section 6 concludes the paper.

2   Asynchronous Task Parallelism and Software Phasers

Phasers were implemented in the Habanero-C research language developed at Rice University. Habanero-C language has two basic primitives, borrowed from X10 [1], for asynchronous task parallel programming: async and finish. The async statement, *async ⟨stmt⟩*, causes the parent task to fork a new child task that may execute ⟨stmt⟩ in parallel with the parent task. Execution of the async statement returns immediately, i.e. the parent task does not wait for the child task to complete. The finish statement, *finish ⟨stmt⟩*, performs a join operation on all the tasks created within ⟨stmt⟩, including transitively spawned tasks.

The async and finish constructs are simpler than the conventional pthread_create and pthread_join APIs, and more flexible than the Cilk spawn and sync keywords [6] and OpenMP task and taskwait directives. For example, the sync or

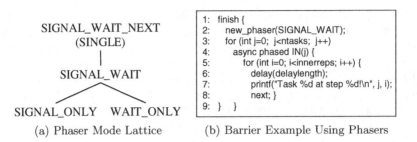

(a) Phaser Mode Lattice   (b) Barrier Example Using Phasers

Fig. 1. Phaser Mode Lattice and Barrier Example

taskwait constructs can only synchronize tasks that are created within the same function scope. Using async and finish as a foundation, we were able to easily experiment with different choices of task parallelism and target platforms.

2.1 Asynchronous Task Synchronization Using Phasers

There are several nice features to use phasers as synchronization constructs with the async and finish task parallel programming model. First, phasers unify collective and point-to-point synchronization in a single set of programming interfaces. The interfaces are ease of use, improving programmer productivity in parallel programming and debugging. Secondly, phasers have two safety properties: deadlock-freedom and phase-ordering [3]. These properties, along with the generality of its use for dynamic parallelism, distinguish phasers from other synchronization constructs in past works including barriers, counting semaphores [7], and X10 clocks [1]. Thirdly, in implementation, phasers have been integrated with a workstealing scheduler that was used in Habanero-C runtime. As a new contribution of this paper, the implementation provided reference solutions to how to map asynchronous tasks with hardware threads when performing synchronization operations. The details of these solutions are discussed in Section 3.

Figure 1(b) shows an example of using phasers to implement a barrier among multiple asynchronously created tasks. The async statement in line 4 and the j-for loop create ntasks child tasks, each registering with the phaser created in line 2 in the same mode as in the master task. The next statement in line 8 is the actual barrier wait; each task waits until all tasks arrive at this point in each iteration of the i-for loop. The first next operation of each task causes itself to wait for the master task to do next operation or to deregister. When the master task reaches the end of the finish scope, it deregisters from the phaser so all child tasks continue and synchronize by themselves in each iteration.

2.2 Software Phasers in Habanero-C

As a synchronization object for dynamic tasks, a phaser has two phases, the signal phase and wait phase, each represented by a counter. Given the mode a task registers with a phaser, a phaser operation could be either or both of a signal and a wait operation, which advances the corresponding phase counter. A task registration is represented by a unique synchronization object, named *sync*, which contains the registration mode and the current signal and wait phase. In order to guarantee deadlock freedom, a child task can only register in a mode that is the same as or below the mode in the parent task according to the phaser mode lattice shown in Figure 1(a). When signaling on a phaser, a task simply increments the signal phase in the *sync* object. The next operation has the effect of advancing each phaser with which a task registers to its next phase, thereby synchronizing all tasks registering with the same phaser. Details operation semantics are described in [3].

Hierarchical Phaser Implementation: The phaser implementation discussed above has used a single master task to advance to its next phase. While the single master approach provides an effective solution for modest levels of parallelism, it quickly becomes a scalability bottleneck as the number of tasks increases. To address this limitation, we have used an approach based on hierarchical phasers [8] for scalable synchronization.

The hierarchical phaser employs a tree of sub-masters, instead of a single master, as in the case of a flat phaser. Tree-based barriers have the advantage that gather operations in the same level (tier) can be executed in parallel by sub-masters. Also, in cases when the hierarchy of sub-masters follows the natural hierarchy in the hardware, each sub-master will leverage data locality among workers in its sub-group. Although the initialization overhead of building a tree is greater than the flat phasers, the runtime of hierarchical phasers outperform the flat phasers heavily on higher number of tasks, as discussed soon in Section 4.

3 Hardware Support in Phasers

The counter-based phaser implementation is a spin-based software approach, also referred to as busy-wait. It consumes both CPU cycles and memory bandwidth, and may quickly become a scalability bottleneck when a large number of tasks are involved in a phaser operations, as in manycores. Recent trends in manycore processor design use tiled architectures to reduce the dependency on the memory bus [9] and to localize synchronizations. In this Section, we explore a phaser implementation that leverages hardware support for synchronization using the IBM Cyclops64 (C64) manycore chip [5] as our evaluation platform.

3.1 Cyclops64 Manycore Architecture

The IBM Cyclops64 is a massively parallel architecture initially developed by IBM as part of the Blue Gene project. As shown in Figure 2, a C64 processor features 80 processing cores on a chip, with two hardware thread units per core that share one 64-bit floating point unit. Each core can issue one double precision floating point Multiply Add instruction per cycle, for a peak performance of 80 GFLOPS per chip when running at 500MHz. The processor chip includes a high-bandwidth on-chip crossbar network with a total bandwidth of 384 GB/s. C64

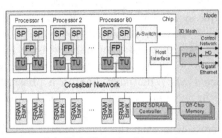 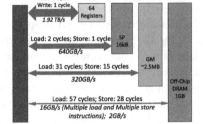

(a) C64 Chip Architecture   (b) C64 Memory Hierarchy

Fig. 2. Cyclops64 Architecture Details

employs three-levels of software-managed memory hierarchy, with the Scratch-Pad (SP) currently used to hold thread-specific data. Each hardware thread unit has a high-speed on-chip SRAM of 32KB that can be used as a cache.

C64 utilizes a dedicated signal bus (SIGB) that allows thread synchronization without any memory bus interference. The SIGB connecting all threads on a chip can be used for broadcast operations taking less than 10 clock cycles, enabling efficient barrier operations and mutual exclusion synchronization. Fast point-to-point signal/wait operations are directly supported by hardware interrupts, with costs on the order of tens of cycles.

The C64 tool chain includes a highly efficient threading library, named TiNy-Threads (TNT) [5], which uses the C64 hardware support to implement threading primitives. Additionally, TNT provides APIs that can be used to access the hardware synchronization primitives to allow for suspension of threads, and including and excluding specific threads from barriers, as shown in Table 1.

Table 1. Cyclops64 TNT APIs for Hardware Synchronization Primitives

Name	Description
tnt_suspend()	Suspend current thread
tnt_awake (const tnt_desc_t)	Awaken a suspended thread
tnt_barrier_include (tnt_barrier_t *)	Join in the next barrier wait operation
tnt_barrier_exclude (tnt_barrier_t *)	Withdraw from the next barrier wait operation
tnt_barrier_wait (tnt_barrier_t *)	Wait until all threads arrive this point

3.2 Optimization Using Hardware Barriers

Barrier operations using phasers can be optimized in manycore architectures that offer direct hardware support for barriers, such as C64. The phaser runtime is able to detect if a phaser operation specified by the user program is equivalent to a barrier operation by checking whether all phasers are registered in SIGNAL_WAIT mode. If so, the underlying hardware support is used directly to perform the barrier operation.

Implementing a hardware barrier in a phaser requires threads to include themselves in the barrier by calling tnt_barrier_include. This requirement is particularly interesting in a workstealing environment due to the fact that the worker that executes the task which is participating in the barrier, has to include itself in the hardware barrier. In workstealing, we cannot include the worker a priori in the barrier. The Habanero-C runtime only includes a worker in the hardware barrier when it is ready to execute a task.

3.3   Optimization Using Thread Suspend and Awake

The TNT API provides functions to suspend a thread and to awake a sleeping thread. A suspend instruction temporarily stops execution in a non-preemptive way, and a signal instruction awakes the sleeping task. Using thread suspend and awake mechanism in place of the busy-wait approach reduces memory bandwidth pressure because all waiting tasks can suspend themselves instead of spinning. The master can collect all the signals from waiting tasks and finally signals the suspended tasks to resume the execution.

The C64 chip provides an interesting hardware feature called the "wake-up bit". When a thread tries to wake up another thread, it sets the "wake-up bit" for that thread. This enables a thread to store a wake-up signal. Hence, if a thread tries to suspend itself after a wake-up signal is sent, it wakes up immediately and the suspend effectively becomes a no-op. This feature is fully utilized by phasers to easily move from phase to phase without worrying about a thread that can execute a suspend after a wake up signal.

3.4   Adaptive Phasers

Adaptability is one of the main features of our phaser implementation. As explained before, the runtime can directly detect the synchronization operation being performed and make a reasonable decision as to how to execute it. A phaser operation can switch to the optimized versions that utilize hardware primitives. These details of how a phaser operation is executed are hidden from the user.

Phaser operations can be implemented in a number of ways to take advantage of the particular characteristics of the underlying hardware. Even when a phaser has all tasks registered in SIGNAL_WAIT mode, it is not guaranteed that a hardware barrier will be used. A task that is registered to support split-phase or fuzzy barriers may signal ahead of its next operation. When a task registers as SIGNAL_ONLY or WAIT_ONLY on a phaser that has been using a hardware barrier, our runtime detects such a scenario and switches to software mode.

The runtime chooses the best mode of operation, depending on the current program state and available features. Each implementation alternately exhibits particular traits: maximum portability and reasonable performance is achieved with a *busy-wait* implementation; low bandwidth and low power usage are featured in the *suspend-awake* implementation.

3.5   Memory Optimizations

Phaser and *sync* objects contain volatile phase counters, and phaser operations involve frequent read and write of those counters in both software based busy-wait approach and hardware-optimized implementations. So low latency and high bandwidth of the memory system are key to the performance of phasers.

The C64's memory hierarchy, as seen in Figure 2, is similar to hardware cache in regular commodity CPUs. The power of using it comes from program manageability as our runtime itself can decide which synchronization objects need to reside on or move to the high-speed SRAM. Yet there is a tradeoff in this software-managed caching approach because the DRAM is limited in its sizes and shared with stack in C64. For a simple DRAM-optimization, the runtime allocates on SRAM, synchronization objects that contain spinning counters. More complex optimizations use heuristic or historical information to identify frequently-accessed data and move them to SRAM. Further memory management by the Habanero-C runtime, such as allocating a list of synchronization objects in a dense array, provide another level of memory optimizations on C64.

4   Implementation and Experiments

Habanero-C includes a workstealing runtime and a compiler for the async and finish task parallel programming constructs. The C64 manycore processor described in Section 3.1 was used as experimental platform for this study. This work is the result of a joint research effort between Rice University and University of Delaware (UDel). Figure 3 shows a description of the infrastructure used for this project as well as the contributions of each institution.

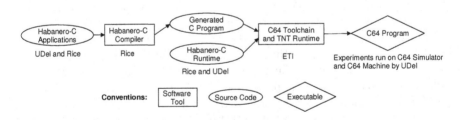

Fig. 3. Collaboration and Software Infrastructure

4.1   Implementation and Experimental Benchmarks

Habanero-C compiler was implemented on top of the ROSE source-to-source compiler framework [10]. The compiler transforms async and finish statements to appropriate library and runtime calls that create and enqueue tasks, and calls to ensure proper task termination within each finish scope.

Habanero-C runtime contains a number of worker threads; each worker thread maintains a double-ended queue (deque). A worker enqueues and dequeues tasks from the tail end of its deque when creating and executing local tasks, respectively. Other workers steal tasks from the head of the deque, when they do not have local tasks to work on. While this approach to the workstealing runtime is similar to the Cilk runtime [6], task creation and enqueuing policy when encountering an async is different from Cilk. In Cilk's "work-first" policy, the code after the async task body (the *continuation*) is pushed onto the deque while the current worker continues the execution of the async body. In our policy, which is referred to as "help first" [4], the async task itself is pushed onto the deque while the current worker continues the execution of the continuation.

The evaluation was conducted using microbenchmarks and common applications. The microbenchmarks include barrier and threadring for evaluating phaser barrier and point-to-point synchronizations. The applications include two-dimensional finite difference time domain (FDTD2D), and Successive Over Relaxation (SOR), to study the performance impact of synchronization overhead using software and hardware approaches, and their tradeoffs.

4.2 Hierarchical Phasers and Memory Optimizations

In Figure 4, we show the barrier overhead of using software flat phasers versus hierarchical phasers, and phasers residing on SRAM versus on DRAM. The dramatic scalability improvements of using hierarchical phasers (4-degree fan-out hierarchy) as compared to flat phasers are obvious. Placing phasers in SRAM results in large (one to two orders of magnitude) overhead reduction for both flat phaser and hierarchical phasers. While this performance does not imply superiority of SRAM over DRAM implementation in general (spin-based solutions may have adverse effects as well), we use the SRAM hierarchical phasers as baseline to compare with other hardware-based implementations in later sections.

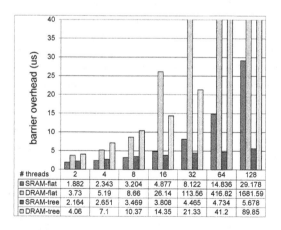

Fig. 4. Hierarchical Phasers and SRAM Optimization

4.3 Barrier and Point-to-Point Microbenchmarks

The barrier microbenchmark was based on the EPCC OpenMP *syncbench* benchmark that was developed for evaluating OpenMP barrier overhead. When using phasers as barriers, barrier wait operations are performed by phaser next operations. A task can dynamically join and leave a barrier wait operation by registering and deregistering with the phaser that is created (with at least SIGNAL_WAIT capability) for this operation. This is different from OpenMP barrier that only allows a fixed number of threads involved in a barrier from the beginning to the end of a parallel region. OpenMP does not permit the use of barriers within parallel loops, either.

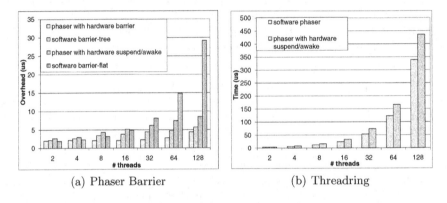

(a) Phaser Barrier   (b) Threadring

Fig. 5. Barrier and Point-to-Point Microbenchmarks

Figure 5(a) shows the barrier overheads using four phaser implementations on C64. The implementation that leverages the C64 hardware barrier incurs much lower overhead than that of the software barrier. The reason behind this is the phaser implementation switches to hardware barriers whenever the tasks registering with the phaser are actually perform the barrier wait operations. The implementation that uses suspend/awake performs worse than software phasers because of the sequentially accumulated cost of hardware interrupt in suspend/awake implementation. For software hierarchical phasers, both signal gathering and wait operations are performed in parallel, thus reducing overhead.

The *threadring* microbenchmark evaluates point-to-point signal-wait operation of two tasks. In this program, a group of tasks form a signal ring; each task waits on the signal from the previous task and signals the next task after receiving the signal. As shown in Figure 5(b), the memory consumption of the software busy-wait approach has little impact on the time required to complete a round of the ring. In fact, the implementation using software phasers performs slightly better than the one using hardware interrupts. These imply the effectiveness of using the portable software-based solution for point-to-point synchronizations.

The high performance obtained using the *busy-wait* implementation is due in part to the high bandwidth and low latency of the local on-chip memory in C64.

Although the other techniques in our experiments use hardware support, they still suffer from overhead in the supporting software required to use the hardware primitives. In contrast, *busy-wait* uses a very simple polling mechanism that does not require complex software support.

4.4   Applications

A simulation of propagation of electromagnetic waves that uses the two-dimensional finite difference time domain (FDTD2D) algorithm was used to test the effectiveness of phasers for commonly used scientific applications. The FDTD algorithm used [11] is an excellent choice to study synchronization and parallelization techniques for manycore architectures; the algorithm has abundant parallelism and its complexity depends on the physical phenomena that it models, ranging from a simple read-modify-write of an array to numerical integration of physical variables. The experiments simulate the propagation of a wave in two dimensions, with an implementation that results in a two dimensional array where each element is updated several times using data from the array elements that surround it. A full description of the FDTD algorithm used here can be found in [12].

The case presented in Figure 6(a) is characterized by a constant amount of computation per array element. Barriers have been successfully used to synchronize multiple threads executing the program, since all threads share approximately the same amount of workload.

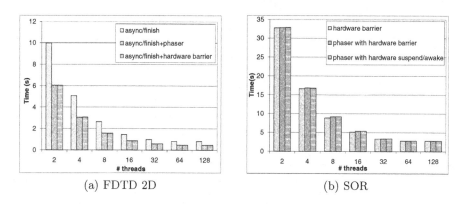

(a) FDTD 2D   (b) SOR

Fig. 6. Applications Performance Using Different Implementations

Figure 6(a) shows FDTD2D performance using following implementations:

1. async/finish: use finish to join tasks as barrier operations; tasks are recreated via async and joined in each iteration. This approach is commonly used in task parallel programming language, such as Cilk.
2. async/finish+phaser: use phaser to perform barrier *wait*; tasks are created once, and then coordinated via phasers in each iteration. Tasks are terminated when the computation completes.

3. async/finish+hardware barrier: similar to async/finish+phaser, but using hardware barrier to perform barrier *wait*.

The implementation using phasers doubles the performance of the one using finish for synchronization. The reason behind this is that finish is a coarse-grained synchronization approach, and it suffers from the runtime overhead for creating and scheduling tasks. Thus algorithms that require fine grained synchronization with large number of iterations should use lower-overhead, finer-grained task coordination mechanism such as phasers. The similar performance between the one using phasers and the one using hardware barriers is because phasers adaptively switch to hardware barrier when it detects a barrier *wait* should be performed.

Another application we used for the evaluation is Red-Black Successive Over-Relaxation (SOR). SOR is a method of solving partial differential equations using a variant of Gauss Seidel method. Task synchronization patterns are similar to FDTD2D, requiring barrier operations to synchronize each iteration. Figure 6(b) shows similar executions time for phasers and hardware barriers, demonstrating the adaptivity of our phaser implementation to the underlying hardware.

5   Related Work

Cilk [6], Cilk++, and OpenMP 3.0 introduced task parallelism at the programming language level. The Cilk's sync and OpenMP's taskwait constructs, related to finish in Habanero-C, are global barrier synchronization points indicating that the execution of current task cannot proceed until all previously spawned tasks have completed. Using this style of synchronization, the runtime efficiency depends heavily on the granularity of parallelism built into the program.

X10 [1] and Chapel [2] provide constructs for dynamic task creation and constructs for task synchronization. X10 allows for the barrier-style phase advancing among all participating tasks using the next operation but it lacks of the point-to-point signal-wait style coordination capability that is available in phasers. Chapel introduce sync variables for programming producer-consumer coordination among tasks. Chapel does not provide direct language construct for barrier operations, or phase-ordered synchronization.

The JUC *CyclicBarrier* class [13] supports periodic barrier synchronization among a set of threads. Unlike phasers, however, *CyclicBarrier* does not support the dynamic addition or removal of threads; nor do they support one-way synchronization or split-phase operations.

6   Conclusions and Future Work

In this paper, we present the design and implementation of *phasers*, a high-level synchronization construct for asynchronous tasks on manycore Cyclops64 processors in the Habanero-C workstealing runtime. We have designed and implemented different techniques for phaser synchronization on C64 that use a combination of software-based busy-wait approach, hardware barriers, and hardware support for thread suspend/awake. Our experiments show that phasers are able to take advantage of hardware primitives on manycore architectures and optimizations for their memory subsystems to provide superior performance to portable software approaches.

In the future, we will experiment with more bandwidth-limited applications on C64 to evaluate the limitations of our busy-wait phaser implementation. We will also investigate more applications for other phasers operations, such as broadcast and reduction.

Acknowledgments. We wish to thank Vincent Cavé and Joshua Landwehr for their hard work on the correctness, performance and efficiency of the Habanero-C runtime. We wish to express our gratitude to ET International for their advice and their logistics support which ultimately boosted the quality and quantity of our experiments. This work was supported by the National Science Foundation through grants CCF-0833122, CCF-0925863, CCF-0937907, CNS-0720531, and OCI-0904534.

References

1. Charles, P., Grothoff, C., Saraswat, V., Donawa, C., Kielstra, A., Ebcioglu, K., von Praun, C., Sarkar, V.: X10: an object-oriented approach to non-uniform cluster computing. In: Proceedings of the 20th Annual ACM SIGPLAN Conference on OOPSLA, pp. 519–538. ACM, New York (2005)
2. Chapel Programming Language, http://chapel.cray.com/
3. Shirako, J., Peixotto, D.M., Sarkar, V., Scherer, W.N.: Phasers: a unified deadlock-free construct for collective and point-to-point synchronization. In: Proceedings of the 22nd ICS, New York, NY, USA, pp. 277–288 (2008)
4. Guo, Y., Barik, R., Raman, R., Sarkar, V.: Work-First and Help-First Scheduling Policies for Async-Finish Task Parallelism. In: IPDPS 2009 (2009)
5. Cuvillo, J.d., Zhu, W., Hu, Z., Gao, G.R.: TiNy Threads: A Thread Virtual Machine for the Cyclops64 Cellular Architecture. In: IPDPS 2005, 265.2 (2005)
6. Frigo, M., Leiserson, C.E., Randall, K.H.: The implementation of the cilk-5 multithreaded language. In: Proceedings of the ACM SIGPLAN Conference on PLDI. Ser. PLDI 1998, pp. 212–223. ACM Press, New York (1998)
7. Sarkar, V.: Synchronization using counting semaphores. In: Proceedings of the 2nd International Conference on Supercomputing, pp. 627–637. ACM, New York (1988)
8. Shirako, J., Sarkar, V.: Hierarchical phasers for scalable synchronization and reductions in dynamic parallelism. In: IPDPS 2010 (2010)
9. Wentzlaff, D., et al.: On-chip interconnection architecture of the tile processor. IEEE Micro. 27(5), 15–31 (2007)
10. ROSE compiler framework, http://www.rosecompiler.org
11. Taflove, A., Hagness, S.: Computational Electrodynamics: The Finite-Difference Time-Domain Method, 3rd edn. Artech House Publishers, Boston (2005)
12. Orozco, D., Gao, G.: Diamond tiling: A tiling framework for time-iterated scientific applications. In: CAPSL Technical Memo 091 (December 2009)
13. Goetz, B.: Java Concurrency In Practice. Addison-Wesley, Reading (2007)

OpenMPspy: Leveraging Quality Assurance for Parallel Software

Victor Pankratius, Fabian Knittel, Leonard Masing, and Martin Walser

Karlsruhe Institute of Technology, IPD
76128 Karlsruhe, Germany
pankratius@kit.edu, {knittel,masing,walser}@student.kit.edu

Abstract. OpenMP is widely used in practice to create parallel software, however, software quality assurance tool support is still immature. OpenMPspy introduces a new approach, with a short-term and a long-term perspective, to aid software engineers write better parallel programs in OpenMP. On the one hand, OpenMPspy acts like an online-debugger that statically detects problems with incorrect construct usage and which reports problems while programmers are typing code in Eclipse. We detect simple slips as well as more complex anti-patterns that can lead to correctness problems and performance problems. In addition, OpenMPspy can aggregate statistics about OpenMP language usage and bug patterns from many projects. Insights generated from such data help OpenMP language designers improve the usability of constructs and reduce error potential, thus enhancing parallel software quality in the long run. Using OpenMPspy, this paper presents one of the first detailed empirical studies of over 40 programs with more than 4 million lines of code, which shows how OpenMP constructs are actually used in practice. Our results reveal that constructs believed to be frequently used are actually rarely used. Our insights give OpenMP language and compiler designers a clearer picture on where to focus the efforts for future improvements.

1 Introduction

Multicore processors are everywhere. Many programmers now use OpenMP [3,4] to write parallel software that exploits the hardware capabilities. Programming with OpenMP does not require programmers to deal with low-level parallelism details; instead, higher-level *#pragma* statements are used to introduce parallelization, e.g., in Fortran and C.

While programming with OpenMP appears to be simple, its approach still has its pitfalls. It is possible to slip and forget important pragma declarations, introduce data races, or use constructs in a way that harms parallel performance. Error feedback from compilers and tools might come too late and thus require significant effort to fix follow-up problems that could have been avoided earlier. In addition, the design of language constructs influences how programmers use those constructs and what kind of mistakes they make. Consequently, information about how OpenMP constructs are used in practice helps assess their usability and make future improvements that reduce error potential.

E. Jeannot, R. Namyst, and J. Roman (Eds.): Euro-Par 2011, LNCS 6853, Part II, pp. 124–135, 2011.
© Springer-Verlag Berlin Heidelberg 2011

This paper takes a new position with a broader view on parallel software quality assurance. Parallel code has good quality if it is correct, easy to understand, performs within defined parameters, and if the potential of introducing errors during extensions is low. Our quality assurance process thus introduces two systematic feedback loops: (1) In a short-term approach, programmers receive feedback on errors in an instant while they are programming. (2) In a long-term approach, we aim to improve parallel language design. To achieve this, we need to learn from individual programmers and projects. Our approach thus establishes the second feedback loop that provides such empirical data back to language designers, so they can improve future versions of OpenMP. In contrast to the status quo where this kind feedback might be sporadically posted on the Web in an unstructured fashion, we are among the first to collect and use it in a structured and systematic way.

In particular, our paper makes the following novel contributions. We present OpenMPspy, a tool that demonstrates the feasibility of our approach. OpenMPspy acts as an online-debugger in the Eclipse environment and gives developers instant feedback. The tool uses static analysis and reports errors affecting correctness (e.g., that might lead to race conditions) and hints on how to improve parallel performance. Furthermore, it collects statistics on matching bug patterns, performance-harming patterns, and construct usage patterns. OpenMPspy can operate in different modes to aggregate such empirical data from many projects. Language designers and compiler writers are provided with a big picture on the usability of OpenMP, so they can make better decisions. We conduct one of the first empirical studies on 46 projects with over 4 million lines of code to determine: (1) what are the most frequently used constructs are and where it makes sense to spend most effort; (2) where to improve syntax and semantics to reduce parallel programming error potential; (3) detect and deal with unused or rarely used language constructs. Generally speaking, such an empirically grounded approach – as opposed to subjective expectations – not only improves parallel software quality, but tailors programming languages to real market needs.

The paper is organized as follows. Section 2 introduces OpenMPspy and explains how it works. Section 3 shows how OpenMPspy works on real projects. We illustrate that OpenMPspy detects previously unreported bugs and discuss how OpenMP language constructs are used in practice on 46 projects. Section 4 presents insights and lessons learned on how to improve OpenMP. Section 5 contrasts related work. Section 6 provides a conclusion.

2 Overview of OpenMPspy

This section introduces OpenMPspy's mode of operations, its code analysis framework, and its analysis features.

2.1 Modes of Operation

OpenMPspy has three modes of operation. In mode (1), it works as a stand-alone online debugger in Eclipse. It provides OpenMP programmers with interactive feedback directly in the IDE. Mode (2) is a batch mode that allows executing OpenMPspy from the command line to analyze a collection of individual projects. OpenMPspy creates aggregate statistics on patterns matching errors, performance problems, and language construct usage. Mode (3) is a client-server mode that implements the data collection in a distributed environment (e.g., over the Internet). In this mode, OpenMPspy locally collects statistics while programmers use the tool during regular program development. Statistics are sent to a server that creates overviews from data of potentially hundreds of thousands of programmers.

2.2   The Code Analysis Framework

OpenMPspy integrates into the Eclipse CDT framework. In particular, it draws upon the abstract syntax tree (AST) generated by Eclipse CDT. However, Eclipse's OpenMP support is incomplete; OpenMP statements are not part of the Eclipse AST and provided separately as a list with preprocessor symbols and code locations. We thus extended the AST data structure to include new nodes for all OpenMP statements, as exemplified below:

```
#pragma omp parallel for        ASTPragma (OpenMP pragma: omp parallel for)
for(int i = 2;i<10;i++)         CPPASTForStatement
{                               CPPASTDeclarationStatement
}                               CPPASTSimpleDeclaration
                                ...
```

We use extensions based on the CODAN framework [1] for static OpenMP code analysis. CODAN and CDT provide functionality to walk our AST and use visitor patterns [6] to perform individual analyses. In the implementation, we define a checker for each issue or problem that we want to analyze. Each checker has one or more associated patterns. A checker is represented by a single class that extends a base checker class from the framework, as outlined below:

```
public class MyChecker extends AbstractIndexExtendedAstChecker { ...
   class CheckStmpVisitor extends ExtendedASTVisitor { ...
      public int visit(...) {...}
   }
}
```

It is possible to employ several visitors at the same time; for example, an outer visitor might visit OpenMP directives and start a nested specialized visitor to work on certain nodes. When a pattern specified within a visitor matches on the AST, results can be displayed in the Eclipse IDE (e.g., as hints, warnings, errors) and provide direct feedback to developers. A pattern detection is usually triggered after each source code modification. This way, we realize an online OpenMP debugger that reports problems while programmers are typing code (see Figure 1).

Our program design based on checkers makes OpenMPspy extensible. For example, it is easy to add an extension and enhance the race detection with additional patterns. The detection capabilities of our tool can be updated by exchanging checker patterns (e.g., by ones downloaded from the Web).

Fig. 1. Instant IDE feedback from OpenMPspy's online debugger

2.3   Analysis Features for OpenMP

OpenMPspy implements three types of code checkers: (1) checkers for error patterns; (2) checkers for performance-harming patterns; (3) checkers for statistics collection of OpenMP construct usage. We describe each category in brief.

(1) Checkers for error patterns. Several checkers are implemented to detect various kinds of parallel programming errors. One of the most complex checkers is the race checker. Using multiple AST visitors, it searches for code patterns that can lead to races. As the entire analysis is static, it can report false positives, however, we use several techniques with specific focus on OpenMP to reduce the number of warnings. For example, the race checker analyzes for each parallel region which variables are private or shared. Then, it looks for unsynchronized variable accesses that can be potentially performed by several threads. In contrast to other tools, our checker pays attention to a variety of special cases that do not lead to races, e.g., for constructs such as *threadprivate, firstprivate, lastprivate,* and variables declared within parallel regions. To avoid unnecessary warnings, the race checker performs analyses on the specific error potential of the most frequently used constructs, such as `omp parallel for`. Further unnecessary warnings are avoided for variables in `reduction` clauses whose write accesses are handled implicitly by OpenMP. Other special cases are considered for `section` blocks that perform parallel work. In `parallel` regions, the checker performs additional analyses on function calls to determine whether variables are copied or passed on by reference. It ensures for referenced variables that no update operations are performed without synchronization.

Another checker controls the wrong usage of the `nowait` clause in work sharing constructs. This clause removes implicit barriers to increase performance, but might introduce data races. The checker analyzes all constructs that have an implicit barrier that is overridden by nowait. It statically follows the control flow until it reaches the next barrier. Along the path so far, it checks all variable read and write accesses for potential races. In particular, it pays attention to special cases such as using `nowait` together with `lastprivate`; this situation might lead to unsynchronized updates on loop variables.

Other checkers detect slips in OpenMP construct usage. For example, they identify inconsistent usage of `omp for` with no associated `parallel`. In addition, warnings are issued for orphaned work sharing constructs, e.g., work sharing declarations with `sections` that don't have a `section` actually defining work. As another example, checkers detect if an `ordered` directive is not within the extent of a `for` or `parallel for` with an `ordered` clause.

OpenMPspy also addresses problems encountered with new constructs from the OpenMP 3 standard, such as *untied tasks*. These tasks can be executed by one thread, halted, and be resumed by another thread. This implicit assumption can cause problems, for example when `threadprivate` variables are accessed in untied tasks. A particular checker reports this kind of problem.

Loop variables are another common source of errors. OpenMPspy detects if code within a `parallel for` loop attempts to modify loop variables and termination conditions, which is not allowed by the OpenMP standard [3].

Other checkers ensure that calls to the OpenMP runtime are used correctly, thus avoiding run-time errors and crashes. For example, specific checks are done to ensure that calls to `set_num_threads` are only done in appropriate locations.

(2) Checkers for performance-harming patterns. Static analysis can detect code that may cause performance problems and provide developers with suggestions for improvement. One of our checkers analyzes critical sections and generates hints if an `atomic` construct could be used instead of a `critical` construct. The `atomic` construct allows more parallelism in certain situations than the `critical` construct. Our tool relieves the programmers from the burden of looking up the language specification about the syntactical details of operators and the data types where `atomic` applies to.

(3) Checkers for language usage patterns. Language usage statistics are collected by special checkers that do not display information in the IDE. For example, these checkers count the number of times a particular construct is used, or the level of nesting. The checkers distinguish between all specific options of a construct. OpenMPspy is able to create statistics showing which syntactical construct variants are actually used.

3   Analyzing with OpenMPspy: A Study of Real Projects

This Section presents results on using OpenMPspy on 46 projects. We sketch the projects, the effectiveness of OpenMPspy to find previously unreported errors, and the statistics collected on OpenMP language construct usage that are relevant to software quality assurance.

3.1   Applications

We study a total 46 OpenMP programs, divided in two categories: real-world programs and OpenMP benchmark programs. As later data will show (Tables 1, 2, and 3), this categorization reveals that real-world programs employ OpenMP differently than benchmark programs.

The benchmark programs are collected from well-known OpenMP benchmarks as presented in Table 1. The real-world projects are collected from the Debian

Repository. We selected all programs that have a dependency to OpenMP *lib-gomp*. Each program was manually checked to ensure it uses OpenMP constructs; a few programs that did not satisfy this condition were pruned. We added to the program set a few more OpenMP programs that were not part of the Debian Repository, and ensured that we used the latest stable version of every programs. Our final set is listed in Table 2.

Tables 1 and 2 include the total number of uncommented lines of code (LOC) for each project and illustrate how many lines contain OpenMP constructs (LOC OpenMP). It is worth noting that on average, OpenMP makes up less than 1% of all lines of code. In particular, benchmarks have a higher percentage of OpenMP (0.21% of LOC) compared to real-world projects (0.022% of LOC).

3.2 Finding Unreported Errors in Real Projects

We illustrate in depth several problems that OpenMPspy was able to detect in real projects, which were not reported so far. The races described next are related to OpenMP language design and suggest that it favors slips and misunderstandings in certain situations.

Error 1: The problem is a race in a video subtitle editor, *aegiSub-2.1.8*, *audio_spectrum.cpp*, line 186. A variable "sample" is declared globally and implicitly shared, and the programmer might have forgotten about this assumption. Inside a **pragma omp for**, the "sample" variable is updated by potentially several threads without synchronization, which can cause a race. This is an error has not been reported so far. It is worth noting that this project has a total of just 2 lines of OpenMP, and the programmer already introduced a race!

Error 2: The problem is a race in a fluid flow tracking application, *libgpiv-0.6.1*, *valid.c*, line 494. Initially, two variables i and j are declared outside a parallel region. Then, a **pragma omp for** is inserted before a nested loop, where the first loop iterates over i and the second over j (without re-declaration inside the **for** parenthesis). No constructs are used to define visibility for i and j, so implicitly i is treated as private and j as shared. The programmer might have wrongly assumed that j is private as well. This can lead to races when the second loop is executed by different threads that each update their counter variable j. This incorrect pattern is used in several places in the code. This problem is serious considering that the project has a total of 25 lines of OpenMP code.

Error 3: The problem is a race in an artificial life simulation application, *critterding-1.0-beta12.1*, *roundworld.cpp*, line 100. In a **pragma omp parallel for ordered shared(freeEnergyc, lmax)**, the programmer includes the *ordered* clause, with the intention to execute loop iterations in the same order as if they were executed on a sequential processor. However, he or she forgets to include a **pragma omp ordered** directive within the loop, which should actually specify what is to be ordered. This causes potentially racy accesses to **freeEnergyc+=...** within the loop, which has no synchronization. This is again an error in a project with just 15 lines of OpenMP code.

Table 1. Programs from OpenMP benchmarks

No.	Project name	Description	LOC total	LOC OpenMP	% LOC OpenMP
B1	EPCC_Microbench 2.0	OpenMP Microbenchmark	886	41	4,63%
B2	NPB3.3 OMP	NASA Parallel Benchmark	4.920	23	0,47%
B3	OmpSCR 2.0	Various OpenMP sources	4.291	126	2,94%
B4	OpenMP Validation Suite	Validates OMP implement.	6.562	799	12,18%
	Parsec_2.1	PARSEC benchmark suite			
B5	/blackscholes	Option pricing	1.262	4	0,16%
B6	/bodytrack	Computer vision app	7.696	6	0,08%
B7	/ferret	Search engine app	10.765	13	0,12%
B8	/freqmine	Data mining app	2.164	18	0,83%
B9	/tbblib	Intel TBB lib	38.319	3	0,01%
	IPP 7.0.1.041	Intel IPP code samples			
B10	/audio video codecs	Codecs samples	371.749	4	0,001%
B11	/data compression	Data compression samples	7.520	4	0,001%
B12	/image codecs	Image codecs samples	112.123	66	0,06%
B13	/realistic rendering	Rendering samples	19.565	85	0,43%
	specomp2001	SPEC OMP benchmarks			
B14	/L2001/321.equake_l	Finite element simulation	1.128	28	2,48%
B15	/L2001/331.art_l	Neural network simulation	1.594	15	0,94%
B16	/M2001/320.equake_m	Finite element simulation	1.102	16	1,45%
B17	/M2001/330.art_m	Neural network simulation	1.594	15	0,94%
B18	/M2001/332.ammp_m	Computational chemistry	9.785	33	0,34%
			603.025	1.299	0,21%

Table 2. Real projects using OpenMP

No.	Project name	Description	LOC Total	LOC OpenMP	% LOC OpenMP
1	3depict 0.0.3	Point cloud visual./analysis	30.816	75	0,24%
2	AegiSub 2.1.8	Video subtitle editor	133.987	2	0,002%
3	aaphoto 0.41	Photo adjusting	3.614	38	1,05%
4	blender 2.49.2	3D content creation	973.291	10	0,001%
5	ccbuild 2.0.1	C++ Source build utility	7.305	9	0,12%
6	coin-or csdp 6.1.1	Operations research	6.875	17	0,25%
7	critterding 1.0 b12.1	Artificial life simulation	81.839	15	0,02%
8	enblend enfuse 4.0	Image blending	17.551	49	0,28%
9	gettext 0.18.1.1	Localization of software	506.366	2	0,0004%
10	gmsh 2.5.0	3D meshing	296.876	9	0,003%
11	gpivtools 0.6.0	Fluid flow tracking	6.804	3	0,04%
12	graphicsmagick 1.3.12	Image processing	233.960	174	0,07%
13	gretl 1.9.3	Econometric analysis	287.215	9	0,003%
14	imagemagick 6.6.7	Image manipulation	299.304	345	0,12%
15	inkscape 0.48.0	Vector graphics editor	396.756	4	0,001%
16	kdegraphics 4.4.5	Gfx apps and libs for KDE	175.507	9	0,01%
17	kipi plugins 1.7.0	KDE Image Plugin Interface	150.388	3	0,002%
18	libcomplearn 1.1.7	Machine learning compressor	2.935	3	0,10%
19	libgpiv 0.6.1	Fluid flow tracking	17.801	25	0,14%
20	libqsearch 1.0.8	Tree search library	2.597	4	0,15%
21	opencv 2.2.0	Computer vision library	426.339	14	0,003%
22	pdf2djvu 0.7.4	Document conversion	5.869	3	0,05%
23	pfstmo 1.4	HDR tone mapping	6.235	27	0,43%
24	projectm 2.0.1	Music visualizer	64.370	41	0,06%
25	sox 14.3.1	Audio file conversion	41.597	7	0,02%
26	tintii 2.4.0	Selective image coloring	4.619	15	0,32%
27	ufraw 0.17	Raw image format importer	34.799	31	0,09%
28	yamas 0.8.5	Genome meta-analysis	2.500	12	0,48%
			4.218.115	955	0,022%

Table 3. OpenMP language usage in real-world and benchmark projects

	(a) LOC Total	(b) %LOC with resp. to R	(c) % of total OpenMP LOC	(d) LOC Total	(e) %LOC with resp. to R	(f)% of total OpenMP LOC
	Real Projects			**OpenMP Bench-**		
				mark Projects		
(A) Synchronization Constructs						
1. critical	246	94.6	25,8	98	49,2	7,5
2. ordered	3	1.2	0,3	9	4,5	0,7
3. atomic	1	0.4	0,1	26	13,1	2,0
4. taskwait	1	0.4	0,1	0	0,0	0,0
5. barrier	0	0	0,0	11	5,5	0,8
6. omp_set_lock	6	2.3	0,6	32	16,1	2,5
7. omp_init_lock	3	1.1	0,3	20	10,1	1,5
8. omp_set_nest_lock	0	0	0,0	1	0,5	0,1
9. omp_test_lock	0	0	0,0	1	0,5	0,1
10. omp_test_nest_lock	0	0	0,0	1	0,5	0,1
R: Reference for col. b & e: sum A1..A10	**260**	**100**	**27,2**	**199**	**100**	**15,3**
(B) Variable Visibility Constructs						
1. shared	279	76.0	29	59	24.1	4,5
2. private	87	23.7	9	175	71.4	13,5
3. threadprivate	1	0.3	0	11	4.5	0,8
R: Reference for col. b & e: sum B1..B3	**367**	**100**	**38**	**245**	**100**	**18,9**
(C) Variable Initialization Constructs						
1. firstprivate	8	100	0,8	10	62.5	0,8
2. lastprivate	0	0	0,0	4	25	0,3
3. copyin	0	0	0,0	2	12.5	0,2
R: Reference for col. b & e: sum C1..C3	**8**	**100**	**0,8**	**16**	**100**	**1,2**
(D) Parallel For Loop Constructs						
(D1)#pragma omp parallel for						
1. sum of for	491	100	51,4	110	100	8,5
2. schedule dynamic	235	48	24,6	34	31	2,6
3. schedule static	88	18	9,2	11	10	0,8
4. schedule guided	1	0	0,1	7	6	0,5
5. schedule runtime	1	0	0,1	0	0	0,0
6. no schedule option	166	34	17,4	58	53	4,5
7. reduction in for	17	3	1,8	26	24	2,0
R: Reference for col. b & e: D1_1	**491**	**100**	**51,4**	**110**	**100**	**8,5**
(D2)#pragma omp parallel{... #pragma omp for						
1. sum of for	26	100	2,7	118	100	9,1
2. schedule dynamic	9	35	0,9	35	30	2,7
3. schedule static	1	4	0,1	12	10	0,9
4. schedule guided	4	15	0,4	3	3	0,2
5. schedule runtime	1	4	0,0	0	0	0,0
6. no schedule option	11	42	1	68	58	5
7. reduction in for	0	0	0,0	17	14	1,3
R: Reference for col. b & e: D2_1	**26**	**100**	**2,7**	**118**	**100**	**9,1**
(E) Tasking Constructs						
1. task	3	75	0,3	0	-	0
2. taskwait	1	25	0,1	0	-	0
R: Reference for col. b & e: sum E1..E2	**4**	**100**	**0,4**	**0**	**-**	**0**
(F) Feedback and Control of Parallelism						
1. get_thread_num	31	74	3,2	102	63	7,9
2. get_num_threads	8	19	0,8	27	17	2,1
3. master	2	5	0,2	15	9	1,2
4. single	1	2	0,1	17	11	1,3
R: Reference for col. b & e: sum F1..F4	**42**	**100**	**4,4**	**161**	**100**	**12,4**
(G) Parallel Section Constructs						
1. section	16	100	1,7	218	100	16,8
2. sections	5	31	0,5	73	33	5,6
R: Reference for col. b & e: G1	**16**	**100**	**1,7**	**218**	**100**	**16,8**

Error 4: The problem is a race in a visual analysis applications, *3depict-0.0.3*, *rdf.cpp*, line 720. The programmer defines a variable *warnBiasCount* outside a parallel region. Within a `parallel for`, the visibility of *warnBiasCount* is not defined explicitly, which means that it is implicitly shared. Incrementing `warnBiasCount` within the loop without synchronization can lead to races. This project has 75 lines of OpenMP code.

OpenMPspy also reports performance issues, which are technically not an error, but which should be fixed to improve performance. Such patterns indeed occur in practice. For example, OpenMPspy reports a *CriticalInsteadOfAtomic* pattern in *graphicsmagick-1.3.12*, file *pnm.c*, line 621. There, a variable status update could be done with `pragma omp atomic` instead of `pragma omp critical`.

Insights. These empirical examples illustrate that even when programmers use just a few lines of OpenMP they still inadvertently introduce races. Implicit assumptions about shared and private variable visibility seem to favor error-proneness. Error potential could have been reduced in the aforementioned examples with explicit visibility declarations for all variables.

3.3   How OpenMP Constructs Are Used in Practice

Table 3 shows OpenMPspy's quantitative results on how OpenMP language constructs are used in all projects. There are seven categories of constructs that are discussed in this Section. The table partitions results by real projects and benchmark projects. Columns (a),(d) show how many lines of code contain a certain construct; columns (b),(e) show the percentage of LOC in relation to the reference value defined for each construct category; columns (c),(f) show the percentage of LOC with a certain construct to the total lines of OpenMP code (see bottom of Tables 1 and 2).

(A) Synchronization constructs. In real projects, `critical` is the most frequently used synchronization construct, which makes up 95.6% of all synchronization constructs. Other constructs such as `atomic`, explicit `barrier`, and explicit locks are almost never used. By contrast, in OpenMP benchmarks, `critical` makes up 49.2% of all synchronization constructs usage. Atomic and locks are used more often in benchmarks to optimize performance.

(B) Variable visibility constructs. In real projects, the `shared` declaration is used in 76% of all visibility declarations and `private` in 23.7%, which is an interesting observation. As OpenMP defines most variables as `shared` by default, one would expect that `private` occurs more frequently as programmers rely on implicit `shared` declarations. It appears that the `shared` declaration is often used for documentation purposes. The situation is reversed in benchmark projects, where `shared` is used in 24.1% of all visibility declarations, and `private` in 71.4%.

(C) Variable initialization constructs. Constructs such as `firstprivate`, `lastprivate`, `copyin`, handle input and output to parallel sections. Real projects, however, hardly use any of these clauses and exchange data mostly over shared variables. In benchmark projects, these clauses are also rarely used.

(D) Parallel for loop constructs. The `#pragma omp parallel for` (D1) is the flagship of OpenMP and the most frequently used directive in real projects (used in about half of all OpenMP lines). It has several options to guide scheduling and improve performance. Looking at all `#pragma omp parallel for`, the most frequently employed option is dynamic (48%), followed by no option (34%), and static (18%). The slightly different syntax (D2) with `#pragma omp for` within a parallel region is rarely used (in less than 3% of all OpenMP lines). In the benchmark projects, both syntactical forms (D1) and (D2) have similar frequency of occurrence, but are not too dominant in relation to the total lines of OpenMP benchmark code. The benchmark projects use no schedule option most frequently, followed by dynamic and static. Surprisingly, `reduction` isn't used a lot – both in real projects and in benchmark projects.

(E) Tasking constructs. Tasking is almost never used in real projects. This is surprising, as tasks were expected to make OpenMP parallel programming easier. It is well possible that programmer don't use tasks because the language standard is too new and the tool chain is immature.

(F) Feedback and control of parallelism constructs. A few real projects use constructs helping with feedback and manual parallelism control, such as `get_thread_num` and `get_max_threads`, `master`, and `single`. About 3% of all OpenMP lines include `get_thread_num`. The other constructs are almost never used. These observations suggest that OpenMP programmers in real projects did not control parallelization too deeply. In the benchmark projects, these constructs are more frequently used (7.9%), which matches the more frequent usage of locks.

(G) Parallel sections constructs. The `section` construct, in combination with the nested `sections` constructs, give programmers more control over what can be run in parallel. Obviously this functionality is almost never used in real programs. By contrast, these constructs are more frequently employed in benchmark projects.

4 Insights for Parallel Software Quality Improvement

OpenMPspy's empirical results teach us important lessons on how OpenMP can be enhanced. In the long run, a better match of OpenMP's syntax and semantics to programmer's intuitions helps improve software quality by: (1) reducing the potential for parallel programming errors and (2) making code easier to understand. Results also show where developers of real-world projects might need more training.

As shown in Section 3.3, programmers specify `shared` variable visibility often, even though it might not be necessary. This can be explained by a need to document the parallel program and make its understanding easier. However, the errors described in Section 3.2 provide evidence that programmers misunderstand when variables are implicitly `shared` and when they are `private`, which is a fertile ground for races. We therefore recommend that each OpenMP variable has a mandatory visibility declaration. In addition, races could be easier to avoid if variables are implicitly `private` by default.

The `atomic` keyword is almost never used. Perhaps most users don't understand how it can improve performance. As `critical` is used for most critical sections, it would make sense to invest in compiler optimizations that replace, where appropriate, `critical` by `atomic` behind the scenes. Locks are also rarely used, which suggests that OpenMP programmers actually want a higher level of parallel programming. This is also supported by the fact that none of the real projects use explicit barriers, which implies that OpenMP's implicit barriers suffice.

Our real program set hardly uses any constructs that give programmers more control over parallelization, such as sections, master, single, get_thread_num. This observation is yet another indication that OpenMP programmers are risk-averse in practice and do not want to get involved in low-level parallelism details.

The `#pragma omp parallel for` is OpenMP's most frequently used construct. Results show that programmers typically chose scheduling options that delegate performance management to the run-time environment. Future run-time environments should therefore emphasize more sophisticated ways to optimize loop performance behind the scenes. Debuggers and race detectors, on the other hand, can refine and perform more detailed analyses to account for the increased error probabilities due to more frequent usage of parallel loop constructs.

The empirical evidence suggests that OpenMP programmers prefer higher-level parallel programming constructs and that there is a clear preference on which constructs are used in practice. Language designers must therefore focus on these issues in the future. Removing unused constructs is another point for discussion in the standardization committee, so compiler and tool developers don't have to invest in unnecessary features.

5 Related Work

An empirical study of parallel programming errors has been presented in [9], but it does not address OpenMP. Recent workshops [2] have begun to tackle usability aspects for programming language design; however, OpenMPspy is the first to present an automated usability checking approach for OpenMP. Debugging parallel programs has been explored in various contexts; [7]

Tool Comparison / Patterns	VisualStudio2008	VisualStudio2010	Eclipse + gcc	Intel Parallel Lint	VivaMP	OpenMPspy
for loop / loop var. modification	x	x				x
for loop / loop-test var. modification	x	x		x	x	x
performance: critical instead atomic						x
'omp_set_num_threads()' in par. region	x	x		x	x	x
data dependency / bad 'nowait' use				x		x
ordered clause / no ordered directive				x	x	x
empty "#pragma omp ordered" region						x
empty "#pragma omp sections" region	x	x	x			x
directly nested parallels				x	x	x
orphaned "#pragma omp for"				x	x	x
orphaned "#pragma omp section"	x	x	x			x
no data-sharing attr. set reminder	x	x	x		x	x
races / modified shared variables				x	x	x
threadprivate vars in "task untied"						x

presents a taxonomy of race detection algorithms and shows the general problem is equivalent to the halting problem, which is why no universal detector exists. Static race detectors such as [8] analyze code without execution; OpenMPspy's static approach detects a larger variety of different patterns, which also include races and performance-harming patterns (see Table). OpenMPspy specializes its code checkers on the particular characteristics of OpenMP constructs to reduce the number of false warnings. In addition, our tool is among the first to also collect and aggregate language usage statistics to enhance language design. Most on-the-fly race detectors require program executions [11] or specialized hardware [10]. Dynamic race detectors such as [5] introduce large run-time overhead, which makes them inappropriate to execute while programmers are typing code.

6 Conclusion

OpenMPspy presents a novel approach to enhance OpenMP software quality. OpenMPspy's online debugger instantly alerts developers to correctness and performance problems. In the long run, OpenMPspy helps language designers improve OpenMP syntax and semantics. Decisions can be based on statistical data from many projects, such as typical errors, performance problems, and language construct usage. The evidence in this paper shows that OpenMP can be adapted in many ways to better match programmer intuitions. Closing this cognitive gap will reduce parallel programming error potential and lead to better code quality.

Acknowledgements. We thank the Excellence Inititative and the Landesstiftung Baden-Württemberg for their support.

References

1. Code Analysis Framework for Eclipse CDT (CODAN) (2010),
   http://wiki.eclipse.org/CDT/designs/StaticAnalysis
2. Evaluation and Usability of Programming Languages and Tools (PLATEAU) Workshops (2010), http://ecs.victoria.ac.nz/Events/PLATEAU
3. The OpenMP API specification for parallel programming (2011),
   http://www.openmp.org
4. Chapman, B., Jost, G., van der Pas, R.: Using OpenMP. Portable Shared Memory Parallel Programming. The MIT Press, Cambridge (2007)
5. Flanagan, C., Freund, S.N.: Fasttrack: efficient and precise dynamic race detection. In: Proc. PLDI 2009, pp. 121–133. ACM, New York (2009)
6. Gamma, E., Helm, R., Johnson, R., Vlissides, J.: Design patterns: elements of reusable object-oriented software, vol. 206. Addison-Wesley, Reading (1995)
7. Helmbold, D.P., McDowell, C.E.: A taxonomy of race detection algorithms. Technical report, UC Santa Cruz, Santa Cruz, CA, USA, September 28 (1994)
8. Intel. Intel parallel lint. (2010), http://software.intel.com
9. Lu, S., et al.: Learning from mistakes: a comprehensive study on real world concurrency bug characteristics. In: Proc. ASPLOS XIII (2008)
10. Nistor, A., et al.: Light64: Lightweight hardware support for data race detection during systematic testing of parallel programs. In: MICRO 2009 (2009)
11. Pozniansky, E., Schuster, A.: Multirace: efficient on-the-fly data race detection in multithreaded C++ programs. Concurr. Comput.: Pract. Exper. 19(3) (2007)

A Generic Parallel Collection Framework

Aleksandar Prokopec, Phil Bagwell, Tiark Rompf, and Martin Odersky

École Polytechnique Fédérale de Lausanne, Lausanne, Switzerland

Abstract. Most applications manipulate structured data. Modern languages and platforms provide collection frameworks with basic data structures like lists, hashtables and trees. These data structures have a range of predefined operations which include mapping, filtering or finding elements. Such bulk operations traverse the collection and process the elements sequentially. Their implementation relies on iterators, which are not applicable to parallel operations due to their sequential nature.

We present an approach to parallelizing collection operations in a generic way, used to factor out common parallel operations in collection libraries. Our framework is easy to use and straightforward to extend to new collections. We show how to implement concrete parallel collections such as parallel arrays and parallel hash maps, proposing an efficient solution to parallel hash map construction. Finally, we give benchmarks showing the performance of parallel collection operations.

1 Introduction

With the arrival of multicore architectures, parallel programming is becoming more widespread. One programming approach is to implement existing programming abstractions using parallel algorithms under the hood. This omits low-level details such as synchronization and load-balancing from the program. Most programming languages have libraries which provide data structures such as arrays, trees, hashtables or priority queues. The challenge is to use them in parallel.

Collections come with bulk operations like mapping or traversing elements. Functional programming encourages the use of predefined combinators, which is beneficial to parallel computations – a set of well chosen collection operations can serve as a programming model. These operations are common to all collections, making extensions difficult. In sequential programming common functionality is abstracted in terms of iterators or a generalized **foreach**. But, due to their sequential nature, these are not applicable to parallel computations which split data and assemble results [18]. This paper describes how parallel operations can be implemented with two abstractions – splitting and combining.

Our parallel collection framework is generic and can be applied to different data structures. It enhances collections with operations executed in parallel, giving direct support for programming patterns such as map/reduce or parallel looping. Some of these operations produce new collections. Unlike other frameworks proposed so far, our solution adresses parallel construction without the aid of concurrent data structures. While data structures with concurrent access

E. Jeannot, R. Namyst, and J. Roman (Eds.): Euro-Par 2011, LNCS 6853, Part II, pp. 136–147, 2011.

are crucial for many areas, we show an approach that avoids synchronization when constructing data structures in parallel from large datasets.

Our contributions are the following:

1. Our framework is generic in terms of *splitter* and *combiner* abstractions, used to implement a variety of parallel operations, allowing extensions to new collections with the least amount of boilerplate.
2. We apply our approach to specific collections like parallel hash tables. We do not use concurrent data structures. Instead, we structure the intermediate results and merge them in parallel. Specialized data structures with efficient merge operations exist, but pay a price in cache-locality and memory usage [20] [17]. We show how to merge existing data structures, allowing parallel construction and retaining the efficiency of the sequential access.
3. Our framework has both mutable and immutable (persistent) versions of each collection with efficient update operations.
4. We present benchmark results which compare parallel collections to their sequential variants and existing frameworks. We give benchmark results which justify the decision of not using concurrent data structures.
5. Our framework relieves the programmer of the burden of synchronization and load-balancing. It is implemented as an extension of the Scala collection framework. Due to the backwards compatibility with regular collections, existing applications can improve performance on multicore architectures.

The paper is organized as follows. Sect. 2 gives an overview of the Scala collection framework. Sect. 3 describes adaptive work stealing. Sect. 4 describes the design and several concrete parallel collections. Sect. 5 presents experimental results. Sect. 6 shows related work.

2   Scala Collection Framework

Scala is a modern general purpose statically typed programming language for the JVM which fuses object-oriented and functional programming [3]. Readers interested to learn more are referred to textbooks on Scala [4].

Its features of interest for this paper are higher-order functions and traits. These language features are not a prerequisite for parallel collections – they serve as a convenience. Our approach can be applied to other general purpose languages as well. Functions are first-class objects – they can be assigned to variables or specified as arguments to other functions. For instance, to find the first even number in the list of integers lst, we write: lst.find(_ % 2 == 0). In languages like Java without first-class functions, anonymous classes can achieve the same effect. Traits are similar to Java interfaces and may contain abstract methods. They also allow defining concrete methods.

Collections form a class hierarchy with the most general collection type Traversable, which is subclassed by Iterable, and further subclassed by Set, Seq and Map, representing sets, sequences and maps, respectively [5]. Some operations (filter, take or map) produce collections as results. They use objects of type **Builder**. **Builder** declares a method += for adding elements to the builder. Its method **result** is called after all the desired elements have been added and it returns the collection. Each collection provides a specific builder.

We give a short example program (Fig. 1). Assume we have two sequences **names** and **surnames**. We want to group names starting with 'A' which have same surnames and print all such names and surnames for which there exists at most one other name with the same surname. The example uses *for-comprehensions* [4] to iterate the sequence of pairs of names and surnames obtained by **zip** and filter those which start with 'A'. They are grouped according to the surname (second pair element) with **groupBy**. Surname groups with 2 or less names are printed. The sugared code on the left is translated to a sequence of method calls similar to the one shown on the right. PLINQ uses a similar approach of translating a query-based DSL into method calls.

We want to run such programs in parallel, but new operations have to be integrated with the existing collections. Data Parallel Haskell defines a new set of names for parallel operations [14]. Method calls in existing programs have to be modified to use corresponding parallel operations. A different approach is implementing parallel operations in separate classes. We add a method **par** to regular collections which returns a parallel version of the collection pointing to the same underlying data. We also add a method **seq** to parallel collections to switch back. Furthermore, we define a separate hierarchy of parallel sequences, maps and sets which inherit corresponding general collection traits **GenSeq**, **GenMap** and **GenSet**.

```
val withA = for {
  (n, s) <- names zip surnames
  if n startsWith "A"           val groups = names.zip(surnames)
} yield (n, s)                    .filter(_._1.startsWith("A"))
val groups = withA.groupBy(_._2)  .groupBy(_._2)
for {                           groups.filter(_._2.size < 3)
  (surname, pairs) <- groups      .flatMap(_._2)
  if pairs.size < 3               .foreach(p => println(p))
  (name, surname) <- pairs
} println(name, surname)
```

Fig. 1. Example program

3   Adaptive Work Stealing

When using multiple processors load-balancing techniques are required. Work is divided to tasks and distributed among processors. Each processor maintains a task queue. Once a processor completes a task, it dequeues the next one. If the queue is empty, it tries to steal a task from another processor's queue. This technique is known as work stealing [8] [2]. We use the Java fork-join framework to schedule tasks [1]. For effectiveness, work must be partitioned into tasks that are small enough, which leads to overheads if there are too many tasks.

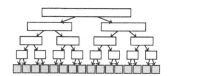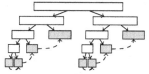

Fig. 2. Fine-grained and exponential task splitting

Assuming uniform amount of work per element, equally sized tasks guarantee that the longest idle time is equal to the time to process one task. This happens if all the processors complete when there is one more task remaining. If the number of processors is P, the work time for $P = 1$ is T and the number of tasks is N, then equation 1 denotes the theoretical speedup in the worst case.

$$speedup = \frac{T}{(T - T/N)/P + T/N} \xrightarrow{P \to \infty} N \qquad (1)$$

In practice, there is an overhead with each created task – fewer tasks can lead to better performance. But this can also lead to worse load-balancing. This is why we've used exponential task splitting [9]. If a worker thread completes its work with more tasks in its queue that means other workers are preoccupied with work of their own, so the worker thread does more work with the next task. The heuristic is to double the amount of work (Fig. 2). If the worker thread hasn't got more tasks in its queue, then it steals tasks. The stolen task is always the biggest task on a queue. Stolen tasks are split until reaching threshold size – the need to steal indicates that other workers may be short on tasks too.

The worst case scenario is a worker being assigned the biggest task it processed so far when that task is the last remaining. We know this task came from the processor's own queue (otherwise it would have been split, enabling the other processors to steal and not be idle). At this point the processor will continue working for some time T_L. We assume input data is uniform, so T_L must be equal to the time spent up to that moment. If the task size is fine-grained enough to be divided among P processors, work up to that moment took $(T - T_L)/P$, so $T_L = T/(P + 1)$. Total time for P processors is then $T_P = 2T_L$. The equation 2 gives a bound on the worst case speedup, assuming $P \ll N$:

$$speedup = \frac{T}{T_P} = \frac{P + 1}{2} \qquad (2)$$

This estimate says that the execution time is never more than twice as great as the lower limit, given that the biggest number of tasks generated is $N \gg P$. To ensure this, we define the minimum task size as $threshold = \max(1, n/8P)$, where n is the number of elements to process.

4   Design and Implementation

4.1   Splitters and Combiners

For the benefits of easy extension and maintenance we want to define most operations (such as `filter` or `flatMap` from Fig. 1) in terms of a few abstractions. We define *splitters* – iterators which have operations `next` and `hasNext` used to traverse. In addition, a splitter has a method `split` which returns a sequence of splitters iterating over disjunct subsets of elements. This allows parallel traversal.

```
trait Splitter[T] extends Iterator[T] {
  def split: Seq[Splitter[T]]
}
```

```
trait Combiner[T, Coll] extends Builder[T, Coll] {
  def combine(other: Combiner[T, Coll]): Combiner[T, Coll]
}
```

Some operations produce collections (e.g. `filter`). Collection parts produced by different workers must be combined into the final result and *combiners* abstract this. Type parameter T is the element type, and Coll is the collection type. Parallel collections provide combiners, just as regular collections provide builders. Method `combine` takes another combiner and produces a combiner containing the union of their elements. Combining results from different tasks occurs more than once during a parallel operation in a tree-like manner (Fig. 2).

The parallel collection base trait `ParIterable` extends the `GenIterable` trait. It defines operations `splitter` and `newCombiner` which return a new splitter and a new combiner, respectively. Subtraits `ParSeq`, `ParMap` and `ParSet` define parallel sequences, maps and sets.

```
class Map[S](f: T => S, s: Splitter[T]) extends Task {
  var cb = newCombiner
  def split = s.split.map(subspl => new Map[S](f, subspl))
  def leaf() = while (s.hasNext) cb += f(s.next)
  def merge(that: Map[S]) = cb = cb.combine(that.cb)
}
```

Parallel operations are implemented within tasks, corresponding to those described previously. Tasks define `split`, `merge` and `leaf`. For example, the Map task is given a mapping function f of type T => S and a splitter s. Tasks are split to achieve better load balancing – the `split` typically calls `split` on the splitter and maps subsplitters into subtasks. Once the threshold size is reached, `leaf` is called, mapping the elements and adding them into a combiner. Results from different processors are merged hierarchically using the `merge` method, which merges combiners. In the computation root `cb` is evaluated into a collection. More than 40 collection operations were parallelized and some tasks are more complex – they handle exceptions, can abort or communicate with other tasks, splitting and merging them is often more involved, but they follow this pattern.

4.2   Parallel Array

Arrays are mutable sequences – class `ParArray` stores the elements in an array.

Splitters. A splitter contains a reference to the array, and two indices for iteration bounds. Method `split` divides the iteration range in 2 equal parts, the second splitter starting where the first ends. This makes `split` an $O(1)$ method.

Combiners do not know the final array size (e.g. `flatMap`), so they construct the array lazily. They keep a linked list of buffers holding elements. A buffer is either a dynamic array[1] or an unrolled linked list. Method `+=` adds the element to the last buffer and `combine` concatenates the linked lists (an $O(1)$ operation). Method `result` allocates the array and executes the `Copy` task which copies the chunks into the target array (we omit the complete code here). When the size is not known a priori, evaluation is a two-step process. Intermediate results are stored in chunks, an array is allocated and elements copied in parallel.

```
class ArrayCombiner[T] extends Combiner[T, ParArray[T]] {
  val chunks = LinkedList[Buffer[T]]() += Buffer[T]()
  def +=(elem: T) = chunks.last += elem
  def combine(that: ArrayCombiner[T]) = chunks append that.chunks
  def result = exec(new Copy(chunks, new Array[T](chunks.fold(0)(_+_.size))))
}
```

4.3   Parallel Rope

To avoid the copying step altogether, a data structure such as a *rope* is used to provide efficient splitting and concatenation [10]. Ropes are binary trees whose leaves are arrays of elements. They are used as an immutable sequence which is a counterpart to the `ParArray`. Indexing an element, appending or splitting the rope is $O(\log n)$, while concatenation is $O(1)$. However, iterative concatenations leave the tree unbalanced. Rebalancing can be called selectively.

Splitters are implemented similarly to `ParArray` splitters.

Combiners may use the append operation for `+=`, but this results in unbalanced ropes [10]. Instead, combiners internally maintain a concatenable list of array chunks. Method `+=` adds to the last chunk. The rope is constructed at the end from the chunks using the rebalancing procedure [10].

4.4   Parallel Hash Table

Associative containers implemented as hash tables guarantee $O(1)$ access with high probability. There is plenty of literature available on concurrent hash tables [13]. We describe a technique that constructs array-based hash tables in parallel by assigning non-overlapping element subsets to workers, avoiding the need for synchronization. This technique is applicable both to chained hash tables (used for `ParHashMap`) and linear hashing (used for `ParHashSet`).

---

[1] In Scala, this collection is available in the standard library and called *ArrayBuffer*. In Java, for example, it is called an *ArrayList*.

Splitters maintain a reference to the hash table and two indices for iteration range. Splitting divides the range in 2 equal parts. For chained hash tables, a splitter additionally contains a pointer into the bucket. Since buckets have a probabilistic bound on lengths, splitting a bucket remains an $O(1)$ operation.

Combiners. Given a set of elements, we want to construct a hash table using multiple processors. Subsets of elements are assigned to different processors and must occupy a contiguous block of memory to avoid *false sharing*. To achieve this, elements are partitioned by their hashcode prefixes, which divide the table into logical blocks. This will ensure that they end up in different blocks, independently of the final table size. The resulting table is filled in parallel.

```
class TableCombiner[K](ttk: Int = 32) extends Combiner[K, ParHashTable[K]] {
  val buckets = new Array[Unrolled[K]](ttk)
  def +=(elem: K) = buckets(elem.hashCode & (ttk - 1)) += elem
  def combine(that: TableCombiner[K]) = for (i <- 0 until ttk)
    buckets(i) append that.buckets(i)
  private def total = buckets.fold(0)(_ + _.size)
  def result = exec(new Fill(buckets, new Array[K](nextPower2(total / lf))))
}
```

Combiners keep an array of 2^k buckets, where k is a constant such that 2^k is greater than the number of processors to ensure good load balancing (from experiments, $k = 5$ works well for up to 8 processors). Buckets are unrolled linked lists. Method += computes the element hashcode and adds it to the bucket indexed by the k-bit hashcode prefix. Unrolled list tail insertion amounts to incrementing an index and storing an element into an array in most cases, occasionally allocating a new node. We used $n = 32$ for the node size. Method `combine` concatenates all the unrolled lists – for a fixed 2^k, this is an $O(1)$ operation.

Method `result` is called in the computation root – the total number of elements `total` is obtained from bucket sizes. The required table size is computed by dividing `total` with the load factor `lf` and rounding to the next power of 2. The table is allocated and the `Fill` task is run, which can be split in up to 2^k subtasks, each responsible for one bucket. It stores the elements from different buckets into the hash table. Assume table size is $sz = 2^m$. The position in the table corresponds to the first m bits of the hashcode. The first k bits denote the index of the table block, and the remaining $m - k$ bits denote the position within that block (Fig. 3). Elements of a bucket have their first k bits the same and are all added to the same block – writes to different blocks are not synchronized. With linear hashing, elements occasionally "spill" to the next block. The `Fill` task records and inserts them into the next block in the merging step. The average number of spills is equal to average collision lengths – a few elements.

4.5   Parallel Hash Trie

A hash trie is an immutable map or set implementation with efficient element lookups and updates ($O(\log_{32} n)$) [11]. Updates do not modify existing tries, but create new versions which share parts of the data structure. Hash tries consist

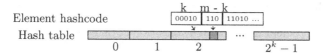

Fig. 3. Hash code mapping

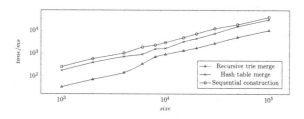

Fig. 4. Hash trie operations

of a root table of 2^k elements. Adding an element computes the hash code and takes the first k bits for the table index i. In the case of a collision a new array is allocated and stored into entry i. Colliding elements are stored in the new array using the next k bits. This is repeated as long as there are collisions. To ensure low space consumption, each node has a 2^k bitmap to index its table (typically $k = 5$) [11]. Hash tries have low space overheads and good cache-locality.

Splitters maintain a reference to the hash trie data structure. Method `split` divides the root table into 2 new root tables, assigning each to a new splitter.

Combiners can contain hash tries. Method `combine` could merge the hash tries (figure 4). The elements in the root table are copied from either of the root tables, unless there is a collision, as with subtries B and E which are recursively merged. This technique turns out to be more efficient than sequentially building a trie – we observed speedups of up to 6 times. We compare the performance recursive merging against hash table merging and sequentially building tries in figure 5. Although it requires less work, recursive merging scales linearly with the trie size. This is why we use the two-step approach shown for hash tables, which results in better performance. Combiners maintain 2^k unrolled lists, holding elements with the same k-bit hashcode prefixes ($k = 5$). The difference is in the method `result`, which evaluates root subtries instead of filling table blocks.

Fig. 5. Recursive trie merge vs. Sequential construction

4.6   Parallel Views

Assume we increment numbers in a collection c, take one half and sum positives:

```
c.map(_ + 1).take(c.size / 2).filter(_ > 0).reduce(_ + _)
```

Each operation produces an intermediate collection. To avoid this we provide *views*. For example, a `Filtered` view traverses elements satisfying a predicate, while a `Mapped` view maps elements before traversing them. Views can be stacked – each view points to its parent. Method `force` evaluates the view stack to a collection. In the example, calling `view` and the other methods on c stacks views until calling `reduce`. Reducing traverses the view to produce a concrete result. *Splitters* call `split` on their parents and wrap the subsplitters. The framework provides a way to switch between strict and lazy on one axis (`view` and `force`), and sequential and parallel on the other (`par` and `seq`).

5   Experimental Results

To measure performance, we follow established measurement methodologies [19]. Tests were done on a 2.8 GHz 4 Dual-core AMD Opteron and a 2.66 GHz Quad-core Intel i7. We first compare two JVM concurrent maps – `ConcurrentHashMap` and `ConcurrentSkipListMap` (both from the standard library) to justify our decision of avoiding concurrent containers. A total of n elements are inserted. Insertion is divided between p processors. This process is repeated over a sequence of 2000 runs on a single JVM invocation and the average time is recorded. We compare against sequentially inserting n elements into a `java.util.HashMap`.

Fig. 6 shows a performance drop due to contention. Concurrent data structures are general purpose and pay a performance penalty for this generality. Parallel hash tables are compared against `java.util.HashMap` in figure 7 I (mapping with a few arithmetic operations) and L (the identity function) – when no time is spent processing an element and entire time spent creating the table (L), hash maps are faster for 1 processor. For 2 or more, the parallel construction is faster.

Microbenchmarks A-L shown in Fig. 7 use inexpensive operators (e.g `foreach` writes to an array, `map` does a few arithmetic operations and the `find` predicate does a comparison). Good performance for fine-grained operators compared to which processing overhead is high means they work well for computationally expensive operators (shown in larger benchmarks M-O). Parallel array is compared against Doug Lea's `extra166y.ParallelArray` for Java.

Larger benchmarks[2] are shown at the end. The Coder benchmark brute-force searches a set of all sentences of english words for a given sequence of digits, where each digit corresponds to letters on a phone keypad (e.g. '2' represents 'A', 'B' and 'C'; '43' can be decoded as 'if' or 'he'). It was run on a 29 digit sequence and around 80 thousand words. The Grouping benchmark loads the words of the dictionary and groups words which have the same digit sequence.

---

[2] Complete source code is available at:
    http://lampsvn.epfl.ch/svn-repos/scala/scala/trunk/

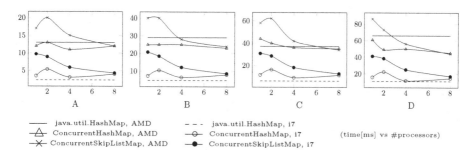

Fig. 6. Concurrent insertion, total elements: (A) 50k; (B) 100k; (C) 150k; (D) 200k

6   Related Work

General purpose programming languages and platforms provide various forms of parallel programming support. Most have multithreading support. However, starting a thread can be computationally expensive and high-level primitives for parallel computing are desired. We give a short overview of the related work in the area of data parallel frameworks, which is by no means comprehensive.

There exists a body of work on data structures which allow access from several threads, either through locking or wait-free synchronization primitives [13]. They provide atomic operations such as insertion or lookup. Operations are guaranteed to be ordered, paying a price in performance – ordering is not always required for bulk parallel executions [18].

.NET langugages support patterns such as parallel looping, aggregations and the map/reduce pattern [6]. .NET Parallel LINQ provides parallelized implementations query operators. On the JVM, one example of a data structure with parallel operations is the Java `ParallelArray` [7], an efficient parallel array implementation. Its operations rely on the underlying array representation, which makes them efficient, but also inapplicable to other data representations. Data Parallel Haskell has a parallel array implementation with bulk operations [14].

Some languages recognized the need for catenable data structures. Fortress introduces conc-lists, tree-like lists with efficient concatenation [17]. We generalize them to maps and sets, and both mutable and immutable data structures.

Intel TBB for C++ bases parallel traversal on iterators with splitting and uses concurrent containers. Operations on concurrent containers are slower than their sequential counterparts [15]. STAPL for C++ has a similar approach – they provide thread-safe concurrent objects and iterators that can be split [16]. The STAPL project also implements distributed containers. Data structure construction is achieved by concurrent insertion, which requires synchronization.

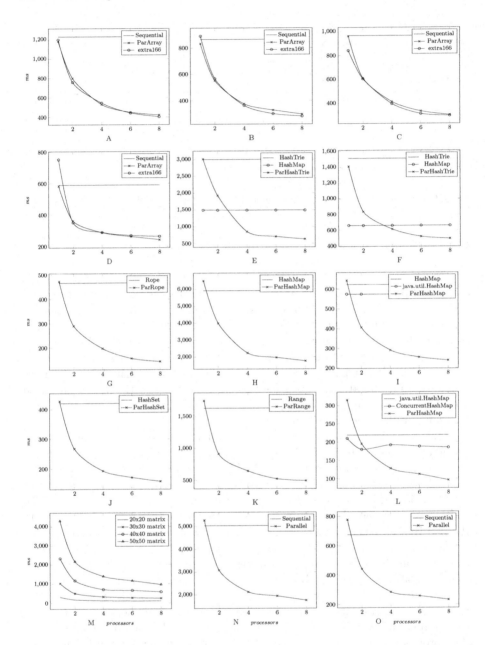

Fig. 7. Benchmarks (running time [ms] vs number of processors): (A) ParArray.foreach, 200k; (B) ParArray.reduce, 200k; (C) ParArray.find, 200k; (D) ParArray.filter, 100k; (E) ParHashTrie.reduce, 50k; (F) ParHashTrie.map, 40k; (G) ParRope.map, 50k; (H) ParHashMap.reduce, 25k; (I) ParHashMap.map, 40k; (J) ParHashSet.map, 50k; (K) ParRange.map, 10k; (L) ParHashMap.map(id), 200k; (M) Matrix multiplication; (N) Coder; (O) Grouping

7   Conclusion

We provided parallel implementations for a wide range of operations found in the Scala collection library. We did so by introducing two divide and conquer abstractions called *splitters* and *combiners* needed to implement most operations.

In the future, we plan to implement bulk operations on concurrent containers. Currently, parallel arrays hold boxed objects instead of primitive integers and floats, which causes boxing overheads and keeps objects distributed throughout the heap, leading to cache misses. We plan to apply specialization to array-based data structures in order to achieve better performance for primitive types [12].

References

1. Lea, D.: A Java Fork/Join Framework (2000)
2. Traore, D., Roch, J.-L., Maillard, N., Gautier, T., Bernard, J.: Deque-free work-optimal parallel STL algorithms. In: Proceedings of the 14th Euro-Par Conference (2008)
3. Odersky, M., et al.: An Overview of the Scala Programming Language. Technical Report LAMP-REPORT-2006-001, EPFL (2006)
4. Odersky, M., Spoon, L., Venners, B.: Programming in Scala. Artima Press (2008)
5. Odersky, M.: Scala 2.8 collections. EPFL (2009)
6. Toub, S.: Patterns of Parallel Programming. Microsoft Corporation (2010)
7. Doug Lea's, Home page, http://gee.cs.oswego.edu/
8. Blumofe, R.D., Leiserson, C.E.: Scheduling Multithreaded Computations by Work Stealing. In: 35th IEEE Conference on Foundations of Computer Science (1994)
9. Cong, G., Kodali, S., Krishnamoorthy, S., Lea, D., Saraswat, V., Wen, T.: Solving Large, Irregular Graph Problems Using Adaptive Work Stealing. In: Proceedings of the 2008 37th International Conference on Parallel Processing (2008)
10. Boehm, H.-J., Atkinson, R., Plass, M.: Ropes: An Alternative to Strings. Software: Practice and Experience (1995)
11. Bagwell, P.: Ideal Hash Trees (2002)
12. Dragos, I., Odersky, M.: Compiling Generics Through User-Directed Type Specialization. In: Fourth ECOOP Workshop on Implementation, Compilation, Optimization of Object-Oriented Languages, Programs and Systems (2009)
13. Moir, M., Shavit, N.: Concurrent data structures. Handbook of Data Structures and Applications. Chapman and Hall, Boca Raton (2007)
14. Jones, S.P., Leshchinskiy, R., Keller, G., Chakravarty, M.M.T.: Harnessing the Multicores: Nested Data Parallelism in Haskell. Foundations of Software Technology and Theoretical Computer Science (2008)
15. Intel Thread Building Blocks: Tutorial (2010), http://www.intel.com
16. Buss, A., Harshvardhan, Papadopoulos, I., Tkachyshyn, O., Smith, T., Tanase, G., Thomas, N., Xu, X., Bianco, M., Amato, N.M., Rauchwerger, L.: STAPL: Standard Template Adaptive Parallel Library. In: Haifa Experimental Systems Conference (2010)
17. Allen, E., Chase, D., Hallett, J., Luchangco, V., Maessen, J.-W., Ryu, S., Steele Jr., G.L., Tobin-Hochstadt, S., et al.: The Fortress Language Specification (2008)
18. Steele Jr., G.L.: How to Think about Parallel Programming: Not! (2011), http://www.infoq.com/presentations/Thinking-Parallel-Programming
19. Georges, A., Buytaert, D., Eeckhout, L.: Statistically Rigorous Java Performance Evaluation. In: OOPSLA (2007)
20. Hinze, R., Paterson, R.: Finger Trees: A Simple General-purpose Data Structure. Journal of Functional Programming (2006)

Progress Guarantees When Composing Lock-Free Objects*

Nhan Nguyen Dang and Philippas Tsigas

Department of Computer Science and Engineering
Chalmers University of Technology
Gothenburg, Sweden
{nhann,tsigas}@chalmers.se

Abstract. Highly concurrent and reliable data objects are vital for parallel programming. Lock-free shared data objects are highly concurrent and guarantee that at least one operation, from a set of concurrently executed operations, finishes after a finite number of steps regardless of the state of the other operations. Lock-free data objects provide progress guarantees on the object level. In this paper, we first examine the progress guarantees provided by lock-free shared data objects that have been constructed by composing other lock-free data objects. We observe that although lock-free data objects are composable when it comes to linearizability, when it comes to progress guarantees they are not. More specifically we show that when a lock-free data object is used as a component (is shared) by two or more lock-free data objects concurrently, these objects can no longer guarantee lock-free progress. This makes it impossible for programmers to directly compose lock-free data objects and guarantee lock-freedom. To help programmability in concurrent settings, this paper presents a new synchronization mechanism for composing lock-free data objects. The proposed synchronization mechanism provides an interface to be used when calling a lock-free object from other lock-free objects, and guarantees lock-free progress for every object constructed. An experimental evaluation of the performance cost that the new mechanism introduces, as expected, for providing progress guarantees is also presented.

1 Introduction

A concurrent data object is lock-free if it guarantees that at least one, among all concurrent operations, finishes after a finite number of steps. Lock-free data objects are immune to deadlocks and livelocks, and typically provide high scalability and performance [12] [11] [20] [22], especially in shared memory multiprocessor architectures. Several lock-free implementations of fundamental data structures have been introduced in the literature, such as queues [15] [21] [9], priority queues [18], linked-lists [23] [19] [18] [10], and hashtables [7] [17] [4]. Moreover, the problem of composing lock-free data objects has been considered recently in an effort to support the use of lock-free objects in the context of complex software development. Composite data

---

* This work was partially supported by the EU as part of FP7 Project PEPPHER (www.peppher.eu) under grant 248481 and the Swedish Research Council under grant 37252706.

E. Jeannot, R. Namyst, and J. Roman (Eds.): Euro-Par 2011, LNCS 6853, Part II, pp. 148–159, 2011.

structures, which are built by nesting multiple basic data structures, were first studied by Cohen and Campell [5]. Recently, Gidenstam et al. [8] and Cederman and Tsigas [3] studied the problem of composing two operations from two different lock-free objects into one compound atomic operation. These results made it possible to perform complex atomic operations such as *moves* that could move an item from one lock-free data object to another lock-free data object in a lock-free way.

Petrank and Steensgaard [16] also studied the problem of composing lock-free programs and services. They provided new formal definitions of lock-freedom, the bounded and unbounded lock-freedom and they extended them to programs and services. These new definitions allowed the authors to formally state and prove the composition theorem. The theorem guarantees lock-free progress for a lock-free program when composing with a service supporting lock-freedom, using the new definitions. This contribution is a step towards formally studying lock-freedom. However, the paper did not consider the case when multiple programs share a service and compete with each other to use it. This way of composing programs and services can affect their progress guarantees.

In this work, we address the lock-free composition problem but from the perspective of object-oriented programming and we do not consider changing the definition of lock-freedom in order to guarantee composition. In object-oriented programs, one lock-free object can be concurrently shared by other lock-free objects. In this setting, composition of several lock-free objects in one object is possible. When examining progress guarantees provided by these objects, we found that they can not provide the lock-free progress guarantee offered by the shared objects that compose them. To help solve this problem, a synchronization mechanism is proposed for a lock-freedom progress guarantee. By applying this mechanism when composing lock-free objects, we can compose as many objects as possible without fear of losing lock-freedom of the individual participants.

The rest of this paper is organized as follows. Section 2 examines the progress guarantees for lock-free objects in a composition. Then, the new synchronization mechanism for composing lock-free objects is proposed in section 3. Section 4 presents a set of experiments to evaluate our synchronization mechanism in practice. A conclusion of our work and discussions about future improvements come last in the section 5.

2   Progress Guarantee When Composing Lock-Free Data Objects

This section examines progress guarantees by lock-free objects used in an object-oriented program. The program can also contain blocking objects. However, since we are considering composing lock-free objects, blocking objects can be taken away without degradation of generality. In the remainder of this paper, all objects mentioned are lock-free.

2.1   Lock-Free Data Objects

Lock-free objects are objects that provide lock-free progress guarantee for their operation executions. The guarantee ensures that some among its concurrent operations succeed after a finite number of steps of their own execution. To provide such a guarantee, lock-free objects usually use non-blocking synchronization primitives to synchronize concurrent accesses to shared memory among the concurrent operations. Two

Algorithm 1. A template of a lock-free object	**Algorithm 2.** Operation Descriptor

```
1  class LF
2    word *ptr
3    public op(args)
4      while (1)
5        oldVal ← *ptr
6        newVal ← calculate(args)
7        if (CAS(ptr, oldVal, newVal))
8          return
```

```
9   struct OpDesc
10    void *oper(void *args)
11    void *args
12    bool done
13    Object src

16
```

synchronization primitives that are commonly used are Compare-And-Swap (*CAS*), Load-Link/Store-Conditional (*LL/SC*). *CAS* [12] takes three arguments: an address, an expected value, and an update value. If the value at the address is equal to the expected value, it is replaced by the update value; otherwise the value is left unchanged. *LL/SC* is a pair of instructions. The *LL* instruction reads from an address. A later *SC* instruction attempts to store a new value at the address. The instruction succeeds if content of the address are unchanged since that thread issued the earlier *LL* instruction to it. The instruction fails if the content has changed in the interval. These instructions are equally powerful since they both have an infinitive consensus number [12].

By observing several lock-free implementation of fundamental data structures such as queues [15] [21], linked-lists [23], and memory allocators [14], we found a common template that most of these implementations followed presented in Algorithm 1. The template object *LF* offers one operation *op*, which takes generalized arguments *args*. This operation computes a *newVal* (line 6) and updates it to *ptr* variable. In a multi-threaded environment, several threads can try to update *ptr* concurrently. Therefore, the *CAS* primitive is used to keep each update atomic. Examples of an *LF* object and an operation *op* that it supports are a lock-free $Queue$ [15] and its *enqueue* operation, respectively. The *enqueue* operation creates a new node containing the new value and inserts it to the *head* of the queue (by a *CAS*) to become the new *head* node.

2.2 Examining Lock-Free Progress Guarantee in Object-Oriented Program

An object-oriented program comprised by three lock-free objects is examined as an example. Among the objects, one, O_{21}, is concurrently shared by the other objects: O_{11} and O_{12}. All are assumed to be implemented by using the above template.

During the executions of O_{11} and O_{12}'s operations, they invoke operations in O_{21} and wait for the returned results. Object O_{21} is lock-free and therefore, always has some executed operations, invoked by O_{11} or O_{12}, finish and return after a finite number of executed steps. But, O_{21} provides no mechanism to ensure fairness among the executions invoked by different objects. As a result, that only executed operations called by one object (e.g O_{11}) succeed while those called by the other object fail to succeed is possible. Consequently, the former object progresses while the latter does not and fails to provide lock-freedom. So, composition causes a lock-free conflict point at O_{21} for O_{11} and O_{12}. When it is the case, lock-freedom of objects that conflict can be violated.

This lock-free conflict concept can be generalized. There can be several objects sharing another object. An object sharing another object can also be shared by other objects and become itself a conflict point. This sharing scenario creates a hierarchy of sharing lock-free objects together with the respective hierarchy of lock-free conflicts.

Our objective is to introduce a new synchronization mechanism enhancing the shared object so that it supports the lock-free property of the sharing objects.

3   A Synchronization Mechanism for Composing Lock-Free Objects

3.1   Our Approach

A new synchronization mechanism for sharing lock-free objects is proposed. Application of this mechanism enhances objects with the capability to maintain fairness among all the objects that invoke its operations. This fairness ensures that any invoking object has at least one operation returned after a finite number of steps. In other words, no object starves because of performing operations at the shared object.

In detail, the proposed synchronization mechanism keeps track of all invocations by sharing objects to the shared object's operations. When those by an object are unsuccessful to execute the instruction(s) at the linearization point many times, the mechanism will announce one of the operations. When such an announcement is made, later invocations help finish the announced operation before performing their expected operations. Completion of the announced operation allows the sharing object to progress.

The description of the proposed synchronization mechanism are introduced in the two next subsections. A correctness proof for the mechanism is also presented.

3.2   The Operation Descriptor

The new synchronization mechanism is introduced so that an unfinished operation can be helped to finish. The operation can be executed by more than one thread but the mechanism guarantees that only at most one execution can successfully complete. To make this helping scheme possible, a description of the operation and its execution status is needed. Any thread can read the description and execute the operation it describes.

The data structure *OpDesc* illustrated in Algorithm 2 is such an operation descriptor. *OpDesc* contains a function pointer **oper* to the operation, along with arguments for the operation; a boolean variable *done* records the status of the operation (finished or unfinished); *src* is a unique identity of the object that invokes this operation.

An *OpDesc* object encapsulates an operation (e.g *enqueue* operation) provided by shared lock-free object. The mechanism introduces a special kind of operation which can help executing other operations. In other words, operations that can read *OpDesc* and execute the operation it described. We call them "super-operations". The term "operation", from this point, refer to an operation representing functionality that other objects want to perform at the shared object, which is described as an *OpDesc* object.

3.3   The Synchronization Mechanism

The implementation of our synchronization mechanism for the lock-free object *LF* is presented in Algorithm 3. The new object *CLF* provides the same interface as that *LF* does to other objects. However each method in the interface is associated with a super-operation instead of an operation.

Any operation *op* in *LF* is re-written into a pair of one public method *op* (a super-operation) and one private one *op_m* (an operation). The operation *CLF.op_m* executes steps to make changes to the *CLF* object similar to that *LF.op* does to the *LF* object. The difference between *CLF.op_m* and *LF.op* is additional steps required by the

---

Algorithm 3. A lock-free object employing the proposed synchronization mechanism

```
17 class CLF
18   word *ptr
19   OpDesc hlps[M], EMPTY;        //EMPTY.done=true

21   public op(src, args)
22     OpDesc me(src, &op_m, (void*)args), hlp

24     for(int i ← 0; i < M; i++) {
25       hlp ← hlps[i];
26       hp_x ← hlp;               //protect hlp with hazard pointer
27       if (hlp != hlps[i]) continue;
28       if (!hlp.done) *hlp.oper(me, hlp)

30     if (¬me.done) op_m(me, me)

32   private op_m(OpDesc me, OpDesc hlp)
33     while (¬hlp.done)
34       for (tries=0; tries < T_MAX ∧ ¬hlp.done; tries++)
35         oldVal ← *ptr
36         newVal ← calculate(hlp.args)
37         tmp ← hlps[hlp.src]
38         if (DCAS(ptr, oldVal, newVal, &hlp.done, false, true))
39           counter[hlp.src] ← 0;
40           CAS(hlps[hlp.src], tmp, EMPTY);
41           break;

43       if (¬hlp.done)
44         if (++counter[me.src] ≥ O_MAX)
45           announce(me)

47   void announce(OpDesc me)
48     curr ← hlps[me.src]
49     if(curr.done)
50       CAS(hlps[me.src], curr, me)
```

synchronization mechanism that will be discussed later. *CLF.op*, is to provide the same interface as that *LF* but the content is totally new. When *CLF.op* is invoked, it is expected to perform modifications on *CLF* similar to functionality of operation *LF.op*. The functionality is now implemented in $CLF.op\_m$. In addition, *CLF.op* can help finish other $CLF.op\_m$ operations that other objects want to perform.

When *CLF.op* is invoked (assuming by object O_i) to perform the operation $CLF.op\_m$, it does not perform the operation immediately. Instead, it first creates an *OpDesc* describing the operation (line 22) which it can perform by itself (line 30) or any thread can help finishing the operation. Then it checks if there are operations of any object needing help to finish (line 24). If there are such operations, the super-operation will execute these operations (line 28). The checking for any object that needs help is performed through a newly introduced array *hlps[]*. When one among the objects needs help, one of the concurrent operations the object performs will be placed in *hlps[]* at a dedicated position for the object. Other concurrent super-operation executions then can help to finish that one. We assume that there are M objects sharing *CLF* object. Therefore, $hlps[]$ can have M elements that one is assigned to an object.

The operation $CLF.op\_m$ introduces two main changes compared to *LF.op*. The first change is that a Double-Compare-And-Swap (*DCAS*) is used instead of a *CAS* in *LF.op* (line 7). *DCAS* atomically compares and exchanges values at two separate memory locations. Lock-free implementations of *DCAS* have been introduced in [6] and [3]. In $CLF.op\_m$, the *DCAS* performs modification of **ptr* and a status variable atomically. The former is similar to *CAS* in *LF.op*. The latter is to set the execution status variable of *OpDesc*. This status variable, which is allowed to be changed only once, makes sure that an *OpDesc* only succeeds once even when multiple threads are executing it.

The second change in $CLF.op\_m$ is the introduction of a counter array $counter[]$ to record the numbers of times invocations by sharing objects try (but fail) to commit the changes to the shared object *CLF*. The counter at position i is increased after a failed *DCAS* execution (line 38) in an operation invoked by object O_i. When this number reaches a threshold, an executed operation invoked by O_i will be announced in *hlps[]* to be helped.

Due to this change, the loop inside this operation is also modified. Our algorithm could have followed the idea of increasing the counter after every failed *DCAS*. In this case, the counter at any position would be shared among several threads and need synchronization for every update which decreases the performance. To avoid this high overhead, in our design, this counter was split into two counters. One local counter *tries* for each operation execution and a shared one ($counter[]$) to record number of tries the executions invoked by the object have made. When *tries* reach a threshold T_{MAX}, an update to *counter[me.src]* is made. And if this counter reaches its threshold O_{MAX}, one of the operation executions whose *src* is the same as *me.src* is announced.

In addition to those changes, a *CAS* is added to remove the reference from the announcement array $hlps[]$ to a successful operation *hlp*. This avoids any unsafe reference to *hlp* in the future when its hazard-pointer protection (line 26) is removed. The memory used by *hlp* can safely be reclaimed later by a memory reclamation scheme.

In short, the synchronization mechanism guarantees that new invocations of *CLF*'s operations helps finish on-going executed operations that need help. Then they executes the operation they are supposed to perform. With this mechanism, objects invoking operations of *CLF* always has one of the invocations finish after a finite number of steps. Therefore, these objects make progress.

3.4 ABA Problem

Similar to other lock-free objects, our mechanism also encounters the ABA problem. The ABA problem happens when the content at an address changes from A to B, and then changes back to A. *CAS* cannot distinguish this case and the case where the content is unchanged. A number of methods have been introduced to tackle with ABA problem such as tagging [1], hazard pointers [13]. In addition, memory words used by lock-free objects must be protected from deletion by concurrent threads when they are in use and reclaimed when they are no more used. Safe Memory Reclamation with hazard pointers introduced in [13] is used for these purposes.

3.5 Linearizability

This section states the lemmas for the linearizability and lock-freedom property of *CLF*. Due to the space limitation, the proofs for these lemmas are not included in this version of the paper.

Lemma 1. *Regardless of the number of threads executing an operation op_m with the same value of hlp argument, only one can succeed.*

Lemma 2. *CLF is linearizable with the linearization point at line 38.*

Lemma 3. *The presented object CLF is lock-free.*

3.6 How Does the Proposed Synchronization Mechanism Resolve Lock-Free Conflicts?

When a lock-free object is concurrently used by other lock-free objects $O_1 \ldots O_M$, it can become a lock-free conflict and block the progress of those objects. This section will prove that when there is such a conflict point at *CLF*, our mechanism can resolve the conflict. Therefore, *CLF* does not block lock-free progress of the objects using it.

A scenario of using *CLF* is a program containing M lock-free objects $O_1 \ldots O_M$ and one *CLF* object. An object O_i can have at most n concurrent invocations (executed by n threads) to *CLF.op* to perform an intended *CLF.op_m* (referred to as *me*). Each invocation creates an execution of operation *CLF.op*. We seek a bound of the maximum number of steps (a step is one execution of *DCAS*) performed by these executions between any two successful operations. If this bound is finite, it guarantees that any object that uses *CLF* progresses. The lemmas and theorem below figure out this bound.

Lemma 4. *An object O_i can make at most n concurrent invocations to super-operation CLF.op. Starting from when the last invocation returns (or when the program starts, if there is no such invocation), if any of these invocations has executed:*

$$U\_BOUND = T_{MAX}.O_{MAX} \tag{1}$$

steps, one of the following condition must hold:

- *at least one invocation finished. Or*
- *one of these concurrent CLF.op_m operations has been announced.*

Lemma 5. *When an operation me is announced in hlps, either me or another operation that has the same src as me.src finishes after it has executed at most*

$$HELP\_BOUND = n(M - 1) + 1$$

steps since when the announcement is made.

Theorem 1. *When CLF is shared by several objects by invoking to CLF's super-operation op, there is always one, among all invocations by one object, finishing after executing a finite number of steps.*

Proof. From lemma 4, there must be one among the invocations from O which finishes before any of them has executed U_BOUND steps. Otherwise, one of the invocations has its operation me announced.

If me is announced, lemma 5 stated that one of the operations whose src is the same as $me.src$ (including me) finishes after it has executed at most $HELP\_BOUND$ steps since the announcement is made. Therefore, one of the invocations from one object returns after executing at most:

$$U\_BOUND + HELP\_BOUND = n(M - 1) + T_{MAX}.O_{MAX} + 1$$

steps; where:

- T_{MAX} is the number of steps executed by an operation before it checks if it should announce itself.
- O_{MAX} is the number of times T_{MAX} was reached by all invocations from one object.
- n is the maximum number of concurrent operations of CLF that can be executed.
- M is the number of objects that are sharing CLF.

4   Experimental Evaluation

For our experimental evaluation we considered the composition scenario where a program containing a number of pseudo objects sharing one queue. The queue is an implementation of the Michael-Scott Queue [15] enhanced with the proposed synchronization mechanism. A set of experiments to evaluate the effectiveness and performance cost of our synchronization mechanism was performed and the results are presented.

In our experiments, the program was executed to perform queue's operations at three contention levels. In high contention, each thread performed one operation right after another. In medium contention, "other work" with a ratio following the normal distribution between 0 and 1 was performed between two consecutive operations. The "other work" was a fixed-times spin loop of a simple calculation. In low contention, "other work" was always performed between two consecutive operations. An exponential back-off was also used after any failed $DCAS$. The program can be run by one to 8

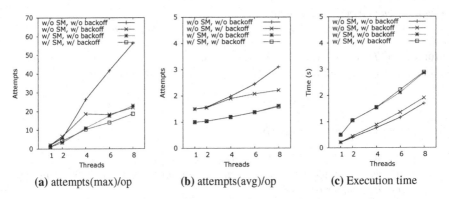

Fig. 1. Measurement results in high contention level

threads and each thread performs 1 000 000 queue operations. Each experiment is the program configured to one contention level and with or without back-off, and set up with a specific number of threads. Each experiment ran five times on a platform with two Intel Core i7 quad-core processors and the average result of the runs was reported. When running the experiments, no other users were using the system.

Three measurements were recorded. The first two were the maximum and average number of attempts between two consecutive successful operations invoked by one object. The maximum number of attempts is an indicator to know whether the proposed synchronization mechanism helped the sharing objects before they starved. The lower this number, the more likely an object is to be helped. On the other hand, the average number of attempts, helps answer a question: does the synchronization mechanism cause the total number of attempts to perform the set of operations increasing? The third measurement was the time it took to finish a run.

Fig. 1 presents the experimental results for the case of high contention. Fig. 1a shows that our synchronization mechanism (w/ SM) significantly reduced the maximum number of attempts to finish one operation when there was no back-off. In the case where no synchronization mechanism was used (w/o SM), the maximum number of attempts when back-off is used (w/ backoff) is much lower than when it is not (w/o backoff). The reason is that back-off reduces the contention among threads and, therefore, lowers the number of attempts. Even though, in this case, there is no lock-free progress guarantee for the sharing objects. The average number of attempts in Fig. 1b shows that when our synchronization mechanism is used, one queue operation needs, on average, about only two thirds of the number of attempts compared to when it is not used. Similar improvements when the synchronization mechanism was used are also observed in medium and low contention levels as shown in Figs. 2a, 2b, 3a, and 3b.

Fig. 1c shows the time to finish all operations at high contention level. Either with or without back-off, the execution time of the runs where our synchronization mechanism was used took about 1.7 of those where the original queue is used. This degradation in performance is because of the overhead cost when applying our synchronization mechanism to achieve the lock-freedom property. In medium and low contention levels, our synchronization performed better which reduced the ratios to 1.5 (Fig. 2c) and 1.2

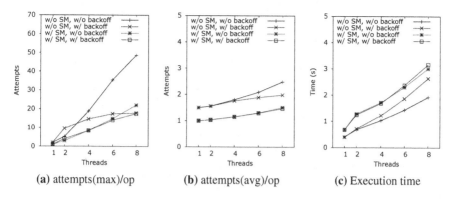

Fig. 2. Measurement results in medium contention level

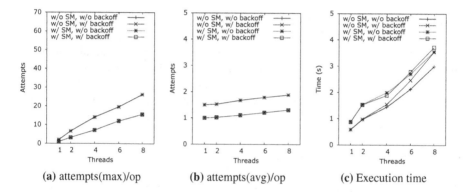

Fig. 3. Measurement results in low contention level

(Fig. 3c) respectively. Especially, in low contention level with back-off, the performance of the queue where our synchronization was used is closer to that when it was not used. Our synchronization mechanism performed better in these contention levels than in high contention levels. This is consistent with the previous result that fewer attempts were performed to finish one queue operation in lower contention level. In addition, when the number of attempts were fewer, the number of cases that the synchronization mechanism was activated to help "unlucky object" were fewer too.

We performed additional experiments to analyze the overhead cost by measuring the performance of *DCAS* comparing to that of *CAS*. The experimental setup was similar to the one described in previous experiments. The only difference was that the queue operations were replaced by an operation containing a simple mathematical calculation and a *DCAS* (or *CAS*). The performance result in Fig. 4 shows that *DCAS* is much more expensive than *CAS* especially in high and medium contention levels. In low contention level, execution time of a *DCAS* operations is quite comparable to that of a *CAS*. These results support a claim that *DCAS* contributes a big portion to the overhead cost of our synchronization mechanism.

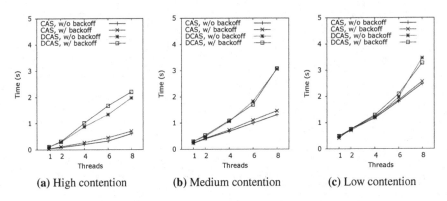

(a) High contention   **(b)** Medium contention   **(c)** Low contention

Fig. 4. Performance of DCAS and CAS

In brief, the experimental results demonstrate that our synchronization mechanism reduces the maximum number of attempts in all the contention level cases. The presented experimental results support the theoretical proofs. The results also show, as expected, that there is a performance overhead cost in order to achieve lock-freedom when composing. The software-implemented *DCAS* mainly contributes to this cost. We expect that with the use of a hardware-supported *DCAS* such as the Advanced Synchronization Facility by Advanced Micro Devices [2], this cost will be reduced significantly.

5   Conclusion

This paper presents our observation on progress guarantees provided by lock-free objects that concurrently share other lock-free objects. We found that these sharing objects can not provide lock-free progress guarantee as expected. A new synchronization mechanism for composing lock-free objects is proposed in order to provide lock-free progress guarantees for each individual. The experimental results show the effectiveness of the new mechanism. A preliminary study for the performance cost introduced by the new mechanism is also presented.

The assumption of the fixed number M of sharing objects should be studied further and if possible removed. Additional experiments can be performed to investigate the influence of choosing T_{MAX} and O_{MAX} on the performance of the mechanism. In addition, an implementation of the mechanism that uses a hardware-supported *DCAS* such as Advanced Synchronization Facility by Advanced Micro Devices is expected to reduce the performance cost.

References

1. IBM System/370 Extended Architecture, Principles of Operations. No. SA22-7085. IBM Publication (1983)
2. AMD: Advanced Synchronization Facility - Proposed Architectural Specification. No. 45432/rev 2.1, AMD (2009)

3. Cederman, D., Tsigas, P.: Supporting lock-free composition of concurrent data objects. In: Conf. Computing Frontiers, pp. 53–62. ACM, New York (2010)

4. Click, C.: A lock-free wait-free hash table, lecture notes in Course EE380 (2006-2007), Stanford University (2007),
http://www.stanford.edu/class/ee380/Abstracts/
070221_LockFreeHash.pdf

5. Cohen, D., Campbell, N.: Automatic composition of data structures to represent relations. In: Proceedings of KBSE 1992, pp. 182–191 (September 1992)

6. Fraser, K., Harris, T.: Concurrent programming without locks. ACM Trans. Comput. Syst. 25(2) (2007)

7. Gao, H., Groote, J., Hesselink, W.: Almost wait-free resizable hashtables. In: Proceedings of IPDPS 2004, p. 50a (2004)

8. Gidenstam, A., Papatriantafilou, M., Tsigas, P.: Allocating memory in a lock-free manner. Algorithmica 58, 304–338 (2005)

9. Gidenstam, A., Sundell, H., Tsigas, P.: Cache-aware lock-free queues for multiple producers/consumers and weak memory consistency. In: Lu, C., Masuzawa, T., Mosbah, M. (eds.) OPODIS 2010. LNCS, vol. 6490, pp. 302–317. Springer, Heidelberg (2010)

10. Harris, T.L.: A pragmatic implementation of non-blocking linked-lists. In: Lecture Notes in Computer Science, pp. 300–314. Springer, Heidelberg (2001)

11. Herlihy, M.: A methodology for implementing highly concurrent objects. ACM Trans. Program. Lang. Syst. 15(5), 745–770 (1993)

12. Herlihy, M., Shavit, N.: The Art of Multiprocessor Programming. Morgan Kaufmann, San Francisco (2008)

13. Michael, M.M.: Hazard pointers: Safe memory reclamation for lock-free objects. IEEE Trans. Parallel Distrib. Syst. 15(6), 491–504 (2004)

14. Michael, M.M.: Scalable lock-free dynamic memory allocation. SIGPLAN Not. 39(6), 35–46 (2004)

15. Michael, M.M., Scott, M.L.: Simple, fast, and practical non-blocking and blocking concurrent queue algorithms. In: Proceedings of PODC 1996, pp. 267–275 (1996)

16. Petrank, E., Musuvathi, M., Steensgaard, B.: Progress guarantee for parallel programs via bounded lock-freedom. In: Proceedings of PLDI 2009, pp. 144–154 (2009)

17. Purcell, C., Harris, T.: Non-blocking hashtables with open addressing. In: Fraigniaud, P. (ed.) DISC 2005. LNCS, vol. 3724, pp. 108–121. Springer, Heidelberg (2005)

18. Sundell, H., Tsigas, P.: Fast and lock-free concurrent priority queues for multi-thread systems. J. Parallel Distrib. Comput. 65(5), 609–627 (2005)

19. Sundell, H., Tsigas, P.: Lock-free and practical doubly linked list-based deques using single-word compare-and-swap 3544, 240–255 (2005)

20. Tsigas, P., Zhang, Y.: Evaluating the performance of non-blocking synchronization on shared-memory multiprocessors. SIGMETRICS Perform. Eval. Rev. 29, 320–321 (June 2001)

21. Tsigas, P., Zhang, Y.: A simple, fast and scalable non-blocking concurrent fifo queue for shared memory multiprocessor systems. In: Proceedings of SPAA 2001, pp. 134–143 (2001)

22. Tsigas, P., Zhang, Y.: Integrating non-blocking synchronisation in parallel applications: performance advantages and methodologies. In: Proceedings of the 3rd International Workshop on Software and Performance WOSP 2002, pp. 55–67 (2002)

23. Valois, J.D.: Lock-free linked lists using compare-and-swap. In: Proceedings of PODC 1995, pp. 214–222. ACM, New York (1995)

Engineering a Multi-core Radix Sort

Jan Wassenberg[1] and Peter Sanders[2]

[1] Fraunhofer IOSB, Ettlingen, Germany
jan.wassenberg@iosb.fraunhofer.de
[2] Karlsruhe Institute of Technology, Karlsruhe, Germany
sanders@kit.edu

Abstract. We present a fast radix sorting algorithm that builds upon a microarchitecture-aware variant of counting sort. Taking advantage of virtual memory and making use of write-combining yields a per-pass throughput corresponding to at least 89% of the system's peak memory bandwidth. Our implementation outperforms Intel's recently published radix sort by a factor of 1.64. It also compares favorably to the reported performance of an algorithm for Fermi GPUs when data-transfer overhead is included. These results indicate that scalar, bandwidth-sensitive sorting algorithms remain competitive on current architectures. Various other memory-intensive applications can benefit from the techniques described herein.

1 Introduction

Sorting is a fundamental operation that is a time-critical component of various applications such as databases and search engines. The well-known lower bound of $\Omega(N \cdot \log N)$ for comparison-based algorithms no longer applies when special properties of the keys can be assumed. In this work, we focus on 32-bit integer keys, optionally paired with a 32-bit (or larger) value. This simplifies the implementation without loss of generality, since applications can often replace large records with a pointer or index [1]. The radix sort algorithm is commonly used in such cases due to its $O(n)$ complexity. In this report, we show a 1.64-fold performance increase over results recently published by Intel [2].

The remaining sections are organized in a bottom-up fashion, with Section 2 dedicated to the basic realities of current and future microarchitectures that affect memory-intensive programs and motivate our approach. We build upon this foundation in Section 3, showing how to speed up counting sort by taking advantage of virtual memory and write-combining. Section 4 applies this technique towards a novel variant of radix sort. The performance of our implementation is evaluated in Section 5. Bandwidth measurements indicate the per-pass throughput is nearly optimal for the given hardware. Its two CPUs outperform a Fermi GPU when accounting for data-transfer overhead.

E. Jeannot, R. Namyst, and J. Roman (Eds.): Euro-Par 2011, LNCS 6853, Part II, pp. 160–169, 2011.

2   Software Write-Combining

We begin with a description of basic microarchitectural realities that are likely to have a serious impact on applications with numerous memory accesses, and show how to avoid performance penalties by means of Software Write-Combining. These topics are not new, but we believe they are often not adequately addressed.

The first problem arises when writing items to multiple streams. An ideal cache with at least as many lines could exploit the writes' spatial locality and entirely avoid noncompulsory misses. However, perfect hit rates are not achievable in practice due to limited ways of associativity a [3]. Since only a lines can be mapped to a cache set, any further allocations from that set result in the eviction of one of the previous lines. If possible, applications should avoid writing to many different streams. Otherwise, the various write positions should map to different sets to avoid thrashing and conflict misses. For current L1 caches with $a = 8$ ways, size $C = 32$ KiB and lines of $B = 64$ bytes, there are $S = \frac{C}{a \cdot B} = 64$ sets, and bits $[\lg B, \lg B + \lg S)$ of the destination addresses should differ (e.g. by ensuring the write positions are not a multiple of $S \cdot B = 4$ KiB apart).

A second issue is provoked by a large number of write-only accesses. Even if an entire cache line is to be written, the previous destination memory must first be read into the cache. While the corresponding latency may be partially hidden via prefetching, the cache line allocations remain problematic due to capacity constraints and eviction policy. Instead of displacing write-only lines that are not accessed after having been filled, the widespread (pseudo-)Least-Recently-Used strategy displaces previously cached data due to their older timestamp. An attempt to avoid these evictions by explicitly invalidating cache lines (e.g. with the IA-32 CLFLUSH instruction) did not yield meaningful improvements. Instead, applications should use *non-temporal streaming store* instructions that write directly to memory. These are guaranteed to avoid cache pollution since they circumvent the cache.

This leads directly to the next concern: single memory accesses involve significant bus overhead. The architecture therefore combines neighboring non-temporal writes into a single burst transfer. However, currently microarchitectures only provide four to ten write-combine (WC) buffers [4]. Non-temporal writes to multiple streams may force these buffers to be flushed to memory via 'partial writes' before they are full. The application can prevent this by making use of Software Write-Combining [5]. The data to be written is first placed into temporary buffers, which almost certainly reside in the cache because they are frequently accessed. When full, a buffer is copied to the actual destination via consecutive non-temporal writes, which are guaranteed to be combined into a single burst transfer.

This scheme avoids reading the destination memory, which may incur relatively expensive Read-For-Ownership transactions and would only pollute the cache. It works around the limited number of WC buffers by using L1 cache lines for that purpose. Interestingly, this is tantamount to direct software control of the transparently managed cache.

We recommend the use of such Software Write-Combining whenever a core's active write destinations outnumber its write-combine buffers. Fortunately, this can be done at a fairly high level, since only the buffer copying requires special vector loads and non-temporal stores (which are best expressed by the SSE2 intrinsics built into the major compilers).

3   Virtual-Memory Counting Sort

We now review Counting Sort of N elements with keys in $[0, M)$ and describe an improved variant that makes use of virtual memory and write-combining.

The naïve algorithm first generates a histogram of the N keys. After computing the prefix sum to yield the starting output location for each key, each value is written at its key's output position, which is subsequently incremented.

Our first optimization goal is to avoid the initial counting pass. We could instead insert each value into a per-key container, e.g. a list of data blocks. However, this incurs some overhead for checking whether the current bucket is full. Preallocating space for M arrays of size N is more efficient, because items can simply be written to the next free position (c.f. Algorithm 1, introduced in [6]). This algorithm only writes and reads each item once, a feat that comes at

---

Algorithm 1: Single-pass counting sort

storage := ReserveAddressSpace$(N \cdot M)$;
for $i := 0$ to $M - 1$ do next $[i] := i \cdot N$;
foreach key,value do
   storage $[$next $[$key$]] :=$ value;
   next $[$key$] :=$ next $[$key$] + 1$;

---

the price of $N \cdot M$ space. While this appears problematic in the Random-Access-Machine model, it is easily handled by 64-bit CPUs with paged virtual memory. Physical memory is only mapped to pages when they are first accessed,[1] thus reducing the actual memory requirements to $O(N + M \cdot \text{pageSize})$. The remainder of the initial allocation only occupies address space, of which multiple terabytes are available on 64-bit systems.

Having avoided the initial counting pass, we now show how to efficiently write values to storage using the write-combining technique described in Section 2. Our implementation initializes the next pointers to consecutive, naturally aligned, cache-line-sized buffers. A buffer is full when its (post-incremented) position is evenly divisible by its size. When that happens, an unrolled loop of non-temporal writes copies the buffer to its key's current output position within storage. These output positions are also stored in an array of pointers.

---

[1] Accesses to non-present pages result in a page fault exception. The application receives such events via signals (POSIX) or Vectored Exception Handling (Microsoft Windows) and reacts by committing memory, after which the faulting instruction is repeated.

4   Radix Sort

After a brief review of radix sorting, we introduce a new variant based on the virtual-memory counting sort described in Section 3.

A radix sort successively examines D-bit 'digits' of the K-bit keys. They are characterized by the order in which digits are processed: starting at the Least Significant Digit (LSD), or Most Significant Digit (MSD).

An MSD radix sort partitions the items according to the current digit, then recursively sorts the resulting buckets. While it no longer needs to move items whose previously seen key digits are unique, this is not especially helpful when the number of passes K/D is small. In fact, the overhead of managing numerous (nearly empty) buckets makes MSD radix sort less suited for relatively small N.

By contrast, each iteration of the LSD variant partitions *all* items into buckets by the current key digit. This amortizes the bucket setup cost over the number of elements and avoids the possibility of load imbalance for parallelization at the price of increased data copying.

To reduce this overhead and also parallel communication, we make use of "reverse sorting" [7], in which one or more MSD passes partition the data into buckets, which are then locally sorted via LSD. This turns out to be even more advantageous for Non-Uniform Memory Access (NUMA) systems because each processor is responsible for writing a contiguous range of outputs, thus ensuring the OS allocates those pages from the processor's NUMA node [8].

Let us now examine the pseudocode of the radix sort (Algorithm 2), choosing $K = 32$ for brevity and $D = 8$ to allow extracting key digits without masking. Each Processing Element (PE) first uses counting sort to partition its items into local buckets by the MSD (digit $= 3$). Note that items consist of a key and value, which are adjacent in memory (ideally within a native 64-bit word, but larger combinations are possible in our implementation via larger user-defined types). After all are finished, the output index of the first item of a given MSD is computed via prefix sum. Each PE is assigned a range of MSD values, sorting the buckets from all PEs for each value. Skewed MSD distributions can cause load imbalance. However, this could be resolved via special treatment of large buckets[2]. The local sort entails $K/D - 1$ iterations in LSD order. The first copies all other PEs' buckets into local memory. The second to last pass also computes the last digit's histogram, thus allowing writing directly to the output positions in the final pass. Note that three sets of buckets are required, which makes heavy use of virtual memory ($3 \cdot 2^D \cdot |\mathrm{PE}| = 6144$ times the input size). While 64-bit Linux grants each process 128 TiB address space, Windows limits this to 8 TiB, which means only about 1.4 GiB of inputs can be sorted[3].

We briefly discuss additional system-specific considerations. The radix 2^D was motivated by easy access to each digit, but is also limited by the cache

---

[2] Sorting buckets larger than $N/|\mathrm{PE}|$ using multiple PEs.

[3] This limitation could be circumvented by estimating bounds for bucket sizes via sampling. In the unlikely case that they are exceeded, a new sample would be drawn and the process repeated.

Algorithm 2: Parallel Radix Sort

parallel foreach item **do**
   $d := \texttt{Digit}(\text{item}, 3)$;
   buckets3 $[d] := \text{buckets3}\,[d] \cup \{\text{item}\}$;
Barrier;
foreach $i \in \left[0, 2^D\right)$ **do**
   bucketSizes $[i] := \sum_{\text{PE}} |\text{buckets3}\,[i]|$;
outputIndices $:= \texttt{PrefixSum}(\text{bucketSizes})$;
parallel foreach bucket3 $\in$ buckets3 **do**
   foreach item $\in$ bucket3 $\forall$ PE **do**
      $d := \texttt{Digit}(\text{item}, 0)$;
      buckets0 $[d] := \text{buckets0}\,[d] \cup \{\text{item}\}$;
   foreach bucket0 $\in$ buckets0 **do**
      foreach item $\in$ bucket0 **do**
         $d := \texttt{Digit}(\text{item}, 1)$;
         buckets1 $[d] := \text{buckets1}\,[d] \cup \{\text{item}\}$;
         $d := \texttt{Digit}(\text{item}, 2)$;
         histogram2 $[d] := \text{histogram2}\,[d] + 1$;
   foreach bucket1 $\in$ buckets1 **do**
      foreach item $\in$ bucket1 **do**
         $d := \texttt{Digit}(\text{item}, 2)$;
         $i := \text{outputIndices}\,[d] + \text{histogram2}\,[d]$;
         histogram2 $[d] := \text{histogram2}\,[d] + 1$;
         output $[i] := \text{item}$;

---

and TLB size. Because of the many required TLB entries, we map the buckets with small pages, for which the Intel i7 microarchitecture has 512 second-level TLB entries. To increase TLB coverage, we use large pages for the inputs. The working set consists of 2^D buffers, buffer pointers, output positions, and 32-bit histogram counters. This fits in a 32 KiB L1 data cache if the software write-combine buffers are limited to a single 64-byte cache line. To avoid associativity and aliasing conflicts, these arrays are contiguous in memory. Interestingly, these optimizations do not detract from the readability of the source code. Knowledge of the microarchitecture can also be applied towards middle-level languages and enables principled design decisions.

5   Performance Evaluation

We characterize the performance of our sorting implementation by its through-put, defined as $\frac{N}{t_1 - t_0}$, where N is the number of items and t_0 and t_1 are the earliest and latest start and finish times reported by any thread. The test platform consists of dual W5580 CPUs (3.2 GHz, 48 GiB DDR3-1066 memory) running Windows XP x64. Our implementation is compiled with ICC 11.1.082 /Ox /Og /Oi /Ot /Qipo /GA /GR- /GS- /EHsc /Qopenmp /QaxSSE4.2. When sorting 350 M uniformly distributed 32-bit keys generated by the WELL512 algorithm [9], the basic algorithm ('VM only') reaches a throughput of 391 M items/s, as shown in the second column of Table 1. After enabling write-combining ('VM+WC'), performance nearly doubles to 657 M/s.

Intel has reported 240 M/s for the same task and a single but identical CPU [2]. For a fair comparison with our dual-CPU system, we double their throughput, which optimistically assumes their algorithm is NUMA-aware, scales perfectly and is not running at a lower memory clock (since our DDR3-1066 is at the lower end of currently available frequencies). We must also divide by the given speedup of 1.2 due to hyperthreads, since those are disabled on our machine. This ('Intel x2') yields 400 M/s; the proposed algorithm is therefore 1.64 times as fast. A separate publication has also presented results [10] for the Many Integrated Cores architecture. The Knights Ferry processor provides 32 cores, each with 4 threads and 16-wide SIMD. The simulation ('KNF MIC') shows a throughput of 560 M/s. Our scalar implementation is currently 1.17 times as fast when running on 8 cores.

Recently, a throughput of 1005 M/s was reported on a GTX 480 (Fermi) GPU [11]. However, this excludes driver and data-transfer overhead. For applications in which the data is generated and consumed by the CPU, we must include at least the time required to read and write data over the PCIe 2.0 bus. Assuming the peak per-direction bandwidth of 8 GB/s is reached, the aggregate throughput ('GPU+PCIe') is 501 M/s. Our implementation, running on two CPUs, therefore outperforms this algorithm on a current top-of-the-line GPU by a factor of 1.31 despite lower transistor counts ($2 \cdot 731$ M vs. 3000 M) and thermal design power ($2 \cdot 130$ W vs. $275 - 300$ W).

Table 1. Throughputs [million items per second] for 32-bit keys and optional 32-bit values

Algorithm	K=32,V=0	K=32,V=32
VM only	391	238
Intel x2	400	307
GPU+PCIe	501	303
KNF MIC	560	(?)
VM+WC	657	452

Similar measurements and extrapolations for the case of 32-bit keys associated with $V = 32$-bit values are given in the third column of Table 1. Since the slowdown is less than a factor of two, the implementations are at least partially limited by computation instead of bandwidth. Intel's algorithm is more efficient in this regard, with only a 1.3-fold decrease vs. our factor of 1.45. The additional data transfers over PCIe render the GPU algorithm uncompetitive.

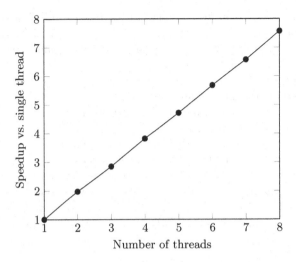

Fig. 1. Linear scalability on two quad-core CPUs with a NUMA factor of 1.5

Since radix sort is bandwidth-sensitive, it is also interesting to examine performance for a varying number of processors. We manually distribute OpenMP threads across CPU packages and cores (in that order) to make use of all available memory controllers. Our NUMA-aware implementation scales linearly with the number of threads, as shown by Figure 1.

To explain the 95% parallel efficiency, we measured the total traffic at each socket's memory controller. Since this information is not available from current profilers such as VTune (which use per-core performance counters), we have developed a small kernel-mode driver to provide access to the model-specific performance counters in the Intel i7 uncore[4]. Uncached writes constitute the bulk of the write combiners' memory traffic and are therefore of particular interest. They are apparently reported as Invalid-To-Exclusive transitions and can thus be counted as the total number of *reads* minus 'normal' reads [12]. We find that 2041 MiB are written, which corresponds to 64 Mi items · 8 bytes per item · 4 passes (slightly less because our final pass cannot use non-temporal writes when the output position is not aligned). Surprisingly, 2272 MiB are read – about 10% more than expected. This amount seems to be influenced by the number of threads. Possible causes may include coherency traffic or page walks and will be investigated in future work. However, we can provide a conservative estimate of the bandwidth utilization. Given the pure read and write bandwidths (38687 MB/s and 28200 MB/s) measured by RightMark [13], the minimum time required for 4 reads and writes of 175 M 8-byte items is 343 ms, which is 89% of the total measured time. This calculation does not include write-to-read turnaround [14, p. 486], so there is even less room for improvement than indicated.

---

[4] The part of the socket not associated with a particular core.

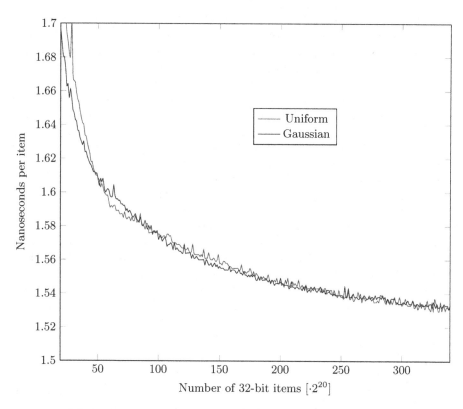

Fig. 2. Time per item for various input sizes and distributions

The previous measurements concern large numbers of items. We now study performance over a wider range of input sizes. The elapsed time per item, shown in Figure 2, varies inversely with the number of items N due to amortization of thread-startup overhead. Performance is within 10% of the best measurement when $N \geq 26 \cdot 2^{20}$, or $N \geq 21 \cdot 2^{20}$ in the case of the approximated Gaussian distribution [15]. It is initially surprising that this distribution does not require more time to sort than uniformly distributed numbers. However, interleaving buckets in the LSD passes (successive buckets are assigned to different threads) avoids load imbalance, and increased occupancy of the central buckets improves locality at the memory page level.

6   Conclusion

We have introduced improvements to counting sort and a novel variant of radix sort for integer key/value pairs. Bandwidth measurements indicate our algorithm's throughput is within 11% of the theoretical optimum for the given hardware. It outperforms the recently published results of Intel's radix sort by a factor of 1.64 and also outpaces a Fermi GPU when data transfer overhead is included.

These results indicate that scalar, bandwidth-sensitive sorting algorithms still have their place on current architectures. However, achieving this level of performance requires awareness of the underlying microarchitecture and some degree of tuning. Our implementation encompasses 5700 lines of C++ (including tests), plus 40,000 lines of shared infrastructure. A demo executable [16] capable of generating or reading 32-bit integers, sorting and efficiently writing them to disk is being made available so that our measurements may be reproduced.

Future Work: While carefully engineered, our implementation is not yet a general solution for all possible sorting applications. Radix sort is limited to relatively small integer keys, and we also assume at least one of the key digits (the MSB) is reasonably equally distributed. Skewed (e.g. constant) distributions currently result in load imbalance. This could be avoided by sorting extremely large buckets from the MSD phase using multiple processors.

We are also interested in testing on larger multi-socket machines with higher NUMA factors and investigating details of the memory subsystem that reduce effective bandwidth. Finally, we believe the general software write-combining technique can provide similar speedups for other memory-intensive applications. In particular, comparison-based sample sort is also expected to benefit from our implementation techniques.

References

1. Bohannon, P., McIlroy, P., Rastogi, R.: Main-memory index structures with fixed-size partial keys. In: SIGMOD Conference, pp. 163–174 (2001),
   http://www.acm.org/sigs/sigmod/sigmod01/eproceedings/papers/
   Research-Bohannon-et-al.pdf
2. Satish, N., Kim, C., Chhugani, J., Nguyen, A., Lee, V., Kim, D., Dubey, P.: Fast sort on CPUs and GPUs: a case for bandwidth oblivious SIMD sort. In: Elmagarmid, A., Agrawal, D. (eds.) SIGMOD Conference, pp. 351–362. ACM Press, New York (2010), http://doi.acm.org/10.1145/1807167.1807207
3. Mehlhorn, Sanders: Scanning multiple sequences via cache memory. Algorithmica 35 (2003)
4. Intel. Intel Architecture Software Developer Manual (2010), System Programming Guide, http://www.intel.com/Assets/PDF/manual/253668.pdf
5. Intel Corporation. Intel 64 and IA-32 Architectures Optimization Reference Manual (November 2007),
   http://www.intel.com/design/processor/manuals/248966.pdf
6. Wassenberg, J., Middelmann, W., Sanders, P.: An efficient parallel algorithm for graph-based image segmentation (June 2009),
   http://algo2.iti.uni-karlsruhe.de/wassenberg/
   wassenberg09parallelSegmentation.pdf
7. Jimenez-Gonzalez, D., Navarro, J., Larriba-Pey, J.: Fast parallel in-memory 64-bit sorting. In: Proceedings of the 2001 International Conference on Supercomputing (15th ICS 2001), Sorrento, Napoli, Italy, pp. 114–122. ACM, New York (2001)

8. an Mey, D., Terboven, C.: Affinity matters! OpenMP on multicore and ccNUMA architectures. In: Parallel Computing: Architectures, Algorithms and Applications, vol. 15, Forschungszentrum Jülich and RWTH Aachen University (Febuary 2008), http://www.compunity.org/events/pastevents/parco07/AffinityMatters_DaM.pdf

9. Panneton, F., L'Ecuyer, P., Matsumoto, M.: Improved long-period generators based on linear recurrences modulo 2. ACM Transactions on Mathematical Software 32 (2006)

10. Satish, N., Kim, C., Chhugani, J., Nguyen, A., Lee, V., Kim, D., Dubey, P.: Fast sort on CPUs, GPUs and intel MIC architectures. Technical report, Intel (2010), http://techresearch.intel.com/userfiles/en-us/FASTsort_CPUsGPUs_IntelMICarchitectures.pdf

11. Merrill, D., Grimshaw, A.: Revisiting sorting for GPGPU stream architectures. Technical Report 3, University of Virginia (February 2010), http://www.cs.virginia.edu/~dgm4d/papers/RadixSortTR.pdf

12. Levinthal, D.: Performance Analysis Guide for Intel Core i7 Processor and Intel Xeon 5500 processors. Intel, http://software.intel.com/sites/products/collateral/hpc/vtune/performance_analysis_guide.pdf

13. Besedin, D.: RightMark memory analyzer, http://cpu.rightmark.org (accessed January 9, 2009)

14. Jacob, B., Ng, S., Wang, D.: Memory systems: cache, DRAM, disk. Morgan Kaufmann, San Francisco (2007)

15. Helman, D., Bader, D., JáJá, J.: A randomized parallel sorting algorithm with an experimental study. J. Parallel Distrib. Comput. 52(1), 1–23 (1998)

16. Wassenberg, J.: Vmcsort demo (May 2011), http://algo2.iti.kit.edu/wassenberg/vmcsort/demo.html

Accelerating Code on Multi-cores with FastFlow

Marco Aldinucci[1], Marco Danelutto[2], Peter Kilpatrick[3],
Massimiliano Meneghin[4], and Massimo Torquati[2]

[1] Computer Science Department, University of Torino, Italy
aldinuc@di.unito.it
[2] Computer Science Department, University of Pisa, Italy
[3] Computer Science Department, Queen's University Belfast, UK
[4] IBM Dublin Research Lab, Ireland

Abstract. FastFlow is a programming framework specifically targeting cache-coherent shared-memory multi-cores. It is implemented as a
stack of C++ template libraries built on top of lock-free (and memory fence free) synchronization mechanisms. Its philosophy is to combine
programmability with performance. In this paper a new FastFlow programming methodology aimed at supporting parallelization of existing
sequential code via offloading onto a dynamically created software accelerator is presented. The new methodology has been validated using a set
of simple micro-benchmarks and some real applications.

Keywords: offload, patterns, multi-core, lock-free synchronization, C++.

1 Introduction

Parallel programming is becoming more and more a *must* with the advent of
multi-core architectures. While up to few years ago faster and faster execution
of programs was mainly the result of increased clock speed and of improvements
in single processor architecture, from now on improvements may only come from
better and more scalable parallel programs.

Here we discuss a semi-automatic parallelization methodology for existing
code which is based on streamization, i.e. on the introduction and exploitation
in the user application of *stream* parallelism. The methodology is based on the
identification of suitable stream parallel patterns within the user application.
Once these patterns have been recognized the computation of a stream of tasks
according to the patterns is delegated to a structured parallel library–FastFlow–
targeting in a very efficient way common cache coherent multi-core architectures.

The proposed methodology is *semi-automatic* as i) the programmer is still in
charge of identifying the appropriate stream parallel patterns, but ii) the stream
parallel pattern implementation is completely and efficiently delegated to the
FastFlow runtime system. This happens by way of offloading onto a software
device behaving as an accelerator (FastFlow software accelerator) which realizes
a parallel pattern (skeleton).

Stream parallelism is the well-known programming paradigm supporting the
parallel execution of a stream of tasks by using a series of *sequential* or *parallel*

E. Jeannot, R. Namyst, and J. Roman (Eds.): Euro-Par 2011, LNCS 6853, Part II, pp. 170–181, 2011.
© Springer-Verlag Berlin Heidelberg 2011

stages [1]. A stream program can be naturally represented as a graph of independent *stages* (kernels or filters) that communicate explicitly over data channels. Parallelism is achieved by running each stage simultaneously on *subsequent* or *independent* data.

As with all kinds of parallel program, stream programs can be expressed as a graph of concurrent activities, and directly programmed using a low-level shared memory or message passing programming framework. Although this is still a common approach, writing a correct, efficient and portable program in this way is a non-trivial activity. Attempts to reduce the programming effort by raising the level of abstraction through the provision of parallel programming frameworks date back at least three decades with a number of significant contributions.

Notable among these is the *skeletal* approach [2] (a.k.a. *pattern-based* parallel programming), which is becoming increasingly popular after being revamped by several successful parallel programming frameworks [3,4,5,6]. Parallel patterns capture common parallel programming paradigms (e.g. MapReduce, ForAll, Divide&Conquer, etc.) and make them available to the programmer as high-level constructs equipped with well-defined functional and parallel semantics. Some of these attempts explicitly include stream parallelism as a major source of concurrency exploitation, such as *pipeline* (running each stage simultaneously on subsequent stream items), *farm* (running multiple independent stages in parallel, each operating on a different task), and *loop* (providing a way to generate cycles in a stream graph). The *loop* skeleton together with the *farm* skeleton can be effectively used to model recursive and Divide&Conquer computations.

The stream paradigm perfectly suits the need for reducing inter-core synchronization overheads in parallel programs for shared cache multi-cores. Therefore, it can be used to build an efficient run-time support for a high-level programming model aimed at the effective design of parallel applications.

The rest of paper discusses the idea of streamization (Sec. 2), outlines the main FastFlow features (Sec. 3), describes the stream acceleration methodology (Sec. 4) and gives experimental results (Sec. 5). Related work (Sec. 6) and Conclusions are then presented.

2   Code Acceleration through Streamization

The parallelization of a sequential code is typically tackled via data dependence analysis [7]. Having fixed a reference grain for the parallelization, the tasks to compute are released from strict sequential order in such a way that the program semantics is preserved. As a result, these objects are organized in a static or dynamically evolving graph of communicating tasks.

Instruction level parallelism is typically exploited at the hardware level within the single core, while coarser grain parallelism is expressed among cores at software level. In the latter case, the primary sources of parallelism exploitation are iterative and recursive tasks since they often model heavy kernels that can be unfolded into fully or partially independent tasks.

Also, if the parallel code is derived from existing sequential code, variable privatization and scalar/array expansion are often applied to further relax *false dependencies* [8,7]. These techniques consist in various levels of duplication of some memory area. Variable privatization nicely couples with stream parallelism making possible *dynamic privatization*. Privatized variables can be copied into a dynamically created data structure (e.g. the stream task type task_t in Fig. 2, right, lines 44–46) and offloaded in a non-blocking fashion to an accelerator.

Computation offloading, which is typically used to feed hardware accelerators (e.g. via OpenCL or CUDA), naturally creates a stream of tasks for the accelerator, provided it is realized via non-blocking mechanisms. As with GPUs, streamization techniques may offer significant opportunities on shared-cache multi-cores.

We classify stream sources in two broad categories: *exo-* and *endo*-streams.

Exo-streams. A stream parallel approach naturally matches the parallelization of applications that manage externally produced (exo) streams of input and output data. These applications are increasingly found in many domains, e.g. multimedia and networking. In many cases, the whole dataset is large and has to be processed online. Moreover, there may be few or no sources of data parallelism to allow use of classical data parallel techniques.

Endo-streams. A stream parallel approach also matches those computations that internally (endo) generate streams. We recognize three distinct sources of endo-streams: recursive computations and iterative computations, with and without dependencies. Recursion (*Recursive kernels*) appears as a natural programming technique in many algorithms working with dynamic data structures such as graphs and trees. In many cases they are data intensive algorithms and require significant computational power. Recursion could be easily modeled as a streaming network using a cyclic graph, whereas it can not readily be modeled by way of a data parallel approach. In this case, stream items are generated *on-the-fly* and represent different invocations of the recursive kernel. *Iterative kernels with independent iterations* represent the simplest case of endo-stream sources and are typically parallelized using a data-parallel approach. Streamization can also be applied in this case (e.g. generating and then processing a stream of items representing the different iterations) and is particularly useful when dynamic loops (i.e. *while*) or *for* loops with conditional jumps in the body (i.e. *break* or *goto* statements) are used. In fact, in all cases when flag variables are used in the code to skip the next code section, classical data parallel techniques are difficult to apply and may lead to poor performance. In the presence of loop-carried dependencies (*Iterative kernels with dependencies*), streamization may lead to more efficient synchronization patterns because it reduces the synchronization overhead due to data sharing in shared memory systems and thus shortens the critical path of execution. In *doAcross* task scheduling, the dependencies across threads are typically cross-iteration dependencies, which means that the underling memory location cannot be privatized. The synchronization overhead must be paid at least once for each iteration of a loop. On the contrary, in a pipeline schedule, loop-carried dependencies can be mapped onto the same thread. The remaining dependencies will still have the same overhead, but privatization will better tolerate the latency.

3   The FastFlow Parallel Programming Framework

FastFlow is a C++ parallel programming framework aimed at simplifying the development of efficient applications for multi-core platforms. The key vision of FastFlow is that ease-of-development and runtime efficiency can both be achieved by raising the abstraction level of the design phase, thus providing developers with a suitable set of parallel programming patterns that can be efficiently compiled onto the target platforms.

FastFlow is conceptually designed as a stack of layers that progressively abstract the shared memory parallelism at the level of cores up to the definition of useful programming constructs supporting structured parallel programming on cache-coherent shared memory multi- and many-core architectures [9].

FastFlow's core is based on efficient Single-Producer-Single-Consumer (SPSC) and Multiple-Producer-Multiple-Consumer (MPMC) FIFO queues, which are implemented in a lock-free and wait-free fashion. On top of its core, FastFlow provides programmers with a set of patterns implemented as C++ templates: *farm*, *farm-with-feedback* (i.e. Divide&Conquer) and *pipeline* patterns, as well as their arbitrary nesting and composition. A FastFlow farm is logically built out of three entities: *emitter, workers, collector*. The emitter dispatches stream items to a set of workers which compute the output data. Results are then gathered by the collector back into a single stream.

Thanks to the lock-free implementation that significantly reduces cache invalidations in core-to-core synchronizations, FastFlow typically demonstrates increased speedup for fine-grained computations over other programming tools such as POSIX, Cilk, OpenMP, and Intel TBB [10]. For more information about the FastFlow implementation and features see [9].

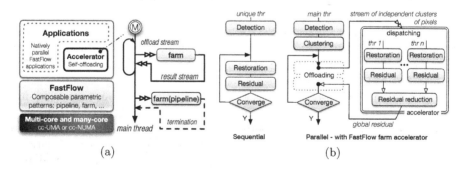

(a)   (b)

Fig. 1. 1(a) FastFlow accelerator architecture with usage examples. 1(b). Flow charts of sequential and a FastFlow accelerated real case algorithm: two-step denoising.

```
1   // Original code
2   #define N 1024
3   long A[N][N],B[N][N],C[N][N];          ①
4   int main() {
5     // < init A,B,C>
6
7     for(int i=0;i<N;++i) {
8       for(int j=0;j<N;++j) {             ②
9
10        int _C=0;
11        for(int k=0;k<N;++k)             ❸
12          _C += A[i][k]*B[k][j];
13        C[i][j]=_C;
14
15      }                                  ④
16    }
17  }                                      ⑤
```

Regions marked with white circled figures ①,②,④,⑤ are copy-pasted.
The region marked with the black circled figure (❸) has been selected to be accelerated with a farm. It is copied with renaming of variables that are concurrently changed, e.g. automatic variables in a loop. A stream of task_t variables is used to keep all different values of these variables.
Grey boxes create and run the accelerator; they are pre-determined according to the accelerator type.
The code marked with ➠ executes the offloading onto the accelerator; the target of the offloading is the svc method ➲ of the Worker class.

```
20  // FastFlow accelerated code
21  #define N 1024
22  long A[N][N],B[N][N],C[N][N];
23  int main() {                          ①
24    // < init A,B,C>
25
26    ff :: ff_farm<> farm(true /* accel */);
27    std :: vector<ff :: ff_node *> w;
28    for(int i=0;i<PAR_DEGREE;++i)
29      w.push_back(new Worker);
30    farm.add_workers(w);
31    farm.run_then_freeze();
32
33    for (int i=0;i<N;i++) {
34      for(int j=0;j<N;++j) {            ②
35        task_t * task = new task_t(i,j);
36        farm.offload(task);              ➠
37      }
38    }                                    ④
39    farm.offload((void *)ff :: FF_EOS);
40    farm.wait();  // Here join
41  }                                      ⑤
42
43  // Includes
44  struct task_t {
45    task_t(int i,int j):i(i),j(j) {}
46    int i; int j;};
47
48  class Worker: public ff::ff_node {
49  public: // Offload target service
50    void * svc(void *task) {             ➲
51      task_t * t = (task_t *)task;
52      int _C=0;
53      for(int k=0;k<N;++k)               ❸
54        _C += A[t->i][k]*B[k][t->j];
55      C[t->i][t->j] = _C;
56      delete t;
57      return GO_ON;
58    }
59  };
```

Fig. 2. Derivation of FastFlow accelerated code from a simple sequential C++ application (matrix multiplication)

4   Self-offloading on the FastFlow Accelerator

A *FastFlow accelerator* is a software device that extends the FastFlow framework with a functional *self*-offloading feature, i.e. offloading from a thread running on the main CPU to other threads running on the main (multi-core) CPUs. The architecture of the accelerator is sketched in Fig. 1(a).

The main aim of self-offloading is to give the programmer an easy and semi-automatic way to introduce parallelism into a C/C++ sequential code by moving parts of the original code into the body of C++ methods, which will be executed in parallel according to the selected FastFlow skeleton (or skeleton composition). As shown in Fig. 2, this requires limited programming effort and may significantly speed up the original code by exploiting efficient stream parallelism.

An accelerator is a collection of threads and has a global life-cycle with two stable states: *running* and *frozen*, plus several transient states. In a running state, all threads of an accelerator are logically able to run (either running or actively waiting on a non-blocking synchronization), whereas in a frozen state they are suspended (at the O.S. level). At any given time, due to non-blocking synchronizations, the total number of threads in running accelerators should typically be smaller (or equal) than core count for performance reasons. This kind of configuration typically benefits from O.S. affinity scheduling. In the case of a higher thread count, running threads will share available cores according to O.S. scheduling policies. The thread-to-core pinning is possible via FastFlow utility functions; automatic thread pinning/mapping are planned for future work.

The accelerator provides the programmer with one (untyped) streaming input channel and one (untyped) streaming output channel that can be dynamically *created* (and *destroyed*) from C++ code (either sequential or multi-threaded) as a C++ object (Fig. 2, right, lines 10–13). Thanks to the underlying shared memory architecture, messages flowing into these channels may carry both values and pointers to data structures.

When an accelerator is created (Fig. 2 lines 26–30), it can be switched on (Fig. 2 line 31): the accelerator threads are created and bound to system cores. A thread of a user can *wait* for an accelerator, i.e. suspend until the accelerator completes its input tasks, and then can put the accelerator into the frozen state. It is also possible to activate the accelerator asynchronously and pop output tasks from the accelerator's output channel using the *load_result* method.

A FastFlow accelerator is defined by a FastFlow skeletal composition augmented with an input stream and an output stream that can be, respectively, pushed and popped from outside the accelerator. Both the functional and extra-functional behaviour of the accelerator are fully determined by the chosen skeletal composition. For example, the *farm* skeleton provides the parallel execution of the same code (within a *worker* object) on independent items of the input stream. The *pipeline* skeleton provides the parallel execution of filters (or stages) exhibiting a direct data dependency. More complex behaviours can be defined by creating compositions of skeletons whose behaviour could be described using a (cyclic or acyclic) graph of tasks with well-defined functional and extra-functional semantics. Clear understanding of accelerator behaviour makes it possible to correctly parallelize segments of code.

The use of a farm accelerator is illustrated in Fig. 2. The code in Fig. 2 (left) shows a sequential program including three loops: simple matrix multiplication. Its accelerated version, shown in Fig. 2 (right), can be semi-automatically derived from the sequential by copy-pasting pieces of code into placeholders on a code template (parts in white background in the left column): for example, code marked with ①,②,④, and ⑤ are copied from left to right. The code that has been selected for offloading, in this case the body of a loop marked with ❸, is copied into the worker body after a suitable *renaming* of variables.

The accelerator shares the memory with its caller. As is well-known, transforming a sequential program into a parallel one requires regulation of possibly concurrent memory accesses. In low-level programming models this is usually done by using critical sections and monitors under the responsibility of the programmer. FastFlow does not prohibit these mechanisms, but promotes a methodology to avoid them. In very general terms, the sequential code statement can be correctly accelerated with FastFlow only mechanisms if the offloaded code and the offloading code (e.g. main thread) instances do not break any data dependency [7]. FastFlow helps the programmer in enforcing these conditions in two ways: *skeletons* and *streams*.

The *skeletal* structure of the accelerator induces a well-defined partial ordering among offloaded parts of code. For example, no-order for farm, a chain of dependencies for pipeline, a directed acyclic graph for farm-pipeline nesting/-composition, and a graph for a farm-with-feedback. The synchronization among threads is enforced by *streams* along the paths of the particular skeleton composition, as in a data-flow graph. True dependencies (read-after-write) are admissible only along these paths. Streams can carry values or pointers, which act as synchronization tokens for indirect access to the shared memory.

Pragmatically, streams couple quite well with the needs of sequential code parallelization. In fact the creation of a stream to be offloaded on the accelerator can be effectively used to resolve anti-dependency (write-after-read) on variables since the stream can carry a copy of the values. For example, this happens when an iteration variable of an accelerated loop is updated after the (asynchronous) offload. This case naturally generalizes to all variables exhibiting a larger scope with respect to the accelerated code. The same argument can be used for output dependency (write-after-write). FastFlow accelerator templates accommodate all variables of this kind in one or more structs or C++ classes (e.g. `task_t`, lines 44–46) representing the input, and, if present, the output stream data type. All other data accesses can be resolved by just relying on the underlying shared memory (e.g. read-only, as with A in line 54, and single assignment as with C in line 55).

It is worth pointing out that the FastFlow acceleration methodology may not be fully automated. It has been conceived to ease the task of parallelization by providing the programmer with a methodology that helps in dealing with several common cases. However, many tasks require the programmer to make decisions, e.g. the selection of the code to be accelerated. In the example code in Fig. 2 there are several choices with different computation granularity: offload only the computations relative to index i, to both i and j, or to all three indices.

The programmer can control the communications among threads at higher level, and in particular can control when the thread reads from the input channel and writes to the output channel, the thread state (running and frozen) at different abstraction levels, the thread termination conditions, and the scheduling and the collection policies in the farm skeleton. In addition, as FastFlow equips the standard OS threads (e.g. POSIX threads) with additional synchronization mechanisms, the user retains the possibility to exploit thread native synchronization mechanisms (e.g. locks) and exploit thread library specific features (e.g. defining and using thread-specific storage). Also, almost any possible nesting of farm and pipeline skeletons is possible.

The low overhead added by the run-time support, together with the flexibility of the framework, widens the parallelization possibilities to a broader class of applications, and especially to programs performing frequent synchronizations.

5   Experimental Evaluation

The proposed self-offloading methodology has been validated on a set of three micro-benchmarks and two novel real-world applications. Several other complex applications —not presented here— have been parallelized using the FastFlow accelerator technique, such as a C.4.5 data classifier, a Gillespie simulator for biological systems and the Smith-Waterman string alignment. We refer back to [9] for an extensive listing.

Two platforms are used in the evaluation: *8-core*) Intel workstation with 2 x quad-core Xeon E5520 Nehalem (16 HyperThreads) @2.26GHz; *48-core*) AMD Magny-Cours 4 x twelve-core Opteron 6174 @2.2GHz. Both run Linux x86_64.

5.1   Micro-benchmarks

We present the results obtained from the parallelization of three simple and well-known algorithms: dense square matrix multiplication, the Quicksort sorting algorithm, and the recursive computation of the n-th Fibonacci number.

The matrix multiplication consists in the parallelization of the "naïve" algorithm; the code is shown fully in Fig. 2; the parallel version achieves a speedup of 7.6 on the 8-core platform (1024x1024 integer matrices). In this case, a stream of submatrix multiplication tasks is created to feed the FastFlow accelerator.

The Quicksort benchmark has been parallelized using a Divide&Conquer parallel pattern (i.e. *farm* plus *loop* patterns). A worker receives a task with two indices describing a partition of a shared array from the emitter and executes a step of the Quicksort algorithm producing two tasks that return to the emitter, which in turn dispatches received tasks to workers. The workers switch to sequential processing at a given partition size threshold. The Fibonacci benchmark behaves similarly; the emitter also accumulates partial results. In both these micro-benchmarks the streams processed through the FastFlow accelerator are generated *on-the-fly* and composed of tasks corresponding to recursive calls to the main procedure. The Quicksort on a 50M integer array and Fibonacci(50) achieve speedups of 6.8 and 9.21, respectively, on the 8-core platform over-provisioned with 16 worker threads. The super-linear speedup achieved by Fibonacci is due to the HyperThreading technology, which, in contrast, brings no benefit to the Quicksort application.

5.2   Applications

We discuss the results achieved when accelerating two applications with Fast-Flow. The applications are representative of two large and significant applications classes: the first performs classical data-parallel (with dependencies) computations and its acceleration is endo-stream parallel, while the second is a classical exo-stream parallel application. In both cases excellent results are achieved on state-of-the-art multi-core architectures.

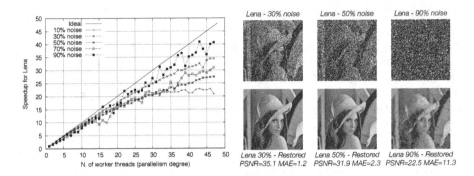

Fig. 3. Left) Speedup for the Lena image on the 48-core platform. Right) Restoration result with PSNR (Peak Signal-to-Noise Ratio) and MAE (Mean Absolute Error).

Edge-Preserving Denoiser. The edge-preserving denoiser is a two-step filter for removing salt-and-pepper noise (see Fig. 3 right). In the first step, an adaptive median filter is used to identify the set of noisy pixels; in the second step, these pixels are restored according to an iterative variational approach up to convergence. The detailed description of the sequential algorithm is beyond the scope of this paper; it ensures state-of-the-art restoration quality and execution time, and it is able to restore also very noisy images (e.g. 90% random noisy pixels) [11]. To ensure a high quality of restoration the algorithm features cross-dependencies among the noisy pixels, which induce a logical data-dependency pattern that cannot be solved with an *a priori* partitioning of the data set, as typically happens in data-parallelism. The idea behind the parallel porting of the proposed algorithm consists in clustering the noisy pixels into independent sets in such a way that cross-dependencies are respected. Following this, independent clusters, which can exhibit very different cardinalities, can be processed in parallel according to a farm paradigm. In particular, the clusters can be streamed (via offloading) to a FastFlow farm accelerator. The porting process, which is sketched in Fig. 1(b), required just a few hours development time. The clustering process, which is not present in the sequential version, has been designed from scratch as sequential code, and thus does not require concurrency skills.

Figure 3 reports the speedup achieved on the 48-core platform for the Lena 256x256 standard test image. The completion time of sequential processing grows linearly with noise ratio: from 9 to 180 seconds with 10% to 90% noise ratio. The parallel version speeds them up to a range of 0.4 to 4 seconds, respectively. Note that restoration quality metrics (PSNR and MAE) are comparable to or better than the best results in the area (e.g. Chan's method [11], while execution time for parallel execution is better than results in the literature [11,9].

Stream File Compressor. This application is a further development of an already parallel application: *pbzip2* [12], i.e. a parallel version of the widely used *bzip2* block-sorting file compressor. It uses pthreads and achieves very good speedup on SMP machines for large files. Small files (less then 1MB) are sequentially compressed. We extend it to manage streams of small files which can be compressed in parallel. In this case, in contrast with previous examples, the stream of data is not created within the application but exists independently of the application (e.g. comes from a POSIX stream, *find* shell command, etc.).

The original pbzip2 application is structured as a farm: the generic input file is read and split into independent parts (blocks) by a splitter thread; then each block is sent to a pool of worker threads which compress the blocks. The pool is hand-coded using pthread synchronizations and extensively hand-tuned. The FastFlow port of pbzip2 (*pbzip2_ff*) was developed by taking the original code of the workers and including it in a FastFlow farm pattern. Then, a second Fast-Flow farm whose workers execute the file compression sequentially was added. The two farms are run as two accelerators and fed by the main thread which selectively dispatches files to the two accelerators depending on the file size. The porting of the pbzip2 application to FastFlow has highlighted two aspects of the approach: (1) by using FastFlow it is possible to parallelize the algorithm using high-level parallel patterns rather than a hand-tuned mutex-based implementation without any performance penalty and with an actual performance improvement; (2) FastFlow non-blocking synchronizations exhibit good performance in comparison to traditional blocking synchronizations based on mutexes, even in worst-case scenarios such as coarse grained CPU-intensive elaborations where non-blocking behaviour might waste CPU cycles.

Table 1 compares the execution times of sequential bzip2, pbzip2 and pbzip2_ff on two different data sets: on the left, on large files shows that pbzip2_ff exhibits no significant slowdown against hand-tuned pbzip2; on the right, on files of various sizes shows the improved speedup of pbzip2_ff against pbzip2.

6   Related Work

The word accelerator is often used in the context of hardware accelerators. Usually accelerators feature a different architecture with respect to standard CPUs and thus, in order to ease exploitation of their computational power, specific libraries are developed. In the case of GPGPUs those (low-level) libraries include *Brook* [13], NVidia *CUDA*, and *OpenCL*. At a higher-level, *Offload* [14] enables offloading of parts of a C++ application, which are wrapped in offload blocks, onto hardware accelerators for asynchronous execution; *OMPSs* [15] enables the

Table 1. On the left, compression and decompression time (S) on a single 1 GBytes file (528 MBytes compressed). On the right, execution time (S) and speedup over bzip2 in the case of a stream of 1078 files: 86% small (0–1 MBytes), 9% medium (1–10 MBytes), 4% large (10–50 MBytes), and 1% very large (50–100 MBytes). pbzip2 uses 16 threads. pbzip2_ff uses 16 threads for each accelerator.

	bzip2	pbzip2			pbzip2_ff				bzip2	pbzip2		pbzip2_ff	
# threads	1	4	8	16	4	8	16		Time (S)	Time (S)	Speedup	Time (S)	Speedup
compres	231	58.8	32.9	26.0	59.3	33.0	25.7	comp.	538	97	5.5	72	7.5
decompres	69	18.0	11.1	8.9	18.5	11.0	8.9	decomp.	126	33	3.8	21	6.0

offloading of OpenCL and CUDA kernels as an OpenMP extension [16]. Fast-Flow, in contrast with these frameworks, does not target specific (hardware) accelerators but realizes a virtual accelerator running on the main CPUs and thus does not require the development of specific code.

Recent work [17] using the Charm++ programming model has demonstrated that accelerator extensions are able to achieve both code portability and speedup. However, in order to exploit the accelerator features, in contrast with FastFlow, the application has to be entirely rewritten using the Charm++ framework.

Streaming applications and patterns are also targeted by StreamIt [18], Intel Concurrent Collections (CnC) [19], and Intel TBB [5]. TBB, in particular, can be used to accelerate C/C++ programs in specific portions of code via parallel patterns (a.k.a. "algorithms") and thread-safe data containers. The only streaming pattern is the *pipeline* which, however, does not support non-linear streaming networks, which therefore have to be embedded in a pipeline with significant programming drawbacks. Farm and Divide&Conquer patterns are not natively provided, even if they can be simulated with lower-level features.

OpenMP [16] supports parallelization of sequential programs via pragmas that are, however, mainly designed to exploit loop-level data parallelism (e.g. *do_independent*) whereas the exploitation of other patterns of parallelism (e.g. farm and Divide&Conquer) may require substantial re-factoring of the code.

A comparative performance study of FastFlow, OpenMP, and TBB on micro-benchmarks and the Smith-Waterman application is reported in [10].

7   Conclusions

In this paper the FastFlow accelerator, which represents an extension of the FastFlow framework specifically designed to support the easy porting of existing sequential C/C++ applications onto multi-cores using stream parallelism, is introduced. We identified exo- and endo-streams, showing that stream parallelism is applicable to a wide range of types of algorithm. The FastFlow accelerator exhibits well-defined functional and extra-functional behaviour represented by a skeleton composition; this helps in ensuring the correctness of the parallelization process. The main vehicle of parallelization is offloading of code kernels onto a number of additional threads running on the same CPU; we call this technique *self-offloading*. Code acceleration is supported by a methodology and by the unique ability of FastFlow to support very fine grain tasks on standard multi-cores.

The effectiveness of the proposed methodology has been demonstrated by a set of codes ranging from very simple kernels to real applications.

References

1. Stephens, R.: A survey of stream processing. Acta Informatica 34(7), 491–541 (1997)
2. Cole, M.: Algorithmic Skeletons: Structured Management of Parallel Computations. Research Monographs in Parallel and Distributed Computing. Pitman (1989)

3. Vanneschi, M.: The programming model of ASSIST, an environment for parallel and distributed portable applications. Parallel Computing 28(12), 1709–1732 (2002)
4. Dean, J., Ghemawat, S.: MapReduce: Simplified data processing on large clusters. In: Usenix OSDI 2004, pp. 137–150 (December 2004)
5. Intel Corp.: Threading Building Blocks (2011),
   http://www.threadingbuildingblocks.org/
6. Asanovic, K., Bodik, R., Demmel, J., Keaveny, T., Keutzer, K., Kubiatowicz, J., Morgan, N., Patterson, D., Sen, K., Wawrzynek, J., Wessel, D., Yelick, K.: A view of the parallel computing landscape. CACM 52(10), 56–67 (2009)
7. Bernstein, A.J.: Program analysis for parallel processing. IEEE Trans. on Electronic Computers EC-15(5), 757–762 (1966)
8. Pop, A., Pop, S., Jagasia, H., Sjodin, J., Kelly, P.H.J.: Improving GCC infrastructure for streamization. In: Proc. of the 2008 GCC Developers' Summit, Ottawa, Canada (June 2008)
9. Aldinucci, M., Torquati, M.: FastFlow website (2009),
   http://mc-fastflow.sourceforge.net/
10. Aldinucci, M., Meneghin, M., Torquati, M.: Efficient Smith-Waterman on multi-core with fastflow. In: Danelutto, M., Gross, T., Bourgeois, J. (eds.) Proc. of Intl. Euromicro PDP 2010: Parallel Distributed and Network-Based Processing, Pisa, Italy, pp. 195–199 (February 2010)
11. Aldinucci, M., Drocco, M., Giordano, D., Spampinato, C., Torquati, M.: A parallel edge preserving algorithm for salt and pepper image denoising. Technical Report 138/2011, Università degli Studi di Torino, Dip. di Informatica, Italy (May 2011)
12. Gilchrist, J.: Parallel data compression with bzip2. In: Proc. of IASTED Intl. Conference on Parallel and Distributed Computing and Systems, pp. 559–564 (2004)
13. Buck, I., Foley, T., Horn, D., Sugerman, J., Fatahalian, K., Houston, M., Hanrahan, P.: Brook for GPUs: stream computing on graphics hardware. In: ACM SIGGRAPH 2004 Papers, New York, NY, USA, pp. 777–786 (2004)
14. Cooper, P., Dolinsky, U., Donaldson, A.F., Richards, A., Riley, C., Russell, G.: Offload – automating code migration to heterogeneous multicore systems. In: Patt, Y.N., Foglia, P., Duesterwald, E., Faraboschi, P., Martorell, X. (eds.) HiPEAC 2010. LNCS, vol. 5952, pp. 337–352. Springer, Heidelberg (2010)
15. Ferrer, R., Planas, J., Bellens, P., Duran, A., González, M., Martorell, X., Badia, R.M., Ayguadé, E., Labarta, J.: Optimizing the exploitation of multicore processors and gPUs with openMP and openCL. In: Cooper, K., Mellor-Crummey, J., Sarkar, V. (eds.) LCPC 2010. LNCS, vol. 6548, pp. 215–229. Springer, Heidelberg (2011)
16. Park, I., Voss, M.J., Kim, S.W., Eigenmann, R.: Parallel programming environment for OpenMP. Scientific Programming 9, 143–161 (2001)
17. Kunzman, D.M., Kalé, L.V.: Towards a framework for abstracting accelerators in parallel applications: experience with cell. In: Proc. of the Conference on High Performance Computing (SC), Portland, Oregon, USA, ACM, pp. 1–12. ACM, New York (2009)
18. Thies, W., Karczmarek, M., Amarasinghe, S.P.: StreamIt: A language for streaming applications. In: Proc. of the 11th Intl. Conference on Compiler Construction (CC), London, UK, pp. 179–196 (2002)
19. Newton, R., Schlimbach, F., Hampton, M., Knobe, K.: Capturing and composing parallel patterns with Intel CnC. In: Proc. of 2nd USENIX Workshop on Hot Topics in Parallelism (HotPar 2010), Berkley, CA, USA (June 2010)

A Novel Shared-Memory Thread-Pool Implementation for Hybrid Parallel CFD Solvers

Jens Jägersküpper[1] and Christian Simmendinger[2]

[1] German Aerospace Center (DLR)
Institute of Aerodynamics and Flow Technology
Center of Computer Applications in Aerospace Science and Engineering ($C^2A^2S^2E$)
38108 Braunschweig, Germany
Jens.Jaegerskuepper @ DLR.de
[2] T-Systems Solution for Research (SfR)
Pfaffenwaldring 38–40,
70569 Stuttgart, Germany

Abstract. The Computational Fluid Dynamics (CFD) solver TAU for unstructured grids is widely used in the European aerospace industry. TAU runs on High-Performance Computing (HPC) clusters with several thousands of cores using MPI-based domain decomposition. In order to make more efficient use of current multi-core CPUs and to prepare TAU for the many-core era, a shared-memory parallelization has been added to one of TAU's solver to obtain a hybrid parallelization: MPI-based domain decomposition plus multi-threaded processing of a domain.

For the edge-based solver considered, a simple loop-based approach via OpenMP FOR directives would – due to the Amdahl trap – not deliver the required speed-up. A more sophisticated, thread-pool-based shared-memory parallelization has been developed which allows for a relaxed thread synchronization with automatic and dynamic load balancing.

In this paper we describe the concept behind this shared-memory parallelization, we explain how the multi-threaded computation of a domain works. Some details of its implementation in TAU as well as some first performance results are presented. We emphasize that the concept is not TAU-specific. Actually, this design pattern appears to be very generic and may well be applied to other grid/mesh/graph-based codes.

1 Intro

The TAU code, which is developed at the Institute of Aerodynamics and Flow Technology of the German Aerospace Center (DLR), is widely used in the European aerospace industry for Computational Fluid Dynamics (CFD), c. f. e. g. [6]. The solver is designed for unstructured grids, yet it may also be used with (block) structured grids. MPI-based domain decomposition allows TAU to be run on HPC clusters withseveral thousands of cores. To make more efficient use of current multi-core CPUs and to prepare TAU for the many-core era a shared-memory parallelization has been implemented for one of the TAU solvers. That

E. Jeannot, R. Namyst, and J. Roman (Eds.): Euro-Par 2011, LNCS 6853, Part II, pp. 182–193, 2011.

solver implements an explicit Runge/Kutta scheme with geometric multigrid acceleration for unstructured grids.

Several approaches for a shared-memory parallelization for TAU have been evaluated, and finally, a novel solution following the thread-pool model has been developed. This thread-pool model allows the concurrent processing of tasks subject to dependencies among the tasks. In addition to temporal data dependencies and the concept of mutual completion of processed data, this solution incorporates also mutual exclusion among tasks to prevent data races. The concept allows for a significantly relaxed thread synchronization compared to bulk-synchronous models. It features an automatic load balancing and allows to implement a straightforward overlap of communication and computation in TAU. The implementation shows a very good performance for the TAU solver. However, the used methodology is not specific to the TAU solver and should be applicable to a wide range of programs that work on unstructured or structured grids/meshes/graphs.

1.1   Motivation

Due to the electric capacity of a CPU chip, the higher the clock frequency, the higher the voltage needs to be – and the higher the thermal power to be dealt with. Since an effective solution of this problem is not available, clock rates no longer increase. Moore's law, however, is still valid: the number of transistors per chip doubles roughly every one and a half years. The CPU manufacturer's policy to further increase the theoretical peak performance of their chips: multi-core chips. Though we see an exponential growth in explicit parallelism, this change in the hardware is not at all reflected in HPC software: With a moderate number of cores per socket there has been simply no need to adapt the parallelization. Due to the large number of cores per socket in modern CPU designs, however, the picture is changing: One MPI process per core has become a problem. For CFD, the more MPI processes are used, the smaller the average computational load per MPI process and the more MPI communication is needed to synchronize the flow variables across the processes. The time for this synchronization is to be considered serial. Hence, simply by Amdahl's law, there is a maximum number of MPI processes (and hence cores) that can be reasonably used.

Shared-memory parallel (multi-threaded) computation of the domains suggests itself as a possible way to increase the scalability. When all cores of a CPU are used to process one domain, the number of domains drops from the number of cores to the corresponding number of sockets. Even though the idea of such a hybrid (2-level) parallelization is straightforward, this approach requires the shared-memory parallel computation of the domains to be sufficiently efficient. It turns out that this is quite a challenge for CFD on unstructured grids.

1.2   Outline

In the following sections we describe how the proposed shared-memory parallel computation of the (MPI-) domains works and give some implementation details and first performance results. We start with a brief introduction to the TAU code in Sec. 2. In particular, the original MPI parallelization of TAU via domain decomposition is explained. In Sec. 3 we describe the concept of the shared-memory implementation. As the tasks to be processed by the threads show not only temporal dependencies, but also data dependencies, a simple loop-based parallelization via OpenMP is precluded. The tasks hence are processed asynchronously using mutual exclusion to prevent data races (asynchronous subject to the dependencies between the tasks). To efficiently handle temporal dependencies we employ the concept of mutual completion.

The relaxed synchronization of the threads improves the scaling considerably: The accumulation of load imbalances, which occur at every synchronization point in the multi-threaded program flow, is drastically reduced since global synchronization is replaced by dynamic local synchronization.

We think that our approach would be applicable to many other numerical codes which use a multi-threaded task parallel approach. Details of how this concept has been actually implemented in the TAU code are given in Sec. 4 and first performance results are presented in Sec. 5. Finally, we conclude in Sec. 6 and give an outlook how to further improve the performance of the shared-memory parallelization presented.

2   The DLR TAU Code

The TAU code is a 3D-flow solver that simulates compressible external flows (steady or time-accurate) on unstructured grids using finite-volume discretization via the Reynolds-averaged Navier-Stokes equations (RANS). TAU features several turbulence models (Spalart/Allmaras, SST, RSM, etc.) as well as hybrid RANS/LES capabilities. It supports central spatial discretization (namely JST) as well as several upwind schemes. Moreover, the user may choose between cell-vertex and cell-centered metric. The most frequently used TAU solvers are Runge/Kutta and LU-SGS. Here we focus on the explicit Runge/Kutta solver with geometric multigrid, cell-vertex metric, central discretization, and a Spalart/Allmaras turbulence model, which requires one equation for the eddy viscosity in addition to the five equations for mass, impulse and energy. As a consequence, the number of degrees of freedom (DoF) is given by 6 times the number of grid points. Each edge in the grid corresponds one-to-one to a face in the so-called dual grid. This dual grid comprises control volumes around the grid points. The calculation of the fluxes between these control volumes, which are also called dual cells, is the main computational task in the scenario considered here. To integrate the fluxes for all dual cells, we could loop over all points and for each point we would loop over the faces of its surrounding dual cell. Recall however that each face in the dual corresponds one-to-one to an edge in the original grid. Thus, in this hypothetical implementation we would touch each edge twice: Once for each of the two end points of the edge.

For a more efficient access to main memory, the current implementation of TAU loops over the edges instead (c. f. [2]): For each edge, i.e., for each face in the dual grid, the respective value for the two points are updated. Using this "edge-based" scheme, the number of point-data loads is halved compared to a loop over the dual cells. Though TAU is edge-based, naturally, there are also point loops. The main computational load, however, is caused by edge loops.

MPI Parallelization via Domain Decomposition

TAU's parallelization is based on a domain decomposition: the grid is cut into several pieces (domains) by means of a partition of the point set, which may be obtained using a graph partitioning software like Chaco ([4]), Zoltan ([3]), (Par)Metis ([5]), etc. Each edge connecting two points in different domains has to be doubled, so that an edge as well as the two incident points exist in both domains (overlap). This adds a number of points to each domain, which are called "ghost points". Furthermore, the total number of edges in all domains equals the number of edges in the original grid plus the number of edges cut during domain decomposition. The flow values at a ghost point must be kept in sync with the corresponding data at the original point. This is done using message passing, namely MPI. MPI-based domain decomposition can be considered the standard parallelization concept for grid-based numerical codes that are run on modern HPC architectures, c. f. [8].

Obviously, the more edges are cut for domain decomposition, the more data has to be passed around via MPI to keep the domains, namely the evolving flow solution at the discretization points, synchronized. The time for this (usually bulk-synchronous) synchronization via MPI plus the inherent load imbalance due to an imperfect partitioning can be considered a serial part of the algorithm ($\rightarrow$Amdahl's law). Nevertheless, TAU's MPI parallelization scales well, c. f. [1].

As a consequence, for any fixed size CFD problem, there is a maximum number of domains that can be effectively used to compute this problem. A further increase of the number of domains eventually results in the parallel efficiency to drop until increasing the number of domains no longer speeds up the calculation at all (limit of scalability). If the number of domains is increased even further, the wall-clock time to solve this problem actually starts to increase. The scalability limit does not only depend on the CFD problem, but also on the cluster used – and on the application, of course.

MPI-synchronized domains + multi-threaded processing of domains

To increase the number of usable compute cores without increasing the number of domains, each domain must be computed on multiple cores. For the shared-memory parallel computation of a single domain with multiple threads, however, data parallelism becomes an issue: If, in an edge loop, two edges incident to a common point are processed concurrently (to update their two points, respectively), the update of the shared point may result in a data race: One of the two updates for this point can get lost. We hence not only need an efficient multi-threaded processing of edges, but also the prevention of data races. We need mutual exclusion among the updates of the same point. In principle this can be achieved in several different ways as we will detail now.

3   The Shared-Memory Parallelization – Generic Concept

We consider a given grid/mesh/graph consisting of a number of points and edges. Each edge connects two points. The number of edges incident to a point may vary, i. e., the connectivity may be regular (like for structured grids), yet it may also be irregular (unstructured girds). Note that whether this is an original graph or one obtained by domain decomposition makes no difference. Usually, the graph is very sparse as there may be several millions of points, yet each point has a very limited number of neighbors, say in the range of tens. Let us consider an algorithm that contains a large number of loops over both, edges as well as points: When passing over the edges, data associated with each edge's two points may be read and also updated. As a consequence, data races are possible when two edges incident to a common point are processed concurrently. Such data races must be prevented to ensure a correct behavior of the program. There are several obvious approaches to do so.

A critical section per point to ensure mutual exclusion of access to data associated with a point. This requires one mutex/lock variable per point, which has to be aquired whenever the point's data is read or written. Pthreads, OpenMP, or system libraries provide adequate functionality. Intrinsic functions for locked memory accesses provided by compilers may be used for a custom implementation.

Atomic updates of point data so that each read-update-write sequence touching a value associated with a point is atomic. For x86 architectures, "lock cmpxchg8b" may be used for an atomic update of a double-precision floating-point value (as it is done by most compilers for "omp atomic" directives).

Obviously, both approaches result in a huge number of locked memory accesses. Nevertheless, these two simple approaches were prototypically implemented in TAU. As expected, the very frequent use of locked memory access turns out a severe performance problem. If all edges incident to a point are exclusively processed by the same thread, no data races are possible for this point. Consequently, one may consider the following approach

Partition of the edge set into as many subsets as threads are concurrently running. Each edge is mapped to a particular thread. For points exclusively touched by edges processed by a single thread, no data races are possible. So mutual exclusion of point data accesses must be provided only for points that are incident to edges processed by two or more different threads.

We call a point "critical" if it is incident to edges processed by different threads so that mutex is necessary. The partition of the edges should be such that the total number of critical points is minimized. In addition, the number of edges processed by each threads should be as balanced as possible to obtain a good load balance. Moreover, the number of critical points touched by each thread should be as balanced as possible. As one might notice, these are quite a number of constraints on the partition of the edge set. In addition to the imbalances due to the edge partition not being perfect, also varying waiting times for locked memory access result in a load imbalance among the n threads when they process the n edge sets in parallel. Despite these load-balancing issues, whenever a point is to be updated within a parallelized edge loop we must know whether this point is critical or not. A prototype implementation in TAU showed that, even though this additional information may be stored without additional memory space (for instance by using signed integers for the point indices and taking the sign bit as an indicator) so that the memory access pattern is not changed, the branching between critical and non-critical points leads to a considerable overhead.

Each edge set spans/induces a subgraph. Note that critical points belong to two or more subgraphs. Assume for a moment that two of the n edge sets are such that the corresponding subgraphs are disjoint. Then these two edge sets can be concurrently processed by two threads without the need to prevent data races. In general, no data races can occur as long as only disjoint subgraphs are concurrently processed. Then each point – in particular the formerly critical ones – is accessed by at most one thread at a time. We thus finally refine the above model towards the following approach:

Asynchronous dependencies-driven parallel processing of a large number of small subgraphs. This model enables a dynamic load-balancing among threads. Moreover, by ensuring mutual exclusion among subgraphs sharing (a) common point(s), no data races can occur. So there is no need to tell between critical points and non-critical ones, which drastically reduces the number of locked memory accesses as well as the overhead of branching.

Assume, just as an example, that there are $10n$ subgraphs so that **on average** each of the n threads processes 10 **dynamically allocated** subgraphs per point/edge loop. Consider the processing of a subgraph s in a loop a task. Then we have task dependencies like "subgraph s must have been processed in the ith loop before s is processed in loop $i+1$". With this, we actually implemented a thread-pool pattern: In each loop the threads process tasks, i. e. subgraphs, that need to be processed at that stage of the program, i. e. for which the temporal data dependencies are met.[1] Furthermore, our task-dispatching logic incorporates the mutual exclusion of tasks for neighboring subgraphs to prevent data races.[2] In other words, the neighborhood structure of the subgraphs induce "mutex dependencies" among the tasks. No explicit thread synchronization is necessary since the threads synchronize automatically via the dispatching of

---

[1] This is somewhat similar to what the "SMP Superscalar" ("SMPSs") programming model/environment provides [7,9], yet in a bottom-up, loop-based approach, rather than the top-down function annotation in SMPSs.

[2] This is dissimilar to SMPSs, as (to our knowledge) SMPSs does not provide explicit locking of neighbouring data segments.

tasks subject to the temporal/mutex dependencies. Finally note that balanced sizes of the subgraphs are no longer that crucial. Instead, the processing of a subgraph should exclude as few as possible other subgraphs from being processed in parallel. So, the neighborhood structure of the subgraphs is most crucial here.

4   The Shared-Memory Parallelization – Implementation Details for TAU

In the TAU solver to be modified, a partitioning of the edge set, namely a **coloring of the edges,** already exists – yet for a different reason than described in the preceding section. Originally, the edge coloring was introduced to prevent data races when a long sequence of edges is processed concurrently on a vector processor. When vector-processor-based supercomputers were superseded by commodity clusters based on the omnipresent x86 architecture, this coloring was repurposed to maximize cache utilization. Even with the memory controller integrated into the CPU this cache optimization is absolutely critical to TAU's (serial) performance.

4.1   Cache Blocking in TAU

The TAU solver considered in this paper is memory-bound on current x86 architectures. To lower the number of loads from main memory per flop, the data layout is optimized with respect to cache utilization: The edges are sorted such that edges incident to the same point follow as closely as possible (temporal blocking of point-data access). Essentially we use space-filling Hilbert curves in this approach. In addition, the points are sorted such that, when accessing the points indirectly while looping over the edges, data associated with subsequently accessed points are close in the memory (spatial blocking). The temporal blocking is supposed to minimize the number of loads from main memory, whereas the spatial blocking is supposed to make use of the prefetching mechanisms of the hardware's memory/cache subsystem. With this strategy, TAU shows very good cache utilization. TAU is still memory-bound, though.

To enable a TAU programmer to split up a large edge loop into several routines (to improve code structure, readability, maintenance, expandability), so-called "colors" exist in TAU. Each color consists of a number of subsequent edges (in optimized order, c. f. the temporal blocking above) such that all the point data touched by these edges (which are spatially blocked, c. f. above) fit into the L2 cache. Note that this coloring is actually a partition of the edge set. A loop over the edges is equivalent to a nested loop over the colors followed by a loop over each color's edges. Code in the body of an edge loop may be split up into several routines on a per-color basis. Then only in the first routine the point data touched by the current color are loaded from main memory. For subsequent routines called for the color, this data is already cached. For commonplace x86 systems, the cache coloring in TAU enables L2-cache-local processing of an edge loop despite the code being split across several routines, possibly in different compilation units.

4.2 Modification of the Colors in TAU to Suite the Hybrid Parallelization Concept

Recall that the coloring in TAU is based on space-filling curves. Unfortunately, this results in a bad neighborhood structure among the subgraphs induced by the colors: there are too many colors with too many neighbors. As our concept for the shared-memory parallel processing of subgraphs described in Sec. 3 requires mutual exclusion among neighboring subgraphs, the coloring needs to be adjusted appropriately – without changing cache performance to the worse.

Recall that the subgraphs should be such that each subgraph touches as few as possible other subgraphs. Furthermore, the subgraphs should be balanced w. r. t. the number of points, whereas the number of edges per subgraph is of minor importance. These requirements perfectly fit graph-partitioning algorithms; for instance Sandia's "Chaco" (cf. [4]) may be used to obtain the colors.

Actually, there are different types of colors since there are three types of points in TAU: points that lie in the physical boundaries of the computational domain, ghost points (forming the halo of a domain obtained by domain decomposition for MPI parallelization), and the rest of the points, which we call "inner points". Correspondingly, there are four color types: inner colors, boundary-touching colors, halo-touching colors, and boundary+halo-touching colors. This enables us to overlap the processing of physical boundaries and the synchronization of ghost points via MPI with the processing of inner colors – subject to mutex and temporal data dependencies, of course. This integrates nicely with the thread-pool model since the processing of domain/physical boundaries at a particular stage in the program can be considered tasks just as well.

4.3 Minimally Invasive Implementation of the Task Dispatching

As TAU is a production code (validated by/for its customers), the numerics cannot be easily changed just to better suite TAU's parallelization. The shared-memory parallelization that has been added (cf. above) is designed such that it would yield exactly the same results as the originally serial code if floating-point arithmetic was exact. With the thread-pool-based parallel processing of the colors, the order in which the edges/points are processed may change from loop to loop. As a consequence, the limited precision of floating-point operations can – at least in principle – result in numerical differences. For the TAU solver considered, however, merely negligible differences are observed, if any.

Besides the consistency of the numerical behavior, the following aspect of software development/engineering has been very important: The shared-memory parallelization was supposed to change the code as little as possible, preferably transparent to the programmers. The proposed model indeed allows for an almost transparent implementation: As a loop over the edges was already split, namely done by a nested loop over the colors and the color's edges, respectively, the task dispatching was easily integrated (using C-code-like syntax): In

```
for(color=colorhead; color != NULL; color=color->next)
  for(eidx=color->start; eidx < color->stop; eidx++)
    {  /* process edge with index eidx */  };
```

merely the first line must be changed into

```
for(color=get_color(grid); color!=NULL; color=get_color(grid))
```

The task dispatching logic is completely encapsulated in the newly introduced `get_color()` function, which returns an appropriate color as long as there are colors left to be processed at that stage, else NULL. Note that `get_color()` may block if, at a given stage, for all colors left at that stage a neighboring color is being processed. To loop over the points, an additional while loop is introduced:

```
for(pidx=0; pidx < grid->npoints; pidx++)
  {  /* process point with index pidx */  };
```

must be replaced by

```
while(get_point_range(grid, &pidx, &pstop))
  for( ; pidx < pstop; pidx++)
    {  /* process point with index pidx */  };
```

The newly introduced `get_point_range()` function simply uses `get_color()` to obtain a color to be processed and with it a point range (each point is associated with exactly one color). The really nice thing with this implementation is that there is no need to change the bodies of point/edge loops – at least in principle. Naturally, when porting a loop, care has to be taken that no data is unintentionally shared. Thread-local storage (TLS) may be necessary. When, for instance, a maximum of a given value at the points is to be computed, TLS is only the half way: The maximum is to be determined as the maximum of the threads' local maxima, necessitating an explicit change in the code. Nonetheless, with this implementation of the concept, only a limited number of changes in the code are necessary to enable multi-threading.

In order to overlap the MPI communication (to synchronize the ghost points) with computation, `this_thread_syncs_halo()` is introduced. This function returns TRUE for exactly one of the threads. In case TRUE is returned, this function may block until no halo-touching color is processed and then keeps tasks for halo-touching colors from being dispatched until the domains' halos are synchronized. Meanwhile the other threads continue to process tasks for non-halo-touching colors. A similar mechanism is used to overlap the single-threaded processing of physical boundaries with the processing of inner colors.

5 First Performance Results

The TAU RANS-solver considered is an explicit 3-stage Runge/Kutta scheme with multigrid acceleration; cell-vertex metric, central discretization (JST), scalar dissipation, Spalart/Allmaras turbulence model are used. We get right to the point: How does the pure shared-memory parallelization face against the pure MPI parallelization. This test was run on a single-CPU machine running SLES 11 with an Intel 6-core Westmere EP X5670 with 2-way SMT enabled. Unfortunately, no Intel compiler was available, so gcc 4.3.3 with full optimization was

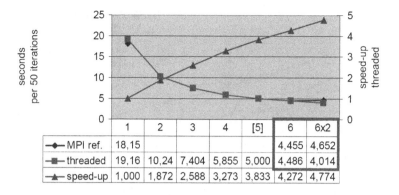

Fig. 1. Pure shared-memory parallelization (green) vs. pure MPI parallelization (blue) for an Intel Westmere 6-core CPU with 2-way SMT (X5670). Wall-clock time for 50 iterations vs. #threads or #domains, respectively; "speed-up" referes to "threaded".

used. Pinning of threads/MPI-processes to physical (logical) cores was applied when running 1 to 6 (12) threads/MPI-processes. The grid has 100,592 points (95,344 tetras and 163,625 prisms), i. e. about 17,000 points per physical core.

As Fig. 1 shows, the modified coloring slightly affects the serial performance (no multigrid). The wall-clock time increases by 5–6 % when comparing the shared-memory version (actually, the hybrid one) running single-threaded vs. the base-line/MPI-only TAU with a single domain. Performance measurements using LIKWID [11] indicate that cache utilization might be the reason. When comparing 6 threads for one domain vs. 6 domains (MPI processes), the shared-memory version is neck and neck with the original MPI version, it is off by less than 1%. As expected, the MPI version fails to utilize the 2-way SMT when running with 12 domains. For the shared-memory version, however, a speed-up of 1.117 is observed when using 12 threads (pinned to the 12 logical cores, respectively). The reason might be a better pipeline utilization and latency-hiding effects. Thus, the shared-memory version outperforms the base-line MPI version by about 10% for this particular setting. This clearly demonstrates the potential of the concept proposed (as well as of its implementation in TAU). According to measurements using LIKWID, the shared-memory parallelized TAU using 12 hardware threads obtains a sustained performance of 8.5 GFlop/s (double precision) for the X5670 (recall that CFD on unstructured grids is considered).

The main reason to add shared-memory parallel processing of domains to TAU, however, was to extend TAU's scalability, i. e., to use more cores more effectively. And indeed, the shared-memory parallelization enables us to effectively utilize more cores. The test depicted in Fig. 2 was run on the $C^2A^2S^2E$ cluster located at the Braunschweig site of the German Aerospace Center, which comprises 648 compute nodes, each with two Intel X5670, connected via QDR Infiniband. The grid used has 13 mio points. Unfortunately, the 2-way SMT of the CPUs is disabled, affecting the performance of the multi-threading, cf. above. The hybrid version of TAU uses one domain (MPI process) per socket, whereas

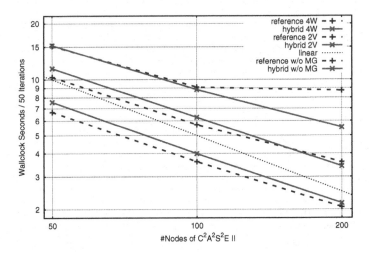

Fig. 2. Strong scaling of TAU. Reference, i. e. MPI only, in blue dashes vs. hybrid in solid red. Grid with 13 mio points on a 2-socket 6-core Westmere (X5670, 2-way SMT disabled) system with QDR Infiniband interconnect (50% blocking). Wall-clock time for 50 iterations vs. #nodes used. 3-stage explict Runge/Kutta with 4W multigrid (top), 2V (middle), and no multigrid (bottom).

for the base-line TAU we have one domain per core, i. e. 12 domains per node. As Fig. 2 shows, the hybrid parallelization significantly increases TAU's scalability – at least when using 4W multigrid, which is preferable practice to do. (The improved scalability can be noticed already for 2V.) For the base-line TAU, for 4W multigrid scalability flattens off at 100 nodes (1200 cores): Using 200 nodes (2400 cores) instead does not result in a noticeable speed-up. In contrast, the hybrid parallel version scales well: The speed-up obtained when using 200 nodes instead of 100 nodes is almost as large as for 50 nodes to 100 nodes. In short words, the hybrid can run 50 iterations in 5.5 seconds, whereas the MPI-only version needs 8.9 seconds – a speed-up of over 1.6 in wall-clock time. Unfortunately, 2-way SMT is disabled, and besides that, more than 200 nodes of the HPC cluster could not be acquired to max out the speed-up attainable for this real-world CFD problem. Nevertheless, this test clearly proves that the hybrid-parallel TAU scales significantly better than its base-line using MPI only.

6   Conclusion and Outlook

As shown by the first, preliminary results described in the preceding section, the concept proposed for shared-memory parallelization of grid-based CFD solvers works well – at least its implementation in the TAU solver considered. Naturally, further tests are needed, in particular for other CPUs. We plan tests on AMD's Magny Cours and on IBM's Power7. It will be interesting to see how these relate to Intel's Westmere considered in the tests presented here.

Even though the implementation here was done for the TAU CFD solver, the concept is not TAU-specific, but rather can be applied to a large number of applications. For example for a stencil based code $A[i] = B[i-1]+B[i+1]$, the concept of dependencies among subgraphs translates to dependencies of thread chunks in OpenMP-parallel FOR loops. Whether or not $A[i]$ of the thread chunk can be calculated, for example, simply depends on whether or not the neighbouring index elements $B[i-1]$ and $B[i+1]$ have already been computed (temporal data dependency). We are currently porting this thread-pool-based parallelization for a stencil-based block-structured CFD turbo machinery code, c. f. [10].

References

1. Alrutz, T.: Investigation of the parallel performance of the unstructured DLR-TAU-code on distributed computing systems. In: Deane, E. (ed.) Parallel Computational Fluid Dynamics, pp. 509–516. Elsevier, Amsterdam (2005)
2. Alrutz, T., Simmendinger, C., Gerhold, T.: Efficiency enhancement of an unstructured CFD-code on distributed computing systems. In: Proc. ParCFD (2009)
3. Devine, K., Boman, E., Riesen, L., Catalyurek, U., Chevalier, C.: Getting started with Zoltan: A short tutorial. In: Proc. Dagstuhl Seminar Combinatorial Scientific Computing, Also Sandia National Labs Tech Report SAND2009-0578C (2009)
4. Hendrickson, B., Leland, R.: A multilevel algorithm for partitioning graphs. In: Proc. 1995 ACM/IEEE Conference on Supercomputing (CDROM), ACM, New York (1995)
5. Karypis, G., Kumar, V.: A fast and high quality multilevel scheme for partitioning irregular graphs. SIAM Journal on Scientific Comp. 20, 359–392 (1998)
6. Kroll, N., Fassbender, J.K. (eds.): MEGAFLOW — Numerical Flow Simulation for Aircraft Design Results of the second phase of the German CFD initiative MEGAFLOW presented during its closing symposium at DLR, Braunschweig, Germany, December 10-11. Notes on Numerical Fluid Mechanics and Multidisciplinary Design, vol. 89. Springer, Heidelberg (2005)
7. Marjanović, V., Labarta, J., Ayguadé, E., Valero, M.: Overlapping communication and computation by using a hybrid MPI/SMPSs approach. In: Proc. 24th ACM Int'l Conference on Supercomputing, pp. 5–16 (2010)
8. Mavripilis, D.: Parallel performance investigation of an ustructured mesh Navier-Stokes solver. The Int'l Journal of High Performance Comp. 2(16), 395–407 (2002)
9. Planas, J., Badia, R., Ayguadé, E., Labarta, J.: Hierarchical task-based programming with StarSs. Int. J. High Perform. Comput. Appl. 23, 284–299 (2009)
10. Simmendinger, C., Kügeler, E.: Hybrid parallelization of a turbomachinery CFD code: performance enhancements on multicore architectures. In: Proc. ECCOMAS-CFD (2010)
11. Treibig, J., Hager, G., Wellein, G.: LIKWID: A lightweight performance-oriented tool suite for x86 multicore environments. CoRR, abs/1004.4431 (2010)

A Fully Empirical Autotuned Dense QR Factorization for Multicore Architectures

Emmanuel Agullo[1], Jack Dongarra[2], Rajib Nath[2], and Stanimire Tomov[2]

[1] LaBRI and INRIA Bordeaux Sud Ouest
[2] University of Tennessee

Abstract. Tuning numerical libraries has become more difficult over time, as systems get more sophisticated. In particular, modern multicore machines make the behaviour of algorithms hard to forecast and model. In this paper, we tackle the issue of tuning a dense QR factorization on multicore architectures using a fully empirical approach. We exhibit a few strong empirical properties that enable us to efficiently prune the search space. Our method is automatic, fast and reliable. The tuning process is indeed fully performed at install time in less than one hour and ten minutes on five out of seven platforms. We achieve an average performance varying from 97% to 100% of the optimum performance depending on the platform. This work is a basis for autotuning the PLASMA library and enabling easy performance portability across hardware systems.

1 Introduction

The hardware trends have dramatically changed in the last few years. The frequency of the processors has been stabilized or even sometimes slightly decreased whereas the degree of parallelism has increased at an exponential scale. This new hardware paradigm implies that applications must be able to exploit parallelism at that same exponential pace. Applications must also be able to exploit a reduced bandwidth (per core) and a smaller amount of memory (available per core). Numerical libraries, which are a critical component in the stack of high-performance applications, must in particular take advantage of the potential of these new architectures. So long as library developers could depend on ever increasing clock speeds and instruction level parallelism, they could also settle for incremental improvements in the scalability of their algorithms. But to deliver on the promise of tomorrow's petascale systems, library designers must find methods and algorithms that can effectively exploit levels of parallelism that are orders of magnitude greater than most of today's systems offer. Autotuning is therefore a major concern for the whole HPC community and there exist many successful or on-going efforts. The FFTW library [1] uses autotuning techniques to generate optimized libraries for FFT, one of the most important techniques for digital signal processing. Another successful example is the OSKI library [2] for sparse matrix vector products. The PetaBricks [3] library is a general purpose tuning method providing a language to describe the problem to tune. It has several applications ranging from efficient sorting to multigrid optimization. In

E. Jeannot, R. Namyst, and J. Roman (Eds.): Euro-Par 2011, LNCS 6853, Part II, pp. 194–205, 2011.
© Springer-Verlag Berlin Heidelberg 2011

the dense linear algebra community, several projects have tackled this challenge on different hardware architectures. The Automatically Tuned Linear Algebra Software (ATLAS) library [4] aims at achieving high performance on a large range of CPU platforms thanks to empirical tuning techniques performed at install time. On graphic processing units (GPUs), among others, [5] and [6] have proposed efficient approaches. FLAME [7] and PLASMA [8] have been designed to achieve high performance on multicore architectures thanks to tile algorithms (see Section 2.1). The common characteristics of all these approaches are that they need intensive tuning to fully benefit from the potential of the hardware.

Tuning a library consists of finding the parameters that maximize a certain metric (most of the time the performance) on a given environment. In general, the term *parameter* has to be considered in its broad meaning, possibly including a variant of an algorithm. The *search space*, corresponding to the possible set of values of the *tunable parameters* can be very large in practice. Depending on the context, on the purpose and on the complexity of the search space, different approaches may be employed. Vendors can afford dedicated machines for delivering highly tuned libraries and have thus limited constraints in terms of time spent in exploring the search space. On the other side of the spectrum, some libraries such as ATLAS aim at being portable and efficient on a wider range of architectures and cannot afford a virtually unlimited time for tuning. Indeed, empirical tuning is performed at install time and there is thus a trade-off between the time the user accepts to afford to install the library and the quality of the tuning. In that case, the main difficulty consists of efficiently pruning the search space. Of course, once a platform has been tuned, the information can be shared with the community so that it is not necessary to tune again the library, but this is an orthogonal problem which we do not address here. Model-driven tuning may allow one to efficiently prune the search space. Such approaches have been successfully designed on GPU architectures, in the case of matrix vector products [2] or dense linear algebra kernels [5,6]. However, in practice, the robustness of the assumptions on the model strongly depends both on the algorithm to be tuned and on the target architecture. There is no clearly identified trend yet but model-driven approaches seem to be less robust on CPU architectures. For instance, even in the single-core CPU case, basic linear algebra algorithms tend to need more empirical search [4]. Indeed, on CPU-based architectures, there are many parameters that are not under user control and difficult to model (different levels of cache, different cache policies at each level, possible memory contention, impact of translation lookaside buffers (TLB) misses, ...) whereas the current generations of GPU provide more control to the user.

In a previous work, we had tackled the issue of maximizing PLASMA performance in order to compare it against other libraries [9]. We first manually pre-selected a combination of parameters based on the performance of the most compute-intensive kernel. We then tried all these combinations for each considered size of matrix to be factorized. This basic tuning approach achieved high performance but required human intervention to pre-select the parameters and days of run to find optimum performance. In the present paper, not only we now tackle the issue of automatically performing the tuning process but we also present new heuristics that efficiently prune the search space so that the whole tuning process is reduced to one hour or so. We illustrate our discussion with the QR factorization implemented in the PLASMA library, which is representative [9] of all three one-sided factorizations (QR, LU, Cholesky) currently available in PLASMA. Because of the trends expose above, we do *not* rely on a model to tune our library (a detailed motivation based on a cased study can be found in Section 2.3 of our corresponding technical report [10]). Instead, we employ a fully empirical approach and we exhibit few empirical properties that enable us to efficiently prune the search space.

The rest of the paper is organized as follows. Section 2 presents the problem and motivates the outline of our two-step empirical approach (Section 3). Section 4 presents the wide range of hardware platforms used in the experiments to validate our approach. Section 5 describes the first empirical step, consisting of benchmarking the most compute-intensive serial kernels. We propose three new heuristics that automatically pre-select (PS) candidate values for the tunable parameters. Section 6 presents the second empirical step, consisting of benchmarking effective multicore QR factorizations. We propose a new pruning approach, which we call "prune as you go" (PAYG), that enables to further prune the search space and to drastically reduce the whole tuning process. We conclude and present future work directions in Section 7.

2   Problem Description

2.1   Tile QR Factorization

The development of programming models that enforce asynchronous, out of order scheduling of operations is the concept used as the basis for the definition of a scalable yet highly efficient software framework for computational linear algebra applications. In PLASMA, parallelism is no longer hidden inside Basic Linear Algebra Subprograms (BLAS) but is brought to the fore to yield much better performance. We do not present tile algorithms in details (more details can be found [8]) but their principles. The basic idea is to split the initial matrix of order N into $NT \times NT$ smaller square pieces of order NB, called *tiles*. Assuming that NB divides N, the equality $N = NT \times NB$ stands. The algorithms are then represented as a Directed Acyclic Graph (DAG) where nodes represent tasks performed on tiles, either panel factorization or update of a block-column, and edges represent data dependencies among them. More details on tile algorithms can be found [8]. PLASMA currently implements three one-sided (QR, LU, Cholesky) tile factorizations. The DAG of the Cholesky factorization is the least difficult to schedule since there is relatively little work required on the critical path. LU and QR factorizations have exactly the same dependency pattern between the nodes of the DAG, exhibiting much more severe scheduling and numerical (only for LU) constraints than the Cholesky factorization. Therefore, tuning the QR factorization is somehow representative of the work to be done

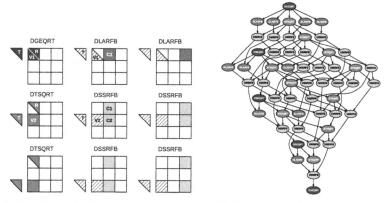

(a) Panel factorization and corresponding (b) DAG when the matrix is split
updates.   in 5 × 5 tiles.

Fig. 1. Tile QR Factorization

for tuning the whole library. In the following, we focus on the QR factorization
of square matrices in double precision statically scheduled in PLASMA.

Similarly to LAPACK which was built using a set of basic subroutines (BLAS),
PLASMA QR factorization is built on top of four serial kernels. Each kernel
indeed aims at being executed sequentially (by a single core) and corresponds
to an operation performed on one or a few tiles. For instance, assuming a 3 ×
3 tile matrix, Figure 1(a) represents the first panel factorization (DGEQRT
and DTSQRT serial kernels [8]) and its corresponding updates (DLARFB and
DSSRFB serial kernels [8]). The corresponding DAG (assuming this time that
the matrix is split in 5 × 5 tiles) is presented in Figure 1(b).

2.2   Tunable Parameters and Objective

The shape of the DAG depends on the number of tiles ($NT \times NT$). For a given
matrix of order N, choosing the tile size NB is equivalent to choosing the number
of tiles (since $N = NB \times NT$). Therefore, NB is a first tunable parameter. A
small value of NB induces a large number of tasks in the DAG and subsequently
enables the parallel processing of many tasks. On the other hand, the serial kernel
applied to the tiles needs a large enough granularity in order to achieve a decent
performance. The choice of NB thus trades off the degree of parallelism with
the efficiency of the serial kernels applied to the tiles. There is a second tunable
parameter, called inner block size (IB). It trades off memory load with extra-flops
due to redundant calculations. With a value $IB = 1$, there are $\frac{4}{3}N^3$ operations
as in standard LAPACK algorithm. On the other hand, if no inner blocking
occurs ($IB = NB$), the resulting extra-flops overhead may represent 25% of the
whole QR factorization (see [8] for more details). The general objective of the
paper is to address the following problem.

Problem 1. *Given a matrix size N and a number of cores ncores, which tile size and internal blocking size (NB-IB combination) do maximize the performance of the tile QR factorization?*

Of course, the performance P we aim at maximizing shall not depend on extra-flops. Therefore, independently of the value of IB, we define $P = \frac{4}{3} \times N^3/t$, where t is the elapsed time of the QR factorization. Note also that we want the decision to be instantaneous when the user requests to factorize a matrix so that the tuning process is to be performed at install time.

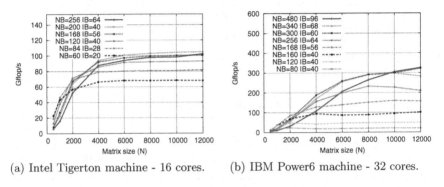

(a) Intel Tigerton machine - 16 cores.   (b) IBM Power6 machine - 32 cores.

Fig. 2. Performance of the PLASMA QR factorization

In a parallel execution of PLASMA, the optimum tile size depends on the matrix size as shown on a 16 cores execution in Figure 2(a). Indeed, if the matrix is small, it needs to be cut in even smaller pieces to provide work to all the 16 cores even if this induces that the serial kernels individually achieve a lower performance. When the matrix size increases, all the cores may evenly share the work using a larger tile size and thus achieving a higher performance. In a nutshell, the optimum tile size both depends on the number of cores and the matrix size, and its choice is critical for performance. Figure 2(b) shows that the impact is even stronger on a 32 cores IBM Power6 machine. The 80-40 combination is optimum on a matrix of order 500 but only achieves 6.3% of the optimum (20.6 Gflop/s against 325.9 Gflop/s) on a matrix of order 12,000.

3   Two-Step Empirical Method

Given the considerations discussed in introduction and further developed in [10], we do *not* propose a model-driven tuning approach. Instead we use a fully empirical method that effectively executes the factorizations on the target platform. However, not all NB-IB combinations can be explored. Indeed, an *exhaustive search* is cumbersome since the search space is huge. For instance, there are more than 1000 possible NB-IB combinations even if we constrain NB to be an even integer lower than 512 (size where the single core compute-intensive kernel reaches its asymptotic performance) and if we impose IB to divide NB. Exploring this search space on a matrix of order $N = 10,000$ with 8 cores on the Intel Core Tigerton machine (described in Section 4) would take several days. Therefore, we need to prune the search space. We propose a two-step approach. In Step 1 (Section 5), we benchmark the most compute-intensive serial kernel. This step is fast since the serial kernels operate on tiles, which are of small granularity ($NB < 512$) compared to the matrices to be factorized ($500 \leq N \leq 10000$ in our study). Thanks to this collected data set and a few well chosen empirical properties, we pre-select (PS) a subset of NB-IB combinations. We propose three heuristics for performing that preliminary pruning automatically. In step 2 (Section 6) we benchmark the effective multicore QR factorizations on the pre-selected set of NB-IB combinations. We furthermore show that further pruning (PAYG) can be performed during this step, drastically reducing the whole tuning process.

4 Experimental Environments

To assess the portability and reliability of our method, we consider seven platforms based Intel EM64T processors, IBM Power and AMD x86_64. **Intel Core Tigerton.** This 16 cores machine is a quad-socket quad-core Xeon E7340 (codename Tigerton) processor, an Intel Core micro-architecture. The processor operates at 2.39 GHz. **Intel Core Clovertown.** This 8 cores server is another machine based on an Intel Core micro-architecture. The machine is composed of two quad-core Xeon X5355 (codename Clovertown) processors, operating at 2.66 GHz. **Intel Core Yorkfield.** This 4 cores desktop is also based on an Intel Core micro-architecture. The machine is composed of one Core 2 Quad Q9300 (codename Yorkfield) processor, operating at 2.5 GHz. **Intel Core Conroe.** This 2 cores desktop is based on an Intel Core micro-architecture too. The machine is composed of one Core 2 Duo E6550 (codename Conroe) processors, operating at 2.33 GHz. **Intel Nehalem.** This 8 cores machine is based on an Intel Nehalem micro-architecture. Instead of having one bank of memory for all processors as in the case of the Intel Core's architecture, each Nehalem processor has its own memory. Nehalem is thus a cache coherent Non Uniform Memory Access (ccNUMA) architecture. Our machine is a dual-socket quad-core Xeon X5570 (codename Gainestown) running at 2.93GHz and up to 3.33 GHz in certain conditions (Intel Turbo Boost technology). The Turbo Boost was activated during our experiments. **AMD Istanbul.** This 48 cores machine is composed of eight hexa-core Opteron 8439 SE (codename Istanbul) processors running at 2.8 GHz. Like the Intel Nehalem, the Istanbul micro-architecture is a ccNUMA architecture. **IBM Power6.** This 32 cores machine is composed of sixteen dual-core IBM Power6 processors running at 4.7 GHz. More details on these platforms can be found in [10]. Note that we do not discuss here the mapping of the tasks onto the cores; this is an orthogonal problem.

5 Step 1: Benchmarking the Most Compute-Intensive Serial Kernel

We explained in Section 2.1 that the tile QR factorization consists of four serial kernels. However, the number of calls to DSSRFB is proportional to NT^3 while the number of calls to the other kernels is only proportional to NT (DGEQRT) or to NT^2 (DTSQRT and DLARFB). Even on small DAGS (see Figure 1(b)), calls to DSSRFB are predominant. Therefore, the performance of this compute-intensive kernel is crucial. DSSRFB's performance also depends on NB-IB. It is thus natural to pre-select NB-IB pairs that allow a good performance of DSS-RFB before benchmarking the QR factorization itself. The practical advantage is that a kernel is applied at the granularity of a tile, which we assume to be bounded by 512 ($NB \leq 512$). Consequently, preliminary benchmarking this serial kernel can be done exhaustively in a reasonable time. *Step 1* thus consists of performing an exhaustive benchmarking of the DSSRFB kernel on all possible NB-IB combinations and then to decide which of these will be kept for further testing in Step 2. Column "Step 1" of Table 1 shows that the total elapsed time for step 1 is acceptable on all the considered architectures (between 16 and 35 minutes). Figure 3(a) shows the resulting set of empirical data collected during step 1 on the Intel Core Tigerton machine. This data set can be pruned a

Table 1. Elapsed time (hh:mm:ss) for Step 1 and Step 2

Machine		Step 1	Step 2		
Architecture	# cores		Heuristic	PS	PSPAYG
Conroe	2	00:24:33	0	14:46:37	03:05:41
			1	09:01:08	00:01:58
			2	07:30:53	**00:34:47**
Yorkfield	4	00:20:57	0	17:40:00	04:48:13
			1	09:30:30	00:05:10
			2	08:01:05	**02:58:37**
Clovertown	8	00:21:44	0	20:08:43	02:56:25
			1	11:06:18	00:13:09
			2	08:52:24	**01:10:53**
Nehalem	8	00:16:29	0	06:20:16	01:51:30
			1	06:20:16	01:51:30
			2	06:20:16	**01:51:30**
Tigerton	16	00:34:18	0	23:29:35	03:15:41
			1	12:22:06	00:08:57
			2	09:54:59	**01:01:06**
Istanbul	48	00:24:23	0	21:09:27	02:53:38
			1	12:25:30	00:11:01
			2	10:04:46	**00:54:51**
Power6	32	00:15:23	0	03:06:05	00:25:07
			1	03:06:05	00:25:07
			2	03:06:05	**00:25:07**

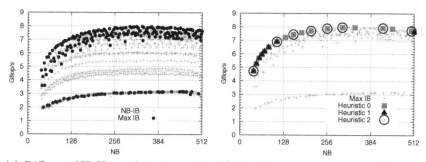

(a) Different NB-IB combinations with a common NB value have the same abscisse; "Max IB" represents the one that achieves the maximum performance among them.

(b) Combinations pre-selected (PS) for each heuristic.

Fig. 3. Performance of the DSSRFB serial kernel depending on the NB-IB combination

first time. Indeed, contrary to NB, which trades off parallelism for kernel performance, IB only affects kernel performance but not parallelism. We can thus perform the following orthogonal optimization:

Property 1 (Orthogonal pruning). *For a given NB value, we can safely pre-select the value of IB that maximizes the kernel performance.*

Applying Property 1 to the data set of Figure 3(a) results in discarding all NB-IB pairs except the ones matching "Max IB", which still represents a large number of combinations. We thus propose and assess three heuristics to further prune the search space. The first considered heuristic is based on the fact that intensive experiments (not reported here) showed the following property.

Property 2 (Convex Hull). *There is consistently an optimum combination on the convex hull of the data set.*

Therefore, **Heuristic 0** consists of pre-selecting the points from the convex hull of the data set (see Figure 3(b)). In general, this approach may still provide too many combinations. Because NB trades off kernel efficiency with parallelism, the gains observed on kernel efficiency shall be considered relatively to the increase of NB itself. Therefore, we implemented **Heuristic 1** that pre-selects the points of the convex hull with a high steepness (or more accurately a point after a segment with a high steepness). The drawback is that all these points tend to be located in the same area as shown in Figure 3(b) corresponding to small values of NB. To correct this deficiency, we consider **Heuristic 2** which first divides the x-axis into iso-segments and pick up the point of maximum steepness on each of these segments (see Figure 3(b) again). Heuristics 1 and 2 are paremetrized to select a maximum of 8 combinations. All three heuristics perform a pre-selection (PS) that will be used as test cases for the second step.

6   Step 2: Benchmarking the Whole QR Factorization

6.1   Discretization and Interpolation

We recall that our objective is to immediately retrieve at execution time the optimum NB-IB combination for the matrix size N and number of cores $ncores$ that the user requests. Of course, N and $ncores$ are not known yet at install time. Therefore, the $(N,ncores)$ space to be benchmarked has to be discretized. We decided to benchmark all the powers of two cores (1, 2, 4, 8, ...) plus the maximum number of cores in case it is not a power of two such as on the AMD Istanbul machine. The motivation comes from empirical observation. Indeed, Figures 4(a) and 4(b) show that the optimum NB-IB combination can be finely interpolated with such a distribution. We discretized more regularly the space on N because the choice of the optimum pair is much more sensible to that dimension (see figures 2(a) and 2(b)). We benchmarked N=500, 1000, 2000, 4000, 6000, 8000, 10000[1]. Each run is performed 6 times to attenuate potential perturbations. When the user requests the factorization of parameters that have not been tuned (for instance N=1800 and ncores=5) we simply interpolate by selecting the parameters of the closest configuration benchmarked at install time (N=2000 and ncores=4 in that case).

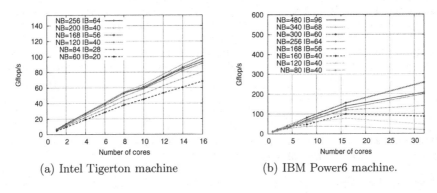

(a) Intel Tigerton machine   (b) IBM Power6 machine.

Fig. 4. Strong scalability - $N = 6000$

6.2   Impact of the Pre-selection on the Elapsed Time of Step 2

Column PS (pre-selection) in Table 1 shows the impact of the heuristics on the time required for benchmarking step 2. Clearly Heuristic 0 induces a very long step 2 (up to 1 day). Heuristic 1 and 2 induce a lower time for step 2 (about 10 hours) but that may be still not acceptable for many users.

---

[1] Except on the IBM Power6 machine where N=10000 was not benchmarked.

6.3   Prune as You Go (PSPAYG)

To further shorten step 2, we can perform complementary pruning on the fly. Indeed, Figures 2(a) and 2(b) show the following property.

Property 3 (Monotony). *Let us denote by $P(NB_1, N)$ and $P(NB_2, N)$ the performances obtained on a matrix of order N with tile sizes NB_1 and NB_2, respectively. If $P(NB_1, N) > P(NB_2, N)$ and $NB_1 > NB_2$, then $P(NB_1, N') > P(NB_2, N')$ for any $N' > N$.*

We perform step 2 in increasing order of N. After having benchmarked the current set of NB-IB combinations on a matrix of order N, we identify all the couples (NB_1, NB_2) that satisfy Property 3 and we remove from the current subset the NB-IB pair in which NB_2 is involved. Indeed, according to Property 3, it would lead to a lower performance than NB_1 on larger values of N which are going to be explored next. We denote this strategy by "PSPAYG" (pre-selection and prune as you go). Column PSPAYG in Table 1 shows that the time for step 2 is dramatically improved with this technique. Indeed, the number of pairs to explore decreases when N increases, that is, when benchmark is costly. For heuristic 2 (values in bold in Table 1), the time required for step 2 is reduced by a factor greater than 10 in two cases (Intel Core Conroe and AMD Istanbul machines).

6.4   Reliability

We employed the following methodology to assess the reliability of the different tuning approaches. We first executed all the discussed approaches on all the platforms with the discretization of the (N,*ncores*) space proposed in Section 6.1. We then picked up between 8 and 16 (N,*ncores*) combinations such that half of them were part of the discretized space (for instance $N = 6000$ and *ncores* $= 32$) and the other half were not part of it (for instance $N = 4200$ and *ncores* $= 30$) so that the reliability of the interpolation is also taken into account. For each combination we performed an (almost) exhaustive search for reference. Table 2 provides a synthesis of the results. Heuristic 2 coupled with the PSPAYG approach is very efficient since it achieves a high proportion of the performance that would be obtained with an exhaustive search (values in bold). The worst case occurs on the Istanbul machine, with an average relative performance of 97.1% (Column "avg"). However, even on that platform, the optimum NB-IB combination was found in seven cases out of sixteen tests (Column "optimum").

Column $\frac{PSAYG}{PS}$ allows to specifically assess the impact of the "prune as you go" method since they compare the average performance obtained with PS-PAYG (where pairs can be discarded during step 2 according to Property 3) compared to PS (where no pair is discarded during step 2). The result is clear: pruning during step 2 according to Property 3 does not hurt performance ($\frac{|PS-PSPAYG|}{PS} < 0.3\%$), showing that Property 3 is strongly reliable. Finally, note that on (N,*ncores*) combinations part of the discretized space, PSPAYG cannot achieve a higher performance than PS since all NB-IB combinations

Table 2. Average performance achieved with a "pre-selection" (PS) method or a "pre-selection and prune as you go" (PSPAYG) method, based on different heuristics (H) applied at step 1. The performance is presented as a proportion of the exhaustive search (ES) or of the prunes search (PS). The column "optimum" indicates the number of times the optimum combination (with respect to the reference method) was found among the number of tests performed.

Machine	H	$\frac{PS}{ES}(\%)$		$\frac{PSPAYG}{ES}(\%)$		$\frac{PSPAYG}{PS}(\%)$	
		avg	optimum	avg	optimum	avg	optimum
Conroe	0	99.67	6/8	99.67	6/8	100	8/8
	1	95.28	0/8	95.28	0/8	100	8/8
	2	99.54	5/8	**99.54**	5/8	100	8/8
Yorkfield	0	98.63	6/12	98.63	6/12	100	12/12
	1	91.53	0/12	91.59	0/12	100.07	10/12
	2	98.63	6/12	**98.63**	6/12	100	12/12
Clovertown	0	98.59	8/16	98.35	7/16	99.76	15/16
	1	91.83	0/16	91.83	0/16	100	16/16
	2	98.49	9/16	**98.25**	8/16	99.76	15/16
Nehalem	0	98.6	8/16	98.9	8/16	100.33	16/16
	1	98.6	8/16	98.9	8/16	100.33	16/16
	2	98.6	8/16	**98.9**	8/16	100.33	16/16
Tigerton	0	97.36	8/16	97.54	5/16	100.21	12/16
	1	91.61	0/16	91.61	0/16	100	16/16
	2	97.51	8/16	**97.79**	7/16	100.31	15/16
Istanbul	0	97.17	7/16	97.17	7/16	100	16/16
	1	94.12	2/16	94.12	2/16	100	16/16
	2	97.23	7/16	**97.1**	7/16	99.87	15/16
Power 6	0	100	16/16	100	16/16	100	16/16
	1	100	16/16	100	16/16	100	16/16
	2	100	16/16	**100**	16/16	100	16/16

tested with PSPAYG are also tested with PS. However, PSPAYG can achieve a higher performance if (N,*ncores*) was not part of the discretized space because of the interpolation. This is why cases where $\frac{PSPAYG}{PS} > 100\%$ may be observed.

7   Conclusion and Future Work

We have presented a new fully empirical autotuned method for tuning dense linear algebra libraries on multicore architectures. Thanks to three strong empirical properties, we showed that the search space can be efficiently pruned. Our tuning process is automatic, fast (less than one hour and ten minutes on five out of seven platforms) and reliable (average performance varying from 97% to 100% of the optimum). We plan to extend our work to the case of non square matrices and to other factorizations. We will then extend our work to the case of hybrid multicore platforms enhanced with multiple GPU accelerators for which heterogeneity will have to be taken into account.

Acknowledgment. The authors would like to thank Jakub Kurzak, Greg Henry and Clint Whaley for their constructive discussions.

References

1. Frigo, M., Johnson, S.: FFTW: An adaptive software architecture for the FFT. In: Proc. 1998 IEEE Intl. Conf. Acoustics Speech and Signal Processing, vol. 3, pp. 1381–1384. IEEE, Los Alamitos (1998)
2. Choi, J.W., Singh, A., Vuduc, R.W.: Model-driven autotuning of sparse matrix-vector multiply on GPUs. In: Proc. ACM SIGPLAN Symp. Principles and Practice of Parallel Programming (PPoPP), Bangalore, India (January 2010)
3. Ansel, J., Chan, C., Wong, Y.L., Olszewski, M., Zhao, Q., Edelman, A., Amarasinghe, S.: Petabricks: A language and compiler for algorithmic choice. In: ACM SIGPLAN Conference on Programming Language Design and Implementation, Dublin, Ireland (June 2009)
4. Clint Whaley, R., Petitet, A., Dongarra, J.J.: Automated empirical optimizations of software and the atlas project. Parallel Computing 27(1-2), 3–35 (2001)
5. Volkov, V., Demmel, J.W.: Benchmarking gpus to tune dense linear algebra. In: SC 2008: Proceedings of the ACM/IEEE Conference on Supercomputing, pp. 1–11. IEEE Press, Piscataway (2008)
6. Tomov, S., Nath, R., Ltaief, H., Dongarra, J.: Dense linear algebra solvers for multicore with gpu accelerators. Accepted for publication at HIPS 2010 (2010)
7. Quintana-Ortí, G., Quintana-Ortí, E., van de Geijn, R., Van Zee, F., Chan, E.: Programming matrix algorithms-by-blocks for thread-level parallelism. ACM Trans. Math. Softw. 36(3) (2009)
8. Buttari, A., Langou, J., Kurzak, J., Dongarra, J.: A class of parallel tiled linear algebra algorithms for multicore architectures. Parallel Computing 35(1), 38–53 (2009)
9. Agullo, E., Hadri, B., Ltaief, H., Dongarra, J.: Comparative study of one-sided factorizations with multiple software packages on multi-core hardware. In: 2009 International Conference for High Performance Computing, Networking, Storage, and Analysis (SC 2009) (2009)
10. Agullo, E., Dongarra, J., Nath, R., Tomov, S.: A Fully Empirical Autotuned Dense QR Factorization For Multicore Architectures. Research Report 7526, INRIA (Febuary 2011)

Parallelizing a Real-Time Physics Engine Using Transactional Memory

Jaswanth Sreeram and Santosh Pande

College of Computing, Georgia Institute of Technology
jaswanth@gatech.edu, santosh@cc.gatech.edu

Abstract. The simulation of the dynamics and kinematics of solid bodies is an important problem in a wide variety of fields in computing ranging from animation and interactive environments to scientific simulations. While rigid body simulation has a significant amount of potential parallelism, efficiently synchronizing irregular accesses to the large amount of mutable shared data in such programs remains a hurdle. There has been a significant amount of interest in transactional memory systems for their potential to alleviate some of the problems associated with fine-grained locking and more broadly for writing correct and efficient parallel programs. While results so far are promising, the effectiveness of TM systems has so far been predominantly evaluated on small benchmarks and kernels.

In this paper we present our experiences in parallelizing ODE, a real-time physics engine that is widely used in commercial and open source games. Rigid body simulation in ODE consists of two main phases that are amenable to effective coarse-grained parallelization and which are also suitable for using transactions to orchestrate shared data synchronization. We found ODE to be a good candidate for applying parallelism and transactions to - it is a large real world application, there is a large amount of potential parallelism, it exhibits irregular access patterns and the amount of contention may vary at runtime. We present an experimental evaluation of our implementation of the parallel transactional ODE engine that shows speedups of up to 1.27x relative to the sequential version.

1 Introduction

The trend towards multi-core and many core processors is pushing more and more applications towards parallelism and is spurring extensive research in concurrent programming models and languages. The potential performance benefits of extracting parallelism and the complexity of specifying efficient concurrent programs are both significant.

Applications that simulate the dynamics and kinematics of rigid bodies or physics engines are examples of applications that are known to have significant amount of parallelism but it this parallelism is often difficult to exploit owing to their complexity. Physics engines that support real-time interactive applications such as games are growing rapidly in sophistication both in their feature-set as

E. Jeannot, R. Namyst, and J. Roman (Eds.): Euro-Par 2011, LNCS 6853, Part II, pp. 206–223, 2011.

well as their design. The popular Unreal 3 game engine is known to consist of over 300,000 lines of code and as described in [12], parallelizing parts of it was a challenging endeavour. Traditional approaches to efficient shared data synchronization such as fine-grained locking are often impractical owing to the size and complexity of the application and the large amounts of hierarchical mutable shared state. On the other hand coarse-grained locking has been found to be too inefficient for maintaining the highly interactive nature of these applications. Further, using fine-grained locks in such applications extracts a significant price in terms of programmer productivity - a factor that deeply affects their commercial development cycle.

Researchers have suggested developing parallel programs in this domain using *transactional memory* to manage accesses to shared state [12]. Software or Hardware Transactional memory has been proposed as a relatively programmer-friendly way to achieve atomicity and orchestrate concurrent accesses to shared data. In this model programmers annotate their programs by demarcating atomic sections (using a keyword such as "atomic" in a language-based TM implementation or specific function calls to a library based TM). The programmer also annotates accesses to shared data within these sections. At run time, these atomic sections are executed speculatively and the TM system continuously keeps track of the set of memory locations each transaction accesses and detects conflicts. This conflict detection step involves checking if a value speculatively read or written has been updated by another concurrent transaction. If so then one of the two speculatively executed transactions is aborted.

Software Transactional Memory systems reduce the burden of writing correct parallel programs by allowing the programmer to focus simply on specifying where atomicity is needed instead of how it is achieved. Further, the benefits of TMs are most apparent when a) the rate of real data sharing conflicts at run time is quite low i.e., most of the concurrent accesses to shared data are disjoint and b) using fine grain locking is difficult either due to the irregularity of the access patterns or the data structures. There has been a substantial amount of interest in hardware and software transactional memory systems recently. However in spite of this recent interest and the significant amount of research most of the studies investigating the use and optimization of these systems have been limited to smaller benchmarks and suites containing small to moderate sized programs [3,4,8,9,6]. Previous studies [18,7] have noted the lack of large real-world applications that use transactional memory without which an effective evaluation of the effectiveness of TM systems in realistic settings becomes difficult.

In this paper we present our experiences in parallelizing and using transactions in the Open Dynamics Engine (ODE), a single-threaded real-time rigid body physics engine [2]. It consists of roughly 71000 lines of C/C++ code with an additional 3000 lines of code for drawing/rendering. In [7] the authors outline a set of characteristics that are desirable in an application using TM. Briefly they are:

1. Large amounts of potential parallelism: As we show in the Section 3, there is a significant amount of data parallelism in the two principal stages in an ODE simulation.
2. Difficult to fine-grain parallelize: ODE exhibits irregular access patterns many structures that can be accessed concurrently.
3. Based on a real-world application: ODE is used in hundreds of open-source and commercial games [2].
4. Several types of transactions: The parallel version of ODE we describe in the rest of this paper has critical sections that access varying amount of shared data, have sizes that vary widely and the amount of contention between them changes during execution.

We started with the single-threaded implementation of ODE and found that the two longest running stages in a time step could be parallelized effectively. While we found many opportunities for fine-grained parallelization at the level of loops in constraint solvers, we choose to focus on a coarser-grained work offloading in order to amortize the runtime overheads. We then modified this parallel program by annotating critical sections and accesses to shared data with calls to an STM library. Our modifications added roughly 4000 lines of code in the ODE.

The rest of this paper is organized as follows: Section 2 presents an overview of collision detection and dynamics simulation in ODE. Section 3 describes the parallelization scheme for ODE and the usage of transactions for atomicity. Section 4 briefly discusses a few issues pertaining to the parallelization. Section 5 presents our experimental evaluation of the application. Related work is presented in Section 6 and Section 7 concludes the paper.

Algorithm 1. Overview of a time step in ODE

```
 1: Create world; add bodies
 2: Add joints; set parameters
 3: Create collision geometry objects
 4: Create joint group for contact points
 5: // Simloop
 6: while (!pause && time < MAX_TIME) do
 7:    Detect collisions; create joints
 8:    Step world
 9:    Clear joint group
10:    time++
11: end while
```

---

2   ODE Overview

At a high level ODE consists of two main components: a collision detection engine and a dynamics simulation engine. Any simulation involving multiple bodies typically uses both these engines. The sequence of events in a typical time step is shown in Algorithm 1. The goal is typically to simulate the movement of one or more bodies in a *world*. Before simulation begins the world and the bodies in it are created and any initial joints are attached. A *contact group* is created for storing the contact joints produced during each collision. During each time step in the simulation loop in line 6, *collision detection* is first carried out which creates contact points/joints which are used in "stepping" or *dynamics simulation* for each body in the world (line 8). After this step all the contact joints are removed from the contact group and the simulation proceeds to the next time step.

2.1   Collision Detection

The collision detection (CD) engine is responsible for finding which bodies in the simulation touch each other and computing the *contact points* for them given the shape and the current orientation of each body in the scene. A simple algorithm would simply test whether each of the "n" bodies collides with any other body in the scene but for large scenes this $O(n^2)$ algorithm does not scale. One solution to this problem is to divide the scene into a number of *spaces* and assign each body to a space. Additionally, the spaces may be hierarchical - a space may contain other spaces. Now, collision detection proceeds in two phases called *broadphase* and *narrowphase* which are as follows:

1. **Broadphase:** In this phase each space $S_1(\in S)$ is tested for collision with each of the other spaces. If S_1 is found to be potentially colliding with space $S_2 \in S$ then S_1 is tested for collision with each of the spaces or bodies inside S_2.
2. **Narrowphase:** In this phase individual bodies that have found to be potentially colliding in the broadphase are tested to check if they are actually colliding.

This approach is similar to the hierarchical bounding box approach used for fast ray tracing and many other problems. If a pair of bodies are found to be colliding the collision detection algorithm finds the points where these bodies touch each other. Each of these contact points specifies a position in space, a surface normal vector and a penetration depth. The contact points are then used to create a joint between these two bodies which imposes constraints on how the bodies may move with respect to each other. In addition to links to the bodies each of these *contact joints* connect, they also have attributes like surface friction and softness which are used in simulating motion in the next step.

By the end of the collision detection step all the contact points in the scene have been identified and the appropriate joints between bodies made. In the dynamics simulation step below, the new positions and orientations of all the bodies in the scene are computed.

2.2   Dynamics Simulation

The joint information computed in the CD step above represents constraints on the movement of the bodies in the scene (for example due to another body in way or due to a hinge). The Dynamics Simulation (DS) engine takes this joint information and the force vectors and computes the new orientation and position for all the active bodies in the scene. It does this by solving a Linear Complementarity Problem (LCP) using a successive over-relaxation (SOR) form of the Gauss-Seidel method. The main output produced in the DS stage are the linear and angular velocities of each body in the scene. These velocities are then used to update the position and orientation of the bodies.

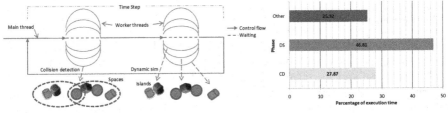

(a) Overview of parallel ODE

(b) Distribution of execution time among phases in single-threaded execution

Fig. 1. ODE overview

3   Parallel Transactional ODE

The broad approach to parallelizing ODE is illustrated in Figure 1a. At a high-level parallelism is achieved by offloading coarse-grained tasks in the CD and DS stages on the main thread onto concurrent worker threads that use transactions to synchronize shared data accesses.

3.1   Global Thread Pool

In order to avoid the overheads of creating and destroying threads, before the simulation begins the main thread creates a global thread pool consisting of t POSIX threads that are initialized to be in a *conditional wait* state. Additionally the pool contains a t-wide status vector that describes each thread's status, a set CM of t mutexes and a set CV of t condition variables. During the course of the simulation the main thread offloads work to a worker thread by scanning the pool for an idle thread, marshalling the arguments and setting the condition variable for the thread to start execution.

3.2   Parallel Collision Detection Using Spatial Decomposition

Detecting collisions between bodies in the world is inherently parallel and indeed the naive $O(n^2)$ algorithm described above can be parallelized by simply performing collision detection for each pair of bodies in a separate thread. However a better scheme would involve a more coarse-grained distribution of work in which a space or a pair of spaces in the world is handled by a separate thread. Before the parallel CD stage starts each of the bodies in the world is assigned to a space S_i. Let S represent the set of spaces in the world i.e., $S = \bigcup_i S_i$. Detecting collisions among bodies contained in the same space can be done independently of (and in parallel with) other spaces. Additionally, detecting collisions between each distinct pair of spaces can be done in parallel. The broadphase stage of parallel CD proceeds as follows.

1. The main thread picks an unprocessed pair of spaces S_1 and S_2 and signals an idle thread $t_{1,2}$ in the thread pool to perform collision detection on them. Additionally the main thread signals idle threads t_1 and t_2 to perform collision detection on bodies contained withing S_1 and S_2 respectively.
2. Thread $t_{1,2}$ first checks if spaces S_1 and S_2 can potentially be touching. It does this by checking if there is an overlap between their axis aligned bounding boxes (AABBs). As described above, the AABB for a space informally is simply the smallest axis aligned box that can completely contain all the bodies in that space. If there is overlap between the AABBs of the two spaces then $t_{1,2}$ has to check if there exist bodies b_1 and b_2 such that $b_1 \in S_1$, $b_2 \in S_2$ and the AABBs of b_1 and b_2 overlap. If they do, b_1 and b_2 are potentially colliding and the narrowphase later on checks if they are actually colliding. After this step thread $t_{1,2}$ marks the space pair $(1, 2)$ as processed.
3. Thread t_1 finds bodies in S_1 that are potentially colliding. This is done again by analyzing the AABBs of bodies in S_1. Thread t_2 does the same for bodies in S_2. Spaces S_1 and S_2 are then marked as processed by their respective threads.
4. All the potentially colliding bodies found above are checked to find actual collisions in the narrowphase. If a pair of bodies do actually collide the appropriate thread computes contact points for the collision (using the positions and orientations of the bodies). These contact points are used by the thread to create contact joints between the pair of bodies.

This approach to assigning collision spaces to threads makes $\left(\binom{n}{2} + n\right)$ thread offloads where n is the number of spaces. An alternate approach is to assign a single thread t_i to each space S_i. This thread computes the collisions for objects within S_i and then performs broadphase and narrowphase collision checking between S_i and all S_j such that $i < j \leq n$. This approach activates only n threads but is likely to be more efficient than the former only if the spaces are *well balanced*. That is all the spaces at each level in the containment-hierarchy contain approximately the same number of subspaces or bodies. Consider a deep space hierarchy with space S_{root} as the root space that contains all other spaces S_i and bodies. In the alternate approach the thread t_{root} has to process collisions between S_{root} and all other spaces/bodies. By definition, S_{root} would collide with every other contained body or space. Thus in general this approach would result in a schedule where threads processing spaces that are high-up in the hierarchy are heavily loaded while threads assigned to spaces that are lower are lightly loaded. However in the former approach, each space-space pair can be processed in parallel - each pair $\{S_{root}, S_j\}$ for $1 < j \leq n$ can be processed in parallel thereby reducing the overall imbalance.

Shared data

Although the collision detection stage described above is quite parallel the participating threads make concurrent accesses to several shared data structures that must be synchronized. The important data structures that are accessed concurrently are the Global Memory buffer that is used to satifsy allocation requests, the joint, contacts and body lists and attributes pertaining to the state of the world and its parameters including the number of active bodies and joints.

We use an STM library to orchestrate calls to these shared data. STM enables efficient disjoint access parallelism - two concurrent threads that do not access the same memory word can execute in parallel. This is in contrast to using more pessimistic coarse-grained locking in which a thread that *could* access/modify shared data (being accessed by some other thread) has to wait to acquire the appropriate lock regardless of whether an actual access takes place or not. The STM library we used is based on the well-known TL2 system described in [1]. In other works such as [18] the authors used an automated compiler-based STM system in which the programmer simply annotates atomic sections and the compiler automatically annotates accesses occurring inside them with calls to the TM runtime. Instead we used the TL2 library based system which means the programmer has to manually identify atomic sections and accesses occurring within them. This choice is because of two reasons. Firstly the TL2 STM has been shown to have lower overheads than other comparable STM systems in several studies [1]. This is especially important since we are using it in the context of a real-time interactive application. Secondly using a library STM offers better flexibility and we are in some cases able to reduce TM overheads by using domain knowledge to elide TM tracking of specific shared data.

3.3   Parallel Island Processing

Island Formation

After the joints in the world have been determined in the CD step the next stage is dynamics simulation or simulating the motion of the bodies under the constraints specified by their shapes and the joints found. This uses the SOR-LCP formulation mentioned above and finding solutions to this problem involves several nested loops that are compute-intensive. However, parallelizing these loops with the work-loading model would result in a very fine-grained parallel system (which is unlikely to scale well [11] and the overheads of synchronization and thread control would likely eliminate any speedups gained. Therefore we choose a more coarse-grained approach in which several connected bodies are processed independently and in parallel with other bodies. All the bodies in the world are assigned to "islands". An island is simply a group of bodies in which each body is connected to one or more bodies in the same island through one or more joints. These islands therefore represent sets of connected bodies that can be processed separately since simulating a body (with some number of joints) does not require accesses to bodies in other islands. In parallel dynamics simulation the main thread first forms islands. The algorithm iterates over all the bodies in the world adding bodies to islands if they haven not already been added. A body is said to be *tagged* when it has been added to some island. Given a body b, the algorithm first finds the untagged neighbors of b and adds adds them to a stack. The algorithm then pops and examines each body in this stack, adding their untagged neighbors. The joints between all these neighbors are collected in a joint list. When the stack is empty, the joint and body lists represent an island of connected bodies that can be processed. The main thread then moves on to the next untagged body in the world in the outermost loop.

Island Processing

While island formation is sequential, processing the bodies in each island can be performed independently of other islands. Immediately after an island is formed, the main thread uses heuristics to check whether the island is suitable to be offloaded to a worker thread. If so, the main thread marshals pointers to body and joint lists for that island, finds an idle thread in the global thread pool and signals it to start processing that island. The main thread then resumes with finding the next island. If the island formed is deemed to be not suitable for offloading, the main thread can process that island itself before continuing with further island formation. A variety of heuristics can be used to decide whether a particular island should be processed in a worker thread or if it should be processed in the main thread. Our system uses a threshold on the number of bodies and number of joints in the island. Because of the overhead of offloading computation to worker threads, if there are very few bodies or joints in the islands then it may be more efficient to process them in the main thread instead. Additionally, if an island is found to have fewer bodies than needed to offload processing to a worker thread, the main thread checks whether the next island in combination with the previous one meets the threshold. If so both these islands are offloaded together to a single worker thread. The main thread chooses and signals a thread from the global thread pool to start island processing. The worker thread uses the body and joint lists and the force vectors to set up a system of equations representing the constraints on the set of bodies and finds. We refer the reader to [2] for details of the constraint solver that is used for finding solutions. The island processing step finishes after computing new values for linear and angular velocity, position and orientation quaternion for each body in the island and atomically updating body with these values.

3.4   Phase Separation

During body simulation in ODE, all the contact joints are typically computed first before dynamics simulation can start since the latter needs these joints to be able to solve the constraint satisfaction problem. In the sequential case this was guaranteed since the dynamics simulation is always preceded by collision detection in each time step. However in the parallel case, the main thread can simply offload the collision detection to worker threads and enter the dynamics simulation step while some of the worker threads are still computing the joints. Therefore there needs to be a thread barrier between the collision detection and dynamics simulation in simulating each time step. The control flow for the main thread is very different from that of the worker threads in our parallelization scheme. Therefore instead of a normal thread barrier that is released when all threads reach a certain program point, in our scheme we use a thread *join point* in the main thread. A *join point* is simply a program point at which the main thread waits for all the active worker threads to finish executing. When the main thread enters the join point, it repeatedly polls the *status* vector and yields its processor if there is at least one worker thread performing collision detection. Note that no lock acquisition is necessary for this polling as the worker thread only ever writes one type of value into its slot in the status vector - the value representing its *IDLE* state. After all worker threads have finished collision detection and have entered the *IDLE* state, the join point is met and the main thread is released. Although it limits parallelism, this join is necessary due to the producer-consumer relation between the stages for joints - the island formation algorithm requires contact joints for all bodies in the world to have been computed.

After island processing has generated new positions and orientations for all the bodies in the world, these new values are used in the collision detection step in the next stage. But after the main thread offloads island processing to worker threads, it could enter the collision detection stage in the next time step while the new body attributes are being computed. This could result in the collision detection stage reading stale position/orientation values for some bodies - the bodies which island processing has not yet updated. Therefore in addition to the dependence between the collision and dynamics simulation steps within a time step there is also a dependence between the dynamics simulation in one time step and the collision detection in the next. We therefore enforce a join point at the end of each time step to make sure that all bodies have been updated. This join point is implemented like the one described above - the main thread simply polls the status vector until all the island processing worker threads have finished.

To see why this join point is needed consider the case of a worker thread with transaction Tx_1 updating the position quaternion R_b of a body b during island processing in time step n. Assume the main thread is allowed to enter the next time step where it offloads collision detection to a worker thread and transaction Tx_1 is reading R_b. If Tx_1 commits after Tx_2 starts but before it finishes then Tx_2 is aborted when the conflict for R_b is detected and the join point would *not* have been necessary. However if Tx_2 commits before Tx_1 does, then Tx_1 is aborted and retried. Thus Tx_1 eventually produces the new value for R_b but Tx_2 ends up using the older value and this phenomenon can adversely affect simulation integrity. Now lets say add a *"last_updated"* field to each body which is updated in Tx_1. So if Tx_2 finds this field for b to be n then Tx_1 is guaranteed to have committed and Tx_2 can read the latest R_b. However if this value is $n-1$ then Tx_2 can be forced to abort to until Tx_1 commit. It may therefore be possible to eliminate the join point at the end of each time step by forcing transactions

reading stale values in the next time step to abort. This could potentially allow
more parallelism by allowing the threads with transactions that only read already
updated bodies to proceed instead of waiting for the other threads.

3.5   Feedback between Phases

A critical factor influencing the amount of effective parallelism achieved during
the CD phase is the assignment of bodies to spaces. Spatial (in the geometric
sense) assignment methods are popularly used in many dynamics simulation
algorithms. In such methods, objects that are geometrically proximal to each
other are assigned to the same space in the containment hierarchy. An important
concern with this approach is that the scene being modelled may evolve to a state
where most of the objects are contained in one or a few spaces. This may in
turn result in the *thread imbalance* problem discussed in Section 3.2. To address
this such methods usually propose a space reassignment step that is invoked
occasionally and reassigns objects such that the threads are once again balanced.
We use a novel method to perform space assignment that reduces imbalance.
Our method is based in the observation that the DS phase in a timestep already
computes entities (islands) of geometrically close bodies - in fact the bodies in
each of these islands are touching each other! After the dynamics simulation
step, the bodies in these islands have been moved so they may not be touching
anymore. However if the simulation timestep is small then in the CD phase in
next iteration these bodies are either still touching each other or are close to
each other. Hence the CD phase bootstraps spaces with clusters of such islands
before performing broadphase checks on these spaces with the result that there
are fewer narrowphase checks to be performed on the contained bodies.

4   Issues

In this section we will discuss a few issues pertaining to using transactions for
synchronization in parallel ODE.

4.1   Conditional Synchronization

Our implementation of parallel ODE makes extensive use of conditional
synchronization for signalling between threads. Indeed constructs such as
`pthread_cond_wait` and `pthread_cond_signal` enable efficient waiting, signalling
and other communication between threads. However these constructs require the
communicating threads to acquire/release locks during doing so. Moreover there
is no direct way to transform these critical sections into transactional atomic
sections. Consider the case of a worker thread t_w waiting for the main thread
t_M to offload work. The thread t_w first acquires a lock on the *waiting mutex* l
and calls `pthread_cond_wait(..,l)`. This call atomically unlocks the mutex and
starts the conditional wait. To signal thread t_w to start execution, the thread t_M
in turn acquires a lock on l, calls `pthread_cond_signal()` and releases the lock

on l. If the critical section protected by the lock acquisition/release in t_M were to be transformed into an atomic section using transactions, then if there is a conflict in the transaction in t_M the transaction cannot roll back since the signal has been set and it is irrevocable. Most STM systems including the TL2 system we used and the compiler-based STM in [10] do no provide transactional methods for conditional synchronization and signalling. Consequently our implementation uses traditional mutex based methods for conditional synchronization.

4.2   Memory Management and Application Controlled alloc/de-alloc.

Dynamic memory allocation is another important programmatic concern for STMs. Most STM systems provide methods for allocating and deallocating memory efficiently from within transactions. Additionally they often implement a large memory buffer from which allocations are made and of course memory that is allocated in a failed transaction is restored back to the buffer. Many of the important classes of objects in ODE are allocated dynamically on the heap. This includes bodies, joints, joint lists, and other shared data. However, ODE implements its own memory allocation/deallocation algorithms that purport to improve locality and to allow objects to be be efficiently garbage collected in addition to implementing its own large stack-shaped buffer from which allocation requests are met. Requests for memory allocations are made using the ODE_Alloc() which simply returns a pointer to the first location in memory that has not previously been allocated. If concurrent transactions in two different threads call ODE_Alloc at the same time, both may receive the address of the same location in memory. And as with all transactional writes to shared data, the modifications they make to this newly allocated memory region will be buffered in their respective private write-buffers. Suppose one of them finishes and commits successfully. At this point its modifications to the heap will actually be written to memory. When a conflict is detected when the second transaction tries to commit it will be aborted. As the TM runtime rolls this transaction back, the memory allocated within it will be freed thereby freeing memory that the first transaction is using. Therefore the memory allocation/deallocation library should be modified to be aware of the revocable nature of allocations. For programs that may make use of such routines from one or more of several external libraries this is a significant problem.

5   Experimental Evaluation

We used the parallel ODE library in to drive an application simulating a scene with approximately 200 colliding rigid bodies (a modified version of the crash program in the ODE distribution). The maximum number of worker threads in the global thread pool was varied from t = 1 to 32 in powers of 2. The number of

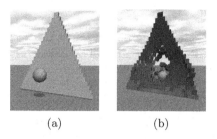

Fig. 2. Scene used in evaluating parallel ODE

threads in the results below therefore represents the maximum number of worker threads available to for offloading and the maximum number of active threads at any instant is (t+1) including the main thread.

We used the TL2 (v0.9.6) STM [1] API and library to provide support for transactions in the ODE library as well as in the driver application program. This version of TL2 is a *word-based write-buffering* STM that uses *lazy version checking* for detecting conflicts and *commit-time locking*. All experiments were carried out on a machine with an Intel Xeon dual processor with two cores per processor and with hyperthreading turned on on all cores (for a total of 8 thread contexts). This in our opinion represents an average platform that may be used to run interactive simulations in ODE. Machines with higher core counts such as (8 or 16) are less common (although they are available) and servers with core counts of 32 and more are less frequently used in running these predominantly desktop oriented simulations. Each core on this machine had a private 32K L1-D cache, 32K L1-I cache, a shared 256KB L2 per processor and a shared 8MB L3 cache and the machine was equipped with 6GB of physical memory. Each thread in our experiments was bound to exactly one core. We compiled all libraries and the driver application with g++-4.3.3 using the default flags and all experiments were run on Ubuntu Linux 2.6.28. All running times were gathered using the `gettimeofday()` call.

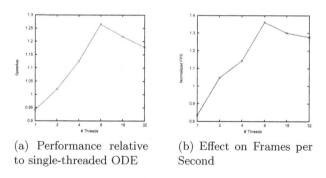

(a) Performance relative to single-threaded ODE

(b) Effect on Frames per Second

Fig. 3. Scalability

5.1   Execution Time

The graph in Figure 3a shows the improvement in execution time as speedup over the single-threaded execution time. The X-axis is the *maximum number of threads available for offloading*. The speedup scales until 8 threads at which point it is roughly 1.27x. At 16 and 32 threads it drops to roughly 1.22x and 1.18x approximately. This means that the heuristics may be too aggressive in offloading work when idle threads are available. This hurts performance since there may not be enough work for a worker thread (not enough joints or bodies in island processing for example) to justify the overhead of offloading. Moreover, at 16 and 32 threads each core is utilized by 2 and 4 threads respectively which means increased contention may also be responsible.

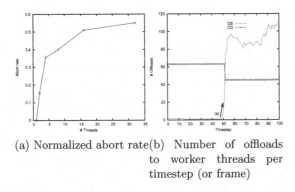

(a) Normalized abort rate (b) Number of offloads to worker threads per timestep (or frame)

Fig. 4. Aborts and Offloads

5.2   Frame Rate

Figure 3b shows the number of frames processed per second (FPS) against the number of threads in the thread pool. In our experiments each time step corresponds to one frame. The frame rate scales in a trend similar to that of execution time speedup. The improvement in frame rate peaks at 1.36x and drops to 1.27x for 32 threads. At more than 8 threads more than one thread is mapped to a processor and contention for shared data also increases reducing the per frame completion time.

5.3   Abort Rate

The abort rate for different number of threads is shown in Figure 4a. The abort rate is defined as the ratio of the number of aborts to the total number of transactions started. Therefore if a, c represent number of aborts and commits, the abort ratio is given by $a/(a + c)$. The abort ratio increases steeply up to 4 threads and continues to rise beyond. The average amount of contention between threads increases as the number of threads increases and the amount of shared data being accessed by these threads remains the same. The abort rate does not

Table 1. Read/Write set sizes

	Reads (bytes)				Writes(bytes)			
Threads	Min	Max	Avg	Total	Min	Max	Avg	Total
1	4	112	112	3094332	4	96	48	1325062
2	4	224	211	5886756	0	192	90	2520386
4	4	2536	596	16620560	0	2036	240	6791206
8	4	2868	1300	36245344	0	2328	530	14775982
16	4	3552	1393	38823380	0	2936	570	15868776
32	4	5184	1504	41912768	0	4196	614	17133684

rise as significantly going from 16 to 32 threads. This is because the average number of *concurrent* threads does not necessarily rise proportionally to the number of threads in the thread pool and therefore the number of aborts increase less steeply.

5.4   Thread Utilization

In contrast to parallelization techniques that purely depend on static decomposition of work, in the scheme for parallel dynamics simulation (DS) described above, only the maximum number of threads in the thread pool is fixed and heuristics are used to dynamically gauge whether to offload island(s) processing to worker threads. The amount of parallelism in the collision detection (CD) stage however remains relatively uniform. The plot in Figure 4b shows the average number of computation offloads occurring in each time-step (or frame) when there are a maximum of 32 threads in the global thread pool. Specifically, the plot shows the number of offloads to worker threads for the first 100 frames of simulation for the scene shown in Figure 2. The number of offloads in the CD stage remain stable and in this stage, a worker thread can be invoked on average roughly 2 times until the point in the simulation noted as *(a)* in the plot. Also, the number of offloads in the DS stage remains low and is also stable until point *(a)*. This is the time step where the stack of bodies in Figure 2 begins to disintegrate as shown in Figure 2(b). While in earlier time steps there was only one island to process, after point *a* there are many smaller islands and therefore there is more parallelism. This is reflected in Figure 4b by the sharp increase in number of offloads in the DS stage after point *(a)*. As mentioned above, the heuristics we used have a relatively low threshold on island count for offloading the work of processing an island to a worker thread. This results in the main thread aggressively offloading work which explains the high number of DS offloads after point (a). The number of offloads in the CD stage remain relatively stable since there the data distribution is based on abstract spaces and not physical artifacts such as joints and islands. Additionally, after point *(a)* the number of offloads in the CD stage are reduced due to contention with the DS stage for worker threads.

5.5   Transaction Read/Write Sets

There are three main types of transactions during execution. The first is the transaction to add a contact joint to the system for a pair of colliding bodies. The second transaction executed during island processing for atomically updating a body's attributes. The third type consists of short transactions to access various shared values such as the number of joints. Table 1 summarizes the characteristics of the read/write sets of all the transactions executed. The average read set sizes are significantly larger than the sizes of the write sets in all cases. This is in line with the average mix of read/write operations in many other transactional programs. Many of the transactions in parallel ODE perform several reads before performing their first write. One commonly occurring transaction for example is atomic insertion into a sorted object list. Here the list is traversed and each element examined to find the right position for insertion before pointer values for the neighboring list elements are updated. The average read and write set sizes remain relatively small for most transactions which shows that hardware transactional memory implementations may also be able to support parallel ODE.

5.6   Scalability Optimizations

Based on the results of the experiments described above, the following observations can be made pertaining to improving scalablity.

1. DS phase offloading: The work offloading algorithm in the island processing phase may be too aggressive in our experimental system. This stems partly from the static threshold used to decide whether processing for a particular island is to be offloaded, inlined or whether it should be combined with another island and then offloaded. The size of the islands changes substantially over the course of the simulation (for example, the one shown in Figure 2a), which results in the threshold becoming too low at several points. A low threshold results in aggressive offloading which in turn results in poor scalability. The processing step for a single island cannot be offloaded to more than one thread in our system. This is because the forces and torques acting on a body are determined by the joints connecting the body to its neighboring bodies and if these bodies were being processed by two separate threads the system of constraints imposed by these joints would have to be communicated between them which we believe would increase the level of synchronization drastically. During the early timesteps of simulating the scene shown above, there are only two islands with one of them containing all the bodies in the world and this large island is then offloaded to a single thread. This restriction therefore has the effect of severely serializing island processing until more islands are formed as a result of collisions.

2. Speculative island formation: The algorithm for discovering islands discussed earlier is sequential - the main thread discovers an island and offloads (or inlines) it before proceeding to discover the next island. This substantially

limits the amount of effective parallelism especially for very large scenes. An algorithm for speculatively discovering islands in parallel and processing them in the worker threads after the speculation has been verified would improve parallel performance greatly (in spite of the additional synchronization costs which are relatively small). Briefly, in this algorithm worker threads speculate on a "seed" body for an island and then "grow" the island. This seed body is picked from a cache of likely candidates built during the island discovery phase in the previous timestep. The worker threads then attempt to verify if the island is valid and was previously undiscovered and if so, continue to the island processing step.

3. Performance of Locks: Coarse-grained locking can be used instead of transactions to protect accesses to shared state and we believe that the performance in both cases would be comparable. Fine-grained locking would be harder to implement given the diversity of both the data structures and the accesses to them. Nevertheless we are in the process of implementing our parallel ODE system with support both coarse-grained and fine-grained locking.

6   Related Work

Several researchers have studied various aspects of parallelizing physics computations for applications from domains ranging from robotics, virtual environments and scientific simulations, to animation [16,13,19,14]. In [19] the authors describe a voxel based parallel collision detection algorithm for distributed memory machines. This algorithm is similar to the abstract space based collision detection scheme discussed in this paper. ParFUM [15] is a framework based on Charm++ for developing parallel applications that manipulate unstructured meshes and supports efficient collision detection. In [6] the authors study the performance of a parallel implementation of the Barnes-Hut algorithm for n-body simulation that uses octree based subdivision for computing particle interactions. In [17] the authors present an algorithm for *continuous collision detection* between deformable bodies that can be executed at interactive rates on present day multicore machines.

Lee-TM [7] is an implementation of Lee's routing algorithm using transactional memory. While the algorithm exhibits large amount potential parallelism the transactional implementation has been shown to have modest scalability. AtomicQuake [18] is an implementation of a parallel Quake game server using transactions. The parallelization is at the level of clients connected to the server - operations for a client are performed on the server by the worker thread that the client is mapped to. Support for transactions is provided by the compiler [10] instead of a library based TM. The programs in STAMP [3] consist of a variety of parallel transactional workloads that represent pieces of larger applications and which can be executed with one of several STM or HTM systems. TMunit [9] is a framework for developing unit tests for evaluating STM systems. RMS-TM [8] is a TM benchmark suite consisting of programs and application kernels. STMBench [5] is a synthetic benchmark that that contains transactions with widely varying characteristics and which operate on non-trivial data structures. Thus while it is very useful for finding problems with specific implementations and stretching the limits of TM designs, it is not representative of any real-world program.

7   Conclusion

In this paper we presented a parallel transactional physics engine for rigid body simulation based on the popular Open Dynamics Engine (ODE). We were able to parallelize the two principal components of ODE - the collision detection engine and the dynamics simulation engine to make use of worker threads from a global thread pool for executing work offloaded from the main thread. We used a software transactional memory for orchestrating concurrent accesses to all shared data. Our approach of coarse-grained parallelization was not only relatively programmer friendly but also helped amortize the cost of the work-offloading. The parallel version of ODE showed speedups of up to 1.27x (for 8 threads) compared to the sequential version. As a continuation of this work we plan to investigate better cost heuristics for making offloading decisions and to investigate techniques for incorporating domain knowledge in optimizing memory transactions in addition to comparing the performance of the transactional implementation with that of versions that use fine-grained and coarse-grained locking.

References

1. Dice, D., Shalev, O., Shavit, N.: Transactional Locking II. In: Proceedings of the 20th International Symposium on Distributed Computing (DISC), Stockholm, Sweeden (September 2006)
2. Open Dynamics Engine, http://ode.org
3. Minh, C.C., Chung, J., Kozyrakis, C., Olukotun, K.: STAMP: Stanford Transactional Applications for Multi-Processing. In: IISWC 2008, pp. 35–46 (2008)
4. Guerraoui, R., Kapalka, M., Vitek, J.: STMBench7: A benchmark for software transactional memory. In: Proceedings of the 2nd European Systems Conference (March 2007)
5. Carey, M.J., DeWitt, D.J., Kant, C., Naughton, J.F.: A status report on the OO7 OODBMS benchmarking effort. In: OOPSLA 1994: Proc. 9th Annual Conference on Object-oriented Programming Systems, Language, and Applications, pp. 414–426 (October 1994)
6. Woo, S.C., Ohara, M., Torrie, E., Singh, J.P., Gupta, A.: The SPLASH-2 Programs: Characterization and Methodological Considerations. In: Proceedings of the 22nd Annual International Symposium on Computer Architecture
7. Ansari, M., Kotselidis, C., Jarvis, K., Lujan, M., Kirkham, C., Watson, I.: Lee-TM: A Non-trivial Benchmark for Transactional Memory. In: Proc. 7th International Conference on Algorithms and Architectures for Parallel Processing (2008)
8. Kestor, G., Stipic, S., Unsal, O.S., Cristal, A., Valero, M.: RMS-TM: A Transactional Memory Benchmark for Recognition, Mining and Synthesis Applications. In: 4th Workshop on Transactional Computing (TRANSACT) (2009)

9. Harmanci, D., Felber, P., Sukraut, M., Fetzer, C.: TMunit: A transactional memory unit testing and workload generation tool Technical Report RR-I-08-08.1, Universite de Neuchatel, Institute Informatique (August 2008)
10. Adl-Tabatabai, A.-R., Lewis, B.T., Menon, V., Murphy, B.R., Saha, B., Shpeisman, T.: Compiler and runtime support for efficient software transactional memory. In: Proc. 2006 ACM SIGPLAN Conference on Programming Language Design and Implementation, pp. 26–37 (June 2006)
11. Reinders, J.: Intel Threading Building Blocks. O'Reilly Media (2007)
12. Sweeney, T.: The Next Mainstream Programming Language: A Game Developers Perspective. Invited Talk at the International Symposium on Principles of Programming Languages (2006)
13. Brown, S., Attaway, S., Plimpton, S., Hendrickson, B.: Parallel strategies for crash and impact simulations. In: Computer Methods in Applied Mechanics and Engineering, vol. 184, pp. 375–390 (2000)
14. Grinberg, I., Wiseman, Y.: Scalable parallel collision detection simulation. In: Proceedings of the Ninth IASTED International Conference on Signal and Image Processing (2007)
15. Lawlor, O.S., Chakravorty, S., Wilmarth, T.L., Choudhury, N., Dooley, I., Zheng, G., Kal, L.V.: ParFUM: a parallel framework for unstructured meshes for scalable dynamic physics applications. In: Engineering with Computers (December 2006)
16. Figueiredo, M., Fernando, T.: An Efficient Parallel Collision Detection Algorithm for Virtual Prototype Environments. In: 10th International Conference on Parallel and Distributed Systems (2004)
17. Tang, M., Manocha, D., Tong, R.: Multi-core collision detection between deformable models. In: SIAM/ACM Joint Conference on Geometric and Physical Modeling (2009)
18. Zyulkyarov, F., Gajinov, V., Unsal, O., Cristal, A., Ayguad, E., Harris, T., Valero, M.: Atomic Quake: Using Transactional Memory in an Interactive Multiplayer Game Server. In: 14th ACM SIGPLAN Symposium on Principles and Practice of Parallel Programming (PPoPP) (Febuary 2009)
19. Lawlor, O.S., Kale, L.V.: A voxel-based parallel collision detection algorithm. In: Proceedings of the 16th International Conference on Supercomputing (2002)

Introduction

Kunal Agarwal, Panagiota Fatourou, Arnold L. Rosenberg, and Frédéric Vivien

Topic chairs

Parallelism permeates all levels of current computing systems. It can be observed in systems as varied as multiple single-CPU machines, large server farms, and geographically dispersed "volunteers" who collaborate over the Internet. The effective use of parallelism depends crucially on the availability of faithful, yet tractable, models of computation for algorithm design and analysis and of efficient strategies for solving key computational problems on prominent classes of computing platforms. No less important are good models of the way the different components/subsystems of a platform are interconnected. With the development of new genres of computing platforms, such as multicore parallel machines, desktop grids, and hybrid GPU/CPU-based systems, new models and paradigms are needed, that will allow parallel programming to advance into mainstream computing. Specific areas of interest within this Topic include, but are not limited to:

- Foundations, models, and emerging paradigms for parallel, distributed, multiprocessor and network computation
- Deterministic and randomized parallel algorithms
- Lower bounds for key computational problems
- Models and algorithms for parallelism in memory hierarchies
- Models and algorithms for real networks (e.g., scale-free, small world, wireless networks)
- Theoretical aspects of routing within networks

This Topic solicited high-quality original papers that contribute new results on foundational issues regarding parallelism in computing and/or propose improved approaches to the solution of specific algorithmic problems. After carefully reviewing each submission, we were able to accept three papers that deal with interesting aspects of the issues outlined above.

"A Bi-Objective Scheduling Algorithm for Desktop Grids with Uncertain Resource Availabilities," by L.-C. Canon, A. Essafi, G. Mounié, and D. Trystram, presents a sophisticated contribution to the theory of scheduling parallel/distributed machines. The paper focuses on scheduling batches of work on a desktop grid in a way that accommodates possible unpredicted lapses in availability of resources on such a platform. The authors provide a scheduling algorithm that prevents unexpected lapses from excessively degrading the makespan of the scheduled job. The quality of the algorithm is demonstrated via simulations based on realistic workflows.

E. Jeannot, R. Namyst, and J. Roman (Eds.): Euro-Par 2011, LNCS 6853, Part II, pp. 224–225, 2011.

"New Multithreaded Ordering and Coloring Algorithms for Multicore Architectures," by M.A. Patwary, A.H. Gebremedhin, and A. Pothen, presents multithreaded vertex ordering and distance-k graph coloring algorithms that are tailored for multicore architectures. The basic challenge here is to overcome the apparent inherent sequentiality of the underlying problems. The authors address this challenge via novel algorithmic devices whose value is demonstrated by experiments on input graphs representing both artificial and real applications, performed on a variety of multicore machines.

"Petri-nets as an Intermediate Representation for Heterogeneous Architectures," by P. Calvert and A. Mycroft, develops a model that allows one to expose what the authors term "performance nondeterminism" in heterogeneous parallel systems. This form of "nondeterminism" is observed when portable programs make implementation decisions (concerning, e.g., resource allocation and algorithmics) in response to the exigencies of different target machines and run-time environments. The authors exemplify and discuss potential uses of the model.

We thank—and acknowledge the efforts of—all of the authors who submitted contributions to this Topic and all of the reviewers who provided useful, insightful comments. Their combined efforts have made this Conference and this Topic possible.

Petri-nets as an Intermediate Representation for Heterogeneous Architectures

Peter Calvert and Alan Mycroft

Computer Laboratory, University of Cambridge
William Gates Building, JJ Thomson Avenue,
Cambridge CB3 0FD, UK
`firstname.lastname@cl.cam.ac.uk`

Abstract. Many modern systems provide heterogeneous parallelism, for example NUMA multi-core processors and CPU-GPU combinations. Placement, scheduling and indeed algorithm choices affect the overall execution time and, for portable programs, must adapt to the target machine at either load-time or run-time. We see these choices as preserving I/O determinism but exposing *performance non-determinism*. We use Petri-nets as an intermediate representation for programs to give a unified view of all forms of performance non-determinism. This includes some scenarios which other models cannot support. Whilst NP-hard, efficient heuristics for approximating optimum executions in these nets would lead to performant portable execution across arbitrary heterogeneous architectures.

Keywords: Petri-nets, parallelism, heterogeneous architectures, scheduling.

1 Introduction

It is becoming clear that modern systems are not only increasingly parallel but also heterogeneous. Common examples include IBM's Cell Broadband Engine and also CPU-GPU combinations. As well as different processing capabilities, the different cores have access to separate memories, with data transfers needing to be managed explicitly. Our research is focused on offering portability for these systems.

On such architectures, achieving optimal performance depends on careful placement of computation and management of the required data transfers. If we want programs to have *portable* performance, this placement must be done automatically. This is an example of *performance non-determinism*, where run-time or load-time decisions affect the overall execution time. A choice between algorithms also causes this (e.g. the fastest sequential algorithm compared to one well suited to parallelisation). We distinguish this from *I/O non-determinism*, where run-time decisions might alter the *result* of the program.

Other work on heterogeneous architectures tends to treat every task as a single unit that must be completed, just giving it a different cost for each processor

E. Jeannot, R. Namyst, and J. Roman (Eds.): Euro-Par 2011, LNCS 6853, Part II, pp. 226–237, 2011.

[14]. However, this does not allow scenarios such as *either perform A once at cost 5 or perform B ten times (possibly in parallel) at cost 1 each*. We wish to encode these possibilities.

In this paper, we use *coloured Petri-nets* as an intermediate representation for parallel programs, and investigate the placement and scheduling problems. In particular, we consider:

- A Petri-net intermediate representation that expresses both parallelism and performance non-determinism, and some example programs (Section 3).
- A simple model of heterogeneous architectures and how this provides a unified model of existing hardware (Section 4).
- How these Petri-nets map onto such hardware, including a cost model and the problem of minimising execution time (Section 5).
- Existing work from both mathematics and computer science that addresses issues that we identify in this minimisation problem (Section 6).
- How compiler optimisations can be applied to these Petri-nets (Section 7).

We also discuss other parallel models and their relation to this work.

2 Notation

Throughout this paper, we make frequent use of *sets*, *multisets* and *ordered lists*. The notation that we will use is as follows:

- $\mathbb{N}_0$ gives the set of non-negative integers—i.e. $\{0, 1, 2, \dots\}$.
- X^∞ gives the set $X \cup \{\infty\}$.
- $\mathbb{R}_+$ gives the set of non-negative real numbers.
- A multiset m over a set X is a function $X \to \mathbb{N}_0$ giving the number of appearances $m(x)$ of $x \in X$ in m. The set of all multisets is written $\mathbf{m}X$. The operations $\cup$, $\setminus$ and $\subseteq$ are defined as liftings of $+$, saturating subtraction and $\leq$ respectively.
- A list l over a set X is a tuple X^n. We will write l_i for the ith element of l (for $i = 1 \dots n$) and $|l|$ for its length n. The set of all lists is written $\ell X = \bigcup_{n \in \mathbb{N}_0} X^n$. Note that we can treat lists as multisets, but not vice-versa.

We also use a very simple type system:

$$\tau ::= \text{int} \mid \text{bool} \mid \text{unit} \mid \tau \times \tau \mid \tau[n] \quad \text{for } n \in \mathbb{N}$$

It is important that each type is of fixed size, given by $\text{sizeof}(\tau)$. Therefore, arrays of different lengths are distinct types. The set of values that a type τ can take is denoted $[\![\tau]\!]$. We do not have higher order types but will refer to transitions of type $\tau \to \tau'$, and write for functions:

$$f : \tau \to \tau' \iff \forall v \in [\![\tau]\!] : (f(v) \text{ defined} \implies f(v) \in [\![\tau']\!])$$

Later, we will abuse $[\![\_]\!]$ to give the operation associated with a transition. When referring to the cost of operations, we will generally use the notation $\langle \_ \rangle$.

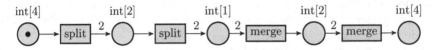

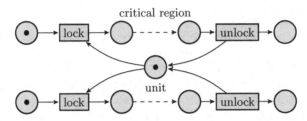

Fig. 1. Petri-net for *merge-sort* of a 4-element list

Fig. 2. Petri-net for a critical region using a *mutual exclusion lock*

3   Petri-net Intermediate Representation

Any intermediate representation for modern architectures must express parallelism, and, we argue, performance non-determinism. The current common formats are *data-flow graphs (DFGs)* and *control-flow graphs (CFGs)*.

Data-flow graphs express parallelism, but not non-determinism. Every computation node in the graph will be performed at some point, and the edges give the dependencies between them. On the other hand, control-flow graphs express no parallelism at all, but branching is often treated as non-determinism (e.g. in static analysis).

Petri-nets are the obvious combination of these two, offering both parallelism and non-determinism. We consider a variant of Jensen's *coloured Petri-nets* or *CP-nets* [8]. *Tokens* (drawn •) containing values are stored at *places* (drawn ◯) which have a type. Execution proceeds by the firing of *transitions* (drawn □). When a transition *fires*, it removes tokens from its *pre-places*, applies a function to them, and adds the results to its *post-places*. Later, we will allow such transitions to take a period of time. For example, a simple merge-sort of a 4-element list can be represented as shown in Figure 1 (the '2's in the diagram show multiplicities where multiple tokens are taken from, or given to, a single place in a firing). Petri-nets also express concurrency primitives well, for example mutual exclusion (Figure 2).

In our nets, the effect of a transition is defined using an existing representation for sequential programs (e.g. functions in LLVM's IR [1] or even individual virtual machine instructions). However, for this work, the exact choice is not important, we simply refer to the set $\mathbb{F}$ of such partial functions[1].

---

[1] We elide the difference between intensional and extensional representations of a function, so for $f_1, f_2 \in \mathbb{F} : (\forall x : f_1(x) = f_2(x)) \not\Rightarrow f_1 = f_2$ since f_1 and f_2 may compute the result in different ways.

The partiality of these functions provides for conditional transitions. If a transition is not defined for a given input, it will not fire. For example, the following transition *cond* of type *bool* → *unit* will generate a token (with *unit* value) only when provided with a token containing the value **true**:

$$[\![\text{cond}]\!](b) = \begin{cases} () & \text{if } b = \text{true} \\ \text{undefined} & \text{if } b = \text{false} \end{cases}$$

We can now define our version of coloured Petri-nets more formally:

Definition 1. *A* **coloured Petri-net** *is a tuple* $N = (S, T, \Gamma, {}^{\bullet}\_, \_^{\bullet}, [\![\_]\!])$ *consisting of:*

1. *A set of places* S.
2. *A set of transitions* T.
3. *A type environment* Γ, *associating a type* τ *to every* $s \in S$.
4. *A pre-place function* ${}^{\bullet}\_ : T \to \ell S$.
5. *A post-place function* $\_^{\bullet} : T \to \ell S$.
6. *A labelling function* $[\![\_]\!] : T \to \mathbb{F}$ *such that transitions are well typed—i.e.*

$$\text{For all } t \in T, \; [\![t]\!] : \Gamma({}^{\bullet}t) \to \Gamma(t^{\bullet})$$

where $\Gamma([s_1, \ldots, s_n]) = \Gamma(s_1) \times \cdots \times \Gamma(s_n)$.

Note that we have defined the pre- and post-places of transitions as *lists* rather than *multisets* to give direct association with argument and result tuples of functions in $\mathbb{F}$.

The state of a coloured Petri-net describes the tokens, and their values, present at each place. This is called a *marking* (although Jensen [8] prefers *token distribution* when tokens carry values) and can be defined as follows:

Definition 2. *A* **marking** *is a function* M *defined on* S *such that* $M(s) \in \mathbf{m}[\![\Gamma(s)]\!]$ *for all* $s \in S$. *We denote the set of all markings as* $\mathbb{M}$. *The operators* $\cup, \setminus$ *and* $\subseteq$ *lift to markings in the obvious manner.*

Given a list of places $i \in \ell S$ (e.g. the pre-places or post-places of a transition), and a list of values $\mathbf{x} \in [\![\Gamma(i)]\!]$ (i.e. which are well typed), the corresponding marking[2] will be written $(i \leadsto \mathbf{x})$.

This allows us to define the *firing rule* $M \to M'$ of the Petri-net as follows:

$$M \cup ({}^{\bullet}t \leadsto \mathbf{x}) \to M \cup (t^{\bullet} \leadsto [\![t]\!](\mathbf{x})) \text{ provided that } [\![t]\!](\mathbf{x}) \text{ is defined}$$

Traditionally, paths through Petri-nets are represented either as *causal nets* [12] or by *pomsets* [10]. When we come to introducing a cost model in Section 5, it will be more convenient to use pomsets.

---

[2] For example, for $A, B, C \in S$ and $\Gamma(s) = \text{int}$ for $s \in \{A, B, C\}$, $([A, B, C, C] \leadsto [10, 4, 5, 2]) = (A \mapsto \{10\}; B \mapsto \{4\}; C \mapsto \{5, 2\})$.

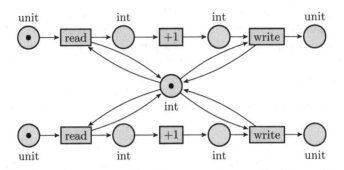

Fig. 3. Petri-net without confluence ($x := x + 1 \parallel x := x + 1$)

Definition 3. *A* **(pomset) path** *is a triple* $(V, \leq, \mu)$ *where* V *is a set of oc-curences labelled with a transition by* $\mu : V \to T$. $\leq$ *defines a partial order on* V. *Two pomset paths are considered equivalent if there is an isomorphism between them.*

We will use $\mathbb{P}$ to give the set of all such paths. Note that for two transiton occurences $v_1, v_2 \in V$, if neither $v_1 \leq v_2$ nor $v_2 \leq v_1$ then the occurences can occur in parallel, whereas otherwise they must fire sequentially.

Just as with traditional flow graphs, not all of these paths are *feasible*, since we cannot check whether the transitions are defined without knowing the values of the tokens. However, any execution trace will be an instance of a path from a distinguished initial place s_0 to a final place s_∞. It must not be possible to fire any transitions from the final place (i.e. $\forall t \in T : s_\infty \notin {}^\bullet t$).

The firing of Petri-nets is non-deterministic, as intended. However, this al-lows all forms of non-determinism to be expressed, not just performance non-determinism (e.g. Figure 3). We will assume that programs are I/O deterministic, and therefore respect *confluence*—i.e. for all markings $M_1, M_2, M_3 \in \mathbb{M}$:

$$(M_1 \to^* M_2) \wedge (M_1 \to^* M_3) \implies \exists M_4 \in \mathbb{M}.((M_2 \to^* M_4) \wedge (M_3 \to^* M_4))$$

4   Simple Hardware Model

We restrict ourselves to a very simple model of heterogeneous architectures. This ignores fine details of the memory system such as caches. We consider a system to consist of *processors*, each with a *local memory*, and *interconnects* between them. The cost of accessing this local memory is low and included in the computation cost of a function. Non-local data must be transferred via interconnects before use, at a cost modelled by latency and bandwidth. This is not dissimilar from the partitioned global address space model (PGAS) that is used elsewhere (e.g. X10). We ignore capacity constraints of memories. Formally, a hardware architecture is defined as follows:

Definition 4. *A* **simple heterogeneous hardware model** H *is a 3-tuple* (P, m, c) *consisting of:*

- *A finite set of* processors P.
- *An* interconnect descriptor *function* $i : (P \times P) \to (\mathbb{R}_+ \times \mathbb{R}_+)^\infty$. *For a pair* (p_1, p_2) *of distinct processors,* $i(p_1, p_2) = (l, b)$ *gives the latency* l *and per-byte cost* b $(= \frac{1}{bandwidth})$ *of the interconnect from* p_1 *to* p_2. *We will refer to the cost of transferring* n *bytes of data with the notation* $\langle p_1 \xrightarrow{n} p_2 \rangle = l + n \cdot b$. *When there is no interconnect from* p_1 *to* p_2, $i(p_1, p_2) = \infty$.
- *A* computation cost *function* $c : (\mathbb{F} \times P) \to \mathbb{R}_+^\infty$, *where* ∞ *indicates that the processor cannot perform the function (e.g. no floating-point support).*

An example model of a multi-core plus GPU architecture is given in Figure 4. The inclusion of small costs, such as ϵ in the example, approximates the effect of cache invalidations, when cores share a memory but have separate caches. Memories not associated with a processor can be modelled as a 'null' processor $p_\perp$ with $c(f, p_\perp) = \infty$ for all $f \in \mathbb{F}$.

Core 1   Core 2

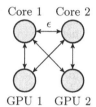

GPU 1   GPU 2

Here ϵ is typically small, and the other costs are based on actual measurements.

Fig. 4. Example model for dual-core CPU with 2 general-purpose GPUs

We assume two sanity constraints: that the memory interconnect is strongly connected[3], and also that all functions can be executed somewhere (i.e. $\forall f \in \mathbb{F} :$ $\exists p \in P : c(f, p) \neq \infty$). These properties ensure that our mapping of software onto hardware is also confluent.

5   Mapping Software to Hardware

Given these two models, we can model all possible executions of a program on an architecture with a single Petri-net. Each feasible path through the net gives a possible execution trace. The intuition behind our construction comes from considering an individual data token x. In program $N = (S, T, \Gamma, {}^\bullet\text{-}, \text{-}^\bullet, [\![\text{-}]\!])$, x must be at some place $s \in S$. However, the architecture $H = (P, m, c)$ on which the software is run, must store x in some memory $p \in P$. Therefore, the location of a *data* token in a running program is described by a pair from the set $S \times P$.

We now consider what might happen to a token x at (s, p). There are two options, either:

- (t, p): The token x is used by transition $t \in T$ executing on processor p (where possible), *or*

---

[3] A graph is strongly connected if for every pair of vertices a and b, there is a path both from a to b, and b to a.

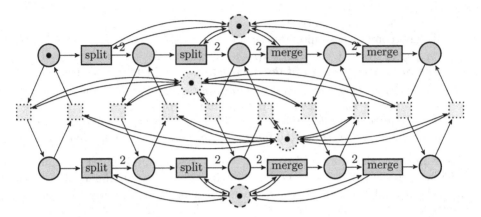

Fig. 5. *Merge-sort* mapped onto a dual core CPU (types are omitted for clarity, and memory transfers are only shown for $n = 1$)

- (s, p, p', n): The token x is transferred to another processor $p' \in P$ via an interconnect (p, p') as part of an n-token transfer.

Therefore, the possible transitions for x must be a subset of $(T \times P) \cup (S \times P^2 \times \mathbb{N})$. Additionally, when we consider the complete system, we require *resource constraints* so that a processor executes only a single transition at any moment, and similarly so that each interconnect is only used for one transfer at a time.

We can encode this as a new Petri-net. To encode the resource constraints, we choose to use mutual exclusion locks similar to Figure 2, although they could also be expressed as restrictions on paths. This results in a net C defined as follows[4]. The Petri-net for our merge-sort example on a dual core CPU is shown in Figure 5 as an example, although we do not intend this model to be a large-scale visual model.

$$C = \Big(\big[\underbrace{(S \times P)}_{\text{Data Places}} \cup \underbrace{P}_{\text{Processor Constraints}} \cup \underbrace{P^2}_{\text{Interconnect Constraints}}\big], \big[\underbrace{(T \times P)}_{\text{Computation Transitions}} \cup \underbrace{(S \times P^2 \times \mathbb{N})}_{\text{Memory Transfers}}\big], \Gamma', {}^{\bullet}\text{-}, \text{-}^{\bullet}, [\![\text{-}]\!]\Big)$$

with[5]:

$${}^{\bullet}(t, p) = [p, ({}^{\bullet}t_1, p), \ldots, ({}^{\bullet}t_{|{}^{\bullet}t|}, p)] \qquad {}^{\bullet}(s, p_1, p_2, n) = [(p_1, p_2), (s, p_1), \ldots, (s, p_1)]$$
$$(t, p)^{\bullet} = [p, (t_1^{\bullet}, p), \ldots, (t_{|t^{\bullet}|}^{\bullet}, p)] \qquad (s, p_1, p_2, n)^{\bullet} = [(p_1, p_2), \underbrace{(s, p_2), \ldots, (s, p_2)}_{n \text{ repetitions}}]$$

---

[4] Excluding the resource constraints, this construction is equivalent to the Cartesian product of hypergraphs, where each Petri-net transition corresponds to a hyperedge.

[5] For these definitions, we use ML-style list syntax (i.e. $[a, b, c]$ is a list, and $1 ::$ $[2, 3, 4] = [1, 2, 3, 4]$).

and:

$$\Gamma'(s') = \begin{cases} \Gamma(s) & \text{if } s' = (s, p) \\ \text{unit} & \text{otherwise (i.e. resource constraint places)} \end{cases}$$

$$[\![(t, p)]\!] = \lambda(r, x_1, \ldots, x_n) \,.\, () :: [\![t]\!](x_1, \ldots, x_n)$$

$$[\![(s, p_1, p_2, n)]\!] = \lambda(r, x_1, \ldots, x_n) \,.\, [(), x_1, \ldots, x_n]$$

In Section 4, we required that our memory be strongly connected. This ensures that data transfers can always be 'undone'. Similarly, since each function in $\mathbb{F}$ can be done on some $p \in P$, we know that the new Petri-net is still confluent. Therefore, the choice of which transition to fire can only affect performance, not correctness.

Since our hardware model gives us costs, we can supplement our 'compiled' Petri-net with a *duration function* $\langle \_ \rangle$. This gives the time taken for each transition to fire. It can be defined as follows:

$$\langle (t, p) \rangle = c([\![t]\!], p)$$

$$\langle (s, p_1, p_2, n) \rangle = \langle p_1 \overset{n \cdot \text{sizeof}(\Gamma(s))}{\longrightarrow} p_2 \rangle$$

Now we have durations associated with each transition, it is reasonable to ask how long a path through C will take to execute. Given a pomset path $\mathfrak{p} = (V, \leq, \mu) \in \mathbb{P}$, this is given by $\langle \mathfrak{p} \rangle = \max_{v \in V}(f(v))$ where the finish time $f(v)$ of an occurence is given by:

$$f(v) = \begin{cases} \max_{\{w \in V | w \leq v\}}(f(w)) + \langle \mu(v) \rangle & \text{if } \exists w \in V : w \leq v \\ \langle \mu(v) \rangle & \text{otherwise} \end{cases}$$

Candidate executions of N on H are given by any paths from an initial place to a final place. Unfortunately, as noted in Section 3, not all paths are feasible. In executing the Petri-net, the aim is to choose a trace with *minimum duration*.

6   Finding Optimal Executions

The problem of finding an optimal execution allows us to consider all performance non-determinism choices together. This is not limited to placement and scheduling, but also programming model specific choices, such as *which thread should get access to the lock first* and *how many times should we perform divide and conquer for our algorithm to best match the available parallelism*.

We can consider the problem in two stages. Firstly, we must be able to find traces of minimum duration where no partial transitions (i.e. conditionals) are used, and therefore all paths are trivially feasible. This corresponds to analysis of straight-line flow graphs. Once solved, extending this to the complete problem will require runtime analysis, since input values will affect which paths can occur. Fortunately, as pointed out in Section 5, we cannot make any *wrong* choices (just slow ones). We might therefore aim to pick transitions that appear in a number of low duration paths.

6.1   Complexity

Even the first part is an NP-hard problem. The proof uses a reduction from the *exact cover* problem, which is very similar to the reduction from *3-dimensional matching* for AND/OR network scheduling [6]. These problems are both known to be NP-complete [9].

Theorem 1. *Checking whether a path exists between two markings is NP-hard.*

Proof. Given a set X, and a set $Y \subseteq \wp(X)$ of subsets, an exact cover is a set $Y^* \subseteq Y$ such that for each element of X it appears in exactly one element of Y^*. Determining whether an exact cover exists for a given X and Y is known to be NP-complete [9].

*We can encode this in a Petri-net $(S, T, \lambda s.unit, {}^\bullet\_, \_^\bullet, \lambda t.(\lambda x.[(), \ldots, ()]))$ as follows. For each $x \in X$, we include a place $s_x \in S$. We also introduce a start place **start**. Now for each subset $y = \{x_1, \ldots, x_n\} \in Y$, we add a transition $t_y \in T$ such that:*

$$^\bullet t_y = [start]$$
$$t_y^\bullet = [start, s_{x_1}, \ldots, s_{x_n}]$$

We can then determine whether an exact matching Y^* exists by checking whether there is a path from $\{start\}$ to $\{start\} \cup X$.

6.2   Similar Problems and Techniques

Scheduling appears to be a similar problem, and despite being NP-complete, effective heuristics do exist for it. This suggests there may also be approximations for our problem. However, there are several key differences between the problems.

In typical scheduling problems, precedence constraints do not allow the concept of a completed task being 'used up' as in Petri-nets, and cycles therefore behave differently. This *dataflow* situation is seen with *streaming models* (e.g. [11]), where the input and output rates of pipeline stages need to be matched. Adaptive runtime approaches have been shown to produce good results [13] for such pipelines.

Flows through graphs also share similarities, with the flow into and out of each node needing to match (our program outputs a single token at s_∞ so tokens cannot build up at intermediate places). We would need to use *hypergraphs*[6] since transitions have multiple pre- and post-places. The *minimum cost flow* problem is most relevant since we know how much data will be input and expected as output. However, this minimises total cost rather than the critical path.

In the standard graph case, there are efficient *cost-scaling* algorithms [7] for this which could be distributed. However, work on hypergraphs [4] appears to be restricted to 'B-hypergraphs' where each edge has only a single head.

Unfortunately, the standard definition of 'hyperflows' fails to describe execution paths (for example, Figure 6). Also a flow would need to be supplemented by an actual schedule for execution.

---

[6] A (directed) hypergraph is here a digraph where each hyperedge e has multiple heads and multiple tails (i.e. $e \in \wp V \times \wp V$).

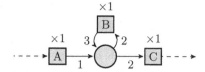

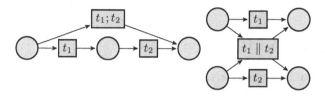

Fig. 6. A hypergraph flow that is not a valid path (A must fire *twice* for B or C to fire)

Fig. 7. Fusing of Petri-net transitions

The non-deterministic choice present in Petri-nets is uncommon in scheduling. *AND/OR networks* do share this characteristic. They are typically solved using list scheduling heuristics [5]. However, it is unclear how this can be combined with the dataflow properties defined above.

Graph flows are again similar since a flow will not necessarily use every edge.

7   Compiler Optimisations

Using the representation that we have described may make parallelism explicit, however it does not consider compiler optimisations apart from within a single transition. The potentially fine-grained nature of transitions means that there will be a limited performance improvement. If we enlarge the transitions in the original program, then some of the explicit parallelism may be lost. The solution to this is to allow transitions to be *fused* at load-time or runtime (Figure 7). For two transitions t_1 and t_2 on a *single processor* p, we can hope that[7] $c([\![t_1; t_2]\!], p) < c([\![t_1]\!], p) + c([\![t_2]\!], p)$ or $c([\![t_1 \parallel t_2]\!], p) < c([\![t_1]\!], p) + c([\![t_2]\!], p)$. For example, with SIMD execution, it may be that $c([\![t \parallel t]\!], p) = c([\![t]\!], p)$.

A fusing can be defined by a path in the original program Petri-net, since a path encodes the ways in which transitions can be combined, including repetitions of a single transition. Any given path is also finite, so we avoid the problem of defining a cost for looping code. We can incorporate this into our mapping by creating a transition for each path $\mathfrak{p} \in \mathbb{P}$ in the program rather than for each transition $t \in T$—i.e. the set of transitions in the mapped Petri-net becomes $((\mathbb{P} \times P) \cup (S \times P^2 \times \mathbb{N}))$.

In practice, we cannot enumerate all paths (an infinite set), so heuristics for specific core types would be used. For example, data parallel devices (e.g. GPUs) would be most effective on paths that compose a calculation in parallel with

---

[7] $t_1; t_2$ and $t_1 \parallel t_2$ give sequential and parallel compositions of t_1 and t_2 respectively.

itself. However, even with heuristics we will generate many transitions. Again looking at data parallel cores as an example, a device that can execute between 1 and N threads will generate a transition for each choice of $n \in \{1, \ldots, N\}$. In the future, we will therefore need to look at how transitions can be parameterised.

Similarly, our current representation does not support any notion of procedures. This becomes impractical for large programs, and we will need to investigate ways of including this. Jensen's hierarchical CP-nets offer one approach [8], but we will also consider techniques from other areas such as hardware description languages (since places behave somewhat like wires).

8   Comparison with Other Models

The best-known bridging model for parallel computation is Valiant's *BSP* [15]. However, this only considers *homogeneous* scenarios. In BSP, computation is grouped onto processors in the program itself. Therefore, any heterogeneous simulation of a BSP program is constrained in its placement and scheduling choices. By choosing to express our programs as dependences, we overcome this and allow full adaptation to the target.

There has also been considerable work on runtimes and languages for parallel systems. StarPU [3,2] uses the HEFT scheduling algorithm [14] to choose an implementation of a task, running it on the relevant processor. However, this framework does not allow choices between sets of tasks.

Coordination languages are perhaps the closest in spirit to our approach. They link together functions from another language to form a complete program. This is similar to how transitions (which are also written in a separate language) are linked together by data places. These include coarse-grain dataflow approaches and are related to streaming languages such as StreamIt [11]. Our model can be seen as a development of both of these.

9   Conclusions

This paper has given a fresh perspective to placement and scheduling on heterogeneous architectures, which we argued in Section 1 are important problems in trying to achieve portable performance. Petri-nets (Section 3) have long been known to provide elegant encodings of concurrency constructs, and we have shown in Section 5 that as an intermediate representation they can also encode the executions of these constructs on our hardware model (Section 4). Whilst the resultant problem is NP-hard, it allows consideration of all performance non-determinism, and there are likely to be efficient heuristics.

In Section 7, we moved away from theory to discuss how compiler optimisations could be performed on our representation.

We feel that the model is promising not just as a theoretical model, but also as the basis of a virtual machine offering portability across heterogeneous architectures. Future work will need to find suitable heuristics for the optimisation problem by further investigation of the related work discussed in Section 6, and also move towards an implementation that allows investigation of behaviour and performance on real programs.

Acknowledgements. We thank the Schiff Foundation, University of Cambridge, for funding this research through a PhD studentship. Thanks are also due to Jonathan Hayman for valuable discussions and feedback.

References

1. Adve, V., Lattner, C., Brukman, M., Shukla, A., Gaeke, B.: LLVA: A Low-level Virtual Instruction Set Architecture. In: Proc. 36th Annual ACM/IEEE International Symposium on Microarchitecture, MICRO-36 (2003)
2. Augonnet, C., Clet-Ortega, J., Thibault, S., Namyst, R.: Data-Aware Task Scheduling on Multi-Accelerator based Platforms. In: Proc. 16th IEEE International Conference on Parallel and Distributed Systems (ICPADS) (2010)
3. Augonnet, C., Thibault, S., Namyst, R., Wacrenier, P.-A.: STARPU: A Unified Platform for Task Scheduling on Heterogeneous Multicore Architectures. In: Sips, H., Epema, D., Lin, H.-X. (eds.) Euro-Par 2009. LNCS, vol. 5704, pp. 863–874. Springer, Heidelberg (2009)
4. Cambini, R., Gallo, G., Scutellà, M.: Flows on Hypergraphs. Mathematical Programming 78, 195–217 (1997)
5. Erlebach, T., Kääb, V., Möhring, R.H.: Scheduling AND/OR-Networks on Identical Parallel Machines. In: Solis-Oba, R., Jansen, K. (eds.) WAOA 2003. LNCS, vol. 2909, pp. 345–346. Springer, Heidelberg (2004)
6. Gillies, D., Liu, J.-S.: Scheduling Tasks with AND/OR Precedence Constraints. In: Proc. 2nd IEEE Symposium on Parallel and Distributed Processing, pp. 394–401 (1990)
7. Goldberg, A.V., Tarjan, R.E.: Finding Minimum-cost Circulations by Successive Approximation. Mathematics of Operations Research 15(3), 430–466 (1990)
8. Jensen, K.: Coloured Petri Nets: A high level language for system design and analysis. In: Rozenberg, G. (ed.) APN 1990. LNCS, vol. 483, pp. 342–416. Springer, Heidelberg (1991)
9. Karp, R.M.: Reducibility Among Combinatorial Problems. In: Complexity of Computer Computations, pp. 85–103. Plenum Press, New York (1972)
10. Pratt, V.: Modeling concurrency with partial orders. International Journal of Parallel Programming 15(1), 33–71 (1986)
11. MIT Computer Architecture Group.: StreamIt,
    http://groups.csail.mit.edu/cag/streamit/
12. Nielsen, M., Plotkin, G., Winskel, G.: Petri nets, Event Structures and Domains, part I. Theoretical Computer Science 13(1), 85–108 (1981)
13. Suleman, M.A., Qureshi, M.K., Khubaib, P.Y.N.: Feedback-directed Pipeline Parallelism. In: Proc. 19th International Conference on Parallel Architectures and Compilation Techniques (PACT), pp. 147–156. ACM, New York (2010)
14. Topcuoglu, H., Hariri, S., Wu, M.-Y.: Performance-effective and Low-complexity Task Scheduling for Heterogeneous Computing. IEEE Transactions on Parallel and Distributed Systems 13(3), 260–274 (2002)
15. Valiant, L.G.: A Bridging Model for Parallel Computation. Communications of the ACM 33, 103–111 (1990)

A Bi-Objective Scheduling Algorithm for Desktop Grids with Uncertain Resource Availabilities

Louis-Claude Canon[1], Adel Essafi[1,2], Grégory Mounié[1], and Denis Trystram[1,3]

[1] Grenoble Institute of Technology, 51 avenue Jean Kuntzmann,
38330 Montbonnot Saint Martin, France
[2] UTIC, Ecole Supérieure des Sciences et Techniques de Tunis
5, Avenue Taha Hussein, B. P. : 56, Bab Menara, 1008 Tunis, Tunisia
[3] Institut Universitaire de France

Abstract. In this work, we consider the execution of applications on desktop grids. Such parallel systems use idle computing resources of desktops distributed over the Internet for running massively parallel computations. The applications are composed of workflows of independent non-preemptive sequential jobs that are submitted by successive batches. Then, the corresponding jobs are executed on the distributed available resources according to some scheduling policy.

However, most resources are not continuously available over time since the users give their idle CPU time only for some time when they are not using their desktops. Moreover, even if the dates of unavailability periods are estimated in advance, they are subject to uncertainties. This may drastically impact the global performances by delaying the completion time of the applications.

The aim of this paper is to study how to schedule efficiently a set of jobs in the presence of unavailability periods on identical machines. In the same time, we are interested in reducing the impact of disturbances on the unavailability periods. This is achieved by maximizing the *stability* that measures the distance between the makespan of the disturbed instance over the initial one. Our main contribution is the design of a new parametrized algorithm and the analysis of its performance through structural properties. This algorithm reduces the impact of disturbances on availability periods without worsening too much the makespan. Its interest is assessed by running simulations based on realistic workflows. Moreover, theoretical results are obtained under the assumption that the size of every availability interval is at least twice the size of the largest job.

Keywords: Scheduling, Availability Constraints, Uncertainty, Stability.

1 Introduction

1.1 Context and Motivation

Today, many kind of parallel platforms are available for running applications. In this work, we focus on desktop grids, which gather idle computing resources

E. Jeannot, R. Namyst, and J. Roman (Eds.): Euro-Par 2011, LNCS 6853, Part II, pp. 238–249, 2011.

of usual desktops distributed over the Internet for running massively parallel computations. Such systems provide a very large computing power for many applications issued from a wide range of scientific domains (including, protein folding [15], gravitational physics [20], etc.).

The applications are composed of workflows of sequential jobs that are submitted by successive batches to a particular user interface machine. Then, the corresponding jobs are transferred to be executed on the distributed available resources according to some scheduling policy. However, usually the resources are not continuously available over time since the users give their idle CPU time only for some time when they are not using their desktops.

Moreover, even if the dates of unavailability periods are estimated in advance, they are subject to uncertainties. This may drastically impact the global performances by delaying the completion time of the application.

In this paper, we study how to schedule efficiently a set of jobs (which corresponds to minimize the makespan) in the presence of unavailability periods. In the same time, we are interested in reducing the impact of disturbances on the unavailability periods. The corresponding objective is the *stability* that measures the ratio between the makespan of the disturbed instance over the initial one [3]. To the best of our knowledge, there is no work studying scheduling with unavailability periods under uncertainties.

1.2 Contributions

The first contribution of this work is to investigate the problem of scheduling with unavailabilities from the view point of studying the impact of uncertainties on the availability periods. Our main contribution is the design of an algorithm and the analysis of its performances through structural properties. It is based on the concept of *slacks* placed just before the unavailability periods that prevent jobs to be delayed. The lengths of the slacks are parametrized by the types of jobs allocated on the available intervals. Then, the good behavior of the algorithm is assessed by running simulations derived from actual workflows of BOINC [1].

The proposed methodology should be useful for solving other scheduling problems with various characteristics like failures or estimated energy consumption.

1.3 Organization of the Paper

The paper is organized as follows. We first recall the most significant related works in Section 2. We distinguished between the works dealing with scheduling under availability constraints and the main existing approaches for studying scheduling in a context with uncertainties. Section 3 is devoted to the description of the computation model and the main notations. Then, we present in Section 5 the algorithm and its worst-case analysis. Before concluding, we present experiments in Section 6 based on simulations on actual workflows and availability constraints.

2   Related Works

In this section, we recall briefly the most significant works related to our problem. We investigate successively each of both sides of the problem, namely scheduling with unavailability constraints and scheduling under uncertainties.

2.1   Scheduling with Unavailabilities

First, notice that most of the approaches used to solve the problem of scheduling with unavailabilities are based on the well-known LPT rule (Largest Processing Times). Lee introduced the problem of scheduling independent jobs with non-simultaneous available times in [16]. This corresponds to scheduling jobs when all the unavailabilities are at the beginning. The main result was to establish that the performance of LPT is bounded by $3/2$. He also proposed a modified version of LPT with an improved performance of $4/3$. A more general problem was studied in [17] for any pattern of availability. Lee showed that the problem cannot be polynomially approximable if no restrictions are done on the availabilities. However, the performance of LPT is bounded by $\frac{m+1}{2}$ when at least one machine is always available and at most one unavailability period per machine is allowed.

In [10], Hwang and Chang analyzed the problem when no more than the half of the machines are unavailable simultaneously. Under this condition, the performance of LPT is bounded by 2. This result was generalized in [11] as follows: if at most λ (ranging from 1 to $m-1$) processors are allowed to be unavailable simultaneously, then LPT generates a schedule whose performance is bounded by $1 + \frac{1}{2}\lceil\frac{m}{m-\lambda}\rceil$.

Liao et al. [19] studied the restriction of the problem on two machines where each machine has one fixed unavailability period. They proposed an optimal exponential-time algorithm. A variant of this particular problem was studied in [21] where the first machine is always available whereas periodic unavailabilities are scheduled on the second. All the unavailabilities have the same duration and all the availabilities have also the same duration. In this case, the performance of LPT is $\frac{3}{2}$ and 2, respectively, for the offline and the online context.

The problem where one machine is always available and with an arbitrary number of unavailabilities on the other processors was analyzed in [5]. It admits no FPTAS, however, a Polynomial Time Approximation Scheme (PTAS) based on the multiple knapsack was designed. A simple list strategy was also proposed.

Notice that all the above approaches are related to sequential jobs. Eyraud et al. studied the problem of scheduling with unavailabilities for parallel rigid jobs [7]. They proved that there is no approximation algorithm in the general case, and they proposed an approximation algorithms for non-increasing unavailability patterns. Moreover, for the problem with restricted unavailabilities, lower and upper bounds were provided for a general list algorithm.

2.2   Scheduling under Uncertainties

Solving scheduling problems with uncertain data has received recently a great attention. There exists a lot of possible approaches depending on the target problem and the desired objectives. A nice survey of scheduling problems was compiled by Billaut et al. [3]. They discuss several complementary approaches from pure pro-active methods (sensitivity analysis), pure on-line strategies and semi on-line methods (flexibility). We focus on this last approach which builds an efficient solution on estimated data and allow simple correction mechanisms at run-time.

Numerous publications are similar to this article, proposing pro-active heuristics based on slacks. In [8], the authors investigate preemption. In [13,14], the authors explore stochastic resource breakdown.

We concentrate on a problem in which interrupted jobs are restarted from the beginning, without migration, but the unavailability constraints can only advance forward in time. Hence, in our case, the reaction and uncertainties are restricted.

3 Models

We present in this section the model of execution that defines the workload and the platform characteristics. Then, without loss of generality, the disturbances are restricted to early shifts of the unavailability periods. Finally, we express formally the problem and define the objectives.

3.1 Model of Execution

In the context of desktop grids, the workload consists of a set of independent jobs. The processing time of the j-th job is p_j. These jobs do not have release or due dates and cannot be preempted.

The platform is composed of m identical machines that are indexed by i. Each of these machines possesses a set of unavailability constraints. We define an interval as an availability period followed by an unavailability period. As intervals are indexed by k, the starting time of unavailability period k on processor i is denoted s_i^k and has a duration u_i^k (see Figure 1). Hence, this period ends at $e_i^k = s_i^k + u_i^k$. Additionally, the duration of the availability period that precedes unavailability k is a_i^k. Finally, the first unavailability period on a processor starts at s_1^i and $e_0^i = 0$.

We denote by λ the number of processors that have no unavailability constraints that are called *free processors*.

3.2 Model of Disturbances

Let δ_i^k be the disturbance that impacts unavailability k on processor i. As we consider that unavailability periods may come earlier, we denote the disturbed unavailability starting time $\tilde{s}_i^k = s_i^k + \delta_i^k$ (by convention, we denote $\tilde{x}$ the disturbed value of the variable x). Unavailability periods cannot overlap, therefore the earliness is limited by the duration of the previous availability period (*i.e.*, $-a_i^k \leq \delta_i^k \leq 0$). Moreover, as it can be seen on Figure 2, only the starting dates are disturbed.

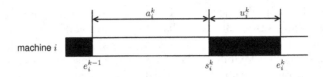

Fig. 1. Representation of interval k on machine i, which contains an availability period followed by an unavailability period with starting dates e_i^{k-1} and s_i^k

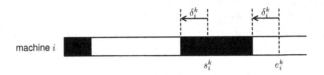

Fig. 2. Unavailability k may start and end earlier due to the disturbance δ_i^k

3.3 Problem Definition

The objective is to generate a schedule given a set of jobs and a set of machines with their unavailability constraints. A schedule is specified by an allocation function $\pi(i, k)$ that gives the set of jobs to be executed during each k-th interval on each i-th processor (jobs are then executed by non-increasing processing times on each interval).

In a disturbed scenario, each unavailability starting date comes early according to our model of disturbance. Moreover, the execution of a schedule is dynamically adapted by using two rules:

- each interrupted job is re-executed as soon as possible without delaying the starting dates of the jobs that follow on the same processor;
- when a processor becomes idle, it starts the execution of its next allocated job.

Assessing the quality of a schedule is done through two objectives: the efficiency and the ability to cope with uncertainties.

We evaluate the first objective by measuring the reference makespan [18] of a schedule, *i.e.*, the makespan when there is no disturbance. It is classically denoted by $C_{\max} = \max_j C_j$ (where C_j is the end date of job j in a given schedule) and the optimal makespan for a given instance is denoted by $C_{\max}^*$.

The second objective is called the *stability*. It is defined as the ratio between the highest disturbed makespan (*i.e.*, the worst makespan among all the possible disturbed scenarios) and the reference makespan, *i.e.*, $S = \frac{\tilde{C}_{\max}}{C_{\max}}$. This objective represents the insensitivity of a schedule to the disturbances. A schedule is said to be *stable* if $S = 1$.

The problem consists in finding a schedule with minimum values of makespan and stability.

4   Analysis of the Stability

In this section, we present the main flexibility mechanism used for coping with uncertain availabilities. The idea is to reserve idle times before the unavailability period to absorb the effect of the disturbances. Idle time, or *slack*, is used for re-executing interrupted jobs in such a way that the reference makespan is delayed the least possible.

Definition 1 (Slack). *The amount of idle time d_i^k preceding unavailability k on processor i is called the* slack.

Definition 2 (Slack rule). *Each slack must be greater or equals to the maximum size of the jobs assigned to its interval, i.e., for each interval k on each machine i:*

$$d_i^k \geq \max_{j \in \pi(i,k)} p_j$$

Proposition 1. *Every schedule based on the slack rule is stable if there is no job scheduled on the last availability period starting before the makespan on each machine (i.e., on the last availability period k for which e_i^k is lower or equals to the makespan).*

Proof. The proof is straightforward since any schedule based on the slack rule is stable within each interval (*i.e.*, in the worst case, the unavailability interrupts the longest job, which can then be absorbed by the slack on this interval).

This is no longer true for the intervals on which a job may be delayed after the makespan. It is the case when an interval starts before the makespan and ends after it. Notice that this condition is strong and may be relaxed in practice.   □

Note that a schedule cannot be stable when there is no free processor: if $\lambda = 0$, the job that terminates at the same time as the makespan may be interrupted and re-executed after the reference makespan. In this case, Proposition 1 is violated because this job is necessarily scheduled on a machine during the last availability that starts before the makespan.

Theorem 1 (Complexity). *Finding a stable schedule with the previous mechanism (flexibility and slack) with minimal $C_{\max}$ is an NP-Hard problem in the strong sense.*

Proof. The proof is based on a reduction from the 3-Partition problem (3-PART in [9, SP15]). The details of the proof are available in [4].   □

5   Bi-objective Algorithm

In this section, we describe a bi-objective algorithm and analyze theoretically its stability. We assumed that there exists at least one free processor (with no un-availability constraint), otherwise it is not possible to generate stable schedules.

5.1 Description

Our bi-objective algorithm uses a compromise parameter β for providing schedules resistant to disturbances (see Algorithm 1). Informally, this parameter indicates at which degree the slack rule is respected (this is called the *relaxed slack rule*). The minimal slack of each interval is proportional to β, *i.e.*, $d_i^k \geq \beta \times \max_{j \in \pi(i,k)} p_j$. Moreover, jobs are never scheduled on the last availability periods that start before the makespan. When $\beta = 1$, the produced schedule is stable because of the slack rule (see Proposition 1). When $\beta = 0$, the slack rule is ignored.

Algorithm 1. Greedy Allocation with Parametrized Slack (GAPS)

Input: a set of jobs J
Output: the allocation function π
 1: Sort intervals by non-decreasing e_i^k (end dates of unavailabilities)
 2: Sort the set of jobs by non-increasing processing times
 3: $S = J$ {Set of unscheduled jobs}
 4: **for all** interval (i,k) **do**   {Consider each interval k on machine i in given order}
 5:   **for all** $j \in S$ **do**   {Consider each job in given order}
 6:     $M = \sum_{j' \in \pi(i,k)} p_{j'}$ {Processing times of the jobs in current interval}
 7:     **if** $e_i^k \leq \frac{\sum_{j' \in S} p'_j - p_j}{\lambda}$ **then**
 8:       **if** $a_i^k - M - p_j \geq \beta \max_{j \in \pi(i,k)} p_j$ **then**   {Relaxed slack rule}
 9:         $\pi(i,k) = \pi(i,k) \cup \{j\}$ {Schedule job j in the current availability}
10:         $S = S \setminus \{j\}$ {Update the set S}
11:       **end if**
12:     **end if**
13:   **end for**
14: **end for**
15: Schedule the remaining jobs using LPT on the λ free processors

---

The first step is to fill greedily the availability periods without violating the relaxed slack rule (Line 8). Note that this step can be seen as a modified version of First Fit Decreasing algorithm for the bin packing problem. In the second step, the λ free processors are treated at once after all the available periods have been filled (Line 15). The transition to the second step occurs when the condition on Line 7 fails. It consists of a lower bound on the time that would be necessary to execute all the unscheduled jobs on the λ free processors. As no job must be scheduled on an availability period that starts just before the makespan, this condition guarantees that the last executed job (such that $C_j = C_{max}$) will be executed on one of the free processors.

The cost of GAPS is low as it only requires jobs and intervals to be sorted. Therefore, its complexity is loglinear in the number of jobs and intervals.

5.2   Theoretical Analysis

In order to schedule at least one job in each interval such that the execution of the job is completed, an assumption is done on the size of the jobs relative to the lengths of the availability periods. These lengths should be greater than twice the maximum size of the jobs, i.e., $2 \times p_{\max} \leq a_{\min}$ (with $p_{\max} = \max_j p_j$ and $a_{min} = \min_{i,k} a_i^k$).

We introduce below the *unavailability ratio* γ that prevents an arbitrarily large approximation ratio for the stability. It characterizes the worst percentage of time during which any machine will stay inactive relatively to its previous availability period.

Definition 3. *Let* $u_{\max} = \max_{i,k} u_i^k$. *The unavailability ratio* γ *is*

$$\gamma = \frac{u_{\max}}{a_{\min}}$$

Intuitively, the larger γ, the longer any rescheduled job will wait before its next execution.

Theorem 2 (Stability). *Under the assumption* $2 \times p_{\max} \leq a_{\min}$, *the stability of the GAPS algorithm is*

$$r_S = \begin{cases} \frac{5}{2} - \beta + \gamma & \text{if } \beta \neq 1 \\ 1 & \text{otherwise} \end{cases}$$

Proof. For any schedule built with GAPS, we determine the amount of jobs that are interrupted and that need to be rescheduled after the makespan in the worst case scenario. We focus on one processor but the argument is general and can be extended easily to any number of processors.

Let K be the number of intervals that finish before the makespan on processor i. Hence, the sum of the slacks is $\sum_{k=1}^{K} \beta \times \max_{j \in \pi(i,k)} p_j$. In the worst case, the K-th unavailability finishes at the same time than the makespan (i.e., $e_K^i = C_{\max}$) and the unavailability periods are arbitrarily small (i.e., $\forall k \in [1..K], u_i^k = \epsilon$). Indeed, it maximizes both the number of jobs that are scheduled and their sizes (which maximizes thus the amount of interrupted work). Then, the sum of the processing times of the scheduled jobs is no more than

$$C_{\max} - \sum_{k=1}^{K} \beta \times \max_{j \in \pi(i,k)} p_j$$

(by discarding the ϵ durations).

In the worst case, each unavailability period undergoes disturbances and come earlier (while being still constant). Moreover, it interrupts a job of maximum duration scheduled in its corresponding availability period just before it can finish its execution. Thus, the sum of the jobs that need to be re-executed is $\sum_{k=1}^{K} \max_{j \in \pi(i,k)} p_j$.

As stated in Section 4, the execution is compact, namely, each job is executed as early as possible. Additionally, each unavailability may only interrupt one job. Therefore, we consider that a fraction of the interrupted jobs are re-executed before the makespan using the time reserved by each slack. Hence, the amount of work that need to be re-executed after the makespan is $(1-\beta)\sum_{k=1}^{K} \max_{j \in \pi(i,k)} p_j$. This amount is maximal when the minimum availability period is maximum, which occurs when each period has the same size (*i.e.*, $a_{min} = \frac{C_{max}}{K}$). Moreover, the largest job has half this size (*i.e.*, $p_{max} = \frac{C_{max}}{2K}$) and each interval has one such job. Therefore, the amount of work to be re-executed becomes $\frac{(1-\beta)}{2} \times C_{max}$.

We separate this amount into two parts. The first corresponds to jobs of maximum sizes while the second corresponds to the remainder (which can also be a job of maximum size that begins before the makespan and finishes after it). The delays due to each of these part are denoted D_1 and D_2, respectively.

There are $\left\lfloor \frac{(1-\beta)}{2} \times \frac{C_{max}}{p_{max}} \right\rfloor = \lfloor (1-\beta)K \rfloor$ jobs of maximum sizes that takes each an entire interval of minimum availability period and maximum unavailability to be re-executed (as it can be interrupted again on this interval). Therefore,

$$D_1 = \lfloor (1-\beta)K \rfloor (a_{min} + u_{max})$$

The second part of this amount (*i.e.*, $\frac{(1-\beta)}{2}C_{max} \mod p_{max}$) either belongs to a job starting its execution before the makespan or is a smaller job. In both cases, it takes a part of an availability period and one complete unavailability period to re-execute it. Therefore,

$$D_2 = (((1-\beta) \times K \mod 1) + \frac{1}{2})a_{min} + \lceil (1-\beta)K \mod 1 \rceil u_{max}$$

Therefore, the worst disturbed maskepan is

$$\tilde{C}_{max} = C_{max} + D_1 + D_2$$
$$\leq C_{max} + (1-\beta)Ka_{min} + Ku_{max} + \frac{a_{min}}{2}$$
$$\leq C_{max} + (1-\beta)C_{max} + \gamma C_{max} + \frac{C_{max}}{2}$$

The stability r_S can directly be derived from this last equation. □

6   Experiments

Simulations are run using data gathered from projects involving BOINC [1]. Traces about availabilities are collected from the project SETI@home [2]. For each processor, the traces provide the starting and ending dates of the availability periods of more than 110,000 processors. These traces were analyzed in [12] and clusters of processors with correlated availabilities were identified.

Workload traces were gathered from project Docking@Home (which was provided to us by Michela Taufer who also modeled the in-progress delay, *i.e.*, the

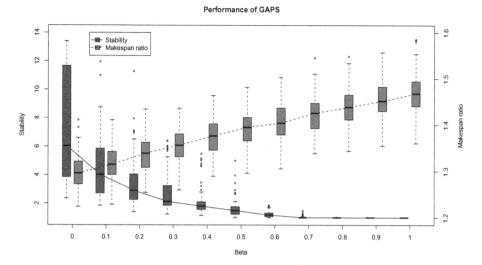

Fig. 3. Effect of the parameter β on the stability and the makespan ratio of GAPS. Each of the 1100 measures represents a simulation with 300 processors and 3000 jobs (GAPS is executed on 100 distinct task and machine instances with 11 values of β for each instance)

computation time required by jobs in several desktop grid projects [6]). These traces report the processing times of more than 150,000 jobs.

A preliminary analysis reports that the traces contain jobs with very short effective execution times (from some seconds to few minutes). In practice, they correspond to jobs that were interrupted during their executions. We remove 382 jobs that are shorter that 30 minutes, assuming that this is a reasonable lower duration that a job should have[1].

Each instance consists of a set of machines and a set of jobs. They are both generated randomly from the traces using a uniform distribution law. Moreover, 20% of processors are free (*i.e.*, with no unavailability period). Indeed, more than 20% of the machines were characterized to be available 95% or more of the time [12].

For each simulation, the inputs of the GAPS algorithm consists of an instance and a parameter β. We measure a lower bound of the makespan, the reference makespan and the disturbed makespan. The latter value is obtained by disturbing the actual schedule 30 times and by getting the median disturbed makespan. Disturbances are generated according to our model using a uniform distribution law. The makespan ratio is obtained by dividing the reference makespan by its lower bound, while the stability is the ratio between the disturbed makespan and the reference makespan.

---

[1] See the statistics reported in
http://www.boinc-wiki.info/Catalog_of_BOINC_Powered_Projects.

Figure 3 depicts the effect of β on the performances of GAPS. In the boxplots, the bold line is the median, the box shows the quartiles, the bars show the whiskers (1.5 times the interquartile range from the box) and additional points are outliers. For easing the reading, a line links each median. As expected, the stability decreases with high values of β whereas the makespan ratio increases. For low values of β, the makespan can increase by an order of magnitude in presence of disturbances. However, increasing β leads to a far better stability for a reasonable degradation of the makespan ratio (around 20%). Note also that it is not necessary to select a high β in order to obtain a good stability (*e.g.*, for $\beta \geq 0.7$, the stability is close to 1).

7   Concluding Remarks

In this work, we have proposed a complete study for scheduling jobs with unavailability periods in an uncertain context. We have introduced a new flexible mechanism based on the concept of adaptive slacks whose sizes are parametrized with the job durations. This leads to a bi-objective algorithm whose principle is to fill the successive intervals by jobs according to non-increasing processing times. The theoretical analysis was assessed by simulations based on actual data for both jobs and availability periods from projects involving BOINC.

Future work is directed towards the implementation of our algorithm for actual large scale applications. Moreover, we plan to extend our model and to adapt our results to uniform machines. Finally, we will derive analogous results for the lateness case in order to develop a more general theoretical framework.

References

1. Anderson, D.P.: Boinc: A system for public-resource computing and storage. In: 5th International Workshop on Grid Computing (GRID), pp. 4–10 (November 2004)
2. Anderson, D.P., Cobb, J., Korpela, E., Lebofsky, M., Werthimer, D.: SETI@home: An Experiment in Public-Resource Computing. Communications of the ACM 45(11), 56–61 (2002)
3. Billaut, J.-C., Moukrim, A., Sanlaville, E. (eds.): Flexibility and Robustness in Scheduling. Wiley-ISTE, Chichester (2008)
4. Canon, L.-C., Essafi, A., Mounié, G., Trystram, D.: A Bi-Objective Scheduling Algorithm for Desktop Grids with Uncertain Resource Availabilities. Research Report RR-LIG-014, LIG, Grenoble, France (2011)
5. Diedrich, F., Jansen, K., Pascual, F., Trystram, D.: Approximation Algorithms for Scheduling with Reservations. Algorithmica 58(2), 391–404 (2010)
6. Estrada, T., Taufer, M., Reed, K.: Modeling Job Lifespan Delays in Volunteer Computing Projects. In: CCGRID 2009: Proceedings of the 2009 9th IEEE/ACM International Symposium on Cluster Computing and the Grid, Washington, DC, USA, pp. 331–338 (2009)
7. Eyraud, L., Mounié, G., Trystram, D.: Analysis of Scheduling Algorithms with Reservations. In: 21st IEEE International Parallel & Distributed Processing Symposium (2007)

8. Fallah, M., Aryanezhad, M., Ashtiani, B.: Preemptive resource constrained project scheduling problem with uncertain resource availabilities: Investigate worth of proactive strategies. In: 2010 IEEE International Conference on Industrial Engineering and Engineering Management (IEEM), pp. 646–650 (December 2010)
9. Garey, M.R., Johnson, D.S.: Computers and Intractability: A Guide to the Theory of NP-Completeness. W. H. Freeman (1979)
10. Hwang, H.-C., Chang, S.Y.: Parallel Machines Scheduling with Machine Shutdowns. Computers & Mathematics with Applications 36(11), 21–31 (1998)
11. Hwang, H.-C., Lee, K., Chang, S.Y.: The effect of machine availability on the worst-case performance of LPT. Discrete Applied Mathematics 148(1), 49–61 (2005)
12. Kondo, D., Andrzejak, A., Anderson, D.P.: On correlated availability in internet-distributed systems. In: Proceedings of the 2008 9th IEEE/ACM International Conference on Grid Computing, Tsukuba, Japan, pp. 276–283 (September 2008)
13. Lambrechts, O., Demeulemeester, E., Herroelen, W.: Proactive and reactive strategies for resource-constrained project scheduling with uncertain resource availabilities. Journal of Scheduling 11(2), 121–136 (2008)
14. Lambrechts, O., Demeulemeester, E., Herroelen, W.: Time slack-based techniques for robust project scheduling subject to resource uncertainty. Annals of Operations Research, 1–22 (2010)
15. Larson, S.M., Snow, C.D., Shirts, M., Pande, V.S.: Folding@Home and Genome@Home: Using distributed computing to tackle previously intractable problems in computational biology. ArXiv e-prints (January 2009)
16. Lee, C.-Y.: Parallel machines scheduling with nonsimultaneous machine available time. Discrete Applied Mathematics 30(1), 53–61 (1991)
17. Lee, C.-Y.: Machine scheduling with an availability constraint. Journal of Global Optimization 9(3), 363–382 (1996)
18. Leung, J.Y.-T. (ed.): Handbook of Scheduling: Algorithms, Models, and Performance Analysis. Chapman & Hall/CCR (2004)
19. Liao, C.-J., Shyur, D.-L., Lin, C.-H.: Makespan minimization for two parallel machines with an availability constraint. European Journal of Operational Research 160(2), 445–456 (2005)
20. LIGO Scientific Collaboration. The Einstein@Home search for periodic gravitational waves in LIGO S4 data. ArXiv e-prints (April 2008)
21. Xu, D., Cheng, Z., Yin, Y., Li, H.: Makespan minimization for two parallel machines scheduling with a periodic availability constraint. Computers & Operation Research 36(6), 1809–1812 (2009)

New Multithreaded Ordering and Coloring Algorithms for Multicore Architectures

Md. Mostofa Ali Patwary[1], Assefaw H. Gebremedhin[2], and Alex Pothen[2]

[1] University of Bergen, Norway
Mostofa.Patwary@ii.uib.no
[2] Purdue University, West Lafayette, IN
agebreme@purdue.edu, apothen@purdue.edu

Abstract. We present new multithreaded vertex *ordering* and *distance-k graph coloring* algorithms that are well-suited for multicore platforms. The vertex ordering techniques rely on various notions of "degree", are known to be effective in reducing the number of colors used by a *greedy* coloring algorithm, and are generic enough to be applicable to contexts other than coloring. We employ *approximate degree* computation in the ordering algorithms and *speculation* and *iteration* in the coloring algorithms as our primary tools for breaking sequentiality and achieving effective parallelization. The algorithms have been implemented using OpenMP, and experiments conducted on Intel Nehalem and other multicore machines using various types of graphs attest that the algorithms provide scalable runtime performance. The number of colors the algorithms use is often close to optimal. The techniques used for computing the ordering and coloring in parallel are applicable to other problems where there is an inherent ordering to the computations that needs to be relaxed for increasing concurrency.

1   Introduction

Multicore platforms with support for multithreading have become commonplace and have reinvigorated the development of shared-memory parallel algorithms. We present new multithreaded algorithms well-suited for such platforms for two inter-related collection of graph problems: vertex *ordering* and *distance-k coloring*. Distance-1 coloring is used (among many others) in parallel scientific computing to discover tasks that can be carried out or data elements that can be updated concurrently [7,8]. Distance-2 coloring is an archetypal model used in the efficient computation of sparse Jacobian and Hessian matrices [5]. We rely on *greedy* algorithms that incorporate a vertex *ordering* stage to solve the coloring problems. The vertex ordering techniques we consider are formulated in a manner independent of a coloring algorithm. They are known to be effective in reducing the number of colors used by a greedy coloring algorithm, but are of interest in their own right with applications in areas outside coloring.

The ordering and coloring algorithms we consider are challenging to parallelize as the computation involved is inherently sequential. We overcome this fundamental challenge using approaches that potentially are useful for other problems as well. For the ordering algorithms, we employ *approximate degree* computation as a mechanism for increasing concurrency. We show that such an approach leads to a scalable performance,

E. Jeannot, R. Namyst, and J. Roman (Eds.): Euro-Par 2011, LNCS 6853, Part II, pp. 250–262, 2011.

whereas an approach that is faithful to the serial behavior of the ordering does not. The approximation-based method does not only lead to scalable performance, but is also far simpler. For the coloring algorithms, we use *speculation* and *iteration* as our primary ingredients for achieving scalable performance. We focus in this work on distance-2 coloring, although the techniques are equally applicable to distance-1 coloring. The algorithms we have developed are implemented using OpenMP. Experiments conducted on an Intel Nehalem machine using a set of graphs designed to cover a wide spectrum of input types show scalable runtime performance. The number of colors the algorithms use is nearly the same as in the serial case, which in turn is often close to optimal.

Like many other graph algorithms, the algorithms we have considered are plagued by several performance impediments besides low concurrency: poor data locality, irregular memory access pattern, and high data access to computation ratio. Our primary focus in this work is on algorithmic techniques and we pay almost no attention to optimization techniques that could further enhance performance.

Preliminaries, Related Work, and Organization. A distance-k coloring of a graph $G = (V, E)$ is an assignment of positive integers, called *colors*, to vertices such that any two vertices connected by a path consisting of at most k edges receive different colors. The objective in the distance-k coloring problem is to minimize the number of colors used, and the problem is known to be NP-hard for every fixed integer $k \geq 1$ (see [5] for pointers to references). Previous work has shown that a *greedy* coloring algorithm—an algorithm that visits vertices sequentially in some *order* in each step assigning a vertex the *smallest* permissible color—is quite effective in practice.

The order in which vertices are processed determines the number of colors used by the algorithm. In an earlier work [6], we identified three ordering techniques, called *Smallest Last* (SL), *Dynamic Largest First* (DLF), and *Incidence Degree* (ID) that are particularly effective in reducing the number of colors used by a greedy coloring algorithm and are generic enough to be useful in other contexts. In particular, the three ordering techniques are characterized (in [6]) purely in terms of *relative* vertex degrees, in a manner decoupled from the coloring algorithm that could use them. Such a characterization makes the orderings of interest in their own right and helps to more easily see their connections with other graph problems. For example, an SL ordering has interesting relationship with such graph concepts as degeneracy, core and arboricity (see [5] for some pointers). In this paper, we present algorithms—which are the first to the best of our knowledge—for *parallelizing* the aforementioned ordering techniques on multithreaded, shared-memory architectures. The algorithms are discussed in Sect. 2.

Using *speculation* and *iteration* as basic ingredients, a framework for effective parallelization of greedy distance-1 coloring on *distributed-memory* architectures was developed in [2]. The framework was extended to distance-2 coloring and related problems in [1]. Recently, a multithreaded algorithm derived from the framework in [2] and tailored for *shared-memory* architectures has been developed for the distance-1 coloring problem in [3]. We present in this paper a similar algorithm for distance-2 coloring on shared memory platforms. The algorithm is described in Sect. 3. We present experimental results in Sect. 4 and conclude in Sect. 5.

Algorithm 1. Template for serial ordering (SL, DLF, ID). Input: graph $G = (V, E)$. Output: An ordered list W of the vertices in V. B is a two-dimensional array used for maintaining *unordered* vertices binned according to their "degrees".

1: **procedure** ORDERINGTEMPLATE($G = (V, E)$)
2:     **for** each vertex $v \in V$ **do**
3:         init $d(v)$
4:         $B[d(v)] \leftarrow B[d(v)] \cup \{v\}$
5:     init i ▷ i is position in W where next vertex in the order is placed
6:     **while** check i **do**   ▷ there remain vertices to order
7:         locate j^*, an *appropriate extreme* index j where $B[j]$ is non-empty
8:         Let v be a vertex drawn from $B[j^*]$
9:         $W[i] \leftarrow v$
10:         $B[j^*] \leftarrow B[j^*] \backslash \{v\}$
11:         **for** each vertex $w \in adj(v)$ such that w is in B **do**
12:             $B[d(w)] \leftarrow B[d(w)] \backslash \{w\}$
13:             update $d(w)$
14:             $B[d(w)] \leftarrow B[d(w)] \cup \{w\}$
15:         update i

Table 1. Table accompanying the ordering template in Algorithm 1

	SL	DLF	ID
L 3: init $d(v)$	$d(v) \leftarrow d(v, G)$	$d(v) \leftarrow d(v, G)$	$d(v) \leftarrow 0$
L 5: init i	$i \leftarrow |V| - 1$	$i \leftarrow 0$	$i \leftarrow 0$
L 6 check i	$i \geq 0$	$i \leq |V| - 1$	$i \leq |V| - 1$
L 7: locate j^*	$j^* = \min_j\{B[j] \neq \emptyset\}$	$j^* = \max_j\{B[j] \neq \emptyset\}$	$j^* = \max_j\{B[j] \neq \emptyset\}$
L 13: update $d(w)$	$d(w) \leftarrow d(w) - 1$	$d(w) \leftarrow d(w) - 1$	$d(w) \leftarrow d(w) + 1$
L 15: update i	$i \leftarrow i - 1$	$i \leftarrow i + 1$	$i \leftarrow i + 1$

2   Vertex Ordering

2.1   The Serial Framework

We give in Algorithm 1 a succinct summary of a *template* for the ordering techniques SL, DLF and ID in the serial setting. Table 1 shows how the template is *specialized* in the three cases. The key idea in the definition (and computation) of these orderings is the use of a *dynamically* changing quantity, the *back* or *forward degree* of vertices. The back degree of a vertex v is the number of vertices that are adjacent to v in G and appear *before* v in the ordering, and the forward degree of v is the number of vertices that are adjacent to v in G and appear *after* v in the ordering. In Algorithm 1 and elsewhere in this paper, the dynamic degree (back or forward) of a vertex v is denoted by $d(v)$, and the *static* degree of the vertex in the input graph G is denoted by $d(v, G)$.

To arrive at an efficient implementation, a two-dimensional array B is used in Algorithm 1 to maintain vertices that are not yet ordered in *bins* according to their dynamic degrees. Specifically $B[j]$ stores a set of unordered vertices where each member vertex u has a current dynamic degree $d(u)$ equal to j. The output of Algorithm 1 is given by the ordered list W of the vertices where $W[i]$ stores the ith vertex in the ordering. In SL, the ordering W is computed right-to-left ($i = |V| - 1$ down to $i = 0$), whereas the ordering in DLF and ID is computed left-to-right ($i = 0$ up to $i = |V| - 1$). The ith vertex in SL ordering is a vertex with the *smallest* back degree among the vertices not yet ordered, in a DLF ordering it is a vertex with the *largest* forward degree among the vertices not yet ordered, and in an ID ordering it is a vertex with the *largest* back degree among the vertices not yet ordered. The rationale behind each of these ordering techniques in the context of a coloring algorithm is to bring vertices that are likely to be highly constrained in choice of colors early in the ordering.

In Line 7 in Algorithm 1, we determine the ith vertex in the ordering in constant time by maintaining a pointer to the last element in the smallest (or largest) index j such that $B[j]$ is non-empty. Once the ith vertex v in the ordering is determined (and removed from B), each unordered vertex w adjacent to v is moved from its current bin in B to an appropriate new bin. With suitable pointer techniques the relocation can also be performed in constant time [6]. Thus the work involved in the ith step of Algorithm 1 is proportional to $d(v, G)$, and the overall complexity of the algorithm is $O(|E|)$.

We point out another interesting connection between the template in Algorithm 1 and an ordering used for an entirely different purpose: an ID ordering obtained by Algorithm 1, when reversed, corresponds to an ordering obtained by the *maximum cardinality search* algorithm [9], which arises in the context of solving sparse linear systems.

2.2 Parallel Ordering

We parallelized the three ordering techniques SL, DLF, and ID employing a common paradigm, but we restrict the presentation in this paper to only SL ordering.

We developed two different approaches for the parallelization. The first approach aims at parallelizing the ordering closely maintaining the serial behavior, while the second approach settles for an approximate solution in favor of increased concurrency. In both approaches, we assume p threads are available and utilized, and we denote by $t(v)$ the thread with which the vertex v is initially associated.

The First Approach—Regular. Algorithm 2 outlines the first approach. The first task Algorithm 2 parallelizes is the population of the global bin array B. To achieve this, with each thread T_k, $1 \le k \le p$, a *local* two-dimensional array B_k is associated. The p local arrays are first populated in parallel (the for-loop in Lines 2–4). Then, the contents are gathered into the global array B, where the parallelization is now switched to run over bins, as shown in the for-loop in Lines 5–8. There and elsewhere in this paper, $\delta(G)$ and $\Delta(G)$ denote the minimum and maximum degree in G, respectively.

The remainder of Algorithm 2 mimics the serial algorithm (Algorithm 1). In the serial algorithm, in each step of the while loop, a *single* vertex—a vertex with the *smallest* current dynamic degree j^*—is ordered and its neighbors' locations updated in B. However, the bin $B[j^*]$ could contain *multiple* vertices. Algorithm 2 takes advantage of this opportunity and strives to order such vertices and update their neighborhoods in parallel. There are a few potential problems that need to be attended while doing so.

– *Problem*: A pair of vertices u and v in $B[j^*]$ are adjacent to each other. In such a case, a thread processing one of the vertices, say u, could try to move the vertex v to

Algorithm 2. A parallel SL ordering algorithm using p threads (the REGULAR variant). Input: graph $G = (V, E)$. Output: An ordered list W of the vertices in V. The array B is as in Algorithm 1, and the arrays B_t, R_t, and A_t are thread-private arrays; the latter two are used to remove or add vertices from or into the global array B.

1: **procedure** SMALLESTLASTORDERING-REGULAR($G = (V, E)$)
2:      **for** each vertex $v \in V$ in `parallel` **do**
3:          $d(v) \leftarrow d(v, G)$
4:          $B_{t(v)}[d(v)] \leftarrow B_{t(v)}[d(v)] \cup \{v\}$
5:      **for** each bin $j \in \{\delta(G), \ldots, \Delta(G)\}$ in `parallel` **do**
6:          **for** $k = 1$ to p **do**
7:              **for** each vertex $v \in B_k[j]$ **do**
8:                  $B[j] \leftarrow B[j] \cup \{v\}$ ▷ note that $j = d(v)$
9:      $i \leftarrow |V|$
10:     **while** $i \geq 0$ **do**
11:         Let j^* denote the *smallest* index j such that $B[j]$ is non-empty
12:         **for** each vertex $v \in B[j^*]$ in `parallel` **do**
13:             **for** each vertex $w \in adj(v)$ such that w is in B **do**
14:                 **if** $w \notin R_{t(v)}$ **then**
15:                     $R_{t(v)}[d(w)] \leftarrow R_{t(v)}[d(w)] \cup \{w\}$
16:                 $r(w) \leftarrow r(w) + 1$ ▷ *atomic* operation
17:             $W[i] \leftarrow v; i \leftarrow i - 1$ ▷ *critical* statements
18:         **for** each bin $j \in \{j^*, \ldots, \Delta(G)\}$ in `parallel` **do**
19:             **for** $k = 1$ to p **do**
20:                 **for** each vertex $v \in R_k[j]$ **do**
21:                     **if** $r(v) > 0$ **then**
22:                         $B[j] \leftarrow B[j] \setminus \{v\}$ ▷ note that $j = d(v)$
23:                         $d(v) \leftarrow d(v) - r(v); r(v) \leftarrow 0$
24:                         $A_{t(v)}[d(v)] \leftarrow A_{t(v)}[d(v)] \cup \{v\}$
25:         **for** each bin $j \in \{j^*, \ldots, \Delta(G)\}$ in `parallel` **do**
26:             **for** $k = 1$ to p **do**
27:                 **for** each vertex $v \in A_k[j]$ **do**
28:                     $B[j] \leftarrow B[j] \cup \{v\}$ ▷ note that $j = d(v)$

---

another bin while another thread at the same time attempts to order v, making the result inconsistent. *Solution*: While ordering the vertex u, we avoid updating the location of the vertex v in B, and instead order v as well in the current step (see Lines 12–17).

– *Problem*: Removal of multiple vertices from the same bin, say $B[j]$. Suppose two vertices u and v from $B[j^*]$ have a common neighbor w in $B[j]$. In the serial case, u and v would be ordered one after another, $d(w)$ would be reduced by 2, and w would be relocated twice. In the parallel case, two threads might try to remove w from $B[j]$ at the same time and the removal of w in constant time will make $B[j]$ inconsistent. Similarly, suppose two vertices u and v in $B[j^*]$ have respective neighbors w and x such that $d(w) = d(x) = j$. In the parallel case, two threads might try to remove w and x from $B[j]$ at the same time while processing u and v in parallel and the removals of w and x in constant time will also make $B[j]$ inconsistent. *Solution*: We let each thread

---

Algorithm 3. A parallel SL ordering algorithm on p threads (the RELAXED variant). Input: graph $G = (V, E)$. Output: An ordered list W of the vertices in V.

1: **procedure** SMALLESTLASTORDERING-RELAXED($G = (V, E)$)
2:    **for** each vertex $v \in V$ in `parallel` **do**
3:        $d(v) \leftarrow d(v, G)$
4:        $B_{t(v)} [d(v)] \leftarrow B_{t(v)} [d(v)] \cup \{v\}$
5:    $i \leftarrow |V|$
6:    **for** $k = 1$ to p in `parallel` **do**
7:        **while** $i \geq 0$ **do**
8:            Let j^* be the *smallest* index j such that $B_k [j]$ is non-empty
9:            Let v be a vertex drawn from $B_k [j^*]$
10:           $B_k [j^*] \leftarrow B_k [j^*] \setminus \{v\}$
11:           **for** each vertex $w \in adj(v)$ **do**
12:               **if** $w \in B_k$ **then**
13:                   $B_k [d(w)] \leftarrow B_k [d(w)] \setminus \{w\}$
14:                   $d(w) \leftarrow d(w) - 1$
15:                   $B_k [d(w)] \leftarrow B_k [d(w)] \cup \{w\}$
16:           $W[i] \leftarrow v;\ i \leftarrow i - 1$ ▷ *critical* statements

---

$T_k, 1 \leq k \leq p$, maintain its own two-dimensional *removal* array R_k, where it stores vertices to be removed from B while the parallel ordering of $B[j^*]$ happens (see the for loop in Lines 13–16). The removal from B takes place once the ordering of vertices in $B[j^*]$ is completed. Since for any two bins $B[j]$ and $B[j']$ the removal from $B[j]$ is independent of the removal from $B[j']$, these could be done in parallel, as shown in Lines 18–24.

– *Problem*: Addition of multiple vertices to the same bin, say $B[j]$. *Solution*: We address this concern by using a similar technique as in the second bullet item. We let each thread maintain its own two-dimensional *addition* array A_k. Again, the addition of vertices to different bins in B can be done in parallel, as shown in Lines 25–28.

The Second Approach—Relaxed. Our second approach for parallelizing the SL ordering algorithm abandons the use of the global array B altogether, and works only with the local arrays B_k associated with each thread T_k. In updating locations of neighbors of a vertex, a thread T_k checks whether or not the vertex w desired to be relocated is in the thread's local array B_k. If w is indeed in B_k it is relocated by the same thread, if not, it is simply ignored. In this manner, only *approximate* dynamic degrees are used while computing the ordering. The approach is formalized in Algorithm 3.

3   Parallel Distance-2 Coloring

The sequential greedy distance-2 coloring algorithm we seek to parallelize iterates over the vertex set V of the graph G, in each step assigning a vertex v the smallest color not used by any of its distance-2 neighbors. It can be implemented such that its complexity is $O(|V| \cdot \bar{d}_2)$, where $\bar{d}_2$ denotes the average number of distinct paths of length at most two edges leaving a vertex [5]. Algorithm 4 shows how we have parallelized the greedy algorithm in this work. The algorithm has two phases, both of which are performed in parallel, and runs in an iterative fashion. In the first phase of each round of the iteration, threads concurrently color their respective vertices in a *speculative* manner (paying attention to already available color information). In this phase, two vertices that are distance-2 neighbors with each other and are handled by two different threads may be colored concurrently and receive the same color, causing a *conflict*. In the second phase, threads concurrently check the validity of colors assigned to their respective vertices in the current round and identify a set of vertices that needs to be re-colored in the next round to resolve any detected conflicts. The algorithm terminates when every vertex has been colored correctly. In the event of a conflict, it suffices to re-color one of the two involved vertices to resolve the conflict. In Algorithm 4 (see Lines 14 and 17), we used the value (id) of vertices to decide the vertex to re-color. Other strategies, such as the use of random numbers associated with vertices, are also possible [2].

Although the tentative coloring and conflict detection phases in each round iterate over the same set U of vertices performing similar operations per vertex visit, the runtime of the conflict detection phase can be significantly reduced by terminating the search for a conflict in the distance-2 neighborhood of a vertex v as soon as the first conflict impacting v is discovered. This is achieved using the **break** statements in Lines 15 and 18. Note that the *cont* boolean variable in Line 12 is used to break out of the for-loop in Line 13 due to a condition in the for-loop in Line 16. Thanks to the use of the early breaks, we observed that the conflict detection phase typically takes roughly around 25% of the overall runtime of the algorithm; without the breaks the conflict detection phase would have taken the same time as the tentative coloring phase.

Scheduling. In the parallel coloring algorithm we just described as well as the parallel ordering algorithms discussed in Sect. 2.2, the runtime performance of the algorithms depends on the manner in which vertices are scheduled on threads. In the results we report in the next section, we used the *dynamic* scheduling option of OpenMP.

4   Experimental Results

In this section we present results on experiments performed on an Intel Nehalem machine equipped with Intel(R) Core(TM) i7 CPU 860 processors running at 2.8GHz. The system has 4 cores with 2 threads on each. The total memory size is 16 GB, with 4×32 KB Instruction and 4×32 KB Data Level-1 cache, 4×256 KB Level-2 cache, and 8 MB shared Level-3 cache. The operating system is GNU/Linux.

Our testbed consists of 20 graphs. Five of them are real-world graphs drawn from various *scientific computing* (sc) applications and are downloaded from the University of Florida Sparse Matrix Collection. The remaining 15 are synthetically generated using the R-MAT algorithm [4]. By combining the four input parameters of the R-MAT algorithm in various ways (the sum of the parameters needs to be equal to one), it is possible to generate graphs with varying properties. We generated three types of graphs: (i) *Erdős-Renyi random* (er) graphs, using the set of parameters $(0.25, 0.25, 0.25, 0.25)$; (ii) *small-world type 1* (g) graphs, using the set of parameters $(0.45, 0.15, 0.15, 0.25)$; (iii) *small-world type 2* (b) graphs, using the set of parameters $(0.55, 0.15, 0.15, 0.15)$.

Algorithm 4. An iterative parallel algorithm for distance-2 coloring using p threads. Input: graph $G = (V, E)$. Output: a vertex-indexed array *color* [] indicating colors of vertices. The vertex set V is assumed to be *ordered*.

1: **procedure** ITERATIVED2COLORING($G = (V, E)$)
2:     $U \leftarrow V$
3:     **while** $U \neq \emptyset$ **do**
4:         **for** each vertex $v \in U$ in parallel **do**   ▷ Phase 1: tentative coloring
5:             **for** each vertex $w \in adj(v)$ **do**
6:                 mark *color* $[w]$ as forbidden to vertex v
7:                 **for** each vertex $x \in adj(w)$ **and** $x \neq v$ **do**
8:                     mark *color* $[x]$ as forbidden to vertex v
9:             Pick the *smallest permissible* color c for vertex v
10:        $R \leftarrow \emptyset$ ▷ R denotes the set of vertices to be recolored
11:        **for** each vertex $v \in U$ in parallel **do**   ▷ Phase 2: conflict detection
12:            $cont \leftarrow true$
13:            **for** each vertex $w \in adj(v)$ **and** $cont = true$ **do**
14:                **if** *color* $[v]$ = *color* $[w]$ **and** $v > w$ **then**
15:                    $R \leftarrow R \cup \{v\}$; **break**
16:                **for** each vertex $x \in adj(w)$ **and** $v \neq x$ **do**
17:                    **if** *color* $[v]$ = *color* $[x]$ **and** $v > x$ **then**
18:                        $R \leftarrow R \cup \{v\}$; $cont \leftarrow false$; **break**
19:        $U \leftarrow R$

These three graph types vary widely in terms of *degree distribution* of vertices and *density of local subgraphs* and represent a wide spectrum of input types posing varying degrees of difficulty for the ordering and coloring algorithms. The er graphs have *normal* degree distribution, whereas the g (for "good") and b ("bad") graphs contain many dense local subgraphs (by good and bad is meant relatively "easy" and "hard" input types). The good and bad graphs differ primarily in the magnitude of maximum vertex degree they contain, the bad graphs have much larger maximum degree. Table 2 provides structural information on all 20 test graphs.

Figure 1 shows scalability results on the two parallel Smallest Last ordering algorithms, SL-Regular (Algorithm 2) and SL-Relaxed (Algorithm 3). The plots show runtimes for various numbers of threads *normalized* by the runtime when 1 thread is used. The raw runtime numbers for the 1 thread case along with the runtime of the pure *sequential* SL ordering and distance-2 coloring algorithms are provided in Table 3. Clearly, the algorithm SL-Regular scaled poorly especially for the sc and rmat-b graphs, whereas SL-Relaxed scaled well across all the graph types tested. We therefore present further results using the better performing algorithm SL-Relaxed.

Figure 2 shows scalability results for the parallel distance-2 coloring algorithm (Algorithm 4) while using the SL-Relaxed algorithm for parallel ordering. The left column shows runtime results considering *only* the coloring stage, whereas the right column shows results on *total* (ordering plus coloring) time. Since distance-2 coloring takes substantially more time than the ordering (recall that the respective sequential complexities are $O(|V| \cdot \bar{d}_2$ and $O(|V| \cdot \bar{d}_1)$), the scalability behavior of just the coloring

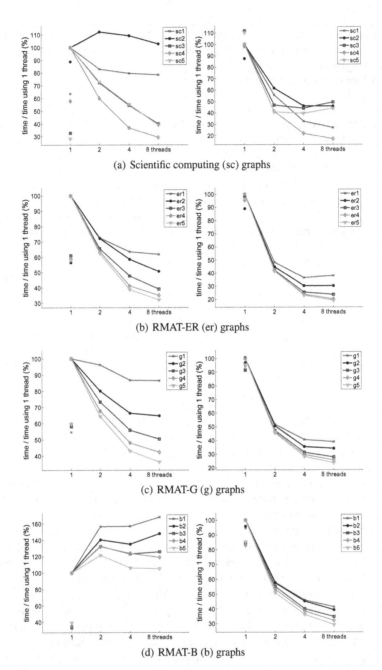

Fig. 1. Scalability results on the two parallel SL ordering algorithms. Left column: Algorithm 2 (SL-Regular). Right column: Algorithm 3 (SL-Relaxed). The plots show runtimes normalized by the runtime when 1 thread is used; the raw numbers for the case of 1 thread are listed in Table 3. Also shown are data points corresponding to runtime of the pure *sequential* algorithm normalized by the runtime of the parallel algorithm run on 1 thread.

Table 2. Structural properties of the various graphs in the testbed: scientific computing (sc), rmat-random (er), rmat-good (g), and rmat-bad (b). Δ denotes maximum degree in G

Name	$|V|$	$|E|$	Δ	Name	$|V|$	$|E|$	Δ
sc1 (bone010)	986,703	35,339,811	80	g1	262,144	2,093,552	558
sc2 (af_shell10)	1,508,065	25,582,130	34	g2	524,288	4,190,376	618
sc3 (nlpkkt120)	3,542,400	46,651,696	27	g3	1,048,576	8,382,821	802
sc4 (er1)	16,777,216	134,217,651	138	g4	2,097,152	16,767,728	1,069
sc5 (nlpkkt160)	8,345,600	110,586,256	27	g5	4,194,304	33,541,979	1,251
er1	262,144	2,097,104	98	b1	262,144	2,067,860	4,493
er2	524,288	4,194,254	94	b2	524,288	4,153,043	6,342
er3	1,048,576	8,388,540	97	b3	1,048,576	8,318,004	9,453
er4	2,097,152	16,777,139	102	b4	2,097,152	16,645,183	14,066
er5	4,194,304	33,554,349	109	b5	4,194,304	33,340,584	20,607

Table 3. Runtime in seconds of the pure sequential algorithms, and of the parallel algorithms when run using one thread. OT shows ordering time, and CT shows distance-2 coloring time

	SL-Seq.		SL-Relaxed		SL-Regular			SL-Seq.		S L-Relaxed		SL-Regular	
	OT	CT	OT	CT	OT	CT		OT	CT	OT	CT	OT	CT
sc1	1.11	20.66	1.18	30.45	1.73	31.18	g1	0.18	2.68	0.18	3.82	0.32	3.84
sc2	0.83	6.83	0.87	10.13	0.91	10.25	g2	0.45	6.31	0.42	8.86	0.74	9.03
sc3	2.05	11.38	1.64	16.45	6.39	28.89	g3	1.02	16.22	1.07	23.25	1.74	23.69
sc4	30.54	306.47	31.19	452.76	51.68	479.81	g4	2.49	43.16	2.54	61.98	4.18	65.64
sc5	5.15	27.51	4.29	39.86	17.68	73.91	g5	5.86	119.20	6.01	168.64	9.59	171.84
er1	0.18	1.45	0.18	2.13	0.32	2.21	b1	0.16	9.20	0.16	12.68	0.44	12.63
er2	0.43	3.30	0.45	5.02	0.71	5.23	b2	0.37	24.10	0.37	32.11	0.95	32.36
er3	1.22	9.24	1.20	12.75	1.84	13.54	b3	0.75	70.26	0.87	94.30	2.09	95.60
er4	2.77	22.48	2.86	33.62	4.54	36.07	b4	1.74	195.60	2.00	280.00	4.48	281.39
er5	6.30	57.13	6.43	83.74	10.51	88.77	b5	4.21	565.59	4.85	785.80	9.87	797.86

stage is nearly identical to that of the overall execution. It can be seen that the coloring algorithm (including the ordering stage) scaled well across all the graphs in the testbed.

Also shown in Figures 1 and 2 is the runtime of a relevant *sequential* algorithm normalized by the runtime of the corresponding parallel algorithm run on 1 thread. This shows the performance advantage (besides functionality) gained by parallelization.

Figure 3 shows the number of colors the parallel distance-2 coloring algorithm (Algorithm 4) used while employing the SL-Relaxed ordering algorithm. In each subfigure, a bar corresponding to the maximum degree (Δ) in a graph, which is a lower bound on the optimal number of colors needed to distance-2 color a graph, is included. It can be seen that the number of colors the parallel algorithm used remained nearly constant as the number of threads is increased for all except the sc graphs. Further, it can be seen that the number in each case is either optimal or very close to optimal.

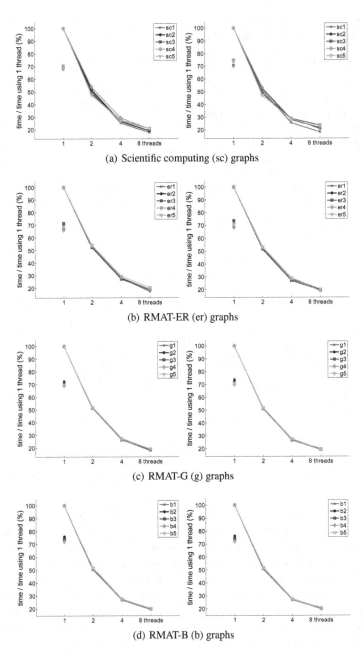

(a) Scientific computing (sc) graphs

(b) RMAT-ER (er) graphs

(c) RMAT-G (g) graphs

(d) RMAT-B (b) graphs

Fig. 2. Scalability results on the parallel distance-2 coloring algorithm (Algorithm 4) while employing the parallel ordering algorithm SL-Relaxed (Algorithm 3). Left column: only distance-2 coloring time. Right column: ordering plus distance-2 coloring time. The plots show runtimes normalized by the runtime when 1 thread is used; the raw numbers for the case of 1 thread are listed in Table 3. Also shown are data points corresponding to runtime of the pure *sequential* algorithm normalized by the runtime of the parallel algorithm run on 1 thread.

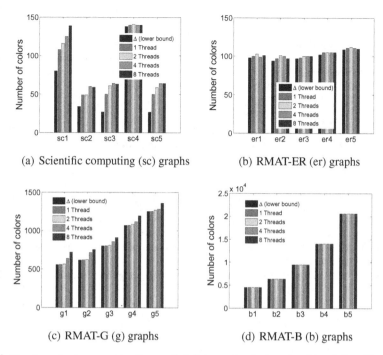

(a) Scientific computing (sc) graphs

(b) RMAT-ER (er) graphs

(c) RMAT-G (g) graphs

(d) RMAT-B (b) graphs

Fig. 3. Number of colors used by the parallel distance-2 coloring algorithm (Algorithm 4) while employing the SL-Relaxed ordering algorithm (Algorithm 3) for various thread counts. The first bar in each subfigure shows the lower bound Δ on the optimal number of colors.

5   Conclusion and Future Work

We presented new parallel ordering and coloring algorithms and a small set of experimental results demonstrating scalable performance on a multicore machine supporting a modest number of threads. Some details and experimental results were omitted for space consideration. In future work, we intend to conduct further studies and provide more extensive results using machines supporting much larger number of threads. One issue we will investigate at large thread count is runtime scalability while maintaining quality of serial solution (to avoid increase in number of colors for some input types as those observed in Figure 3 a). The color choice strategy (see Line 9 of Algorithm 4) used in all of the results reported in this paper is *First Fit*, i.e., each thread searches for a permissible color for a vertex starting from *color 1*. We intend to investigate the merits of alternative color choice strategies (such as Staggered First Fit, Least Used, Random etc [2]) that could reduce the likelihood of conflicts.

Acknowledgements. We thank Fredrik Manne and Duc Nguyen for helpful discussions, and the referees for their helpful comments. One of the referees pointed out the relevance of reference [9]. This research was supported by the U.S. Department of Energy through the CSCPAES Institute grant DE-FC02-08ER25864 and by the National Science Foundation through grant CCF-0830645.

References

1. Bozdağ, D., Catalyurek, U.V., Gebremedhin, A.H., Manne, F., Boman, E.G., Ozgunner, F.: Distributed-memory parallel algorithms for distance-2 coloring and related problems in derivative computation. SIAM J. Sci. Comput. 32(4), 2418–2446 (2010)
2. Bozdağ, D., Gebremedhin, A.H., Manne, F., Boman, E.G., Catalyurek, U.V.: A framework for scalable greedy coloring on distributed-memory parallel computers. Journal of Parallel and Distributed Computing 68(4), 515–535 (2008)
3. Catalyurek, U., Feo, J., Gebremedhin, A.H., Halappanavar, M., Pothen, A.: Multithreaded algorithms for graph coloring. Submitted for Journal Publication (2011)
4. Chakrabarti, D., Faloutsos, C.: Graph mining: Laws, generators, and algorithms. ACM Comput. Surv. 38(1), 2 (2006)
5. Gebremedhin, A.H., Manne, F., Pothen, A.: What color is your Jacobian? Graph coloring for computing derivatives. SIAM Review 47(4), 629–705 (2005)
6. Gebremedhin, A.H., Nguyen, D., Patwary, M.M.A., Pothen, A.: ColPack: Graph coloring software for derivative computation and beyond. Submitted for Journal Publication (2010)
7. Jones, M.T., Plassmann, P.E.: Scalable iterative solution of sparse linear systems. Parallel Computing 20(5), 753–773 (1994)
8. Saad, Y.: ILUM: A multi-elimination ILU preconditioner for general sparse matrices. SIAM J. Sci. Comput. 17, 830–847 (1996)
9. Tarjan, R.E., Yannakakis, M.: Simple linear-time algorithms to test chordality of graphs, test acyclicity of hypergraphs, and selectively reduce acyclic hypergraphs. SIAM J. Comput. 13(3), 566–579 (1984)

Introduction

Jesper Larsson Träff, Brice Goglin, Ulrich Bruening, and Fabrizio Petrini

Topic chairs

The Euro-Par topic on high-performance networks and communications is devoted to communication issues in scalable compute and storage systems, such as tightly coupled parallel computers, clusters, and networks of workstations, including hierarchical and hybrid designs featuring several levels of possibly different interconnects. All aspects of communication in modern compute and storage systems are of interest, for example advances in the design, implementation, and evaluation of interconnection networks, network interfaces, system and storage area networks, on-chip interconnects, communication protocols and interfaces, routing and communication algorithms, and communication aspects of parallel and distributed algorithms.

The papers submitted for the topic were reviewed by the four chairs and their selected subreviewers. Bar two (who had three reviews), the papers received 4, hopefully useful reviews (that in some cases admittedly could have been more extensive). The several submitted papers all fitted well to the call for papers as outlined above and the specific list of themes. Based on the reviews, the quality aspirations and the overall balance of Euro-Par, only two contributions were accepted for presentation at the conference, making for a selective topic. The topic chairs thank all submitting authors, the presenters, and the audience who will be listening and participating in the discussions. High-quality submissions to the topic also in the coming years are encouraged.

The first paper presented at the conference titled *Kernel-Based Offload of Collective Operations - Implementation, Evaluation and Lessons Learned* by Timo Schneider, Sven Eckelmann, Torsten Hoefler and Wolfgang Rehm deals with issues in offloading collective communication algorithms to the communication network layer. A kernel-based architecture for implementing a framework for offloading such algorithms is described, implemented and experimentally evaluated with specific microbenchmarks on a standard cluster by comparing to traditional implementations of (non-blocking) collective operations with progress in user-space. Especially reduced CPU overhead and improvement in the capability to overlap communication with computation are shown.

The second paper by Alexandre Denis on *A High Performance Superpipeline Protocol for InfiniBand* discusses improved pipeline schemes for point-to-point communication, and in particular gives a more detailed analysis than usual in a *LogP* inspired performance model. Benchmarks show that in particular the costs of memory registration can be eliminated, making for significantly better performance even on "first touch" of a user-space communication buffer.

E. Jeannot, R. Namyst, and J. Roman (Eds.): Euro-Par 2011, LNCS 6853, Part II, p. 263, 2011.
© Springer-Verlag Berlin Heidelberg 2011

Kernel-Based Offload of Collective Operations – Implementation, Evaluation and Lessons Learned

Timo Schneider[1], Sven Eckelmann[1], Torsten Hoefler[2], and Wolfgang Rehm[1]

[1] TU Chemnitz, Germany
{timos,ecsv,rehm}@hrz.tu-chemnitz.de
[2] University of Illinois at Urbana-Champaign, IL, USA
htor@illinois.edu

Abstract. Optimized implementations of blocking and nonblocking collective operations are most important for scalable high-performance applications. Offloading such collective operations into the communication layer can improve performance and asynchronous progression of the operations. However, it is most important that such offloading schemes remain flexible in order to support user-defined (sparse neighbor) collective communications. In this work, we describe an operating system kernel-based architecture for implementing an interpreter for the flexible Group Operation Assembly Language (GOAL) framework to offload collective communications. We describe an optimized scheme to store the schedules that define the collective operations and show an extension to profile the performance of the kernel layer. Our microbenchmarks demonstrate the effectiveness of the approach and we show performance improvements over traditional progression in user-space. We also discuss complications with the design and offloading strategies in general.

1 Introduction

The Message Passing Interface (MPI) standard [12] is the de-facto standard for implementing today's large-scale high-performance applications. Part of MPI's success is it's high portability, not only from a correctness, but also from a performance perspective. This is achieved by defining several high-level collective communication operations that specify communication primitives on groups of processes instead of process pairs. The implementation of such collective operations can now be optimized to the particular machine architecture and network topology. Several non-trivial algorithms have been developed to optimize those group communications, e.g., [2, 16].

A recent addition to the MPI standard (in the upcoming MPI-3.0 standard) builds upon this success and introduces nonblocking versions of all MPI collective operations [7]. Nonblocking collective operations allow the application to perform computations while the communication (and synchronization) is performed "in the background".

E. Jeannot, R. Namyst, and J. Roman (Eds.): Euro-Par 2011, LNCS 6853, Part II, pp. 264–275, 2011.

Different software implementation options have been explored for different network architectures [6] but the major problem, how to progress the collective algorithm efficiently, remains open. This problem exists because most advanced collective algorithms have multiple stages where a stage can only be started if some preconditions are satisfied. A simple example is a binary tree where an inner node can only send the message to its children after it has received it from its parents. Checking if a message was received and conditionally starting new transmissions requires to transfer the program control from the application to the collective implementation. Hoefler and Lumsdaine analyzed in [5] different schemes to progress the communication subsystem. Their study showed that, without loosing CPU power (cores), the application needs to progress the library by calling it regularly (e.g., with MPI_Test()). This solution is of course not feasible in the general case due to long calls to libraries that are not MPI-aware (e.g., Level 3 BLAS).

As systems grow larger, collective operations on the whole set of processes might not be feasible. Even though, many collective operations scale logarithmically with the number of processes for small input sizes, frequent communication can inhibit scalability. Thus, algorithm-design needs to address this issue and localized communications (e.g., nearest neighbor) become most important. Nevertheless, most algorithms are still written in a bulk synchronous model [17] with iterative communication and computation phases and the computation phases of many such applications communicate within a fixed neighborhood (e.g., each process has four neighbors in a two-dimensional five-point stencil computation). Such neighbor exchanges can be viewed as a localized (or *sparse*) collective group communication and optimized with similar principles as traditional collective operations [10]. The addition of a set of calls to support such a communications is considered for MPI-3.0. The communication topology of such sparse collective communications is expressed by the user at runtime and their nonblocking variants suffer from similar progression issues as traditional (we call them *dense*) collective operations.

1.1   Related Work

Several communication systems offer direct (offloaded) hardware support for some MPI collective operations [1,14]. However, such implementations often fail to support the full spectrum of collective operations and cannot express user-defined sparse collectives.

Several works propose to offload an abstract definition of a collective operation into the communication layer (e.g., a network interface card). Portals 4 [15] specifies triggered operations where new messages can be sent based on arriving messages. InfiniBand ConnectX-2 [4] specifies chained Queue Pair operations which can trigger new messages inside the HCA. GOAL allows to specify communication schedules as complete dependency graphs that can be downloaded into the communication layer [9]. All offload techniques allow nearly fully asynchronous execution of nonblocking collective operations with minimal impact on the running application. Akihiro and Ishikawa show a possible design for kernel-level asynchronous operations in [13].

Open-MX [3] offers fast point-to-point communication for Ethernet networks. It is similar to ESP (cf. Section 3.2) in that it uses the Linux skb mechanism to transmit data. However, large parts of the protocol (for example reliability) are handled in user-space. Thus, it is not possible to use it for reliable communication from kernel-space yet.

In this work, we discuss a possible implementation of the flexible Group Operation Assembly Language (GOAL) framework in a general purpose operating system. GOAL allows to express arbitrary communication patterns and dependencies. The operating system acts as the resource broker on each end-node, it can immediately react to incoming messages (interrupts) from the hardware and progress the collective communication and thus solve the progression issue. In Section 2.1, we will discuss optimized design options for collective operation schedules, kernel-level execution limitations, and an extension for performance profiling of the kernel layer. In Section 3 we discuss our experimental design of the kernel-level in detail. Results are presented in Section 4 followed by a discussion of issues in our design and conclusions.

2   Expressing Collective Operations

GOAL allows to specify communication as a local dependency graph on each process [9]. The basic set of supported vertex types are sends, receives, and local operations. Dependencies (edges) can be added to enforce a certain execution order (i.e., an edge A→B means that operation A needs to complete before operation B is started). The matching send/receive statements across processes form a global communication graph that can be transformed during a compilation phase. GOAL allows the specification and transparent optimization of complex communication patterns.

Lower-level APIs, such as ConnectX-2 or Portals 4 would act as a concrete machine language, something that abstract GOAL graphs could be compiled into. However, both interfaces are only available on certain hardware platforms. In this work, we define a scheme which enables the execution of GOAL graphs within an operating system on standard hardware, such as Ethernet.

2.1   The GOAL Interpreter

The task of the GOAL interpreter is to take an optimized representation of the dependency graph and execute the primitive operations which are defined by it. Each primitive operation should be executed as early as possible but without violating the specified dependencies. GOAL graphs are serialized in traversal order and are stored as a cache-friendly binary format called *schedule*. In our implementation[1], the binary schedule is stored in the format described in Figure 1.

---

[1] http://www.tu-chemnitz.de/informatik/RA/dw/doku.php?id=en:espgoal:study

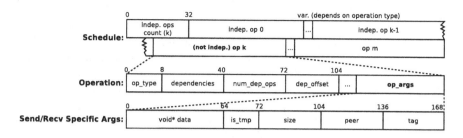

Fig. 1. Schedule Binary Format

This representation has the advantage that the whole dependency graph is stored in a contiguous memory block. This enables fast copying of the graph when transferring it from userspace to kernelland. If there is a dependency between the two operations u and v, which would be represented as edge $u \to v$ in the dependency graph, the dependency counter (*dependencies*) of the operation v will be at least one. The offset of the operation v will be listed in the adjacency list (num_dep_ops, dep_offset_i) of u.

When the interpreter starts to execute a schedule the *indep_ops_count* operations that have no incoming dependencies are executed first. Because each operation has a different set of arguments, the type of operation is encoded in an eight bit value at the beginning of each operation, so that the interpreter knows how much data has to be read. The *dependencies* value specifies how many incoming edges this operation has. When this counter reaches zero the operation will be executed. The scheduler is notified by the underlying network protocol whenever an operation is finished. When the scheduler starts an operation it's offset in the schedule binary is passed to network protocol layer. This address is also present in the information the scheduler gets upon completion. The adjacency list num_dep_ops, dep_offset_i of the finished operation will be traversed. It contains the offsets of all operations that depend on this (now finished) operation. For each operation in that list the scheduler will decrease the *dependencies* counter by one. If such a counter reaches zero, the corresponding operation is executed.

GOAL supports three types of primitive operations: send, receive and local operations. Each of those operations either operates on a single contiguous block of data or on scatter/gather lists. Send and receive operations are non-blocking. An operation completes if all specified buffers can be read and modified. That implies that a send operation can be finished as soon as the data has been copied into a temporary buffer in the case of eager send. The schedule execution is non-blocking, thus, all send and receive operations are implicitly nonblocking. Local operations are predefined arithmetic (add, sub, mult, div, max, min) and binary operations (and, or, xor) on all signed, unsigned and floating point datatypes from one to 64 bit width, a copy operation to copy data between local buffers, as well as a timing operation which records a timestamp at the time it is executed by the GOAL scheduler.

2.2   User vs. Kernel Level Design

The GOAL API allows the user to specify arbitrary dependency graphs. Each node in such a graph represents a single send, receive or local calculation operation. Therefore nodes can be created by the user by calling, for example, GOAL_Send() or GOAL_Recv(). For each input or output buffer that is given to these functions, there is a corresponding argument which can be either GOAL_USERSPACE or GOAL_SCRATCHPAD. The reason for this is that schedules can be defined in a different place than they are executed. If one would write a function that creates a tree based gather schedule with GOAL, this function would have to allocate a temporary buffer. But this function can not contain the corresponding call to free(), unless the schedule is also executed, waited on, and destroyed in that function — which would make it impossible to overlap the communication with computation. Therefore GOAL has a primitive memory management functionality. For each schedule, the user specifies how much temporary space is needed. GOAL will allocate such a *scratchpad buffer* before it starts to execute the schedule. If the user wants to reference data in the scratchpad buffer, he can do this by passing the byte offset (relative to the start of the scratchpad) and set the memory type argument to GOAL_SCRATCHPAD.

Edges $t \rightarrow h$ in the dependency graph can be created with the function GOAL_Requires(). After the graph is complete it can be compiled into the binary representation using GOAL_Compile() and run with GOAL_Run. The functions GOAL_Test() and GOAL_Wait() can be used to test and wait for completion of the handle returned by GOAL_Run. The GOAL API is implemented as a userspace library, while the actual interpreter is a kernel module. The GOAL_Run() function will hand over the binary schedule to the kernel module by doing an ioctl(). The complete control flow is depicted in Figure 2.

In our particular example implementation, we use the kernel-based Ethernet Streaming Protocol (ESP) [8], but we remark that any kernel-level communication mechanism will suffice. The ESP network protocol uses MAC addresses and device ids to identify peers. We collect this information during the definition phase of the schedule: If the user adds a send operation to rank 7 to a dependency graph on rank 8 we use MPI_Isend/Irecv to exchange the MAC addresses

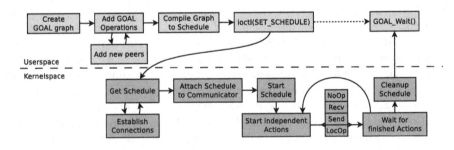

Fig. 2. ESPGOAL Control Flow

between both peers. To keep the amount of out of band communication low, we cache that information in userspace. During GOAL_Run we pass the schedule as well as the list of all MAC addresses and device ids of the peers we will communicate with in that schedule to the GOAL interpreter. The interpreter will update the peer list for the active communicator by opening new connections (if they don't exist yet) before the schedule is started. Upon completion, the GOAL interpreter will change a memory location in the process address space, so the GOAL_Wait()/-Test() functions can poll that value to gather information about the status of a schedule in progress.

3   Integration into the Operating System

3.1   Anatomy of the Linux Kernel Network Stack

The Linux kernel network stack consists of multiple layers, each tries to provide a different level of abstraction. The Linux network stack is shown in Figure 3(a).

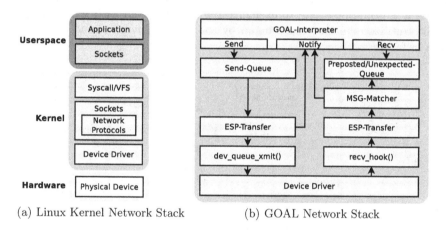

(a) Linux Kernel Network Stack   (b) GOAL Network Stack

Fig. 3. Comparison of the default Linux with the GOAL network stack

A network interface card typically has a hardware buffer to temporarily store incoming network packets. When a new packet arrives the Linux kernel is notified with an interrupt from the network card. The device driver retrieves the newly received packets and stores them in so called socket buffers (*skbs*).

Incoming packets (skbs) are handled by "receive hook" functions, registered with dev_add_pack() in Linux. Full skbs (including all Ethernet packet data, such as destination, ethertype, etc.) are sent with dev_queue_xmit(). This is the lowest layer of abstraction which is offered by the Linux kernel to send and receive data in a device independent manner. Our implementation uses this interface to the driver layer inside the kernel. The benefit of this approach is that the functions mentioned above do not sleep and therefore they can be called in an irqhandler or tasklet.

Another possibility how to implement network communication in the Linux kernel is to use the kernel socket API. Utilizing kernel sockets is very similar to userspace socket programming, however, in the kernel one has to employ mutual exclusion strategies to prevent race conditions. Most network protocols supported by the Linux kernel, such as TCP are implemented with the kernel socket API. One disadvantage of the socket API is that certain functions, for example, sending data via `kernel_sendmsg()` can not be performed in an interrupt handler or tasklet. If such functionality is required it has to be implemented in a separate kernel thread or a workqueue element. Thus, we used workqueues to implement GOAL over ESP (ESPGOAL). Other network protocols such as TCP do not have to use workqueues or an extra kernel thread as the problematic socket API function which might sleep are usually called from userspace.

This raises the question if the scheduling overhead implied by using workqueues has a negative impact on ESPGOALs performance, compared to the other possible approaches to send and receive data in the kernel. If the overhead required to start a new work item in a workqueue is significant it is desirable to have an upper bound on its performance impact so that we can decide if it would be useful to exchange the ESP protocol with something that directly utilizes the functions offered by the device driver to send and receive data in future work.

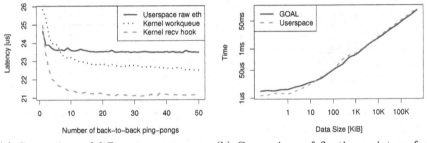

(a) Comparison of different ways to communicate over Ethernet

(b) Comparison of floating point performance in user- and kernelspace

Fig. 4. Microbenchmark Results

We implemented a microbenchmark to assess the overheads affiliated with the different choices. Our benchmark consists of three different implementations of a pingpong scheme, one using the raw socket API from userspace and the other two run in kernelspace. The benchmark performs multiple round-trips for each measurement to amortize the startup overheads. Figure 4(a) shows the benchmark results for two CHiC nodes, see Section 4.1. We observe that the overhead for inserting and scheduling a workqueue element adds about 1.6 μs of latency to each transmission.

3.2 The Ethernet Streaming Protocol

The Ethernet Streaming Protocol (*ESP*) [8] is a connection-oriented, port mul-
tiplexed and reliable protocol on top of Ethernet with optimized congestion
control for static, switched networks. It can be used through standard sockets
from kernel- and userspace. This makes it ideal to utilize it in a kernel based version of the GOAL interpreter. As mentioned before, Open-MX cannot be used
from inside the kernel directly at this stage.

The ESP protocol is transfer based. There are special flags to signal the start
(TXS) and finish (TXF) of a transfer. After the initial TXS, the receiver requests
more data, until he receives a TXF packet. This has the advantage that the
receiver handles flow control and adapts it to the number of streams. The GOAL
scheduler is only invoked for packets with the TXF flag set.

3.3 Asynchronous Progression

The GOAL interpreter is activated (run as a kernel workqueue element) in two
conditions: Either ESP received a packet that had the TXF flag set or ESP could
not receive more data into an skb because there is not enough memory available
(memory pressure). The TXF flag indicates that a transfer is completed and the
GOAL scheduler will try to match the received message against the preposted
receive queue or put the message in the unexpected receive queue. If the message
matched, the scheduler will mark the corresponding node in the schedule as
completed and decrease the dependency counters of all nodes (operations) that
depend on it. The scheduler will then immediately start all operations where the
dependency counter reached zero.

If the scheduler was called because of memory pressure, it will also try to
process messages that are not completely transferred yet. The header that is
needed to perform message matching is transferred in the first 28 bytes of each
transfer so the interpreter can perform message matching and partially copy the
payload to the final destination for every socket that contains at least 28 bytes
of data and holds a message that belongs to a preposted receive.

A workqueue item is implemented as a function pointer that will be executed
in a special kernel thread. A modern Linux kernel (i.e., 2.6.36) will run one
workqueue execution thread per core and decide which workqueue item to run
based on a number of flags that can be set when allocating the workqueue
structure. Currently the GOAL interpreter is run as a high-priority workload,
which means that available workqueue items are to be scheduled by the kernel
as soon as possible. Also our workqueue items are marked as unbound, which
means they can be run on any core available, to maximize the chance that they
are executed immediately.

3.4 Performing Reduction Operations in Kernel Space

Performing floating point operations inside the kernel space is supported by the
macros kernel_fpu_begin()/end(), which save and recover the FPU state and
disables preemption.

In order to assess the performance impact of a kernel-based reduction, we implemented a simple benchmark that computes the maxima of two 32 bit floating point vectors. We ran the benchmark in userspace and in the kernel with a GOAL local operation. We excluded the schedule startup overhead from all GOAL measurements. As shown in Figure 4(b) the kernel implementation is slightly slower than the userspace implementation for small datasizes, but outperforms the userspace implementation for larger vectors.

4   Benchmark Results

We implemented several collective operations with GOAL on top of ESP. It was shown that test-based schemes achieve reasonable overlap for large messages [7]. However, overlapping small-message communications remains hard due to the high ratio between control overhead and message-sending. Thus, we focus especially on small-message operations because they are most important at large-scale and are hardest to overlap. We implemented several optimized collective algorithms for small-message collectives. For all-to-all communication, we used the scheme proposed by Bruck [2] and for barrier and allreduce we implemented the well-known dissemination algorithm. Both schemes use $\log_2(P)$ stages in P processes and have a relatively complex dependency structure.

4.1   Experimental Setting

We conducted all experiments on the CHiC Cluster System at the University of Technology Chemnitz. CHiC consists of 530 compute nodes with two Opteron 2218 Dual-Core 2.6 GHz CPUs running Linux. Each node is equipped with two Tigon3 BCM95704A6 rev 2100 network cards which are connected to an 48 port Gigabit Ethernet Switch (SMC 8848M). We used an MTU of 1500. The NIC used supports interrupt coalescing. With the default coalescing parameters the latency was very high. If we disabled interrupt coalescing completely our systems became unstable. Therefore we optimized the interrupt coalescing settings with a genetic algorithm and the omx-pingpong tool included in the Open-MX distribution. The coalescing settings used were:

rx-usecs	rx-frames	rx-usecs-irq	rx-frames-irq	tx-usecs	tx-frames	tx-usecs-irq	tx-frames-irq
1	1	996	95	32	94	724	128

We used Open MPI 1.4.2 and Open-MX 1.3.4 in all experiments. We compare LibNBC over MPI with the different transports: TCP/IP (TCP), Open-MX (OMX), and ESP (used from user-level) with out kernel-based ESPGOAL implementation. The latency of a blocking execution (initiation call immediately followed by a wait) of the GOAL nonblocking collective operations and the LibNBC nonblocking collective operations is very similar. Figure 5 shows barrier as an example. It can be seen that the latency of the ESPGOAL Barrier is between the Open MPI implementation with the TCP and MX BTL. This can be attributed to the well tuned Open-MX implementation, which uses a lot of optimizations that have not been done in ESP. Also note that minimizing latency for blocking collectives was not the goal for this work — we just want to ensure that our implementation is not substantially slower, which could invalidate our overlap results shown in Section 4.2. If an implementation spends a lot of time waiting for IO it would be easy to overlap the collective and the CPU overhead would be low.

4.2   Asynchronous Progress and Overlap

We now analyze the ability of ESPGOAL to asynchronously progress messages. For this, we use NBCBench [7] without any explicit progression. NBCBench uses a work-loop, which is calibrated at the beginning of the benchmark, to determine the overlap. This means that all interruptions by the kernel will "steal" time from the work loop and show up as overhead, see [5] for a detailed description.

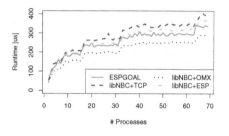

Fig. 5. Barrier Latency

NBCBench reports the share of the communication that can be overlapped with computation, a number between 0 and 1 (higher is better). Figure 4.2 shows the results for all-to-all of size 8 bytes per process and barrier. As expected, we

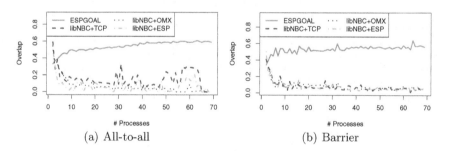

(a) All-to-all   (b) Barrier

Fig. 6. NBCBench Overlap Benchmarks

see high overlap with the ESPGOAL implementation while the unprogressed nonblocking collective operations exhibit very low overlap due to missing asynchronous progression (all but the first stage of the algorithm will be performed during the wait call).

4.3   CPU Overheads

In this section, we assess the absolute CPU overheads, i.e., the absolute non-overlappable time of the communication. We showed that asynchronous progression works well for the investigated operation. Reducing the absolute CPU

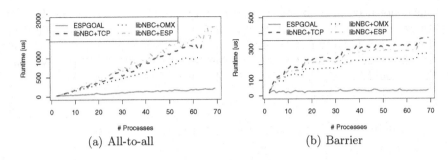

Fig. 7. NBCBench CPU Overhead

overhead of the operations is most important to "free" the CPU for the user application. Figure 4.3 shows the absolute CPU overhead for each configuration. ESPGOAL causes a significantly lower CPU overhead than LibNBC in all configurations.

5   Conclusions and Future Work

We implemented a dependency driven communication framework that offers true asynchronous progress without an extra progression thread. We defined an API to use such a framework that supports simple sends and receives, vector sends and receives, and local operations. Our framework shows significant improvements in terms of host overhead over existing userland implementations of non-blocking collectives. Our work shows that it is possible to implement dependency driven communication schemes as a Linux kernel module without placing constraints on the user. For example our GOAL scheduler does not require the user to pin the memory used for communication buffers.

In future work this implementation should be tuned further so that it can compete with state of the art low overhead Ethernet protocols such as Open-MX. One possible way to tune ESPGOAL even further would be to replace the ESP protocol with another low overhead Ethernet protocol that shows better performance in point to point latency benchmarks, for example it could be investigated if ESP can be replaced with the kernel part of Open-MX.

Another interesting optimization would be the use of the memory subsystem on multi-core nodes. An optimized GOAL implementation could directly push or pull the data into other processes memory similarly to kernel-level zero copy mechanisms such as KNEM [11].

Acknowledgments. This work was supported in part by the DOE Office of Science, Advanced Scientific Computing Research X-Stack Program.

References

1. Almasi, G., et al.: Optimization of MPI collective communication on BlueGene/L systems. In: ICS 2005, pp. 253–262. ACM Press, New York (2005)
2. Bruck, J., Ho, C.T., Upfal, E., Kipnis, S., Weathersby, D.: Efficient algorithms for all-to-all communications in multiport message-passing systems. IEEE Trans. Parallel Distrib. Syst. 8, 1143–1156 (1997)
3. Goglin, B.: High-Performance Message Passing over generic Ethernet Hardware with Open-MX. Elsevier Journal of Parallel Computing (PARCO) 37(2), 85–100 (2011)
4. Graham, R.L., et al.: ConnectX-2 InfiniBand management queues: First investigation of the new support for network offloaded collective operations. In: IEEE International Symposium on Cluster Computing and the Grid, pp. 53–62 (2010)
5. Hoefler, T., Lumsdaine, A.: Message Progression in Parallel Computing - To Thread or not to Thread? In: Proceedings of the 2008 IEEE International Conference on Cluster Computing. IEEE Computer Society Press, Los Alamitos (2008)
6. Hoefler, T., Lumsdaine, A.: Optimizing non-blocking Collective Operations for InfiniBand. In: Proceedings of the 22nd IEEE International Parallel & Distributed Processing Symposium (IPDPS) (April 2008)
7. Hoefler, T., Lumsdaine, A., Rehm, W.: Implementation and Performance Analysis of Non-Blocking Collective Operations for MPI. In: Proc. of the SC 2007. IEEE Computer Society/ACM (November 2007)
8. Hoefler, T., Reinhardt, M., Mietke, F., Mehlan, T., Rehm, W.: Low Overhead Ethernet Communication for Open MPI on Linux Clusters CSR-06(06) (July 2006)
9. Hoefler, T., Siebert, C., Lumsdaine, A.: Group Operation Assembly Language - A Flexible Way to Express Collective Communication. In: ICPP-2009 - The 38th International Conference on Parallel Processing, IEEE, Los Alamitos (2009)
10. Hoefler, T., Traeff, J.L.: Sparse collective operations for MPI. In: Proceedings of the 23rd IEEE International Parallel & Distributed Processing Symposium (IPDPS), HIPS Workshop (May 2009)
11. Moreaud, S., Goglin, B., Goodell, D., Namyst, R.: Optimizing MPI Communication within large Multicore nodes with Kernel assistance. In: CAC 2010, IEEE Computer Society Press, Atlanta (2010)
12. MPI Forum: MPI: A Message-Passing Interface Standard. Version 2.2 (June 23rd 2009), www.mpi-forum.org
13. Nomura, A., Ishikawa, Y.: Design of kernel-level asynchronous collective communication. In: Keller, R., Gabriel, E., Resch, M., Dongarra, J. (eds.) EuroMPI 2010. LNCS, vol. 6305, pp. 92–101. Springer, Heidelberg (2010)
14. Petrini, F., Frachtenberg, E., Hoisie, A., Coll, S.: Performance Evaluation of the Quadrics Interconnection Network. Journal of Cluster Computing 6(2), 125–142 (2003)
15. Riesen, R.E., Pedretti, K.T., Brightwell, R., Barrett, B.W., Underwood, K.D., Hudson, T.B., Maccabe, A.B.: The Portals 4.0 message passing interface (April 2008), Sandia National Laboratories, Tech. Rep., SAND2008-2639 (April 2008)
16. Sanders, P., Speck, J., Träff, J.L.: Two-tree algorithms for full bandwidth broadcast, reduction and scan. Parallel Comput. 35, 581–594 (2009)
17. Valiant, L.G.: A bridging model for parallel computation. Commun. ACM 33(8), 103–111 (1990)

A High Performance Superpipeline Protocol for InfiniBand

Alexandre Denis

INRIA Bordeaux Sud-Ouest / LaBRI, France
`Alexandre.Denis@inria.fr`

Abstract. InfiniBand high performance networks require that the buffers used for sending or receiving data are registered. Since memory registration is an expensive operation, some communication libraries use caching (rcache) to amortize its cost, and copy data into pre-registered buffers for small messages. In this paper, we present a software protocol for InfiniBand that always uses a memory copy, and amortizes the cost of this copy with a superpipeline to overlap the memory copy and the RDMA. We propose a performance model of our protocol to study its behavior and optimize its parameters. We have implemented our protocol in the NewMadeleine communication library. The results of MPI benchmarks show a significant improvement in cache-unfriendly applications that do not reuse the same memory blocks all over the time, without degradation for cache-friendly applications.

1 Introduction

INFINIBAND NETWORKS are nowadays the leading technology for high performance networks in clusters. Parallel applications usually exploit this network through an MPI library that makes their usage seamless for the end-user. Under the hood the MPI implementations access the InfiniBand network cards through an API called *verbs*. Unlike API used to program other networking technologies, the *verbs* API is very low-level. It means that a lot of things have to be done by hand by the MPI library programmer; on another hand, the programmer has a lot of control on how to exploit the network interface.

Network transfers are based on RDMA and are executed by the DMA engine on the network card. The card sees the system from the PCIe bus, thus works with physical addresses. The application, MPI library, and InfiniBand software stack run in user space, with no system call involved thanks to OS bypass. Since they run in user space, they use virtual addresses. Thus, when sending data from user space through InfiniBand, translation has to be done from virtual address space to physical address space. The network card can do the translation if it has been told previously the mapping from virtual to physical space. This process is called *memory registration* and in InfiniBand it has to be performed explicitly by the user. Actually, the memory registration is comprised of both the communication of the translation table to the network card, and memory pinning to prevent swapping. All memory involved in sending and receiving

E. Jeannot, R. Namyst, and J. Roman (Eds.): Euro-Par 2011, LNCS 6853, Part II, pp. 276–287, 2011.
© Springer-Verlag Berlin Heidelberg 2011

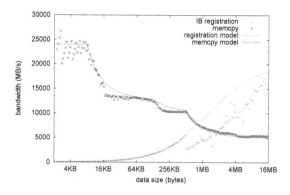

Fig. 1. Registration and memory copy performance comparison on cluster `graphene`

operations in InfiniBand must be registered. Two approaches are possible to satisfy this constraint: (1) register memory blocks on the fly; (2) register a buffer at application startup, then copy data into this pre-registered buffer. Memory registration has a significant cost [1] and both approaches have an impact on the overall network performance.

In this paper, we present a software protocol for InfiniBand that copies data through a pre-registered buffer and amortizes the cost of the memory copy by using a superpipeline to overlap copy and RDMA transfer. We propose a performance model of our protocol to study its behavior and optimize its parameters.

The remainder of this paper is organized as follows. In Section 2 we analyze the performance of memory copy and registration. In Section 3 we present an analysis of a pipeline for memory copy. In Section 4, we describe our superpipeline protocol. Section 5 gives benchmark results. Section 6 compares our work to related works. Section 7 concludes our paper.

2   Performance Analysis

In this section, we analyze the performance of memory registration, memory copy, and network transfer, and we propose a performance model. We run our tests on multiple InfiniBand clusters. Cluster `graphene` features ConnectX DDR (MT26418) cards on quad-core nodes equiped with Intel Xeon X3440. Cluster `infini` has ConnectX2 QDR (MT26428) cards on quad Intel Xeon X5570. To visualize closely what happens, all our graphs use a 5 % increment for message size (i.e. powers of 1.05), not only powers of 2 that hide a lot of details.

Performance of registration. Memory registration in InfiniBand is an expensive operation. We have conducted benchmarks to measure the time consumed to register memory on several InfiniBand clusters. For example, the results obtained on cluster `graphene` are depicted in Figure 1, represented as a bandwidth. Let L be the length of a given message, we model the registration time

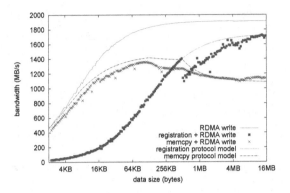

Fig. 2. Impact of registration and memcpy on communication performance on cluster **graphene**

as a linear function in the form $T_{reg}(L) = \lambda_{reg} + \frac{L}{B_{reg}}$ with λ_{reg} the latency of memory registration, and B_{reg} its bandwidth. On this cluster, we have measured $\lambda_{reg} = 68\,\mu s$. and $B_{reg} = 20\,GB/s$ (actually, $200\,ns$ per $4\,KB$ page). We observe the same order of magnitudes on other DDR and QDR InfiniBand boards.

The cost of registration may have a huge impact on actual communication performance. A naive protocol to send a block of data would consist in dynamically registering the memory region, send the data on the network, then deregister the memory region; the receiver has to perform registration/deregistration too. The performance of such a protocol is depicted in Figure 2 for cluster **graphene**. We observe that the overhead introduced by memory registration lowers the bandwidth by as much as 60 % for packets of roughly 64 KB, and is far from negligible even for larger sizes of messages, with an apparent bandwidth converging asymptotically to $\frac{1}{\frac{1}{B_{net}} + \frac{1}{B_{reg}}}$, which is 91 % of the network bandwidth on our cluster.

Performance of memory copy. Performance of a raw `memcpy` is depicted in Figure 1. The apparent bandwidth decreases when the size of data increases, as a result of cache effects. We roughly model its behavior with four different bandwidth figures for L1, L2 and L3 caches, and memory. It would require to know actual cache policy and associativity to get a more precise model.

A naive copy-based protocol would copy data on the sender side into a registered memory zone, send the data, then copy the data from the registered memory zone into its final destination in the receiver side. Since memory copies are fast for small messages, this copy-based protocol is usually used for small messages sent eagerly. Larger messages are usually [2] sent with a *rendezvous* protocol to avoid copies that would lower the available bandwidth.

The apparent bandwidth of the naive copy-based protocol is depicted in Figure 2, converging asymptotically to $\frac{1}{\frac{1}{B_{net}} + \frac{2}{B_{copy}}}$, which is 57 % of the network bandwidth on our cluster.

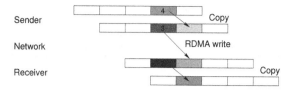

Fig. 3. Pipeline for memory copy: sender copies chunk 3 while sending chunk 4; receiver copies chunk 2 while receiving chunk 3

Real behavior of *rcache*. To amortize the cost of memory registration, a *pin-down cache* [3] has been proposed, or as commonly called today, a *registration cache* (in short: *rcache*). It means that the sender does not unregister the memory zone after a message is sent, in case the same zone is sent again. However it requires a lot of care to be correct [4]. One must use `malloc` hooks, `libc` symbol interception through `LD_PRELOAD` or kernel patches, to invalidate the cache (unregister memory) when memory is deallocated. These mechanisms are not quite portable and may break in subtle ways when interacting with various versions of `libc`, Fortran or OpenMP runtimes, or any runtime that supplies its own memory allocator.

It must be noted that *rcache* does not increase performance by itself. The first send exhibits the same performance as the naive registration-based protocol. Only the subsequent sends of the *same* memory zone will be faster, at the nominal speed of the network. The real world performance of *rcache* depends on the buffer reuse scheme of the application, and obviously varies from one application to another. For example in NAS Parallel Benchmarks, SP and CG have 99 % cache hits, IS has less than 5 % cache hits, and LU sends mostly small messages not covered by *rcache*. Our goal is to improve performance of IS without degrading performance of other benchmarks.

3   Pipelining Memory Copy

In this section, we study an InfiniBand software protocol which manages memory registration by copying data into a pre-registered buffer instead of dynamically registering data in place.

Since copy and RDMA may be overlapped, we use a *pipeline* to amortize the cost of the memory copy. Both operations share the same memory bus, but experiments show that copies have a negligible impact on an overlapped RDMA, while the copy is slowed down by no more than the bandwidth used by the network. As depicted in Figure 3, each message is divided into *chunks* of a given size. Then on the sender side, we overlap the RDMA transfer of one chunk with the memory copy of the next chunk. Since we use RDMA write, on the receiver side nothing has to be done to make the overlapping happen.

Cost analysis. Let L be the message length, and C the chunk size. For convenience, we assume L is a multiple of C. To model the network with multiple chunks and overlap, we use a model close to LogP [5] with the following notations. Let λ_{net} be the network latency (the L of LogP) and B_{net} the network bandwidth, then we have $T_{net}(L) = \lambda_{net} + L/B_{net}$ as end-to-end transfer time for a raw RDMA write, assuming data is already registered. Let g be the *gap* between messages, then we have $T_{net}(L_1, L_2) = \lambda_{net} + \frac{L_1}{B_{net}} + g + \frac{L_2}{B_{net}}$ as transfer time for two packets of length L_1 and L_2. Let o be the *overhead* for sending messages, namely the CPU time needed to initiate the RDMA operation, that will not be available for overlapping.

In Section 2, we have shown that `memcpy` bandwidth depends on message length. Let $B_{copy}(L)$ be the copy bandwidth for a message of length L. We must notice that since the bandwidth depends on cache effects, the length to take into account is the whole data set, namely L, not C. We assume $\lambda_{copy} = 0$. Then we have $T_{copy}(L) = L/B_{copy}(L)$ as time to copy data of length L.

Then the time for the full pipelined transfer is comprised of: copy of the first chunk: $\frac{C}{B_{copy}(L)}$; steady state with $\frac{L}{C}$ chunks copied and sent: $\frac{L}{C} \times \left(g + \frac{C}{B_{net}}\right)$; the network latency: λ_{net}; copy of the last chunk: $\frac{C}{B_{copy}(L)}$. Therefore the total time of pipelined transfer is:

$$T_{pipeline}(L, C) = \frac{2 \times C}{B_{copy}(L)} + \frac{L}{C} \times g + \lambda_{net} + \frac{L}{B_{net}} \qquad (1)$$

Compared to a raw RDMA write, the overhead for the pipeline is $\frac{2 \times C}{B_{copy}(L)} + \frac{L}{C} \times g$. It is comprised of the copy of the first chunk on the sender side, the copy of the last chunk on the receiver side, and the gaps.

Optimal pipeline. We can find the optimal value for C the chunk size. When we draw a graph of $C \to T_{pipeline}(L, C)$ for any fixed L and realistic values of B_{copy} and g, we can see this function admits a minimum. If we assume C to be real instead of integer to make the function differentiable, the derivative with respect to C for a given message length L is:

$$T'_{pipeline}(C) = \frac{2}{B_{copy}(L)} - \frac{L}{C^2} \times g \qquad (2)$$

Let C_{opt} be the the optimal chunk size for a given message length L. It corresponds to the zero of the derivative. We solve the equation and get:

$$C_{opt}(L) = \sqrt{\frac{L \times g \times B_{copy}(L)}{2}} \qquad (3)$$

Using our performance models for network and copy, we estimate the performance of the pipeline with optimal chunk size, depicted in Figure 4. The performance increase compared to naive protocols is huge, but bandwidth is still lower than raw InfiniBand RDMA and may still be improved. We can see that for messages smaller than 16 KB, the naive copy-based protocol is faster than the optimal pipeline; when computing the optimal chunk size for these messages, we get an optimal with less than one chunk per message, which is wrong. Our hypothesis of C being real instead of integer works only for messages large enough.

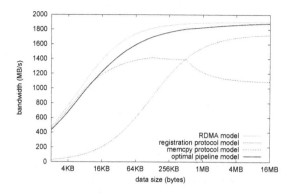

Fig. 4. Bandwidth model for pipeline using parameters from cluster `graphene`

4   Optimizations beyond Pipeline: Superpipeline

In this Section, we present various mechanisms to improve the performance of our protocol, beyond the vanilla pipeline with optimal chunk size presented in the previous Section.

Super-pipeline to lower the number of gaps. Memory copy has a higher bandwidth than RDMA write over the network, as measured in Section 2. Therefore, when pipelining, for each chunk `memcpy` finishes earlier than the RDMA write for the previous chunk. Thus we propose to increase the chunk size from chunk to chunk while the pipeline is running. We call this mechanism a *super-pipeline*, like superpipelines in CPU architecture. This superpipeline mechanism is depicted in Figure 5. It is expected to have a lower number of gaps than the plain pipeline, thus reducing the overhead of the protocol. We must define a suitable progression rule for the chunk size. Let C_i be the chunk size at step i. We may compute the sequence that enables a full overlap of `memcpy` and RDMA; such a sequence will have the fewest gaps. It is defined as:

$$\frac{C_{i+1}}{B_{copy}(L)} = \frac{C_i}{B_{net}} + g - o \tag{4}$$

in other words the time to copy chunk C_{i+1} may be as high as the time to send C_i on the network, including g the gap between packets, but excluding the non-overlapable overhead o, with o and g as defined in Section 3. Since o and g are of the same order of magnitude, $g - o$ is at most in the order of $100\,ns$ and may be neglected compared to the other terms when C_i is several kilobytes. Then Equation 4 simplifies as $C_{i+1} = C_i \times \frac{B_{copy}(L)}{B_{net}}$. Therefore, the general term of the sequence is:

$$C_i = C_0 \times q^i \quad \text{with } q = \frac{B_{copy}(L)}{B_{net}} \tag{5}$$

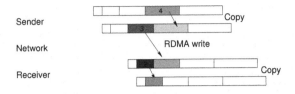

Fig. 5. Super-pipeline for memory copy: a pipeline with a variable chunk size

To compute the protocol overhead, we need to compute the number of gaps. Let n be the number of gaps for a given message of length L. Then L, as the sum of all chunks, is a finite geometric series:

$$L = \sum_{i=0}^{n-1} C_i = C_0 \times \sum_{i=0}^{n-1} q^i = C_0 \times \frac{1 - q^n}{1 - q} \qquad (6)$$

We then solve this equation to get n the number of gaps:

$$n = \log_q \left(1 + \frac{L}{C_0}(q - 1) \right) \qquad (7)$$

Therefore the total transfer time of our superpipeline protocol is:

$$T_{superpipeline}(L) = \frac{C_0 + C_n}{B_{copy}(L)} + n \times g + \lambda_{net} + \frac{L}{B_{net}} \qquad (8)$$

It is very similar to the cost of plain pipeline given in Equation 1, except that the number of gaps is $O(L)$ for fixed-chunk pipeline, $O(\sqrt{L})$ for pipeline with the optimal chunk size C_{opt} as defined by Equation 3, and is lowered to $O(\log(L))$ for superpipeline.

Sub-blocking to lower last chunk copy overhead. The transfer time for the superpipeline given in Equation 8 includes $\frac{C_n}{B_{copy}(L)}$ the time needed to copy the last chunk at the receiver side. Given the general term of the sequence given in Equation 5, C_n is expected to be quite large.

To amortize the cost of the copy at the receiver side, we propose a *sub-blocking* mechanism — namely, a pipeline in the pipeline — to overlap the RDMA and the memcpy of the *same chunk*. Among the possible strategies [6] of the receiver side to detect the arrival of RDMA data, we chose to poll a flag at a known memory location. The receiver sets it to zero; the sender writes a 1 through RDMA.

Our *sub-blocking* mechanism consists in dividing each chunk into blocks of a given size b. Every block is comprised of data payload and a flag indicating the presence of data. All the blocks that form a chunk are sent through a single RDMA write. The receiver is then able to detect and consume blocks as they arrive, only one block behind the one being written by the NIC.

With such a method, the cost of the copy at the receiver side is at most the copy of a block $\frac{b}{B_{copy}(L)}$. This methods adds flags in every blocks, which increases the size of packets sent on the network and must be taken into account. However, if we take for example $b = 4$ KB (page size and multiple of MTU) and a flag on a 64-bit word (to avoid atomicity issues depending on endianness), then the overhead is less than 0.2%.

Overlap *rendezvous* to lower first chunk copy overhead. The overhead of our superpipeline protocol includes the cost of the memory copy for the first chunk C_0. We propose to overlap this copy with the *rendezvous* to lower its impact on performance.

We have shown in Section 3 that the optimal number of chunks is 1 for small messages. The fastest option to send small messages is the naive copy-based protocol. It means that we will use pipeline (or superpipeline) protocols only for large messages, that are sent using a *rendezvous* mechanism to ensure that data is received in place. It must be noted that, although our protocol involves a copy, the *rendezvous* is still relevant because our protocol copies data on the fly and needs to know where to store data.

The cost of a *rendezvous* is twice the latency, namely $2 \times \lambda_{net}$. We propose that the sender copies C_0 the first chunk of the message while the *rendezvous* round-trip takes place. If $\frac{C_0}{B_{copy}} < 2 \times \lambda_{net}$, then the copy of C_0 is free. To maximize overlap in the common case, we chose to use: $C_0 = B_{copy} \times 2\lambda_{net}$. With contemporary hardware, we get values from 8 KB to 16 KB for C_0.

Pipeline folding using N-buffering. In our previous descriptions of our superpipeline, we have assumed that the preregistered buffer is as large as the message. However, registered memory is a finite resource and cannot be arbitrarily large. Therefore we *fold* our superpipeline to fit statically-allocated buffers. Since at any given time, one buffer is copied while another buffer is sent (or received), we can then make our superpipeline a *double-buffering* algorithm. A flow control mechanism is needed to make sender and receiver synchronize their buffer swaps. Since such synchronization through network has a significant latency, we loosen the coupling between the sender and the receiver with triple-buffering. Before it may send chunk C_i, the sender does not have to wait for the C_{i-1} acknowledgment — that may arrive late because of network latency — but for the C_{i-2} acknowledgement. Moreover, folding the superpipeline in a smaller workspace than the full message length improves cache reuse, although the precise impact is hard to model and to predict.

Discussion. With all the heuristics and optimizations applied, the total transfer time of our superpipeline protocol is:

$$T_{superpipeline}(L) = \frac{b}{B_{copy}(L)} + g \times \log_q \left(1 + \frac{L}{C_0}(q-1)\right) + \lambda_{net} + \frac{L}{B_{net}} \quad (9)$$

The overhead for the copy of the last sub-block b is low, and the number of gaps is $O(log(L))$.

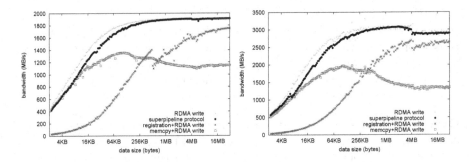

Fig. 6. Raw superpipeline protocol performance on cluster **graphene** (left) and **infini** (right)

The plain *pipeline* is easy to model and we have determined analytically its optimal chunk size, as we did in Section 3. However, the performance of *super-pipeline* depends on a sequence, not a single value, which makes it difficult to solve analytically. Optimization depends on a lot of parameters, that's why we used heuristics (maximize overlap) to determine q and C_0.

Our model is not so precise and some behaviors are hard to predict (cache effects depend on cache policy, associativity, data alignment). The theoretical optimal is thus not necessarily optimal once implemented. However when we trace $C_0 \rightarrow T_{superpipeline}(L, C_0)$ for a given L, variations are *small* around the optimal. We conclude that not-so-precise tuning works well with superpipeline, which is confirmed by experience. Hardwired $C_0=12$ KB and $q = 1.5$ gives results almost as good as optimal values obtained through auto-tuning.

5   Benchmarks

In this Section, we present benchmarks of our superpipeline protocol and compare it against other protocols and implementations.

Raw protocol benchmark. We have implemented our superpipeline protocol in a test program to evaluate its behavior regardless of any other implementation artefact. The results of a ping-pong bandwidth test on clusters **graphene** and **infini** are depicted in Figure 6. We observe the superpipeline is much faster than naive registration and memcpy-based protocols, and is very close to the raw RDMA performance. The overhead compared to raw RDMA is 15 % for 16 KB messages, and less than 5 % for messages larger than 64 KB. However this test program does not implement *rendezvous*, thus cannot overlap C_0 copy and *rendezvous*; real-life overhead is then expected to be lower.

MPI micro-benchmarks. We have then implemented our superpipeline protocol as a driver in our NEWMADELEINE [7] communication library, which already has an *rcache* method for InfiniBand, and run MPI benchmarks using its Mad-MPI [8] interface. NEWMADELEINE has a 32 KB *rendezvous* threshold, and uses

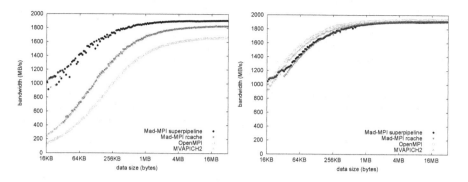

Fig. 7. MPI performance on cluster **graphene** on first touch (left), and best time (right)

plain copy for small messages (no pipeline, no *rcache*). We compare Mad-MPI against OpenMPI-1.4.3 and MVAPICH2-1.6rc2. The results of a MPI ping-pong bandwidth test on cluster **graphene** are depicted in Figure 7. The benchmarks performs 100 round-trips for each message size; we draw separate graphs for the *first* and for the *best* round-trip. We observe the superpipeline gets roughly the same performance as Mad-MPI *rcache best* time, very similar to OpenMPI and MVAPICH2 *best* time. However, when we compare the performance of the *first* round-trip, we observe that Mad-MPI *rcache*, OpenMPI and MVAPICH2 all get low performance because of registration, whereas Mad-MPI superpipeline is unaffected. The superpipeline gets its best performance already at the first send; the others, relying on *rcache*, get poor performance on first send.

MPI NAS Parallel Benchmarks. We have run some benchmarks from the NAS Parallel Benchmarks 3.3.1 on cluster **graphene**. On tests sp.B.8, lu.B.8 and cg.C.8, Mad-MPI superpipeline and *rcache* get the same performance, 3 % slower than OpenMPI and MVAPICH2, explained by the fact that Mad-MPI has a slightly higher latency. However, on is.C.8 Mad-MPI superpipeline is 9 % faster than Mad-MPI *rcache* (respectively 3.05 s and 3.32 s), but slightly slower than MVAPICH2 (3.01 s) and OpenMPI (2.92 s). It means that superpipeline actually improves performance over *rcache* on IS which is cache-unfriendly, but some work has to be done in Mad-MPI to improve latency for small message to get overall better performance.

6 Related Works

People working on MPI implementations and other communication libraries have already studied InfiniBand memory registration and proposed solutions to amortize its cost. Memory registration performance has been analyzed [1] and communication performance modeled [9] without proposing solution to improve performance. Various caching strategies [3,10,4] have been proposed, as well as protocols to overlap *rendezvous* and registration [11] in case of cache miss; however, all these solutions exhibits the pitfalls of cache-based strategies. It has been proposed in OpenMPI [12] to pipeline registration; our model shows that pipelining copies gives better performance than pipelining registration. For the InfiniBand device for MPICH2 [2] and the BCopy mode of SDP [13], it has been investigated to use a copy pipeline with fixed-size chunks. Performance was not convincing, because at that time memory bandwidth was not significantly higher than network bandwidth; our proposal goes further with a superpipeline rather than a flat pipeline, and our theoretical study shows that it works because $B_{copy} > B_{net}$, which has become the common case nowadays with contemporary CPUs and their integrated memory controlers.

7   Conclusion and Future Works

Memory registration has a major impact on performance for InfiniBand networks. In this paper, we have proposed performance models for InfiniBand network, copies, and registration, and used them to analyze and optimize the performance of a protocol that uses pipelined copy to bring data into registered memory. We have proposed an alternative called *superpipeline* that reduces the number of gaps, and some optimization mechanisms to reduce the cost of the first and last chunk copy. We have implemented and benchmarked our superpipeline protocol, and observed that it gets roughly the same performance as *rcache*-based protocol on cache-friendly communication patterns, and better performance on cache-unfriendly patterns.

As a future work, we will study adaptive strategies to automatically tune the protocol parameters to the machine it is running, and to monitor *rcache* misses/hits to dynamically choose between strategies depending on observed application behavior.

Acknowledgements. This work was supported in part by the ANR-JST project FP3C and the ANR project COOP. Experiments presented in this paper were carried out using the Grid'5000 experimental testbed, being developed under the INRIA ALADDIN development action with support from CNRS, RENATER and several Universities as well as other funding bodies (see https://www.grid5000.fr).

References

1. Mietke, F., Rex, R., Baumgartl, R., Mehlan, T., Hoefler, T., Rehm, W.: Analysis of the memory registration process in the mellanox infiniband software stack. In: Nagel, W.E., Walter, W.V., Lehner, W. (eds.) Euro-Par 2006. LNCS, vol. 4128, pp. 124–133. Springer, Heidelberg (2006)
2. Liu, J., Jiang, W., Wyckoff, P., Panda, D.K., Ashton, D., Gropp, W., Buntinas, D., Toonen, B.: Design and implementation of MPICH2 over infiniband with RDMA support. In: Proceedings of the International Parallel and Distributed Processing Symposium, IPDPS 2004 (2004)

3. Tezuka, H., O'Carroll, F., Hori, A., Ishikawa, Y.: Pin-down cache: a virtual memory management technique for zero-copy communication. In: Proceedings of the First Merged International Parallel Processing Symposium and Symposium on Parallel and Distributed Processing (IPPS/SPDP), pp. 308–314 (April 1998)
4. Ohio, P.W., Wyckoff, P., Wu, J.: Memory registration caching correctness. In: IEEE International Symposium on Cluster Computing and the Grid(CCGrid). IEEE Computer Society, Los Alamitos (2005)
5. Culler, D., Karp, R., Patterson, D., Sahay, A., Schauser, K.E., Santos, E., Subramonian, R., von Eicken, T.: Logp: towards a realistic model of parallel computation. In: ACM SIGPLAN Symposium on Principles and Practice of Parallel Programming. PPOPP 1993, pp. 1–12. ACM, New York (1993)
6. Gupta, R., Tipparaju, V., Nieplocha, J., Panda, D.K.: Efficient barrier using remote memory operations on via-based clusters. In: Cluster 2002 (September 2002)
7. Aumage, O., Brunet, E., Furmento, N., Namyst, R.: Newmadeleine: a fast communication scheduling engine for high performance networks. In: CAC 2007: Workshop on Communication Architecture for Clusters, Held in Conjunction with IPDPS 2007 (March 2007)
8. Trahay, F., Denis, A., Aumage, O., Namyst, R.: Improving reactivity and communication overlap in MPI using a generic I/O manager. In: EuroPVM/MPI. LNCS, Springer, Heidelberg (2007)
9. Hoefler, T., Mehlan, T., Mietke, F., Rehm, W.: Logfp – a model for small messages in infiniband. In: IPDPS (2006)
10. Bell, C., Bonachea, D.: A new dma registration strategy for pinning-based high performance networks. In: Proceedings of the International Symposium on Parallel and Distributed Processing, p. 10 (2003)
11. Ou, L., He, X., Han, J.: An efficient design for fast memory registration in RDMA. Journal of Network and Computer Applications 32, 641–642 (2009)
12. Woodall, T.S., Shipman, G.M., Bosilca, G., Graham, R.L., Maccabe, A.B.: High performance RDMA protocols in HPC. In: Mohr, B., Träff, J.L., Worringen, J., Dongarra, J. (eds.) PVM/MPI 2006. LNCS, vol. 4192, pp. 76–85. Springer, Heidelberg (2006)
13. Goldenberg, D., Kagan, M., David, R., Tsirkin, M.S.: Zero copy sockets direct protocol over infiniband – preliminary implementation and performance analysis. In: Symposium on High-Performance Interconnects, HOTI 2005 (2005)

Introduction

Eric Fleury, Qi Han, Pedro Marron, and Torben Weis

Topic chairs

The tremendous advances in wireless networks, mobile computing, sensor networks along with the rapid growth of small, portable and powerful computing devices offers opportunities for pervasive computing and communications. Topic 14 deals with cutting-edge research in various aspects related to the theory or practice of mobile computing or wireless and mobile networking, including architectures, algorithms, networks, protocols, modeling and performance, applications, services, and data management.

The accepted papers discuss very interesting issues about wireless ad-hoc networks, mobile telecommunication systems and sensor networks. As a highlight, the paper "FEW Phone File System: Proving Ubiquitous Access to the User's Data" by João Soares and Nuno Preguiça from CITI / DI/FCT univeristy of Lisabon, deals with the use of multiple computing devices and, more concretely, with the presentation of the FEW phone file system, that allows for the high availability of data across multiple devices. This paper supports beautifully the notion of ubiquitous computing and allows the use of these devices as personal, portable servers with high availability of data and with an automatic synchronization to external data sources that could be in the cloud.

We would like to take the opportunity to thank all authors who submitted a contribution, the Euro-Par Organizing Committee, and all reviewers for their hard and valuable work. Their efforts made this conference and this topic possible.

E. Jeannot, R. Namyst, and J. Roman (Eds.): Euro-Par 2011, LNCS 6853, Part II, p. 288, 2011.
© Springer-Verlag Berlin Heidelberg 2011

ChurnDetect: A Gossip-Based Churn Estimator for Large-Scale Dynamic Networks

Andrei Pruteanu, Venkat Iyer, and Stefan Dulman

Embedded Software Group, Delft University of Technology, The Netherlands
{a.s.pruteanu,v.g.iyer,s.o.dulman}@tudelft.nl

Abstract. With the ever increasing scale of dynamic wireless networks (such as MANETs, WSNs, VANETs, etc.), there is a growing need for performing aggregate computations, such as online detection of network churn, via distributed, robust and scalable algorithms. In this paper we introduce the ChurnDetect algorithm, a novel solution to the distributed churn estimation problem. Our solution consists in a gossiping-based algorithm, which incorporates a periodic reset mechanism (introduced as DiffusionReset). The main difference with existing state-of-the-art is that ChurnDetect does not require nodes to advertise their departure from the network nor to detect neighbors leaving the network. In our solution, all the nodes are interacting with each other wirelessly, by using a gossip-alike approach, thus keeping the message complexity to a minimum. We only use easy accessible information (i.e., about new nodes joining the network) rather than presuming knowledge on nodes leaving the system since that is highly unfeasible for most distributed applications. We provide convergence proofs for ChurnDetect, and present a number of results based on simulations and implementation on our local testbed. We characterize the performance of the algorithm, showcasing its distributed light-weight characteristics. The analysis leads to the conclusion that ChurnDetect is an attractive alternative to existing work on online churn estimation for dynamic wireless networks.

1 Introduction

Recent technological advances have led to a tremendous increase in the number of embedded devices having processing and wireless communication capabilities. Large-scale networks of resource-limited devices are already in operation: wireless sensor networks (WSNs), mobile ad-hoc networks (MANETs) and vehicular networks (VANETs). Following this trend, current research projects show that significantly larger networks could be envisaged (e.g., programmable matter - claytronics [11], swarm robots [16], amorphous computing [1], etc.). Overall, devices tend to become smaller, networks increase in size and mobility becomes the basic assumption.

As such, there is a growing need for performing aggregate computations via distributed, robust and scalable algorithms. For example, the *online estimation* of network churn in dynamic scenarios is of crucial importance for a large number

E. Jeannot, R. Namyst, and J. Roman (Eds.): Euro-Par 2011, LNCS 6853, Part II, pp. 289–301, 2011.

of applications. While the churn rate can be computed offline from network traces, the online estimation of this quantity is a problem of increasing interest (e.g. for MANETs and WSNs). As we show in Section 1.1, the current state of the art presents unfortunately a quite limited set of alternatives. The particular aspects raising the difficulty of the problem are dynamic multihop architectures involving mobile nodes and failures at both node and communication levels.

A direct use-case for such an algorithm is the detection of traffic congestion on a highway. The potential imbalance between the inflow and the outflow of cars passing through a section of the road caused by an accident for instance, translates directly into a change of the churn level. Since our approach is fully distributed and convergences fast to a good estimate, we are able to prevent the drivers of a potentially hazardous situation faster than a centralized technique.

We approach the problem of estimating the network churn size by means of *diffusion algorithms* (also known as *gossiping*) [13]. This class of algorithms allows easy dissemination of information in a network, and has been already used to compute network aggregates such as averages, sums and aggregates, perform random sampling, compute quartiles, etc. (see [17]).

Inspired by real-world deployments of WSNs, where periodic resets of nodes are a known failure mode [3], we propose a new diffusion algorithm, by incorporating resets into a gossiping algorithm. We show that the new mechanism, called *DiffusionReset*, retains the properties of gossiping in terms of message complexity for achieving convergence exponentially fast. Although our algorithms are derived from gossiping algorithms sensitive to *mass conservation* [15, 17], our approach specifically exploits the property that total mass varies in a dynamic network. Our algorithm is able to track churn level evolution, even when it changes with time.

Based on *DiffusionReset*, we develop the *ChurnDetect* algorithm which we propose as a solution for the online estimation of the network churn rate (defined as the percentage of nodes that join/leave a network in a period of time). We are *not* assuming that the nodes advertise their departure from the network *nor* that nodes can detect when neighbors leave the network. In short, new nodes joining a network need to use a different reset value than "older" nodes. The results of the gossiping algorithm is an average aggregate value, available at all nodes, which can be used to compute the online churn estimate. To the best of our knowledge, this is the first work for the online estimation of churn that is addressing an arbitrary mobile multihop topology while still offering very good churn percentage estimates in a fully distributed manner.

We validate our work with both simulation and experiments on our wireless testbed. For the analysis of our algorithm we considered different mobility and network density scenarios that cannot be matched with corresponding traces from real deployments due to their scarcity, especially for mobile ad-hoc networks. The paper is structured as follows: Section 1.1 describes existing state-of-the-art. In Section 2 we introduce the underlying diffusion mechanism, while in Section 3 we present the network churn estimation algorithm. The algorithms are analyzed in Section 4 and we draw the conclusion in Section 5.

1.1   Related Work

Several definitions exist for the term *network churn*, most of them coming from the peer-to-peer context (see [10] for an overview). In the following, we denote as *churn* the changes in the set of networked nodes due to joins, graceful leaves, and failures. The churn is thus the percentage of the nodes in the network that changes during a given time period.

The problem of estimating churn in large-scale networks has been mostly studied in the context of Internet peer-to-peer systems. For these applications, the importance of knowing how many nodes enter and exit the system at any given time is fundamental. Due to the highly distributed nature of these systems, gossip-based protocols [15] have emerged as one of the most used techniques for the estimation of churn ratios. The main assumption these algorithms make is that *detection of nodes leaving a network* is feasible, either by advertisement or by nodes periodically checking their neighborhood. Unfortunately, for the case of a MANET or WSN, these approaches are not feasible due to the difficulty of discovering and maintaining neighborhood information within a finite amount of time and with a reduced energy budget.

The gossip-based algorithm presented in [12], was designed for estimating churn in arbitrary topologies (that can be represented as undirected graphs) built on top of a peer-to-peer network. The algorithm cannot be used in the presented form on a multihop wireless network, suffering from the shortcomings described above. Even if assuming node departure detection is feasible, the convergence of the algorithm is dependent on the network diameter, rapidly degrading when the network size exceeds a certain threshold.

Aside from gossip-based methods, there are algorithms that estimate the level of churn based on the amount of time a peer spends while being connected to the system (online time) or while being disconnected from the system (offline time) [4, 9]. While for some distributed applications (i.e., peer-to-peer - where clients are *not* behind firewalls) it is feasible to presume that they can signal their departures, or the peers can ping each other at regular time intervals, for networks where most users are behind firewalls (cannot be checked by others for availability) or have other restrictions (i.e., in our case: energy consumption and unavailability of a cheap ping function on a multihop topology) this assumption cannot be made. Furthermore, although these papers claim to provide a churn estimate, they actually focus on a slightly different definition for churn, and showcase the estimation of the online time and not specifically on the amount of nodes that constantly enter or exit the system. We acknowledge the difference in terminology with the peer-to-peer community, and notice that the results presented there are not directly applicable in our case.

A large amount of work is actually targeted at reducing the effects of churn (in all the above mentioned communities: P2P, MANETs and WSNs). Most of them assume existence of network traces, and estimate churn offline [10,8,18]. We believe that to be able to enable algorithms to adapt at run time to a dynamic environment, online estimation of churn is an important building block.

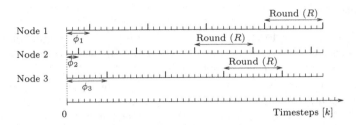

Fig. 1. Discrete time model for three nodes (ϕ_i - reset phase)

2   Diffusion Algorithms

We introduce the *DiffusionReset* algorithm under the same assumptions as used in [17] for the Push-Sum algorithm: i) communication operates in discrete time; ii) nodes do not need to have globally unique IDs (although at the lowest communication layer we need to be able *to distinguish* between neighbors); iii) the network does not become partitioned with time.

We make use of the discrete time assumption in order to simplify the description of the algorithm and ease the intuitive grasp of the concepts. The term *communication round* is being used in the sense described in [14] - it captures the fact that each node performs, in a given (large) time interval, an equal amount of actions. The beginning of rounds need not be synchronized (see Figure 1), in fact *DiffusionReset* is actually relying on this. Rounds are considered to be orders of magnitudes longer than the clock drift of the devices, thus, usual communication networks can be modeled as such. The last assumption (unpartitioned network or, equivalently, network as *a single* multihop cluster) should be interpreted from the perspective of large periods of time - it may be invalidated for the case of mobile scenarios for short moments (as in single nodes having no neighbors at a particular instance of time) but still holds for large time spans, so the assumption of the network not being disconnected does hold. Nevertheless, mobility actually helps by significantly accelerating the convergence of diffusion algorithms [19]. When introducing the *ChurnDetect* algorithm in Section 3, for the simulations we consider perfect radio communication between nodes. On the other hand, the usage of acknowledgments for messaging is *not* required. The effects of these two assumptions are addressed in Section 4.

Notations - $\mathcal{S}$ denotes the set of all n nodes in the network. The neighborhood of a node i, *including* the node itself is defined by $\mathcal{S}_i^+$ and has n_i nodes. We use i and j as node indexes, k to index time steps and r to index time rounds.

2.1   The DiffusionReset Algorithm

In this section we introduce the first contribution of this paper, the *Diffusion-Reset* algorithm (see Algorithm 1), which is the foundation of our solution to the churn estimation problem. We build upon a basic *diffusion algorithm* (lines

Algorithm 1. *DiffusionReset*(μ, ϕ_i)

1: ▷ *state update step*
2: $m_i[k] \leftarrow \sum_{j \in \mathcal{S}_i^+[k-1]} \lambda_{j,i}[k-1]m_j[k-1]$

3: $\omega_i[k] \leftarrow \sum_{j \in \mathcal{S}_i^+[k-1]} \lambda_{j,i}[k-1]\omega_j[k-1]$

4: ▷ *reset step*
5: **if** rem (k, R) == ϕ_i **then**
6:     $\{m_i[k]; \omega_i[k]\} \leftarrow \{\mu, 1\}$
7:     Choose values $\lambda_{i,j}$
8: **end if**

9: ▷ *communication step*
10: **for all** neighbors j **do**
11:     Send j: $\{\lambda_{i,j}m_i[k]; \lambda_{i,j}\omega_i[k]\}$
12: **end for**

13: ▷ *return value*
14: $\{m_i[k]; \omega_i[k]\}$

Algorithm 2. *ChurnDetect*(ϕ_i)

1: ▷ *initialization step*
2: **if** node i just entered the network **then**
3:     Update phase: $\phi_i \leftarrow$ rem(k, R)
4:     Reset mass value: $\mu_i \leftarrow 0$
5: **end if**

6: **if** node i inside network longer than R **then**
7:     Reset mass value: $\mu_i \leftarrow 1$
8: **end if**

9: ▷ *diffusion step*
10: $\{m_i[k], \omega_i[k]\} \leftarrow$ *DiffusionReset* (μ_i, ϕ_i)

11: ▷ *return value*
12: $\frac{m_i[k]}{\omega_i[k]}$

1–3 and 9–14 in Algorithm 1), adding the novel feature that each node periodically (albeit asynchronously), *resets its local variables to a default value* (i.e., the tuple $\{\mu; 1\}$ - lines 4–8 in Algorithm 1). The rationale for this mechanism is that we can model churn if a large number of nodes enter and exit the network constantly with time – actually, this is identical to having a number of nodes periodically reset (detailed in Section 3). The inspiration for this algorithm comes from a very common failure pattern met in real-world WSNs deployments – where nodes reset randomly [3] (see Figure 2 for the expected behavior).

Basic diffusion mechanism – *DiffusionReset* borrows parts of the *Push-Sum* and *Push-Vector*, introduced in [17] (lines 1–3 and 9–14 in Algorithm 1). In short, these work as follows: each node i holds a local state variable (given by the tuple of values $\{m_i[k]; \omega_i[k]\}$) at the beginning of the communication time step k (m_i is usually referred to as "mass"). During the time step, each node splits its local variable in several shares that get distributed to its neighbors. At the end of the time step, the node adds all the shares of received variables and updates to the new state value. The effect of this mechanism is that, with time, all local variables converge to the same value (the average of the original variable set) *regardless of the synchronization model* [17] (allowing us to relax the synchronous communication assumption).

Let i indicate the current node and j be the index of a neighbor $j \in \mathcal{S}_i^+[k]$. During each time step, node i defines a *share vector* $\mathbf{\Lambda}_i[k]$ of size $n_i[k]$, with elements corresponding to the share of local variables to be distributed to each neighbor. Let $\lambda_{i,j}[k]$ be the share assigned by node i to a neighbor j in time step k. The shares are chosen such that, at any time step k, $\sum_{j \in \mathcal{S}_i^+[k]} \lambda_{i,j}[k] = 1$ holds. During each time step k, each node i sends to all its neighbors a weighted vector: $\{\lambda_{i,j}[k]m_i[k]; \lambda_{i,j}[k]\omega_i[k]\}$ and receives the sets $\{\lambda_{j,i}[k]m_j[k]; \lambda_{j,i}[k]\omega_j[k]\}$ from its neighbors. At the time step $k+1$, the node updates its m_i (ω_i is updated similarly) value as follows: $m_i[k+1] = \sum_{j \in \mathcal{S}_i^+[k]} \lambda_{j,i}[k]m_j[k]$.

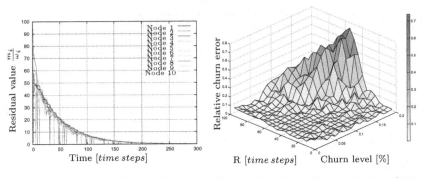

Fig. 2. *DiffusionReset* (200 nodes, tr. range 0.15 *units*, R(reset interval) = 50, $\mu = 0$)

Fig. 3. Influence of reset period (500 nodes, tr. range 0.1 *units*)

In matrix form, ($\mathbf{M}$ and $\mathbf{\Omega}$ being column vectors with the m_i and respectively ω_i elements), we have $\mathbf{M}[k + 1] = \mathbf{\Lambda}^T \mathbf{M}[k]$, $\mathbf{\Omega}[k + 1] = \mathbf{\Lambda}^T \mathbf{\Omega}[k]$. As shown in [17], if no errors occur and *the set of nodes remains the same*, the sums $\sum_{i \in S} m_i[k]$ and $\sum_{i \in S} \omega_i[k]$ remain constant over time. The usage of the share vector $\mathbf{\Lambda}_i[k]$ allows a great flexibility in the algorithm design: if all the elements in $\mathbf{\Lambda}_i[k]$ are zero, except for two entries (corresponding to i and a random neighbor j) equal to 0.5 each, the algorithm maps onto the classic definition for gossiping using unicasts. If all the entries in $\mathbf{\Lambda}_i[k]$ are taken to be $\frac{1}{n_i[k]}$ then we model a local broadcasting mechanism.

Reset mechanism – The reset mechanism (lines 4–8 in Algorithm 1) works as follows: every R time steps, a node resets its state value to $\{\mu; 1\}$. The reset phase of each node is ϕ_i (see Figure 1). Let $\delta[k]$ be the discrete Dirac function. The moment k when node i resets is signaled by $t_i[k] = 1$, where $t_i[k] = \delta\left[\text{rem}\,(k - \phi_i, R)\right]$ (rem(a, b) gives the remainder of the division of a to b). Let $\mathbf{x}_i[k]$ be the local state variable on node i (i.e., the vector $[m_i[k], \omega_i[k]]$). The state transition can be written as

$$\mathbf{x}_i[k + 1] = (1 - t_i[k]) \sum_{j \in S_i^+} \lambda_{j,i}[k] \mathbf{x}_j[k] + t_i[k][\mu, 1]. \tag{1}$$

We define the vector $\mathbf{X} = [\mathbf{x}_1[k], \mathbf{x}_2[k], ..., \mathbf{x}_n[k]]^T$. Let $\mathbf{A}[k]$ be the adjacency matrix and $\mathbf{I}$ the identity matrix. We define the square matrix $\mathbf{\Delta}[k]$ with the terms $t_i[k]$ on its diagonal. Let $\mathbf{D}$ be a $n \times 2$ matrix with elements μ on the first column and 1 on the second column. The algorithm can be written in matrix form as $\mathbf{X}[k + 1] = (\mathbf{I} - \mathbf{\Delta}[k])(\mathbf{I} + \mathbf{A}[k])\mathbf{\Lambda}^T[k]\mathbf{X}[k] + \mathbf{\Delta}[k]\mathbf{D}$, where the first term on the right side maps to the basic diffusion mechanism and the second term on the right side maps to the asynchronous resets.

2.2   Convergence of DiffusionReset

As shown by Dimakis et. al [19] mobility enables the construction of "short" routes between all pairs of agents, accelerating the diffusion process. If the entire network becomes mobile, the speed of information diffusion approaches the one of a fully connected network.

In our case, the influence of mobility and multihop topology is captured by $\mathbf{A}[k]$ and $\mathbf{\Lambda}[k]$ matrices - that change at each moment in time. Since a closed form solution for this type of equation is not available [5], we propose the following approach: we determined the convergence values and speed for the case of a fully connected network. Based on the results presented in [19], our convergence results will hold for a multihop mesh network *in which at least a small fraction of nodes is mobile*.

We assume that nodes use broadcast communication ($\lambda_{j,i} = \frac{1}{n}$). Each node resets after R time steps and the reset phase for each node is random and follows an uniform distribution. This results in an approximately constant number of nodes resetting in a time step. The mass value to which all nodes reset is equal to μ. The expected values are $E\left\{\sum_{i=1}^{n} t_i[k]\right\} = \frac{n}{R}$, $E\left\{\sum_{i=1}^{n} t_i[k]^2\right\} = \frac{n}{R}$. where we used the fact that $t_i[k]$ can be either 0 or 1, leading to $t_i[k] = t_i[k]^2$. Let $f = \left(1 - \frac{1}{R}\right)$. The error on each node is defined as $|m_i[k] - \mu|$. We can prove the following two lemmas:

Lemma 1 (Convergence of Mass for *DiffusionReset*). *With time, the total mass of the system converges to:* $\lim_{k\to\infty} M[k] = n\mu$.

Proof. The total mass in the network at time $k+1$ is: $M[k+1] = \sum_{i=1}^{n} m_i[k+1]$. From Equation 1,

$$M[k+1] = \sum_{i=1}^{n}(1 - t_i[k])\sum_{j=1}^{n}\frac{m_j[k]}{n} + \frac{n\mu}{R} =$$

$$= \sum_{j=1}^{n} m_j[k]\left(1 - \frac{1}{n}\sum_{i=1}^{n} t_i[k]\right) + \frac{n\mu}{R} = M[k]f + \frac{n\mu}{R}$$

$$M[k] = M[0]f^k + \frac{n\mu}{R}\left(f^{k-1} + ... + f^0\right) = M[0]f^k + n\mu\left(1 - f^k\right) \quad (2)$$

As $f < 1$, we obtain $\lim_{k\to\infty} M[k] = n\mu$.   □

Lemma 2 (Convergence Speed of *DiffusionReset*). *The overall error* $v[k] = \sum_{i=1}^{n}(m_i[k] - \mu)^2$ *decreases exponentially fast in the squared norm form:* $v[k+1] = v[k]f^2$.

Proof. $v[k+1] = \frac{1}{n^2}(M[k] - n\mu)^2 \sum_{i=1}^{n}(1 - t_i[k])^2 = \frac{1}{n}(M[k] - n\mu)^2 f$.

Using Equation 2 to expand $M[k]$ leads to $v[k+1] = \frac{1}{n}(M[0] - n\mu)^2 f^{2k+1}$, and then to $v[k+1] = v[k]f^2$.   □

For a static multihop network, standard gossiping is very expensive in terms of message complexity, requiring $O(n^2 \log e^{-1})$ messages [6] to compute the average within accuracy e. When even a small fraction of nodes are mobile, the communication complexity drops significantly to $O(n \log e^{-1})$ messages, the same order as a fully connected graph [19], this being the basis of our reasoning. Our algorithm has the same message complexity as standard gossiping and while node mobility improves the convergence speed, *ChurnDetect* does not require the nodes to be mobile. As Lemma 1 shows, the average of the distributed variable $\overline{M}_i = \frac{m_i[k]}{\omega_i[k]}$, will converge to $\{\mu; 1\}$ with time, regardless of the initial values $\{m_i[0], \omega_i[0]\}_{i \in S}$. Intuitively we can think of it as if the network "forgets" the initial value exponentially fast - see Figure 2. This property extends also to disturbances in the network: if a node local values become arbitrary, the system will converge back to $\{\mu; 1\}$ exponentially fast.

3   Churn Detection Algorithm

In this section we introduce the *ChurnDetect* algorithm, as a solution to the problem of online network churn estimation – i.e., the percentage of nodes that are entering/leaving the network in a given amount of time (see Algorithm 2). The main idea is that a network comprises two sets of nodes: ones that "are fresh" (entered less than R time steps) and the ones that already "belong" to the network (entered more than R time steps ago). The periodic reset mechanism in *DiffusionReset* is used with one change: the "fresh" nodes reset to $\{0, 1\}$ and the "old" nodes reset to $\{1, 1\}$. The value to which the algorithm converges (available readily at each node) is a function of the churn rate - thus each node can estimate it directly. The novelty in our approach is the fact that nodes need not advertise leaving the system. The algorithm automatically tracks their departure through the change of the global shared variable M. This is a fundamentally different when compared to classic approaches. The elegance of our approach comes from the fact that we drop the assumption of nodes advertising leaving the network.

The intuitive explanation for why this works is the following: say that at each moment in time, a number n_{new} nodes enter the network (initialized with the value $\{0, 1\}$). At the same moment, a number of $n_{old} = n_{new}$ nodes leave the network taking with them the values $\{m_i, \omega_i\}$. This is equivalent to having *a network not changing its set of nodes*, but instead having *a subset of n_{old} nodes reset to $\{0, 1\}$ at each moment in time* (the subset needs not be the same at each moment in time). This equivalence maps churn directly onto *DiffusionReset* using two distinct reset values for the nodes. Please note that we are *not* assuming that the nodes advertise their departure from the network *nor* that nodes can detect when neighbors leave the network. The power of the algorithm is that only new nodes in the network are asked to behave slightly differently.

Lemma 3. *Convergence of* DiffusionReset *for various μ_i Assume that n_1 nodes reset to $\{\mu_1, 1\}$ and n_2 nodes reset to $\{\mu_2, 1\}$ ($n_1 + n_2 = n$). Then $\lim_{k \to \infty} M[k] = n_1\mu_1 + n_2\mu_2$.*

Proof. We rewrite Equation 1 under the current assumptions. The formula is derived similarly to Lemma 1.   □

Modeling-wise, the churn mechanism is equivalent to forcing each node to reset more than once per time round, to $\{0,1\}$ instead of $\{1,1\}$, the ratio being a function of the churn rate. Let ψ represent the churn rate, defined as the percentage of nodes entering/exiting the network at each *time step*. ψ can be interpreted as well as the probability with which a node needs to reset to $\{0,1\}$ at any given time step.

In *ChurnDetect* the nodes are forced to reset more often than once per period. The probability that a node *does not reset to* $\{0,1\}$ for R consecutive time steps is: $P_{noreset} = (1 - \psi)^R$. In this case, $\overline{M}$ equals the ratio between the number of nodes resetting to 1, and the total number of nodes (resetting to either ($\{1,1\}$ and $\{0,1\}$ - see Lemma 3). From this definition, it follows actually that $\overline{M} = P_{noreset}$. This leads to each node being able to estimate ψ (making abstraction of the low-pass filter needed for a smooth estimate) as $\psi_e = 1 - \sqrt[R]{\overline{M}}$.

An important aspect of the algorithm is the selection of the reset period. Although synchronization is not needed, there is a dependency between the network dynamics (churn ratio) and the reset intervals.

As the shape of the graph in Figure 3 does not depend on the actual network size, nodes could check at run time if their reset period and the computed estimate are "in the safe zone", adapting otherwise. Due to size constraints this extension is not presented in this paper.

4   Analysis of ChurnDetect Algorithm

We base our evaluation by conducting simulations on Matlab. The mobile nodes are assumed to be deployed in a square space of 1 $units^2$. A circular disk communication model is assumed and the transmission range of the nodes is set to 0.1 $units$. The nodes move through space with a speed ranging from a minimum of 0.01 $units/time\ step$ to a maximum of 0.2 $units/time\ step$ (using the *Random Walk* [2] mobility model). Each experiment consisted of simulations running for 500 *time steps*. The maximum speed, the reset period and the number of nodes were varied across simulations to achieve different characteristics for mobility.

4.1   Experimental Evaluation via Simulations

Influence of Network Density and Mobility on Accuracy - As shown in Figure 4, the variation in network density from 5 to 25 nodes per squared unit affects the accuracy of the estimation as expected. This is in line with the generally agreed intuition that, for most distributed systems, increasing network density increases the information diffusion speed. On the other hand, as predicted in [19], even a small percentage of mobile nodes tremendously accelerates the information diffusion. In Figure 4 the difference between the cases with high and low mobility cannot be distinguished. One explanation for this graph is that the effects presented in these figures are dampened by the uniform spatial distribution of the churn nodes in the simulation. Nevertheless, Figure 4 confirms that an increased density decreases exponentially the relative churn error.

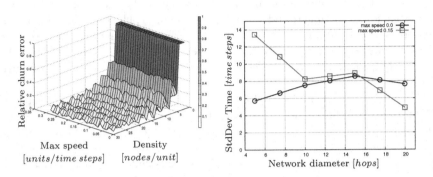

Fig. 4. Influence of density and speed. ChurnRatio = 0.1.

Fig. 5. Convergence speed. ChurnRatio = 0.1.

Influence of Network Diameter on Convergence Speed - One of the most interesting results is the influence of the network diameter and maximum node speed on the convergence time. As seen in Figure 5 where we showcase the standard deviation for the convergence time, the network diameter does show its role. The higher the diameter, the faster the algorithm converges to 90% estimation accuracy if the maximum nodes speed is non zero. Again, as predicted, having a percentage of node mobile, increases the convergence time even for the case of a high network diameter.

Influence of Network Diameter and Mobility on Accuracy - A trade-off exists between the network diameter and the speed of the nodes. While the former aspect is well understood [7], the latter is still a subject of active research [19]. Given a network diameter, if the mobility of the nodes is above a certain threshold, then the mobile multihop network becomes actually equivalent to a single hop network from the convergence speed perspective. Intuitively, a large network diameter increases the number of diffusion steps needed to spread the information, while node mobility helps the nodes "mix" faster, reducing the number of diffusion steps. The results confirm our hypothesis: as shown by the standard deviation in Figure 6, for the largest network diameter, the algorithm performs badly for the static case in comparison with the mobile one. The increased node speed accelerates the diffusion. For networks with low diameter, however, the increased speed of the nodes has relatively little effect. This confirms the results derived in Equation 1 - although derived for a fully connected graph, it holds for determining the churn ratio in a mobile case.

Communication Costs - A point of interest for a practical implementation of the *ChurnDetect* algorithm is the amount of communication that has to take place to ensure a good churn estimate. Although gossiping has a low message complexity, when the amount of nodes that constantly enter and exit the system increases, we have to reduce the reset intervals and thus linearly increase the amount of transmitted messages to fasten up the diffusion of information.

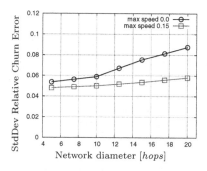
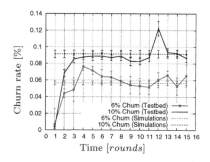

Fig. 6. Influence of network diameter. ChurnRatio = 0.1.

Fig. 7. Comparison between experiments and simulations

Overall, the low complexity is maintained, making *ChurnDetect* an attractive distributed protocol.

4.2   Experimental Evaluation on the Testbed

In order to validate our results, we have implemented *ChurnDetect* on our wireless sensor network, consisting of 108 *GNode* nodes statically deployed across the floor of our department, using the TinyOS-2.x as the software platform. The *GNodes* are sensor nodes built around the MSP430-microcontroller combined with a Chipcon-CC1101 transceiver. Our experimental objective was two-fold; i. e., (1) To study the accuracy of the *ChurnDetect* algorithm on a real network, and (2) To recommend guidelines or implementation practices for protocols that operate on communication *rounds*. Figure 7 shows the comparison between the testbed runs (that had a network diameter of at most 3 communication hops) and simulation results. We tested *ChurnDetect* for varying values of churn, namely 6% and 10%. We implemented the network churn using a lookup table of nodeids, indexed by gossiping round number. Typically, for a gossiping round, every nodeid in the lookup table resets its value to 0. Each data point in the graph represents the average churn estimate over all nodes for a communication round. We note that apart from minor outliers, the testbed results are not only comparable with simulations, but also found to be within 10% of the actual churn value.

There was one notable issue we have had to face with real world experimentation, namely communication failures. Particularly, packet acknowledgment collisions early on within a gossiping round tends to increase the $\{m_i, \omega_i\}$ values on a node. This has an effect of an overestimated value of $\overline{M}$, which in consequence, underestimates the churn ratio. We alleviated this issue by allowing nodes to message randomly within a time interval (5 seconds in our experiments). We believe that in practice, every node should follow a probabilistic approach to messaging that adapts itself depending upon the number of channel contenders.

5   Conclusions

Online computation of network churn, in a distributed and reliable manner, is a common research interest for several communities such as MANET, WSN, VANET and peer-to-peer. In this paper, we introduced *ChurnDetect*, an algorithm for the online estimation of churn in dynamic networks. To the best of our knowledge, this is one of the first algorithms specifically targeted at multihop, mobile networks. The solution we proposed is derived from gossiping algorithms and incorporates the notion of periodic asynchronous resets for being able to provide at each node an estimation of the churn percentage. We analyze the *ChurnDetect* algorithm and validate our contribution analytically, through simulations and experiments on a wireless sensor network. As future work, we plan to address some of the shortcomings that we encountered with respect to the implementation of the algorithm on the testbed platform.

References

1. Abelson, H., Allen, D., Coore, D., Hanson, C., Homsy, G., Knight Jr. T.F., Nagpal, R., Rauch, E., Sussman, G.J., Weiss, R.: Amorphous computing. Commun. ACM 43(5), 74–82 (2000)
2. Bettstetter, C.: International workshop on modeling analysis and simulation of wireless and mobile systems. In: MSWIM 2001, pp. 19–27 (2001)
3. Beutel, J., Römer, K., Ringwald, M., Woehrle, M.: Deployment techniques for sensor networks. In: Ferrari, G. (ed.) Sensor Networks. Signals and Communication Technology, pp. 219–248. Springer, Heidelberg (2009)
4. Binzenhöfer, A., Leibnitz, K.: Estimating churn in structured p2p networks. Managing Traffic Performance in Converged Networks, 630–641 (2007)
5. Boyd, S., Ghosh, A., Prabhakar, B., Shah, D.: Randomized gossip algorithms. IEEE/ACM Trans. Netw. 14(SI), 2508–2530 (2006)
6. Dimakis, A.D.G., Sarwate, A.D., Wainwright, M.J.: Geographic gossip: Efficient averaging for sensor networks. IEEE Transactions on Signal Processing 56(3), 1205–1216 (2008)
7. Dimakis, A.G., Sarwate, A.D., Wainwright, M.J.: Geographic gossip: efficient aggregation for sensor networks. In: Proceedings of IPSN 2006, pp. 69–76. ACM, New York (2006)
8. Friedman, R., Gavidia, D., Rodrigues, L., Viana, A.C., Voulgaris, S.: Gossiping on MANETs: the Beauty and the Beast. ACM SIGOPS Operating Systems Review 41(5), 67–74 (2007)
9. Giuffrida, C., Ortolani, S.: A Gossip-based Churn Estimator for Large Dynamic Networks. In: Proceedings of ASCI 2010 (2010)
10. Brighten Godfrey, P., Shenker, S., Stoica, I.: Minimizing churn in distributed systems. SIGCOMM Comput. Commun. Rev. 36, 147–158 (2006)
11. Goldstein, S.C., Campbell, J.D., Mowry, T.C.: Programmable matter. IEEE Computer 38(6), 99–101 (2005)
12. Gramoli, V., Kermarrec, A.M., Le Merrer, E.: Distributed churn measurement in arbitrary networks. In: Proceedings of the Twenty-Seventh ACM Symposium on Principles of Distributed Computing, p. 431. ACM, New York (2008)

13. Hedetniemi, S.M., Hedetniemi, S.T., Liestman, A.L.: A survey of gossiping and broadcasting in communication networks. Networks 18(4), 319–349 (1988)
14. Iwanicki, K., Van Steen, M.: On hierarchical routing in wireless sensor networks. In: Proceedings of IPSN 2009, pp. 133–144. IEEE, Los Alamitos (2009)
15. Jelasity, M., Montresor, A., Babaoglu, O.: Gossip-based aggregation in large dynamic networks. ACM Trans. on Computer Systems 23(3), 219–252 (2005)
16. Karpelson, M., Wei, G.-Y., Wood, R.J.: Milligram-scale high-voltage power electronics for piezoelectric microrobots. In: ICRA 2009 (2009)
17. Kempe, D., Dobra, A., Gehrke, J.: Gossip-based computation of aggregate information. In: FOCS 2003 (2003)
18. Qadri, N.N., Alhaisoni, M., Liotta, A.: Mesh based P2P streaming over MANETs. In: Proceedings of MOMM 2008, pp. 29–34. ACM, New York (2008)
19. Sarwate, A.D., Dimakis, A.G.: The impact of mobility on gossip algorithms. In: INFOCOM 2009, pp. 2088–2096. IEEE, Los Alamitos (2009)

Introduction

Olivier Coulaud, Kengo Nakajima, Esmond G. Ng, and Mariano Vazquez

Topic chairs

As demand in high-resolution and high-fidelity modeling and simulation increases, desktop computers or small clusters of processors have proven to be insufficient to carry out the calculations in many scientific, engineering, and industrial applications. Indeed, many such applications typically require a significant amount of computing time or need to process a large amount of data. The High-Performance and Scientific Applications Topic highlights recent progress in the use of high-performance parallel and scientific computing, with an emphasis on successes, advances, and lessons learned in the development and implementation of novel scientific, engineering, and industrial applications.

Today's large computational solutions are often required to operate in complex information and computation environments, where data access can be as important as computational methods and performance, so the technical approaches in this topic span a wide range of areas, which include, but are not limited to, high performance parallel computing, data access, and the associated problem-solving environments that compose and manage advanced solutions. This is reflected by the papers accepted. The papers cover areas such as parallel numerical methods for sparse matrix computation and block-structured adaptive mesh refinements; multicore implementations of numerical optimization algorithms and cellular automata; computer science issues such as fault tolerance and tree search algorithms; and applications such as real-time contingency analysis for power grid and data analysis for radio telescopes. We invite you to read and enjoy this collection of papers.

E. Jeannot, R. Namyst, and J. Roman (Eds.): Euro-Par 2011, LNCS 6853, Part II, p. 302, 2011.

Real Time Contingency Analysis for Power Grids

Anshul Mittal[1], Jagabondhu Hazra[1], Nikhil Jain[2], Vivek Goyal[3],
Deva P. Seetharam[1], and Yogish Sabharwal[1]

[1] IBM Research - India,
New Delhi, India
{mittal.anshul,jaghazra,dseetharam,ysabharwal}@in.ibm.com
[2] University of Illinois at Urbana-Champaign,
Illinois, USA
nikhil@illinois.edu
[3] IIT Delhi,
New Delhi, India
cs1070191@cse.iitd.ernet.in

Abstract. Modern power grids are continuously monitored by trained
system operators equipped with sophisticated monitoring and control
systems. Despite such precautionary measures, large blackouts, that affect more than a million consumers, occur quite frequently. To prevent
such blackouts, it is important to perform high-order contingency analysis in real time. However, contingency analysis is computationally very
expensive as many different combinations of power system component
failures must be analyzed. Analyzing several million such possible combinations can take inordinately long time and it is not be possible for
conventional systems to predict blackouts in time to take necessary corrective actions.

To address this issue, we present a scalable parallel implementation of
a probabilistic contingency analysis scheme that processes only most severe and most probable contingencies. We evaluate our implementation
by analyzing benchmark IEEE 300 bus and 118 bus test grids. We perform contingency analysis up to level eight (contingency chains of length
eight) and can correctly predict blackouts in real time to a high degree of
accuracy. To the best of our knowledge, this is the first implementation
of real time contingency analysis beyond level two.

1   Introduction

Electric power systems are prone to various kinds of faults or disturbances. To
withstand such disturbances, trained operators rely on computer simulations
to continuously monitor the system and take corrective actions. However, large
blackouts continue to occur across the globe[1]. For example, even in the US
power grid having sophisticated controls, the frequency of blackout, which was
about 7 per year until 1995, has grown to 36 per year in 2006[2]. Due to these
blackouts, both the utilities and the consumers incur massive losses. According

E. Jeannot, R. Namyst, and J. Roman (Eds.): Euro-Par 2011, LNCS 6853, Part II, pp. 303–315, 2011.
© Springer-Verlag Berlin Heidelberg 2011

to the US Department of Energy, 2003 US blackouts resulted in losses amounting to 6 billion USD[3]. Increasing frequency of severe blackouts indicate the need for tools that can reliably predict and prevent blackouts in real time.

One such tool is Contingency Analysis (CA), which assesses the ability of a grid to withstand cascading component failures/contingencies. The results of contingency analysis provide the basis for preventive and corrective operation actions against blackouts[1]. CA uses the current state reported by SCADA[1] or EMS[2] to identify possible series of component failures and check for collapse cases. The CA schemes are usually referred to as $(N-x)$ CA, where N is the total number of components (could be lines, generators and transformers) in the grid under consideration and x is the level/order. $(N-x)$ CA represents checking all possible permutations of x or less components (out of the total N) for a collapse. For example, a $(N-5)$ CA would evaluate all possible combinations of up to five components failing together in a cascade.

As the number of components (N) and number of levels (x) increase, the number of possible combinations that need to be evaluated increases exponentially $(\sum_{i=1}^{x} {}^{N}P_i)$. Due to this computational complexity, contingency analysis has been traditionally limited to $(N-1)$ CA. However, post event analysis of major blackouts has shown that failing of a component leads to additional component outages in its vicinity. Moreover, the current trend of operating power grids closer to their capacity and integrating intermittent renewable energy sources has increased the probability of multiple component failures. Therefore performing higher order $(N-x)$ CA has become important. In fact, the North American Electricity Reliability Corporation (NERC) has recommended higher order CA as part of its Transmission Planning standards.

However, performing higher order CA for practical grids in real time (a few minutes) is not feasible using conventional techniques. Typically, a practical grid consists of a few thousands of components and even performing level 5 contingency analysis will involve a few billions of contingencies. Each contingency analysis takes about 50-100 ms on an ordinary computer. Hence, it is obvious that the computational workload is beyond what a single personal computer can achieve for real-time operation. This has lead researchers to turn to high performance computing platforms in order to accelerate power grid contingency analysis. The contingency analysis problem involves a large number of small independent computations. The challenge is not merely in parallelising it, but in doing so in real time. An important aspect here is to devise a load balancing scheme which scales to a large number of processors so that the full capabilities of a parallel system can be realised.

Our Contribution: In this paper, a parallel implementation of a probabilistic contingency analysis scheme, that processes only the most severe and most probable contingencies, has been developed. We have adopted search space reduction techniques to reduce the computational burden. We have also proposed a novel

---

[1] Supervisory Control and Data Acquisition.
[2] Energy Management System.

load balancing scheme which scales to thousands of processors. To the best of our knowledge, this is the first effort that goes beyond $(N-2)$ CA and scales well up to $8k$ processors.

The rest of this paper is organized as follows. Section 2 discusses previous work in this domain. Section 3 describes the risk based probabilistic approach and the algorithm used in this paper and the search space reduction techniques adopted. In Section 4, we introduce our novel load balancing scheme and provide a detailed comparison with the previous schemes. In Section 5, we present the results of our experiments. Finally in Section 6, we conclude the paper and propose some future work.

2 Previous Work

Contingency analysis in power systems was first proposed by Ejebe *et al*[4] in 1979. Since then several CA methods have been developed, each varying in methodology and complexity. However, they either employ approximate solution techniques, or use approximate models of the grid. Moreover, these methods are not suitable for higher order $(N-x, x > 1)$ CA.

Recently, for higher order contingency analysis, Monte-Carlo simulation[5], Importance Sampling[6], and Risk Index (RI)[7] have been proposed. Monte-Carlo simulation and Importance Sampling techniques are not so efficient as they simulate the same set of contingencies repeatedly, delaying the convergence. RI approach, on the other hand, avoids repeated simulation and is much faster than Monte-Carlo simulation and Importance Sampling techniques.

Researchers have proposed several schemes to improve the computational speed of CA. Alves *et al*[8] proposed a parallel and distributed computing architecture for CA. For fast CA, Santos *et al*[9] developed a socket based client-server model where dynamic load balancing scheme is implemented for improved performance. Morante *et al*[10] developed a pervasive grid middleware which uses a broker system for reserving on-demand computational resources and for automatically splitting the contingency analysis task into sub-tasks and to allocate them to reserved resources based on a master-slave computing model. However, these and other methods focussed solely on $(N-1)$ analysis with a small set of cases. For massive higher order $(N-x; x \geq 2)$ contingency analysis, Huang *et al*[11] and Chen *et al*[12] proposed dynamic load balancing schemes to perform $(N-x)$ CA. However, their scheme naively selects either all or, a random subset of contingencies. Moreover, scalability remains to be an issue when more processors are used and more cases are analyzed. Most recently, Jin *et al*[13][14] proposed a CA approach using parallel betweenness centrality for contingency selection. This method identifies the most important lines based on base case power flow through lines and restricts higher order contingency analysis to those lines. This approach is overly conservative because it always considers the same set of contingencies as being critical. However, in case of multiple component outage, sets of critical contingencies dynamically change as power flow changes with each outage.

3   Risk Based Algorithm

In power grids, cascading failures happen in a chronological sequence. Therefore, it is convenient to model them with an event tree as shown in Fig. 1, where each node represents a state of the system and the branch between any two nodes represents a contingency. In the tree, the root node represents the pre-fault state of the system whereas a node with no forward branch represents an end node, i.e. a node which is either on the last level (level x), or represents a cascade leading to a blackout in the system.

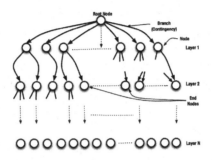

Fig. 1. Event tree

For any real grid, it is very difficult to explore the full event tree due to inordinately large number of possible paths. Therefore, CA schemes try to identify and traverse only those paths which may lead to a system collapse. In this paper, an intelligent search space reduction technique based on Risk Index (RI) is implemented. RI associated with each node is computed by multiplying the severity and probability of any contingency. Severity of any contingency is computed based on voltage instability, load loss, overload, available power margin, and frequency deviation as proposed in[7]. During event tree exploration, less relevant, low risk nodes are discarded at each node. This process is continued until the desired level is reached. RI captures both local and global information of the system, the probability part is computed based on local information whereas the severity part is computed based on global information.

We now propose a parallelization technique to make this algorithm suitable for execution in real-time (Algorithm 1). We compute N event trees (N being the number of lines), one for each line. Initially, the lines are (almost) equally divided amongst the processors. The processor responsible for an event tree begins the computation by simulating the tripping of the corresponding line; this corresponds to a real-life scenario wherein a natural event causes the line to trip, thereby triggering the breakdown process. A processor, say x, then determines the next set of elements (lines, generators, transformers) in the vicinity of this element that are most likely to get affected by the failure of this element. These are referred to as *exposed elements*. It creates a new child node in the event tree for each of the exposed elements. The processing of these child nodes is

---

Algorithm 1

Compute base case load flow for the current system state.
Select child processor, send first lines and base case matrix to it.
MPI_Irecv(RI response from children)
MPI_Irecv(A new contingency to evaluate)
while(1)
    Check end condition (whether all jobs completed or not)
        If yes, exit program.
    Test if RI response received from all children for a contingency, if yes,
        if(RI value is selected)
            MPI_Isend(go-ahead message to that child)
        else if(RI value is rejected)
            MPI_Isend(abort message to that child)
            goto beginning of while loop.
    Test if a new contingency received for evaluation, if yes,
        compute RI.
        If there is a system collapse
            report and put $RI = \infty$.
        MPI_Isend(RI value to parent)
        MPI_Irecv(go-ahead/abort message from parent)
        MPI_Irecv(A new contingency to evaluate)
    Test if (go-ahead/abort) message received from any parent, if yes,
        if(abort received or max level reached)
            goto beginning of while loop.
        else if(go-ahead received)
            find exposed elements; select children processors
            send new contingency and base case solution to the children.
            MPI_Irecv(RI response from children)
            goto beginning of while loop.

---

distributed to new processors. If there are m exposed elements, x selects m new processors and sends them an "RI-evaluation" request, handing over to them, the responsibility of performing the processing for each of these nodes. It then waits for the child nodes to compute and return the RI value for the nodes allocated to them. Once it receives the RI values from all the nodes, it selects the most risky elements based on a certain criteria (explained later) dependent on the RI values. It then sends a "go-ahead" message to the processors corresponding to nodes that are selected for further exploration and an "abort" message to the remaining processors. This completes the processing of the current node for x.

A processor may receive four types of messages, "RI-response", "RI-evaluation" request or, "abort" message or "go-ahead" message. Note that at any point of time, a processor may have requests corresponding to multiple event-tree nodes pending with it. These are queued by MPI; he processor receives one request at a time from MPI and handles it as shown in Algorithm 1. The messages are processed in the given order so as to give priority to those messages which free up the memory. The end condition and process of selecting the children varies according to the load balancing scheme used and is discussed in detail in the next section.

The decision of whether or not to further explore a node in the event tree is based on the RI values. The objective here is to maximise the risk coverage, defined as follows:

$$RC_l = \frac{\sum_{i=1; i \in N}^{k} RI_i}{\sum_{i=1}^{N} RI_i} \times 100 \qquad (1)$$

where RC_l is the percentage risk coverage up to layer l, N is the number of possible blackouts up to layer l, k is the number of blackouts identified by the proposed method and RI_i is the risk associated with the blackout sequence i. RI_i is calculated as follows:

$$RI_i = SI_i * p_{fault} \times \prod_{j \in tripped} p_{cj} \qquad (2)$$

$$p_{cj} = p_j \times \prod_{k \in \ exposed \& not \ tripped; k \neq j} p_k \qquad (3)$$

where SI_i and Pr_i are severity and probability of the blackout i respectively, p_{fault} is the probability of the initiating fault, p_{cj} is the conditional tripping probability of exposed equipment j in the blackout sequence i, and p_j is the tripping probability of equipment j.

The risk coverage, and hence, the effectiveness of the entire analysis, is heavily dependent on the choice of the number of event tree nodes to explore further. Some strategies to determine this parameter are (i) select all the child nodes to explore further, (ii) select a fixed percentage of the child nodes or (iii) select all the child nodes above a certain threshold of RI value. While it is desirable to explore all the child nodes, this leads to a significant computation load. It is also very difficult to obtain a single RI value to use as a thumb rule for the threshold as the RI values vary significantly depending on the specific contingency and the test case under consideration. While the option of selecting a fixed percentage looks promising, our experiments show that this strategy does not result in good risk coverage.

To address this issue, we devise a new strategy that combines a novel extension of the percentage selection strategy and the threshold based strategy. The idea is based on the observation that as the event tree grows exponentially with increasing levels (depth), we can afford to explore more nodes at lower levels (towards the top of the event tree) but fewer nodes at higher levels (towards the bottom of the tree). We therefore apply a linear function to determine the percentage of nodes to select for exploration; this linear function is set up so that it returns 100% at the top level and about 20% at the bottom-most level. Along with this, we also use a threshold value to rule out contingencies with very small RI values. This heuristic results in very good risk coverage (around 80%) and reduces the computation significantly (see Figure 2). 80% percentage risk coverage is fairly good for our application because remaining unidentified

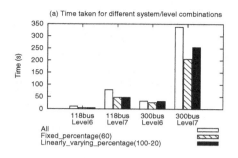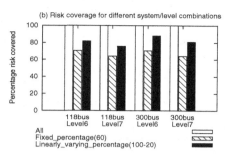

Fig. 2. Comparison of different next-element selection schemes

cascades are of very low probability and consist of longer chain of events. Hence, operator will get enough time to take preventive/corrective controls.

Another optimization is based on the observation that in most of the contingency chains that lead to a collapse, there was a significant jump (at least a 15%) in the RI value before the tripping of the last element in the contingency sequence. For instance, consider a contingency sequence L1-L2-L3-L4-L5 that leads to a collapse. Then, there would be an at least a 15% jump in the sequence RI1, RI2, RI3, RI4 corresponding to the risk indices for the contingencies L1, L1-L2, L1-L2-L3, L1-L2-L3-L4, respectively. This is not surprising since the tripping of the first few lines may cause some lines to become highly overloaded or a few generators may reach their limit leading to considerable worsening of the grid health. Therefore, at the penultimate level, we give a go-ahead to only those sequences that exhibit such a jump in the RI values. Since more than 90% of the contingencies are in the last level only, incorporating this optimization in the algorithm significantly reduces the computation load.

4   Load Balancing Schemes

The ratio of communication to computation per task should be minimised for a scheme to achieve good performance. In case of higher order CA, the number of tasks is huge (of the order of millions) and the task granularity (relative amount of computation per task) is very low (of the order of milliseconds). Therefore, any good load balancing scheme for this problem should involve minimum book keeping and communication amongst the processors, as even a small delay can reduce the communication to computation per task considerably.

Load balancing schemes can be broadly classified into two categories: (i) centralized and (ii) decentralized schemes. While centralized schemes offer better control over the load balance as all the information is available at a single node, decentralized schemes, in contrast, are less prone to congestion, particularly when the number of processors is very large.

Huang et al.[11] have implemented and compared several centralized load balancing schemes. However, these schemes are not directly applicable to the algorithm discussed in this paper due to the modular nature of our contingency analysis scheme, where contingency selection and contingency evaluation are performed separately (Section 2). We have implemented a modification of these schemes for our algorithm. We first discuss them and then propose a new decentralized scheme. Our decentralized scheme does not require any book keeping and scales linearly with increasing number of processors. It is thus highly suited for contingency analysis when there are a large number of processors and the analysis is performed for higher levels.

4.1   Centralized Load Balancing Schemes

In the centralized schemes, a processor designated as master allocates tasks to others processors.Whenever a processor needs to spawn some tasks (corresponding to the child nodes in the event tree), it sends a request to the master indicating the number of tasks to be spawned. The master then, based on the scheme being used, assigns a set of processors and sends a list of these processors back to the requesting node. The master keeps track of the number of jobs spawned in the system and sends an end signal to all processors when all jobs have completed. We now discuss some of these schemes investigated in prior work.

Static allocation: In this scheme, the master ensures that every processor gets equal number of tasks by doing round robin based allocation. The amount of bookkeeping done at master is minimal. This scheme should work well for small system sizes as shown by Huang et al[11]. For larger system sizes, this scheme is expected to underperform as the advantages gained due to the minimal bookkeeping diminish.

Dynamic allocation: This scheme tries to equalize the computation load across the processors. In order to do this, the master maintains a list of active jobs on every processor and assigns request for new tasks to processors that are least loaded. This list is updated as tasks begin and finish on the processors. This involves a considerable amount of bookkeeping and therefore results in an increase in the computation time at the master. For large levels and large number of processors, this has the affect of increasing communication delays between the processors and the master as intermittently requests tend to get queued up at the master. While, this scheme should outperform the static allocation scheme on account of better load balancing, its performance deteriorates for large levels and system sizes.

Two master allocation: Chen et al[12] proposed a variant which they referred to as the multi counter based dynamic allocation. In this scheme, the inital task list is divided into multiple masters which allocate processors for new tasks using the dynamic approach. If the task list of any processor becomes empty, it steals them from the other master(s). The observed performance of this scheme has been found to be similar to previous scheme.

Though these schemes are expected to do well when the number of tasks is not too large and the levels are few, they suffer from congestion issues at the master nodes and hence do not scale to a large number of tasks and larger levels of analysis.

4.2   Decentralized Load Balancing Scheme

Chen et al[12] suggest that if the allocation queries can be serviced instantaneously by the master then ideal speedup can be achieved in the centralized scheme with dynamic allocation. However, as number of tasks increase for larger levels, the congestion at the master causes the network queues to build up and the service time cannot be ignored anymore. In order to address this, we propose a decentralized load balancing scheme that aims at reducing the service time for new task requests while continuing to balance the computation load across all the processors. The master performs the bookkeeping primarily for two purposes; the first is to balance the load amongst the processors and the second is to declare completion of the processing. In our new scheme, we eliminate bookkeeping alltogether in order to enable decentralized control.

To handle the load balancing, we handle task allocation as follows. Whenever new tasks have to be spawned corresponding to the child nodes in the event tree, the processor handling the current (parent) node selects as many processors as the number of child nodes uniformly at random from the set of available processing nodes. It then sends the information regarding the task to be performed directly to the corresponding processors. As every processor makes local decisions regarding the set of processors to allocate the tasks to, the queries are serviced locally and hence instantaneously. There is no master involved in this scheme. If all the processors start with distinct initial seeds for the random number generation, it can be shown that when the number of tasks spawned is very large, the tasks are distributed over the processors uniformly with very small deviation. Hence for higher levels of contingency analysis, considering the granularity of the tasks and the number of tasks involved, the load imbalance is not expected to be high. This scheme is therefore expected to scale linearly with the number of processors as well as the levels of contingency analysis performed.

To detect completion of processing, the nodes perform a collective Allreduce operation at regular interval to determine the number of unfinished tasks. Completion is declared when all the processors report that there are no unfinished tasks remaining. The Allreduce is performed at an interval of $100t$ units where t represents the units taken to perform an Allreduce. This ensures that the overheads of completion detection are no more than 1% of the processing time.

5   Results

In this section, we evaluate our algorithm and compare it with previously studied algorithms.

Hardware setup. All implementations are on Blue Gene/P - IBM's massively parallel supercomputer; Each node of the Blue Gene/P system consists of four

850 MHz PowerPC 450 processor cores. Torus network handles the bulk of the communication data from an application and offers the highest bandwidth in the system. Each node supports 850 MBps bidirectional links to each of its nearest neighbors for a total of 5.1GB/s bidirectional bandwidth per node.

Test Cases. We evaluate the performance using the IEEE Standard test cases[15] comprising of 118 bus system containing 186 lines and 300 Bus System containing 411 lines.

We present the number of contingency chains generated using the RI based selection technique for varying levels on IEEE 118 and IEEE 300 bus systems in Figure 3(1). These results conform to the expected exponential increase in the search space with increase in the number of levels explored. We observe a factor 10x increase in the number of contingency chains with every level (branching factor in event tree). The number of contingency chains runs in tens of millions for level 7 and hundreds of millions for level 8.

In Figure 3(2), results for comparative study of execution time of various load balancing schemes with varying levels are presented. The results show that the increase in execution time for our scheme is commensurate to increase in problem size. In contrast, for centralized schemes the execution time increases super-linearly for large problem sizes.

We study the scalability of our scheme with increase in system size in Figure 3(3). Our scheme outperforms both the centralized schemes for large system sizes. The increase in system size has a negative effect on scaling of both the centralized schemes. Our scheme, on the other hand, scales almost linearly. The gap in performance increases as the problem size increases (from level 4 to level 5). For level 5, the performance of our scheme is an order of magnitude better than the centralized schemes. These gains can be attributed to absence of wait queues at the master node. However, for small system sizes, static scheme outperforms both the dynamic scheme (due to large turnaround time) and our scheme (due to inefficient allocation of jobs). For level 6, the centralized schemes fail to complete successfully in certain cases; this is primarily attributed to the requests piling up on the master causing the processor to run out of memory.

In Figure 3(4), we present strong scaling results for our scheme to show its scalability to very large levels and very large system sizes. We report results for level 6 and 7 for which real time analysis has been made possible by our scheme even for medium system sizes like 512 and 1024 processors. Figure 3(4) shows that our scheme scales nearly linearly for large levels for large systems with up to $2k$ processors. It can also be seen that the scheme scales very well up to $8k$ processor systems. We obtain a factor 12 speedup for $8k$ processor system relative to 512 processor system. Along with the good speed up, it is worth noticing the fact that all these runs up to level 6 and 7 can be done in real time; in contrast, as of today no system goes beyond level 2 for online calculations.

To test the real time nature and scalability of our scheme to highest level, we also ran the code for *level 8 on IEEE 118 bus system* and it took only *259 seconds on 8k processors*. The number of contingencies evaluated in this case is nearly 150 million. A serialized, or parallel centralized scheme based, contingency analysis

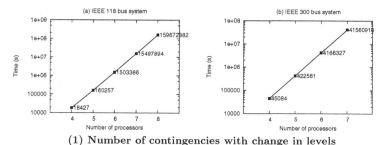

(1) Number of contingencies with change in levels

Level	Time (s)		
	Static	Dynamic	Decentralized
3	0.070	0.074	0.092
4	0.640	0.395	0.247
5	56.970	15.837	1.199

(2) Level wise comparison of load balancing schemes on 118 bus system

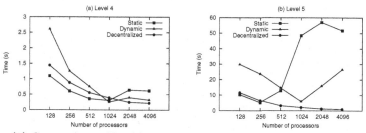

(3) Comparison of load balancing schemes on IEEE 118 bus system

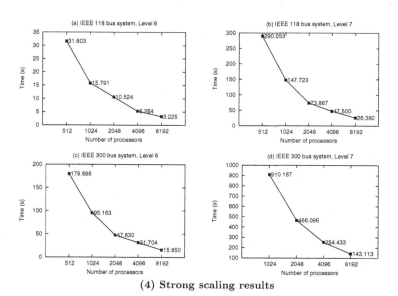

(4) Strong scaling results

Fig. 3. Results with varying levels and system sizes

version will take several days to complete this analysis. The results indicate that our scheme scales well for large number of processors and outperforms the other load balancing schemes. The difference between the schemes becomes more prominent with increasing levels, due to the increasing load on the system and with increasing system sizes, due to the increasing difficulty in load balancing.

6   Conclusions and Future Work

We presented a parallel implementation of probabilistic real time contingency analysis scheme which could be used for blackout prediction in power grid. We evaluated up to 150 million contingencies and showed real time results up to level 8. We also presented a novel load balancing scheme achieving good scalability up to 8k processors. Future work includes incorporating transient stability analysis into this implementation and analyzing the performance on different machines.

References

1. Knight, U.G.: Power Systems in Emergencies: From Contingency Planning to Crisis Management. Wiley, New York (2000)
2. http://edition.cnn.com/2010/TECH/innovation/08/09/smart.grid/index.html
3. Bill Parks: Transforming the Grid to Revolutionize Electric Power in North America. In: U.S. Department of Energy, Edison Electric Institutes Fall 2003 Transmission, Distribution and Metering Conference (2003)
4. Ejebe, G.C., Wollenberg, B.F.: Automatic contingency selection. IEEE Trans. Power Apparatus and Systems PAS-98 (1), 92–104 (1979)
5. Xingbin, Y., Singh, C.: A practical approach for integrated power system vulnerability analysis with protection failures. IEEE Trans. Power Systems (2004)
6. Thorp, J.S., Phadke, A.G., Horowitz, S.H., Tamronglak, S.: Anatomy of power system disturbances: importance sampling. Int. J. Electr. Power Energy Syst. (1997)
7. Hazra, J., Sinha, A.K.: Identification of catastrophic failures in power system using pattern recognition and fuzzy estimation. IEEE Trans. Power System (2009)
8. Alves, A., Monticelli, A.: Parallel and distributed solutions for contingency analysis in EMS. In: Proc. of the Midwest Symposium on Circuits and Systems (1995)
9. Santos, J.R., Exposito, A.G., Ramos, J.L.M.: Distributed Contingency Analysis: Practical Issues. IEEE Trans. on Power Systems 14(4), 1349–1354 (1999)
10. Morante, Q., Ranaldo, N., Vaccaro, A., Zimeo, E.: Pervasive Grid for Large-Scale Power Systems Contingency Analysis. IEEE Transactions on Industrial Informatics 2(3) (August 2006)
11. Huang, Z., Chen, Y., Nieplocha, J.: Massive Contingency Analysis with High Performance Computing. In: Proc. PES General Meeting, Canada (July 2009)
12. Chen, Y., Huang, Z., Chavarra-Miranda, D.: Performance Evaluation of Counter-Based Dynamic Load Balancing Schemes for Massive Contingency Analysis with Different Computing Environments. In: Proc. of the 2010 IEEE PES General Meeting (2010)

13. Jin, S., Huang, Z., Chen, Y., Chavarra-Miranda, D., Feo, J.T., Wong, P.C.: A Novel Application of Parallel Betweenness Centrality to Power Grid Contingency Analysis. In: IPDPS 2010, pp. 1–7 (2010)
14. Gorton, I., Huang, Z., Chen, Y., Kalahar, B., Jin, S., Chavarra-Miranda, D., Baxter, D., Feo, J.T.: A High-Performance Hybrid Computing Approach to Massive Contingency Analysis in the Power Grid. In: Proc. of Fifth IEEE International Conference on e-Science (2009)
15. http://www.ee.washington.edu/research/pstca/

CRSD: Application Specific Auto-tuning of SpMV for Diagonal Sparse Matrices*

Xiangzheng Sun, Yunquan Zhang, Ting Wang, Guoping Long,
Xianyi Zhang, and Yan Li

Lab. of Parallel Software and Computational Science,
Institute of Software, Chinese Academy of Sciences.
Graduate University of Chinese Academy of Sciences
{xiangzheng08,yunquan,wangting,guoping,xianyi,liyan08}@iscas.ac.cn

Abstract. Sparse Matrix-Vector multiplication (SpMV) is an important computational kernel in scientific applications. Its performance highly depends on the nonzero distribution of sparse matrices. In this paper, we propose a new storage format for diagonal sparse matrices, defined as Compressed Row Segment with Diagonal-pattern (CRSD). We design diagonal patterns to represent the diagonal distribution. As the diagonal distributions are similar within matrices from one application, some diagonal patterns remain unchanged. First, we sample one matrix to obtain the unchanged diagonal patterns. Next, the optimal SpMV codelets are generated automatically for those diagonal patterns. Finally, we combine the generated codelets as the optimal SpMV implementation. In addition, the information collected during auto-tuning process is also utilized for parallel implementation to achieve load-balance. Experimental results demonstrate that the speedup reaches up to 2.37 (1.70 on average) in comparison with DIA and 4.60 (2.10 on average) in comparison with CSR under the same number of threads on two mainstream multi-core platforms.

Keywords: CRSD, Auto-tuning, SpMV, Diagonal-pattern, Application Specific Optimization.

1 Introduction

The Sparse Matrix-Vector multiplication (SpMV) is one of the most important computational kernels in sparse linear algebra. Algorithms based on Compressed Sparse Row(CSR) format often perform poorly on modern computer systems. The performance highly depends on nonzero distribution, which determines the memory access pattern and varies significantly among different applications.

In this paper, we study the optimization for diagonal sparse matrices, in which the nonzeros mainly distribute along diagonals. Diagonal sparse matrices are

---

* This paper is supported by the National 863 Plan of China (No.2006AA01A125, No. 2009AA01A129, No. 2009AA01A134), the China HGJ Significant Project (No. 2009ZX01036-001-002), the Knowledge Innovation Program of the Chinese Academy of Sciences (No.KGCX1-YW-13), the Ministry of Finance (No. ZDYZ2008-2).

E. Jeannot, R. Namyst, and J. Roman (Eds.): Euro-Par 2011, LNCS 6853, Part II, pp. 316–327, 2011.

universal. As far as we know, the Finite Difference Method(FDM) is widely used to solve the numerical problems. Once the FDM is used, the coefficient matrix of discrete Partial Differential Equations(PDEs) is usually the diagonal sparse matrix. The numerical solution to the PDEs is an approximation to its exact solution by using a discrete representation to the PDEs on the $m \times n \times l$ mesh points (x_i, y_j, z_k), where $1 \leq i \leq m, 1 \leq j \leq n, 1 \leq k \leq l$. In the finite difference scheme, the unknown's value $U_{i,j,k} = U(x_i, y_j, z_k)$ is related to $U_{i \pm t, j \pm p, k \pm q}$ (where t, p, q may normally be 1 or 2). As long as the difference scheme is fixed, t, p and q remain unchanged. When we modify m, n and l to change the problem size, the diagonal distribution remains similar.

The Diagonal format(DIA) [1] is designed to store the diagonal sparse matrix. All nonzeros on the same diagonal share the same index. However, a large number of zeros should be filled when there are many scatter points or the diagonal is broken by a long zero section. We define the long zero section as *idle section*.

To address this problem, we propose a novel storage format CRSD. In order to represent the diagonal distribution, we design the diagonal pattern, which divides diagonals into different groups. Furthermore, the matrix is split into row segments. In each row segment, nonzeros on the diagonals of the same group are viewed as the unit of storage and operation. We store those nonzeros contiguously and organize the operation on them in one loop. Simultaneously, the scatter points are also detected in each row segment. The number of filled zero for idle section can be controlled according to the application.

Because the diagonal distribution remains similar in one application of different problem sizes, most diagonal patterns remain unchanged. We define the unchanged diagonal pattern as application specific diagonal pattern(detailed in section 2.2). For any given application, we analyze one matrix, named the sample matrix, to obtain application specific diagonal patterns. Next, the optimal SpMV codelets for those diagonal patterns are generated automatically. Finally, we combine those codelets as the optimal SpMV implementation. In addition, the information collected during auto-tuning process can also be utilized for parallel implementation to achieve load-balance. As the unchanged diagonal patterns vary across diverse applications, the optimization is application specific.

The rest of this paper is organized as follows: section 2 describes the diagonal pattern and CRSD storage format; section 3 presents the process of automatic performance tuning; in section 4, the experiment results are provided and analyzed; the related works are given in section 5. At last, conclusion is summarized in section 6.

2   CRSD Storage Format

2.1   Diagonal Pattern

For any two diagonals in the matrix, if the absolute value of difference of their offset [1] is 1, they are adjacent. We can group a sequence of diagonals by the following steps: if two diagonals are adjacent, put them into an *adjacent (AD) group*; after removing the diagonals within the adjacent groups, the original diagonal sequence is broken up into pieces. We assign the diagonals of each piece into a *nonadjacent(NAD) group*. The *diagonal pattern* is defined as the way that the AD group(s) and the NAD group(s) are organized.

When the group is represented by group type(AD or NAD) and the number of diagonals in it, then

group = (group type, the number of diagonals)

According to the definition, the diagonal pattern is represented as follows:

diagonal-pattern = {$group_1$, $group_2$, ... $group_m$}

If the whole matrix contains several diagonal patterns, then

matrix = {dia-pattern$_1$, dia-pattern$_2$, ... dia-pattern$_n$}.

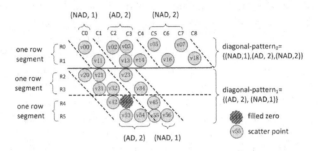

Fig. 1. Example of diagonal sparse matrix

For example, there are two diagonal patterns in the matrix shown in Fig. 1 except nonzero v55. The matrix is represented as follows:

matrix = {{(NAD,1),(AD,2),(NAD,2)}, {(AD,2), (NAD,1)}}.

With diagonal pattern, we can process idle section: if there are few zeros in the idle section, we can fill zeros to maintain the diagonal structure; otherwise, if a large number of zeros are needed, we believe that the diagonal is broken and the diagonal pattern should be changed. For example, a zero is filled at v43 position to maintain the diagonal structure, while the main diagonal is broken. The application developer can set the maximum number of filled zeros according to the property of application and the problem size.

2.2   Application Specific Diagonal Pattern

The diagonal patterns that remain unchanged among different problem sizes are abstracted as **Application Specific Diagonal Pattern** and stored into one group with group type *Application Specific (AS)*. As there are more than one application specific diagonal patterns, a tag is needed to identify them. In this way, the group is represented as (AS, tag). We analyze one matrix, named the sample matrix, from a given application to obtain application specific diagonal patterns. For the matrix listed in Fig. 1, if the second diagonal pattern is viewed as application specific diagonal pattern with tag 1, then

$$(AS, 1) = \{(AD, 2), (NAD, 1)\}$$
$$matrix = \{\{(NAD,1),(AD,2),(NAD,2)\}, \{(AS, 1)\}\}.$$

2.3 Storage Format

We have grouped the diagonals using diagonal pattern. Furthermore, the matrix is split into row segments. The number of rows in each row segment is defined as **row segment size** and represented by the token **mrows**. In this way, the whole matrix is split in two dimensions, as the dotted lines show in Fig. 1. In each row segment, nonzeros on the diagonals of the same group are the **storage unit** of CRSD. Additionally, if only one nonzero is on a diagonal within one row segment, the nonzero is viewed as scatter point, such as v55 in Fig. 1.

In CRSD storage format, the scatter points and the nonzeros in diagonal are stored separately. In order not to change the order of floating point operations, the whole row where the scatter point locates is stored together. The row number, the number of nonzeros in this row and the column index of each nonzero are used as the indices and stored in array **scatter_index**. The nonzero values are stored in array **scatter_val**.

Except scatter points, the whole matrix is represented by diagonal patterns. All nonzeros in the same diagonal pattern share the same index: the diagonal pattern, the start row number of the diagonal pattern, the number of row segments, and the column indices of diagonals. The column index of each diagonal is needed for nonadjacent group, while only the column index of first diagonal in adjacent group needs to be recorded. The diagonal pattern is stored in array **matrix** and the remaining of index value is stored in array **crsd_dia_index**. The nonzero values in each storage unit are stored contiguously in array **crsd_dia_val**, such as v20, v31, v21 and v32.

The number of diagonal patterns and rows that contain the scatter point are assigned to **num_dia_patterns** and **num_scatter_rows** respectively. An example is shown in Fig. 2 for the matrix in Fig. 1, when row segment size is 2.

---

num_scatter_rows=1
num_dia_patterns=2

matrix={{(NAD,1),(AD,2),(NAD,2)}, {(AS, 1)} }

crsd_dia_index = {R0, 1, C0, C2, C5, C7, | R2, 2, C0, C4}
crsd_dia_val = {{(v00,v11),(v02,v13,v03,v14),(v05,v16,v07,v18)}, {(v20,v31,v21,v32),(v23,v24)}, {(v42,v53,0,v54),(v45,v56)} }

scatter_index = {R5, 4, C3, C4, C5, C6}
scatter_val = {v53, v54, v55, v56}

Fig. 2. The CRSD storage format for matrix shown in Fig. 1 when mrows=2

2.4 SpMV Implementation for CRSD

In the SpMV implementation for CRSD, the storage unit is also the *operation unit*, for the reason that all SpMV operations on the elements in each storage unit are organized together. When we set the upper limit of the number of diagonals in (non)adjacent group, it is practical to enumerate the SpMV operations for all kinds of groups. Once the number of diagonals in one group exceeds the upper limit, it will be split into many sub-groups until the number of diagonals is less than the upper limit.

```
for each row-segment in diagonal pattern
    for each group of the diagonal pattern in one row segment
        switch group type
        case (NAD, 1): operation for group (NAD, 1)
        case (AD, 2) etc: operation for the enumerated groups
        case (AS, 1): // for application specific diagonal pattern
            for each row-segment in diagonal pattern // Loop inside
                // the generated SpMV codelet for (AS, 1)={(AD,2),(NAD,1)}
```

```
            done
        end switch
    done
done
```

Fig. 3. SpMV code fragment for matrix shown in Fig. 1 when mrows=2

As application specific diagonal patterns are only available after sampling the sample matrix, a code generator is designed to generate codelets for those diagonal patterns. Because application specific diagonal patterns are inherent to an application, the SpMV for CRSD becomes application specific. In addition, it is not reasonable to generate all application specific diagonal patterns, since some patterns cover few nonzero values. A *threshold* is set to determine the least number of nonzeros that an application specific diagonal pattern should cover.

A SpMV code fragment is given in Fig. 3. In processing the adjacent group, the elements of x can be reused. For this reason, we can load the element into a register and use it repeatedly, such as register x1. The group with type *AS* represents application specific diagonal pattern and describes diagonal distribution of entire row segment. Then operations for entire row segment are organized in one loop.

3 Application Specific Automatic Performance Tuning

In order to improve the performance as well as portability of the generated codelets, we apply auto-tuning to select the optimal implementation. For the reason that the SpMV for CRSD is application specific, the auto-tuning process is also application specific.

We optimize the generated codelets by applying SSE intrinsics and explicit prefetching. The SSE intrinsics allow simultaneous operations on a vector of two double precision values. Explicit prefetching is implemented via compiler intrinsic _builtin_prefetch. It sets prefetch distance to determine which elements to be preloaded. Meanwhile, it can change the temporal locality of the preloaded elements by modifying the prefetch locality, ranging from 0 to 3. The bigger the prefetch locality is, the higher the temporal locality is. Furthermore, we reschedule the SpMV operation via modifying the latency between data read and data available(LAT_RD) as well as the latency between data operation and result available(LAT_OP)(as shown in Fig. 3).

For different diagonal patterns, the performance is affected by the following parameters on different hardware platforms and the value ranges are determined according to the experimental statistics results:

- mrows. It determines the number of nonzeros to be processed in one loop, ranging from 2 to 8;
- prefetch distance and prefetch locality. The prefetch function is applied to nonzeros and vector x. The range of prefetch distance is from 30 to 300.
- LAT_RD and LAT_OP. If the number of SSE registers is defined as num_SSE_regs. Their range is from 1 to $num\_SSE\_regs/3$;

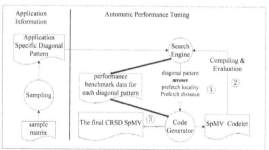

Fig. 4. Application specific auto-tuning

Table 1. Experimental Platforms

platform #	AMD	Intel
CPU	AMD Phenom TM II X4 940, 3.0 GHz	Intel Xeon X5550, 2.67GHz
MEM	8GB	8GB
Sockets	1	2
Compiler	GCC 4.3.3	GCC 4.4.3
Compiler option	-msse2 -O3 -fopenmp	

After obtaining the application specific diagonal patterns, the whole automatic performance tuning process is described as follows (shown in Fig 4):

Step 1. Search engine reads each application specific diagonal pattern, determines value set of parameters and passes them to code generator.

Step 2. The SpMV Codelet is generated, compiled and executed. The performance information is sent back to search engine and recorded until all application specific diagonal patterns are measured.

Step 3. The optimal performance and the corresponding parameter values for all application specific diagonal patterns are sent to the code generator to produce the final SpMV implementation.

The matrix, which is used to evaluate the generated codelet in step 2, is extended according the indices of the corresponding diagonal pattern, since the performance of SpMV is affected by the input sparse matrix.

The search engine uses orthogonal search method[6] to determine the parameter value set. The search order is *prefetch distance, prefetch locality, LAT_RD, LAT_OP* and *mrows*. Moreover, the entire process is completed during the building phase rather than at runtime.

3.1   The Final CRSD SpMV Implementation

As the auto-tuning records show, the row segment size is not same for different diagonal patterns when the performance of generated codelet is optimal. When we split the matrix in row direction, we chose different row segment size for different diagonal patterns. The SpMV codelet for each diagonal pattern should be generated according the corresponding row segment size. Then we combine those generated codelets to produce the final CRSD SpMV implementation.

3.2   Parallelization

When one matrix is stored in CRSD storage format, the diagonal pattern and corresponding number of row segments are obtained. Given the performance for processing each diagonal pattern, we can estimate the execution time. Then we can split the matrix into sub-matrices and keep the estimated time for processing each sub-matrix equal. The diagonal patterns may be split in the process when necessary. The parallelization is implemented using OpenMP. specifically, we distribute the scatter points according to the row range of each sub-matrix to avoid write confliction of destination vector y.

4   Evaluation

In this section, we present the performance improvement of CRSD on two platforms(Table 1) and 13 matrices(Table 2). Those matrices are categorized in five classes: the first three classes come from [14]; the last two classes are from an astrophysics application [15]. The coverage represents the percentage of number of nonzeros in this diagonal pattern and threshold 15%, mentioned in section 2.4, is used to identify the application specific diagonal pattern.

We choose CSR and DIA storage formats to compare with our CRSD storage format. Owing to that SpMV is not available in ACML, we select Intel MKL, with version 10.2.6.038, on the two x86-based architectures.

We also select OSKI-1.0.1h[7] for the comparison. When we set hint that all the matrices are diagonal sparse matrices, it fails to tune the matrices and return the input matrices.The same situations occur in earlier work[4][5].Thus the performance of CSR can be viewed as the result of OSKI.

Table 2. Matrix Set and Application Specific Diagonal patterns

#	name	row	nnz	picture	#	content	Coverage(%)
1	atmosmodd*	1270423	8814880				
2	atmosmodj	1270423	8814880		Dp1	{(NAD, 2), (AD, 3), (NAD, 2)}	95.3
3	atmosmodm	1270423	8814880				
4	cell1*	7055	30082				
5	cell2	7055	30082		Dp2	{(AD, 2), (NAD, 1), (AD, 2)}	79.9
6	kim1*	38415	933195				
7	kim2	456976	11330020		Dp3	{(AD,5),(AD,5), (AD, 5), (AD, 5) (AD, 5)}	98.0
8	A1	620000	4917600				
9	A2	1080000	8578800		Dp4	{(NAD,2), (AD,3), (NAD,3)}	44.0
10	A3*	320000	2532800		Dp5	{(NAD,3), (AD,3), (NAD,2)}	45.9
11	B1	620000	4917600				
12	B2	1080000	8578800		Dp6	{(NAD,2), (AD,5),(NAD,2)}	97.5
13	B3*	320000	2532800				

(Matrix Information; Application Specific Diagonal pattern)

* the sample matrix

4.1 The Auto-Tuning Records

The auto-tuning records on two platforms are given in Table 3. From the records we can conclude that the parameter values are different when the performance is optimal for the same diagonal pattern on different platforms.

Table 3. The Auto-tuning Records for Application Specific Diagonal patterns

Dp#	Segment size		Locality for x		Locality for nonzeros		Prefetch distance		LAT_RD		LAT_OP	
	AMD	Intel	AMD	Intel	AMD	Intel	AMD	Intel	AMD	Intel	AMD	Intel
Dp1	3	8	2	3	0	0	50	30	2	3	4	2
Dp2	8	8	2	3	1	0	150	30	2	4	3	4
Dp3	5	4	2	3	1	0	150	90	3	4	3	2
Dp4	3	6	3	3	0	0	50	150	2	2	2	3
Dp5	2	6	2	3	0	0	150	90	3	3	3	3
Dp6	4	4	3	3	0	0	40	90	3	3	3	3

The *prefetch locality* for x is almost 3, the highest temporal locality, whereas that for nonzeros is almost 0. The effect of *prefetch locality* for matrix B3 on two platforms is shown in Fig 5. The maximum difference of performance reaches up to 44.8% and 24.0% on platform AMD and Intel respectively. The major reason is the different access behavior of nonzeros and vector x: the elements of x may be accessed repeatedly, which is determined by nonzero distribution, whereas the elements in nonzeros are only accessed once.

We also collect the performance data affected by *prefetch distance* for matrix B3. The maximum difference of performance is only 11.8% and 2.2% on platform AMD and Intel respectively. This phenomenon exists for other matrices. We can conclude that prefetch locality plays a more important role than prefetch distance on SpMV performance for diagonal sparse matrices.

The performance improvement for the automatic performance tuning is given in Fig 6. The final CRSD SpMV uses the variable row segment and other optimization methods, such as SSE intrinsic. The performance using only variable row segment size is viewed as the basic CRSD implementation. The performance

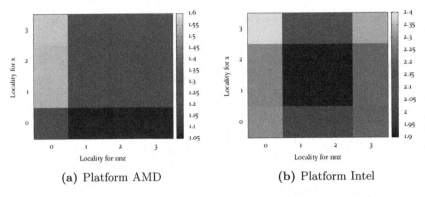

(a) Platform AMD   (b) Platform Intel

Fig. 5. The performance effect of prefetch locality on AMD and Intel platforms

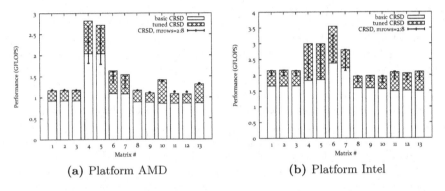

(a) Platform AMD   (b) Platform Intel

Fig. 6. The perfomance improvement of auto-tuning

gain differs among the matrices and the platforms. The average performance improvement is 36.3% and 36.6% and maximum is 65.9% and 63.1% on platform AMD and Intel respectively.

In Fig 6, the performance range of CRSD with different identical *mrows* is also given. The final CRSD SpMV outperforms those implementations for all matrices on platform Intel. On platform AMD, the performance difference between the final CRSD SpMV and upper range bound is less than 4%. This verifies that the final CRSD SpMV is efficient.

4.2   Serial Performance Improvement

The performance comparison with CSR and DIA is given in Fig 7. The maximum speedup compared with CSR reaches 3.83 and 4.38 on platform AMD and Intel respectively. And the average of speedup reaches 2.44 and 3.05.

In comparison with DIA, we find that the performance of DIA for cell1 and cell2 is very poor. For the reason that the number of filled zeros is almost 33 times larger than the number of nonzeros. The nonzeros distributes on 169 distinct diagonals. Therefore large number of idle sections exist. Even CSR is faster

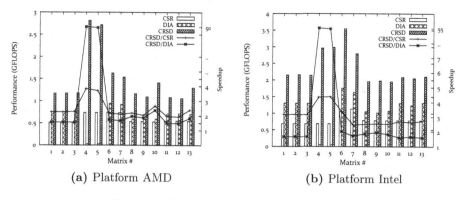

Fig. 7. Serial performance comparison results

than DIA by the factor of 23.50 and 11.87 on AMD and Intel respectively. However, using diagonal pattern CRSD is suitable for the two matrices, especially application specific diagonal pattern covers 79.9% of the nonzeros. In comparison with CSR, the speedups reach 3.69 and 4.34 on platform AMD and Intel respectively. Except cell1 and cell2, the maximum speedups reach 2.37 and 2.02 on platform AMD and Intel respectively. The average reaches 1.73 and 1.74.

4.3 Parallel Performance Improvement

Since the implementation based on the DIA is not parallelized in Intel MKL, only the CSR format is used for the comparison. The comparison results are shown in Fig 8. CRSD outperforms CSR under different number of available threads on two platforms. The maximum and average speedups are listed in Table 4.

Table 4. Speedup of parallel CRSD compared with parallel CSR

# of threads		2	4	8
AMD	Max	4.07	4.64	X
	Average	2.51	2.32	X
Intel	Max	4.51	4.34	4.61
	Average	2.42	2.16	2.17

The performance of CRSD for matrices with prefix cell and kim are relatively high, especially for matrices whose sizes are small enough to be fitted into cache. In those matrices, a large percent of diagonals are adjacent. Hence the elements of x are reused frequently.

5 Related Work

Im and Yelick et al. propose register blocking, cache blocking and reordering techniques. Register blocking[3][7][8] is based on BCSR format. BCSR is suitable for matrices, in which nonzeros primarily distribute in dense blocks. This

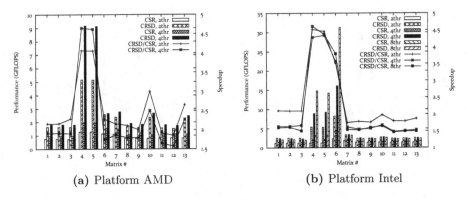

(a) Platform AMD   **(b)** Platform Intel

Fig. 8. Performance comparison results between parallel CSRD and parallel CSR

property is universal for the matrices produced by Finite Element Method(FEM, another major method for PDEs). Vuduc et al. estimate the performance bounds for the register blocking and propose a new approach to choose the register block size[9]. However, excessive zeros have to be filled to maintain the block format in BCSR, which wastes the computation and memory resources. To reduce the number of filled zero, Vuduc et al. in [2][10] exploit variable block structure rather than identical block size; Belgin et al. explore the distribution pattern of nonzeros in dense block and propose PBR to store matrices without zero filling[4]. Cache blocking[3] is used to increase the temporal locality by reordering the memory access, Nishtala et al. present a new performance models, which takes TLB misses into account, and a criteria to determine when to apply the cache blocking [11]. Samuel Williams [5] sums up all those optimization methods on the emerging multi-core platforms. To mitigate the memory bandwidth pressure, Willcock[12] and Kourtis et al. [13] utilize data compression to reduce the index. Furthermore Kourtis also introduce CSR-VI to compress the nonzero value when most of nonzero values are identical.

6   Conclusion

In this paper, we propose CRSD for the diagonal sparse matrix. We design diagonal pattern to describe the diagonal distribution, making CRSD more suitable than DIA. Furthermore, as diagonal distributions are similar for different problem sizes, we introduce the idea of application specific diagonal pattern to optimize SpMV implementation. During the building phase, the optimal codelets for application specific diagonal patterns are generated automatically. It differs from OSKI, for OSKI chooses the optimal implementation at runtime. The autotuning records are also utilized to achieve load-balance for parallelization.

The results from our experiments demonstrate that CRSD is efficient for processing the diagonal sparse matrices from one application, when there are several major diagonal patterns and diagonal distribution remains similar among the matrices. We are transplanting the CRSD to the GPU. Our preliminary tests on

GPU indicate a strong potential for better performance. In the future, we will study more types of nonzero distributions and optimize the SpMV specifically on distinct architectures.

References

1. Bell, N., Garland, M.: Implementing sparse matrix-vector multiplication on throughput oriented processors. In: Supercomputing (2009)
2. Vuduc, R.W.: Automatic Performance of Sparse Matrix Kernels. The dissertation of Ph.D, Computer Science Division, U.C. Berkeley (2003)
3. Im, E.: Optimizing the performance of sparse matrix-vector multiplication. PhD thesis, University of California, Berkeley (2000)
4. Belgin, M., Back, G., Ribbens, C.J.: Pattern-based sparse matrix representation for memory-efficient SMVM kernels. In: International Conference on Supercomputing, NY, USA (2009)
5. Williams, S., Oliker, L., Vuduc, R., Shalf, J., Yelick, K., Demmel, J.: Optimization of sparse matrix-vector multiplication on emerging multicore platforms. In: Proceedings of the 2007 ACM/IEEE Conference on Supercomputing, Reno, Nevada, November 10-16 (2007)
6. Kulkarni, M., Pingali, K.: An experimental study of self-optimizing dense linear algebra software. Proceedings of the IEEE 96(5), 832–848 (2008)
7. Vuduc, R., Demmel, J., Yelick, K.: OSKI: A library of automatically tuned sparse matrix kernels. In: Proceedings of SciDAC 2005, Journal of Physics: Conference Series (2005)
8. Im, E.-J., Yelick, K.A.: Optimizing sparse matrix computations for register reuse in SPARSITY. In: Alexandrov, V.N., Dongarra, J., Juliano, B.A., Renner, R.S., Tan, C.J.K. (eds.) ICCS-ComputSci 2001. LNCS, vol. 2073, pp. 127–136. Springer, Heidelberg (2001)
9. Vuduc, R., Demmel, J., Yelick, K., Kamil, S., Nishtala, R., Lee, B.: Performance optimizations and bounds for sparse matrix-vector multiply. In: Supercomputing, Baltimore, MD (2002)
10. Vuduc, R.W., Moon, H.-J.: Fast sparse matrix-vector multiplication by exploiting variable block structure. In: Yang, L.T., Rana, O.F., Di Martino, B., Dongarra, J. (eds.) HPCC 2005. LNCS, vol. 3726, pp. 807–816. Springer, Heidelberg (2005)
11. Nishtala, R., Vuduc, R., Demmel, J.W., Yelick, K.A.: When cache blocking sparse matrix vector multiply works and why. Applicable Algebra in Engineering, Communication, and Computing (2007)
12. Willcock, J., Lumsdaine, A.: Accelerating sparse matrix computations via data compression. In: ICS 2006: Proceedings of the 20th Annual International Conference on Supercomputing, pp. 307–316. ACM Press, New York (2006)
13. Kourtis, K., Goumas, G., Koziris, N.: Optimizing sparse matrix-vector multiplication using index and value compression. In: Proceedings of the 5th Conference on Computing Frontiers, Ischia, Italy, May 5-7 (2008)
14. Boisvert, R., Pozo, R., Remington, K., Miller, B., Lipman, R.: NISTMatrixMarket, http://math.nist.gov/MatrixMarket/index.html
15. Chana, K.H., Li, L., Liao, X.: Modelling the core convection using finite element and finite difference methods. Physics of the Earth and Planetary Interiors 157(2), 124–138 (2006)

The LOFAR Beam Former:
Implementation and Performance Analysis

Jan David Mol and John W. Romein

Stichting ASTRON (Netherlands Institute for Radio Astronomy)
Oude Hoogeveensedijk 4, 7991 PD Dwingeloo, The Netherlands
{mol,romein}@astron.nl

Abstract. Traditional radio telescopes use large, steel dishes to observe radio sources. The LOFAR radio telescope is different, and uses tens of thousands of fixed, non-movable antennas instead, a novel design that promises ground-breaking research in astronomy. The antennas observe omnidirectionally, and sky sources are observed by signal-processing techniques that combine the data from all antennas.

Another new feature of LOFAR is the elaborate use of *software* to do signal processing in real time, where traditional telescopes use custom-built hardware. The use of software leads to an instrument that is inherently more flexible. However, the enormous data rate (198 Gb/s of input data) and processing requirements compel the use of a supercomputer: we use an IBM Blue Gene/P.

This paper presents a collection of new processing pipelines, collectively called the beam-forming pipelines, that greatly enhance the functionality of the telescope. Where our first pipeline could only correlate data to create sky images, the new pipelines allow the discovery of unknown pulsars, observations of known pulsars, and (in the future), to observe cosmic rays and study transient events. Unlike traditional telescopes, we can observe in hundreds of directions simultaneously. This is useful, for example, to search the sky for new pulsars. The use of software allows us to quickly add new functionality and to adapt to new insights that fully exploit the novel features and the power of our unique instrument. We also describe our optimisations to use the Blue Gene/P at very high efficiencies, maximising the effectiveness of the entire telescope. A thorough performance study identifies the limits of our system.

1 Introduction

The LOFAR (LOw Frequency ARray) telescope is the first of a new generation of radio telescopes. Instead of using a set of large, expensive dishes, LOFAR uses many thousands of simple antennas. Every antenna observes the full sky, and the telescope is pointed through signal-processing techniques. LOFAR's novel design allows the telescope to perform wide-angle observations as well as to observe in multiple directions simultaneously, neither of which are possible when using traditional dishes. In several ways, LOFAR will be the largest telescope in the world, and will enable ground-breaking research in several areas of astronomy and particle physics [1].

Another novelty is the elaborate use of software to process the telescope data in real time. Previous generations of telescopes depended on custom-made hardware to combine data, because of the high data rates and processing requirements. The availability

E. Jeannot, R. Namyst, and J. Roman (Eds.): Euro-Par 2011, LNCS 6853, Part II, pp. 328–339, 2011.

of sufficiently powerful supercomputers however, allow the use of software to combine telescope data, creating a more flexible and reconfigurable instrument. Because LOFAR is driven by new science, flexibility in the design is essential to explore the possibilities and limits of our telescope.

For processing LOFAR data, we use an IBM BlueGene/P (BG/P) supercomputer. The LOFAR antennas are grouped into stations, and each station sends its data (up to 198 Gb/s for all stations) to the BG/P. Inside the BG/P, the data are processed using both real-time signal-processing routines as well as two all-to-all exchanges. The output data streams are sufficiently reduced in size to be able to stream them out of the BG/P and store them on disks in our storage cluster.

In this paper, we will present the LOFAR *beam former*: a collection of software pipelines that allow the LOFAR telescope to be pointed at hundreds of sources simultaneously. A *beam* consists of a 1D stream of data representing the signal from a certain area in the sky, and thus is different from a correlator, that creates 2D snapshot images of the sky. Simplified, a beam former performs a weighted addition of the input signals, while a correlator multiplies the input signals.

It is LOFAR's unique design that allows us to point at many sources at once. Traditional telescopes use dishes that have a narrow field-of-view: they are only sensitive to a small region around the source they are pointed at. LOFAR's antennas are omnidirectional. Groups of antennas (*stations*) are sensitive to a wide field-of-view around the source. These views, or *station beams*, are sent to the BG/P, that generates weighted additions of the station input data, called *tied-array beams*. Each tied-array beam represents an offset pointing within the wide field-of-view of the stations.

The primary scientific use case driving the work presented in this paper is pulsar research [2]. A pulsar is a rapidly rotating, highly magnetised neutron star, which emits electromagnetic radiation from its poles. Similar to the behaviour of a lighthouse, the radiation is visible to us only if one of the poles points towards the Earth, and appears to us as a very regular series of pulses, with a period as low as 1.4 ms. Pulsars are weak radio sources, and their individual pulses often do not rise above the background noise that fills our universe. Our beam former can track several pulsars at LOFAR's full observational bandwidth. Alternatively, the beam former is capable of efficiently performing sky surveys to discover new pulsars (or other radio sources) by covering the sky with hundreds of tied-array beams at a reduced observational bandwidth.

The main contributions of this paper are threefold. First, we demonstrate the power of a *software* telescope; its flexibility allows us to add new functionality with modest effort and we show how the use of supercomputer technology enables new science in astronomy and particle physics. Second, we describe the first system which allows a telescope to be pointed in hundreds of directions. Third, we elaborately analyse the performance of our application and the effectiveness of our optimisations.

This paper is organised as follows. First, we will describe the key characteristics of the IBM BlueGene/P supercomputer in Sec. 2. Then, we describe LOFAR and beam forming in more detail in Sec. 3. Section 4 describes the implementation of our pipelines, followed by the performance analysis in Sec. 5. We briefly discuss related work in Sec. 6, and conclude in Sec. 7.

2   IBM BlueGene/P

We use an IBM BlueGene/P (BG/P) supercomputer for the real-time processing of station data. We will describe the key features of the BG/P; more information can be found elsewhere [8]. Furthermore, we will describe how our BG/P is connected to its input and output systems, and how we perform real-time processing using a BG/P.

2.1   System Description

Our system consists of 3 racks, with 12,480 processor cores that provide 42.4 TFLOPS peak processing power. One chip contains four PowerPC 450 cores, running at a modest 850 MHz clock speed to reduce power consumption and to increase package density. Each core has two floating-point units (FPUs) that provide support for operations on complex numbers. The chips are organised in *psets*, each of which consists of 64 cores for computation (*compute cores*) and one chip for communication (*I/O node*). Each compute core runs a fast, simple, single-process kernel, and has access to 512 MiB of memory. The I/O nodes consist of the same hardware as the compute nodes, but additionally have a 10 Gb/s Ethernet interface connected. They run Linux, which allows the I/O nodes to do full multitasking. One rack contains 64 psets, which is equal to 4096 compute cores and 64 I/O nodes.

The BG/P contains several networks. A fast *3-dimensional torus* connects all compute nodes and is used for point-to-point and all-to-all communications over 3.4 Gb/s links. The torus uses DMA to offload the CPUs and allows asynchronous communication. The *collective network* is used for communication within a pset between an I/O node and the compute nodes, using 6.8 Gb/s links. In both networks, data is routed through compute nodes using a shortest path.

2.2   External I/O

We customised the I/O node software stack [9] and run a multi-threaded program on each I/O node that is responsible for the handling of both the input and the output. Unfortunately, the I/O nodes cannot saturate their 10 Gb/s Ethernet interfaces, because the 850 MHz cores do not have enough computational power to handle the overhead caused by IRQs, IP, and UDP/TCP. An I/O node can output at most 3.1 Gb/s, unless it has to handle station input (3.1 Gb/s per station), in which case it can output at most 1.1 Gb/s. We implemented a low-overhead communication protocol called FCNP [6] to efficiently transport data between the I/O nodes and the compute nodes. The compute nodes perform the signal processing. The I/O nodes forward the results to our storage cluster, which can sustain a throughput up to 80 Gb/s.

2.3   Real-Time Processing

Radio telescopes, including LOFAR, can observe for 24 hours per day: Rayleigh scattering, which causes optical sun light to dominate the sky during the day, is nearly nonexistent at radio frequencies. A LOFAR observation typically runs for several minutes to several days, and requires a single rack for real-time processing. Our other two

Fig. 3. Tied-array beams (hexagons) formed within two station beams (ellipse)

Fig. 2. The left antenna receives the wave later

Fig. 1. Locations of the stations

racks are used for development, and as hot spares in case of unexpected hardware failures, which happens a few times per year. The BG/P is not a hard real-time system: almost all variance occurs in the networks within the BG/P due to clashes caused by scheduling intricacies, which can force our software to discard station input. To keep post-processing tractable, a lost input sample causes all output samples that depend on it to be discarded. We tolerate at most 0.1% of data loss, but loss is typically a lot rarer.

3 LOFAR and Beam Forming

The LOFAR antennas are grouped in *stations*. The stations are strategically placed, with 20 stations in the centre (the *core*) and 24 stations at increasing distances from the core, spanning five nations (see Fig. 1). A core station can act as two individual stations in some observational modes, resulting in a total of 64 stations. A station is able to produce 248 frequency subbands of 195 kHz in the 10 – 250 MHz sensitivity range. Each sample consists of two complex 16-bit integers, representing the amplitude and phase of the X and Y polarisations of the antennas.

Even though the antennas are omnidirectional, they can be pointed due to the fact that the speed of electromagnetic waves is finite. Signals emitted by a source reach different antennas at different times (see Fig. 2). A process called *delay compensation* delays the signals such that they align (are *coherent*) for the desired source. Beam forming subsequently adds the aligned signals. The stations perform delay compensation and beam forming to combine the antenna signals into a station beam with a wide field-of-view. The BG/P subsequently combines the signals from different stations to form tied-array beams within the sensitive area of the station beams (see Fig. 3). In the BG/P, the samples from different stations are shifted with respect to each other to compensate delay at a sample-level granularity. Sub-sample delay compensation is performed by a complex multiplication per sample, which shifts the phase of each sample. The weights used in the complex multiplication depend on the location of the stations, the observational frequency of the sample, and the sky coordinates of the tied-array beam. The beam former thus creates tied-array beams by adding the station signals using different complex weights for each beam.

Our beam former supports several pipelines. The *complex voltages* pipeline stores the tied-array beams as is (X and Y polarisation samples). The *Stokes IQUV* pipeline transforms the complex voltages into Stokes parameters, which are a different representation of the signal. Finally, the *Stokes I* pipeline stores just the signal strength for each beam, and can be integrated in time to reduce the output data rate and to increase the number of tied-array beams that can be formed. Finally, our software can produce the Stokes parameters of an *incoherent* beam, which is an accumulation of unweighted station signals. The incoherent beam is less sensitive than a coherent beam, but it maintains the wide field-of-view of the stations. The incoherent beam is typically formed in parallel with other pipelines, and is used to detect the presence of pulsars, but does not reveal their location within the station beams.

4   Beam Former Pipelines

In this section, we will describe in detail how the full signal-processing pipelines operate, in and around the beam former. The use of a software pipeline allows us to reconfigure the components and design of our standard imaging pipeline, described elsewhere [7]. Due to the flexibility of software, we can run several pipelines in parallel on the same data, as long as resource limits are not exceeded. Figure 4 gives an overview of our system. Our software is written in C++, with core routines ported to assembly to obtain maximal performance.

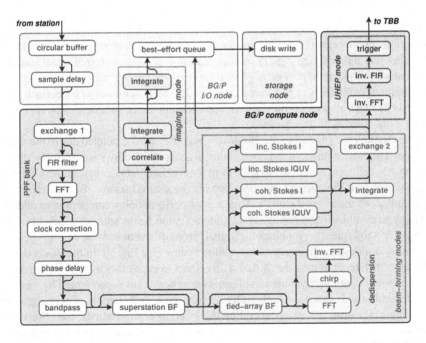

Fig. 4. The on-line pipelines of LOFAR. The imaging and UHEP pipelines are outside the scope of this work.

4.1   Input from Stations

Each station sends data to a different I/O node. The beam former, however, needs data from all stations together to form tied-array beams. The station data thus have to be rearranged inside the BG/P, to collect the data from different stations but also to split it along different dimensions in order to distribute the workload. At the I/O nodes, the station data are split into chunks of one subband and 0.25 seconds. The chunk size is chosen such that the compute cores have enough memory to perform all of the necessary processing. Due to the BG/P design, an I/O node sends chunks to its own compute cores using the collective network. The compute cores then exchange these chunks over the torus network using an all-to-all exchange, shown in Fig. 5.

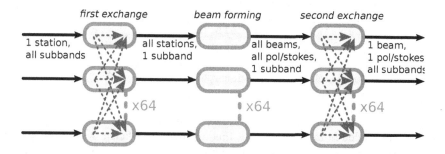

Fig. 5. The data flow and data ordening in our pipelines

4.2   First All-to-All Exchange

The first all-to-all exchange allows the compute cores to distribute the chunks from a single station, and to collect all the chunks of the same subband from all of the stations. The exchange is performed over the fast torus network, but with up to 198 Gb/s of station data to be exchanged, special care still has to be taken to avoid network bottlenecks. It is impossible to optimise for short network paths due to the physical distances between the different psets across a BG/P rack. Instead, we optimised the data exchange by creating as many paths as possible between compute cores that have to exchange data. Within each pset, we employ a virtual mapping such that the number of possible routes between communicating cores in different psets is maximised.

The all-to-all exchange is asynchronous. Once a compute core receives a complete chunk from a single subband, it performs a sequence of processing steps on it. The first step is a conversion from 16-bit little-endian integers into 32-bit big-endian floats, to be able to use the BlueGene's powerful FPUs. Figure 4 shows which steps are performed before the tied-array beam forming occurs. Note the Fast Fourier Transform (FFT) that divides the 195 kHz subbands into (typically) 12 kHz channels. We use the efficient *Vienna* version of FFTW [5]. The superstation beam former is a simplified version of our beam former, used to combine multiple stations as if it were one, and is used in our imaging pipeline to reduce the workload. Once the chunks from all stations are received and processed asynchronously, the processed data are ready to be beam formed.

4.3 Beam Forming

The beam former combines the chunks from all stations, producing a chunk for each tied-array beam. Each beam is formed using different complex weights for the frequency of the channel, the locations of the stations, and the beam coordinates. The positional weights are precomputed by the I/O nodes and sent along with the data to avoid a duplicated effort by the compute nodes. The delays are applied to the station data through complex multiplications and additions.

All time-consuming pipeline components are written in assembly, to achieve maximum performance. The assembly code minimises the number of memory accesses, minimises load delays, minimises FPU pipeline stalls, and maximises instruction-level parallelism. We learnt that optimal performance is often achieved by combining multiple iterations of a multi-dimensional loops:

```
FOR Channel IN 1 .. NrChannels DO
  FOR Station IN 1 .. NrStations STEP 6 DO
    FOR Time IN 1 .. NrTimes STEP 128 DO
      FOR Beam IN 1 .. NrBeams STEP 3 DO
        BeamForm6StationsAnd128TimesTo3BeamsAssembly(...)
```

This is much more efficient than to create all beams one at a time, due to better reuse of data loaded from main memory. Finding the most efficient way to group work is a combination of careful analysis and, unfortunately, trial-and-error. The coherent beam former achieves 86% of the FPU peak performance, not as high as the 96% of the correlator [7], but still 16 times more than the C++ reference implementation.

4.4 Channel-Level Dedispersion

Another major component in the pulsar-observation pipeline is real-time dedispersion. Since light of a high frequency travels faster through the interstellar medium than light of a lower frequency, the arrival time of a pulse differs for different wave lengths. To combine data from multiple frequency channels, the channels must be aligned (shifted in time). Otherwise, the pulse will be smeared or even overlap with the next pulse, causing many details to be lost. This process, called *dedispersion*, is done by post-processing software that runs after the observation has finished. However, to observe at the lowest frequencies, or to observe fast-rotating millisecond pulsars, dedispersion must also be performed *within* a channel, since our channels (typically 12 kHz) are too wide to ignore dispersion.

Figure 6 shows pulses of pulsar J0034-0534 at four frequencies. The pulse period is 1.88 ms. On the left is the original dispersed signal, which results in a smeared pulse when the frequencies are collapsed into a 12 kHz channel. On the right is the dedispersed signal, which results in a sharp pulse profile when collapsed.

Dedispersion is performed in the frequency domain, by doing a 4096-point FFT that splits a channel into 3 Hz subchannels. The phases of the observed samples are corrected by applying a chirp function, i.e., by multiplication with precomputed, channel-dependent, complex weights. These multiplications are programmed in assembly, to reduce the computational costs. A backward FFT is done to revert to 12 kHz channels.

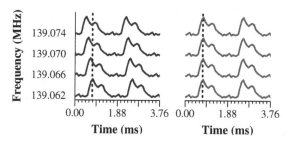
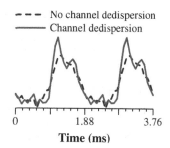

Fig. 6. Pulse arrival times within a 12 kHz channel before (left) and after (right) channel-level dedispersion

Fig. 7. Pulse profiles with and without channel dedispersion

Figure 7 shows the observed effectiveness of channel-level dedispersion, which improves the effective time resolution from 0.51 ms to 0.082 ms, revealing a more detailed pulse and a better signal-to-noise ratio. Dedispersion contributes significantly to the data quality, but it also comes at a large computational cost due to the two FFTs it requires. The channel-level dedispersion demonstrates the power of using a *software* telescope: the component was implemented, verified, and optimised in only one month time.

4.5 Stokes Calculations

The beams are optionally converted into Stokes IQUV or Stokes I parameters, again using assembly routines to achieve optimal performance. The Stokes parameters are calculated through $I = X\overline{X} + Y\overline{Y}, Q = X\overline{X} - Y\overline{Y}, U = 2 \cdot \text{Re}(X\overline{Y}), V = 2 \cdot \text{Im}(X\overline{Y})$, with $\overline{X}$ as the complex conjugate of X. Although the formulas are simple, the Stokes parameters are expensive to calculate. The required operations for I and Q do not map well onto the FPU instruction set of the BG/P, even though the instruction set is extended with support for operations on complex numbers.

4.6 Second All-to-All Exchange

Even though the beams are formed and optionally converted into Stokes parameters, they are still distributed as chunks across the BlueGene. Because the compute nodes cannot send their data directly to the I/O node that sends it to storage, a second all-to-all exchange is required to rearrange the chunks for output. Only chunks that are sent to the same I/O node can be sent to storage as a single data stream.

Unfortunately, the output bandwidth available at each I/O node can be less than the bandwidth required by the beams. An I/O node can output 3.1 Gb/s, and only 1.1 Gb/s if the I/O node also has to process station input at the same time. The bandwidth required for a complex voltages, Stokes IQUV, or (unintegrated) Stokes I beam however is 6.2 Gb/s, 6.2 Gb/s, and 1.5 Gb/s, respectively. We therefore split the beams and send the polarisations or Stokes parameters to different I/O nodes and store them in different files in our storage cluster. In some cases, it is necessary to split the beams further.

Due to memory constrains on the compute cores, the cores that performed the beam forming cannot be the same cores that receive the beam data after the second exchange.

We assign a set of cores (*output cores*) to receive the chunks. The output cores are chosen before an observation, and are distinct from the *input cores* which perform the earlier computations in the pipeline.

The output cores receive the chunks asynchronously, which we overlap with computations. For each chunk, the data are reordered into their final ordering. Reordering is necessary, because the data order that will be written to disk is not the same order that can be produced by our computations without taking heavy cache penalties. Once all of the chunks are received and reordered, they are forwarded to the I/O node.

For the distribution of the workload over the output cores, three factors are considered. First, all of the data belonging to the same beam has to be processed by output cores in the same pset, to ensure that one I/O node can concatenate all of the 0.25 second chunks that belong to the beam. Second, the maximum output rate per I/O node has to be respected. Finally, the presence of the first all-to-all exchange, which uses the same network at up to 198 Gb/s. The second exchange uses up to 81 Gb/s. Even though each link sustains 3.4 Gb/s, it has to process the traffic from four cores, as well as traffic routed through it between other nodes. The network links in the BG/P become overloaded unless the output cores are scattered sufficiently.

4.7 Transport to Disks

Once an output core has received and re-ordered all of its data, the data are sent to the core's I/O node. The I/O node forwards the data over TCP/IP to the storage cluster. To avoid any stalling in our pipeline due to network congestion or disk issues, the I/O node uses a best-effort buffer which drops data in the unusual case that it cannot be sent.

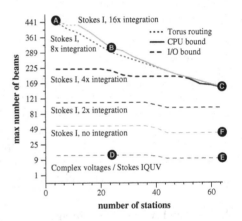

Fig. 8. The number of beams that can be formed

5   Performance Analysis

We will focus our performance analysis on the most challenging cases that are of astronomical interest. We present measurements for a single BG/P rack.

5.1   Overall Performance

Figure 8 shows the maximum number of beams that can be formed when using a various number of stations, in each of the three pipelines: complex voltages, Stokes IQUV, and Stokes I. Both the complex voltages and the Stokes IQUV pipelines are I/O bound. Each beam is 6.2 Gb/s wide. We can form up to 13 beams without exceeding the available 81 Gb/s to our storage cluster. If 64 stations are used, the available bandwidth is 70 Gb/s due to the fact that an I/O node can only output 1.1 Gb/s if it also has to process station data. The granularity with which the output can be distributed over the I/O nodes, as

Table 1. Several highlighted cases (CD = channel dedispersion, IF = integration factor)

Case	Mode	CD	IF	Stations	Beams	Input	Output	Bound	Used for
Ⓐ	Stokes I	N	16	4	450	12 Gb/s	44 Gb/s	Torus	Surveys
Ⓑ	Stokes I	N	16	24	310	74 Gb/s	30 Gb/s	CPU	Surveys
Ⓒ	Stokes I	N	8	64	155	198 Gb/s	30 Gb/s	CPU	Surveys
Ⓓ	Stokes IQUV	Y	-	24	13	74 Gb/s	81 Gb/s	I/O	Known sources
Ⓔ	Stokes IQUV	Y	-	64	10	198 Gb/s	62 Gb/s	I/O	Known sources
Ⓕ	Stokes I	Y	1	64	42	198 Gb/s	65 Gb/s	I/O	Known sources

well as scheduling details, determine the actual number of beams that can be formed, but in all cases, the beam former can form at least 10 beams at full observational bandwidth.

In the Stokes I pipeline, we applied several integration factors (1, 2, 4, 8, and 16) in order to show the trade-off between beam quality and the number of beams. Integration factors higher than 16 does not allow significantly more beams to be formed, but could be used in order to further reduce the total output rate. For low integration factors, the beam former is again limited by the available output bandwidth. At 8x integration, the number of beams is limited by the virtual mapping we applied to optimise both of the all-to-all exchanges (see Sec. 4.2): the high number of routes causes more collisions than the compute cores have spare time for to handle. With higher integration factors, a few more beams can be formed before the compute cores run out of computational resources. For observations for which a high integration factor is acceptable, the beam former is able to form 155–450 tied-array beams, depending on the number of stations used. For observations that need a high time resolution and thus a low integration factor, the beam former is still able to form at least 42 tied-array beams.

5.2 System Load

We analyse the workload of the compute cores by highlighting a set of cases, summarised in Table 1. We will focus on case Ⓐ, which creates the highest number of beams, and on CPU-bound cases useful for performing surveys, with either 24 stations (Ⓑ) or 64 stations (Ⓒ) as input. Cases Ⓓ and Ⓔ represent high-resolution observations of known sources, and are I/O bound configurations with 24 and 64 stations, respectively. Case Ⓕ focusses on the observations of known sources as well, using Stokes I output, which allows more beams to be formed. Channel-level dedispersion is applied for all cases that observe known sources.

The average workload of the compute cores for each case is shown in Fig. 9. For the CPU-bound cases Ⓑ and Ⓒ, the average load has to be lower than 100% to recover from small delays in the processing, that can occur since the BG/P is not a real-time system. These fluctuations typically occur due to clashes within the BG/P torus network which is used for both all-to-all-exchanges, and cannot be avoided in all cases.

In the cases where we create many beams (Ⓐ Ⓑ Ⓒ), most of the cycles are spent on beam forming and on calculating the Stokes I parameters. The beam forming scales with

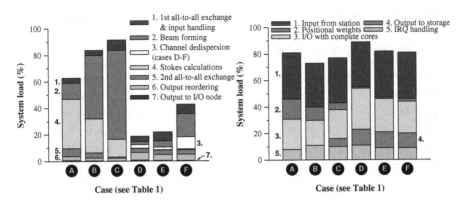

Fig. 9. The load of the compute cores   **Fig. 10.** The load of the busiest I/O nodes

both the number of stations and the number of beams, while the Stokes I calculation costs depends solely on the number of beams. Case Ⓐ has to beam form only four stations, and thus requires most of its time calculating the Stokes I parameters. Cases Ⓑ and Ⓒ use more stations, and thus need more time to beam form. The costs for both all-to-all exchanges are mostly hidden due to overlaps with computation. The remaining cost for the second exchange is proportional to the output bandwidth.

For the I/O-bound cases Ⓓ Ⓔ Ⓕ, only a few tied-array beams are formed and transformed into Stokes I(QUV) parameters, which produces a lot of data but requires little CPU time. Enough CPU time is therefore available to include channel-level dedispersion, which scales with the number of beams and is an expensive operation.

Figure 10 shows the workload for the busiest I/O nodes in each case, including the system time spent to handle IRQs. The processing of station data and the communication with the compute cores cause most of the load. In cases Ⓐ Ⓑ, the output is handled by I/O nodes that do not process station data. In both cases, a significant amount of time is spent computing the positional weights (see Sec. 4.3). A similar amount of time is required in cases Ⓒ Ⓓ Ⓔ Ⓕ to process the output.

6   Related Work

The LOFAR beam former is the only beam former capable of producing hundreds of tied-array beams. A radio dish can be extended to focus on multiple sources by placing additional receivers in its focal point (a *focal plane array*) [4], but such a solution does not scale. The Murchison Widefield Array (MWA) uses a design similar to LOFAR [3], and has far fewer antennas but groups them into more stations. The MWA will be able to form 16 tied-array beams, reducing 320 Gbit/s of input to 10 Gbit/s of output.

7   Conclusions

We have shown the capabilities of our beam former pipelines, running in software on an IBM BlueGene/P supercomputer. Our system can form 13 tied-array beams at LOFAR's full observational bandwidth before our output limit of 81 Gb/s is met. Alternatively, it can form hundreds of beams at a reduced resolution, the exact number depending on the number of stations and the pipeline used. Finally, an incoherent beam can be formed, which retains the wide field-of-view offered by our stations. None of these feats are possible with any other telescope.

The use of a software solution on powerful interconnected hardware is a key aspect in the development and deployment of our pipeline. Because we use software, rapid prototyping is cheap, allowing novel features to be tested to aid the exploration of the design space of a new instrument. The resulting pipelines retain the flexibility that software allows. The control flow and bookkeeping have become complex while remaining manageable through software abstraction. We can run the same station data through multiple pipelines in parallel, and even multiple independent observations in parallel, as long as there are enough resources. The science which drives LOFAR, and which is driven by it, is accelerated through the use of an easily reconfigurable instrument.

The BG/P supercomputer provides us with enough computing power and powerful networks to be able to implement the signal processing and all-to-all-exchanges that we require, without having to resort to a dedicated system which inevitably curbs the design freedom that the supercomputer provides. As with any system, platform-specific parameters nevertheless become important when maximal performance is desired. Although a C reference implementation allowed us to quickly develop and test features, we needed handcrafted assembly to keep the double FPUs of each compute core busy. The architecture of the BG/P makes some tasks more difficult as well. We cannot freely schedule the workload, because an I/O node can only communicate with its own compute cores. Instead, we have to manually route the data using two all-to-all exchanges to stream the data from and to the right I/O nodes. To achieve maximum performance, we tuned the distribution of the workload over the cores to avoid network collisions.

References

1. de Bruyn, A.G., et al.: Exploring the Universe with the Low Frequency Array. A Scientific Case (2002), http://www.lofar.org/PDF/NL-CASE-1.0.pdf
2. Stappers, B.W.: Observing pulsars and fast transients with LOFAR. Astronomy & Astrophysics (to appear, 2011)
3. Lonsdale, C.J., et al.: The Murchison Widefield Array: Design Overview. Proc. of the IEEE 97(8), 1497–1506 (2009)
4. Staveley-Smith, L., et al.: The Parkes 21cm Multibeam Receiver. Publications of the Astronomical Society of Australia 13(3), 243–248 (1996)
5. Lorenz, J., et al.: Vectorization Techniques for the Blue Gene/L Double FPU. IBM Journal of Research and Development 49(2/3), 437–446 (2005)
6. Romein, J.W.: FCNP: Fast I/O on the Blue Gene/P. In: Proc. of PDPTA, pp. 225–231 (2009)
7. Romein, J.W., Broekema, P.C., Mol, J.D., van Nieuwpoort, R.V.: The LOFAR Correlator: Implementation and Performance Analysis. In: Proc. of ACM PPoPP, pp. 169–178 (2010)
8. IBM Blue Gene team. Overview of the IBM Blue Gene/P Project. IBM Journal of Research and Development 52(1/2) (2008)
9. Yoshii, K., Iskra, K., Naik, H., Beckman, P., Broekema, P.C.: Performance and Scalability Evaluation of "Big Memory" on Blue Gene Linux. International Journal of High Performance Computing (to appear)

Application-Specific Fault Tolerance via Data Access Characterization

Nawab Ali[1], Sriram Krishnamoorthy[1], Niranjan Govind[1],
Karol Kowalski[1], and Ponnuswamy Sadayappan[2]

[1] Pacific Northwest National Laboratory, Richland, WA
{nawab.ali,sriram,niri.govind,karol.kowalski}@pnl.gov
[2] The Ohio State University, Columbus, OH
saday@cse.ohio-state.edu

Abstract. Recent trends in semiconductor technology and supercomputer design predict an increasing probability of faults during an application's execution. Designing an application that is resilient to system failures requires careful evaluation of the impact of various approaches on preserving key application state. In this paper, we present our experiences in an ongoing effort to make a large computational chemistry application fault tolerant. We construct the data access signatures of key application modules to evaluate alternative fault tolerance approaches. We present the instrumentation methodology, characterization of the application modules, and evaluation of fault tolerance techniques using the information collected. The application signatures developed capture application characteristics not traditionally revealed by performance tools. We believe these can be used in the design and evaluation of runtimes beyond fault tolerance.

Keywords: Fault tolerance, Data access characterization, NWChem.

1 Introduction

The increasing component counts in modern supercomputer designs, coupled with a decrease in micro-architectural feature size, and considerations of power envelope predict a significant decrease in the mean time between failures (MTBF) of the next generation of leadership-class machines [27]. Long-running scientific applications should expect multiple failures, both hard and transient, during execution. This has increased the need for applications to incorporate capabilities to identify and make forward progress in the presence of faults.

Making a large-scale scientific application fault tolerant is an arduous task. The first step involves evaluating different fault tolerance approaches and quantifying their impact in terms of space and time overhead, the amount of work lost in the event of a fault, and the feasibility of incorporating the fault tolerance approaches into the application. In this paper, we present our approach to evaluating key modules of NWChem [32,17], a large computational chemistry application consisting of close to two million lines of code. NWChem is a widely used computational chemistry suite shown to scale on the largest systems.

Understanding the key characteristics of such a large application through study of the source code is a daunting task. This has long been recognized by performance tools

E. Jeannot, R. Namyst, and J. Roman (Eds.): Euro-Par 2011, LNCS 6853, Part II, pp. 340–352, 2011.

researchers, who have developed several ways of characterizing applications [29,2,9]. While useful in identifying scalability bottlenecks and performance inefficiencies, performance tools are not always suited for evaluating the feasibility of a particular approach to fault tolerance. In this regard, we identify critical application characteristics and construct signatures that are valuable from a fault tolerance perspective. Rather than characterize computation and communication behavior, we study the application modules in terms of their constituent data structures and accesses to them. This complements profiling provided by traditional performance tools.

We characterized all key NWChem modules and, due to space constraints, selectively present our results, which arose from considering a suite of fault tolerance techniques and evaluating feasibility for the different modules. The contributions of this paper are:

- A data-structure-oriented instrumentation methodology
- Data access characterization of key modules in a large application
- An *incremental checksum* approach to fault tolerance that combines the features of incremental checkpointing and checksum-based fault tolerance
- Evaluation of a broad class of fault tolerance techniques in the context of NWChem.

This work represents an early effort in making such large applications fault tolerant. To the best of our knowledge, we are not aware of a data access characterization approach for studying application behavior and its use in evaluation of a suite of fault tolerance techniques. Beyond fault tolerance, this methodology provides a means for joint understanding of macro-scale application behavior by both computer scientists and domain experts. It also encourages further investigation into causes of the exhibited behavior beyond fault tolerance, including performance optimization.

2   Related Work

Application profiling can provide useful insights into runtime behavior. This information can be used to pinpoint performance bottlenecks, optimize algorithms and data structures, design application-specific fault tolerance techniques, and fine-tune data access mechanisms. The GNU gprof is a call graph execution profiler [14] that performs dynamic program analysis and lists the frequency and duration of all function calls. HPCToolkit [2] and TAU [29] are another set of tools used for measuring and analyzing the performance of high-performance computing (HPC) applications. These tools are used primarily on multicore machines and large supercomputers. While useful in capturing performance-related information, they do not focus on data structures employed by the application and the associated access patterns.

Profiling message-passing libraries such as Message Passing Interface (MPI) [3] allows users to characterize the communication patterns [8] of applications. PMPI, the standard profiling interface for MPI, allows developers to gather diagnostic data by implementing custom wrappers to MPI calls. P^NMPI [28] extends the PMPI interface to include multiple, concurrent tool stacks.

Often, I/O bandwidth is considered a bottleneck for scientific applications. As such, profiling the I/O patterns of applications can provide avenues for performance improvement. Darshan [9] is an I/O characterization tool used to discern interesting patterns in application I/O behavior. Other studies [21,26] also have profiled the data access patterns of scientific workloads for analysis and tuning purposes.

While applications typically are profiled to identify performance bottlenecks, we characterize the data access patterns of a computational chemistry application to study various fault tolerance techniques that would safeguard the application against failures.

3   Background

This section provides background information on the applications and programming models presented in this paper.

3.1   NWChem

NWChem [32,17] is a massively parallel computational chemistry application developed and maintained by the Environmental Molecular Sciences Laboratory (EMSL) at Pacific Northwest National Laboratory (PNNL). The package, consisting of nearly two million lines of code, provides a variety of ground- and excited-state methods for quantum mechanical calculations, as well as classical simulation methods. In this paper, we provide a detailed analysis of access patterns of various data structures of the Gaussian basis set-based Hartree-Fock (HF) and Coupled Cluster (CC) modules. NWChem employs Global Arrays (GA) to manage and manipulate its key data structures.

3.2   Global Arrays

Global Arrays [20], a library-based implementation of the Partitioned Global Address Space (PGAS) programming model, provides applications a shared-memory, multidimensional view of data distributed among the physical memories of processors. Applications can create, destroy, and manipulate matrices using one-sided communication primitives such as GA_Get, GA_Put, and GA_Accumulate. In addition to the ease of the shared-memory abstraction, GA, which is fully interoperable with MPI, allows users to query locality information to further optimize their code. This programming model has proven to be highly scalable and can simplify array-based computations in large codes, including NWChem.

4   Instrumentation Methodology

The typical approach to profiling an application is to trap its function invocations. A profiling interface such as PMPI uses name-shifted weak bindings to instrument the application code by intercepting MPI calls made by the application. Since NWChem employs GA as its underlying data management and communication substrate, we intercept the GA calls made by NWChem to collect relevant performance data.

To profile NWChem's data access behavior, we instrumented the underlying GA library to keep a record of the data structures and the operations being performed on them. For example, a call to GA_Create or GA_Destroy stores the operation identifier, GA identifier, size, and timestamp of the operation to a file. GA primitives that manipulate data using one-sided communication, such as GA_Get, GA_Put, and GA_Accumulate, track the matrices being operated on during a phase. At the end of each phase (signified by a GA_Sync operation), the list of GA identifiers, operation identifier, and associated data volume are written to a file along with a timestamp of the GA_Sync operation.

Maintaining a summarized log of GA operations allows replay of the execution for post-processing analysis without the overhead associated with conventional communication tracing approaches that track each communication call. During the post-processing phase, we can build elaborate data access models of the NWChem modules. These models paint a detailed picture of the application's data access patterns and provide insight into effective fault tolerance techniques tailored for the application. Section 7 provides a detailed analysis of NWChem's runtime behavior.

5   Fault Tolerance Techniques

In this section, we briefly describe the fault tolerance techniques evaluated and considerations that led to the construction of the data access characterization. We focus on the GA data structures throughout the application. While additional state might be crucial to recovering from failures, the total state of the application being evaluated is dominated by the GA matrices alive at any point in execution.

Checkpoint-restart: This widely studied approach [12,13] involves backing up the GA matrices to stable storage periodically and restoring them in the event of a failure. As suggested in prior work, we store redundant copies in-memory at periodic intervals. We consider two variants of this approach. In full checkpointing, all GA matrices are backed up at every checkpoint interval. In incremental checkpointing, we track the updates to the matrices. At each checkpoint interval, only the modified arrays are backed up. While more efficient, incremental checkpointing requires tracking all changes, including potential *out-of-band* accesses to data. Checkpointing is done in a collective fashion, at every GA_Sync.

Redundant data communication: In this approach, duplicate copies of all data structures are maintained throughout execution. This is an extension of the disk-based Redundant Array of Independent Disks (RAID) approach to in-memory data. We consider the technique evaluated earlier [4], where each communication operation is repeated on a shadow copy. This scheme minimizes the work lost at the expense of increased communication overhead.

Checksum-based fault tolerance: This approach exploits the fact that read-only data can be restored from checksums, which incur much lower space overhead. This is an extension of recent efforts in fault tolerant linear algebra [10,7,5]. When a data structure is modified, the checksums no longer help in its recovery. Therefore, each GA matrix is duplicated before it is modified. The duplicate is discarded, and the checksums are recomputed once the changes to the array are complete. By only requiring duplicate storage for arrays being modified at any point, this approach improves upon the space requirements of the aforementioned schemes.

Incremental checksums: This approach combines the features of incremental check-pointing and the checksum-based approach. Checksums conserve space by minimizing duplicate storage. On the downside, checksums need to be computed repeatedly when a data structure frequently transitions between read-only and modified states. In this scheme, all GA matrices are periodically backed up using checksums. The first transition of an array into modified state is intercepted, and the array to be modified is duplicated. When an array remains read-only between checkpoint intervals, the space overhead is minimized. We are unaware of prior work in evaluating this approach—inspired by copy-on-write policy in process management—in the context of fault tolerance.

6 Application Evaluation Axes

This section presents the metrics used to characterize applications. We describe the metrics and discuss how they are plotted in the graphs presented in subsequent sections.

(a) Array creation and destruction: This metric provides a distribution of the creation and destruction of GA matrices as a function of execution time. We identify the sizes of arrays involved and runtime phases that could serve as bottlenecks in terms of space. The size of each individual creation and destruction is plotted with respect to the associated time. The y-axis represents the GA size in megabytes (MB).

(b) Number and (c) space consumed by GAs: This is a cumulative representation of the creation and destruction of arrays. Together with the impact on total space consumed, these axes depict trends in space utilization. This data directly relates to the space overhead of the various checkpointing schemes. In addition to measuring the total number and size of all GAs at any given time, we also measure those that are modified. A large fraction of modified data implies more state to be duplicated. The y-axis represents the number of GAs and the total GA size in gigabytes (GB), respectively.

(d) Reuse factor: Reuse factor is measured as the ratio of the total data volume associated with an array in a particular phase to the size of the array. A large reuse factor implies a high degree of reuse with potential improvements through locality optimizations. More importantly, a high reuse factor favors checkpointing approaches because it increases the cost associated with redundant data communication. The reuse factor for each array is plotted at the time of each GA_Sync. The y-axis represents the reuse factor of the GA matrices.

(e) Data liveness: This metric measures the period between creation and destruction of arrays. Short-lived arrays often correspond to temporary data structures, while long-lived arrays correspond to key data structures. A long checkpoint interval can avoid storage of temporary variables, reducing the time required to perform a checkpoint. Data liveness is presented as a cumulative distribution of the percentage (in terms of size) of all GA matrices ever created that are alive for less than a given percentage of the total execution time. Plots to the left of the graph correspond to temporary arrays, while those to the right correspond to long-lived arrays. Note that short-lived arrays

do not always correspond to a small impact on checkpointing overhead. In particular, short-lived arrays could be spread in such a way as to always constitute a significant overhead. This information is partly revealed by the metrics (b) and (c). Array liveness complements these metrics by providing an additional perspective. The y-axis represents the data liveness of the GA matrices as a percentage.

(f) Read-only window: As mentioned earlier, long durations in which an array is read-only benefit checksum-based approaches. To evaluate this metric, we measure the cumulative distribution of the percentage of data that are read-only for contiguous durations greater than a certain percentage of the total execution time. The total data is measured in terms of all data alive at any point, ignoring the modification phases for any array. In essence, the total data size is the area under the curve in metric (c). This is plotted with reverse key along the x-axis. A computation with long read-only windows will present a fast-growing graph, while several modifications will result in a slow-growing graph. The y-axis represents the read-only window as a percentage.

7   Data Access Characterization of NWChem

This section discusses data access characterization of the different NWChem modules. The experiments were conducted on Chinook, a 163 teraflops HP supercomputer available at EMSL. Chinook consists of 2310 HP DL185 nodes. Each node contains two 64-bit, quad-core AMD 2.2 gigahertz (GHz) Opteron processors. The nodes are also equipped with 32 GB of main memory and 365 GB of local disk space. A single rail InfiniBand interconnect provides high-bandwidth communication between the nodes. The cluster runs a customized version of Red Hat Linux Advanced Server. The experiments were conducted on 2048 processor cores using NWChem v6.0.

7.1   Hartree-Fock/Density Functional Theory

The HF method is a single-determinant theory [30] that forms the basis for higher-level electronic structure theories, such as Møller-Plesset perturbation theory (MP), CC theory, and other post-HF approaches. Density functional theory (DFT) also is a single-determinant approach and affords an alternate approach to the many-electron problem [18,22,23]. Both HF and DFT are similar in structure and are typically solved using iterative approaches involving basis set expansions [18,16,19].

We performed a DFT calculation with pure HF exchange on the C_{240} molecule. Pure HF exchange involves the calculation of two-electron integrals. This is performed via four-center integrals and results in an overall scaling of $O(N^4)$. The calculation was performed without symmetry using the 6-31G basis set [1] with a total of 3600 basis functions for the whole system. All integral evaluations were performed using the direct method, and the Fock matrix was constructed using the distributed data approach [15].

Data access characterization. Fig. 1 shows the data access characterization of a pure HF calculation using the NWChem DFT module. The creation and destruction of arrays are shown in Fig. 1(a). All arrays created, except at the end of the calculation, are of

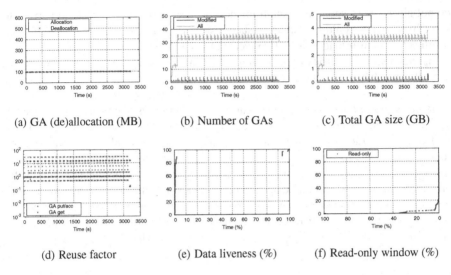

(a) GA (de)allocation (MB)   (b) Number of GAs   (c) Total GA size (GB)

(d) Reuse factor   (e) Data liveness (%)   (f) Read-only window (%)

Fig. 1. NWChem HF/DFT data access signature. The y-labels are provided in the captions and explained in Section 6

the same size and on the order of $O(N^2)$, where N is the number of basis functions. At the end of the calculation, the computed results are analyzed via molecular orbital processing. This is an optional analysis phase that might be omitted in certain execution scenarios. Even more importantly, the execution time involving the largest array is a small fraction of the total execution time. Hence, this array can effectively be ignored in designing a fault-tolerant scheme for the HF/DFT module.

As the first plot in Fig. 1 shows, arrays are created and destroyed throughout program execution. Fig. 1(b) shows that a batch of 10 arrays are created first, followed by 20 more arrays. The initial batch mostly corresponds to arrays that persist through the calculation—Fock, density, exchange correlation, eigenvector matrices, etc. The 20 subsequent arrays created correspond to the direct inversion of the iterative subspace (DIIS) arrays, the default number chosen to ensure quick convergence. As evident from Figs. 1(b) and 1(c), most of the matrices are read-only. A small number of GA matrices transition between read-only and modified states at any point in time. The periodic blips on the two curves correspond to matrices being created and destroyed.

Fig. 1(c) measures the size of the application state as a function of the wall time. The total GA size tops at about 3.2 GB and is constant (barring the periodic blips mentioned earlier) for the duration of the application execution. The notable trend evident from the figures reflects an application state that consists predominantly of read-only data structures. For this calculation, modified matrices form only about 10% of the total state. The modified GAs correspond to matrices involved in the self-consistent field iterative cycle (e.g., Fock, density, potentials, and eigenvector matrices).

The data reuse factor of HF/DFT varies from less than 1 to more than 20 for different GA matrices (see Fig. 1(d)). HF is inherently a non-linear computation with $O(N^4)$ operations being performed on $O(N^2)$ data. As such, we expect the data reuse factor to be high for all arrays. However, the key matrices—Fock matrices—are organized into blocks. If a given block does not contain any element larger than the threshold (due to Schwarz screening), its interactions with other blocks are ignored. Note that an overall reuse factor of less than 1 does not preclude any single block of data from being reused more than once. For data-dependent calculations, the reuse factor again relies on the problem at hand. In general, we observed the reuse factors were typical of calculations of chemical interest.

Almost 90% of the arrays, in terms of size, end up being live for less than 5% of the execution time. Thus, a large part of the application state does not significantly impact the checkpointing schemes for moderate checkpointing intervals or, through careful design, can potentially be ignored while still tolerating faults (see Fig. 1(e)).

The read-only windows for HF/DFT are small for most of the calculation. Approximately 80% of the read-only windows are shorter than 5% of the execution time, and less than 5% have a duration of 5%–35% of the wall time.

7.2   Coupled Cluster Theory

Many aspects of computational chemistry and physics require accuracies that can only be achieved by higher-order post-HF computational methods that account for the instantaneous interactions or correlations between electrons in molecules [6]. Among the many methods that describe correlation effects systematically, the CC formalism has evolved into a widely used and accurate method for solving the electronic Schrödinger equation. Compared with other wavefunction-based formalisms, the main advantage of CC methods lies in the fact that the correlation effects are efficiently encapsulated in the exponential form of the wavefunction. A simple consequence of this Ansatz is the size-extensivity of the resulting energies.

In most cases, due to quickly growing numerical complexity, the cluster operator is approximated by low-rank contribution. For example, the numerical complexity of the most rudimentary CC approximation—the CCSD approach (CC with singles and doubles) [25,11]—is $O(N^6)$, while the storage requirements are $O(N^4)$, where N refers to the system size. Using the currently available NWChem module, we can routinely perform CCSD calculations on systems consisting of about 1000 orbitals.

Data access characterization.   Fig. 2 shows the data access characterization of the NWChem CCSD module. The initialization phase of the CCSD module is small in terms of time and application state. The total number of GA matrices in the system is relatively small ($\sim$10), although the total application state during execution is $\sim$35 GB, signifying that the individual arrays are fairly large. The square wave pattern in Fig. 2(c) represents the individual tensor contractions, where the matrices are produced and consumed as part of the contraction.

The CCSD calculation consists of HF followed by the iterative part. As evident from the various graphs shown in Fig. 2, the first 1000 seconds are spent in the HF calculation. Immediately following the HF, the wavefunctions are transformed from the atomic to the molecular orbital basis using a procedure referred to as the *four-index transform*. The iterative part of the calculation is clearly visible from the periodic features. There also is an observable, large *plateau* in each iteration. This corresponds to the most expensive tensor contraction in the iterative procedure.

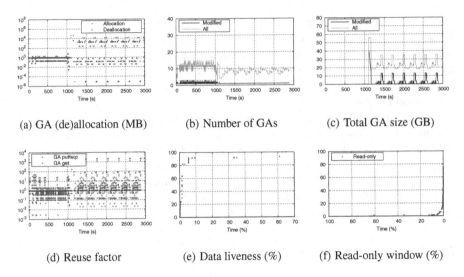

(a) GA (de)allocation (MB)   (b) Number of GAs   (c) Total GA size (GB)

(d) Reuse factor   (e) Data liveness (%)   (f) Read-only window (%)

Fig. 2. NWChem CCSD data access signature. The y-labels are provided in the captions and explained in Section 6.

Fig. 2(a) shows the four-index transform, while inexpensive in terms of execution time, involves the largest array in the calculation. In fact, Fig. 2(c) demonstrates this phase serves as the space bottleneck for the entire CCSD calculation. This has led to an ongoing effort to restructure the calculation to compute the array corresponding to atomic integrals to be produced and consumed on-the-fly, without storing the entire array at any point in time.

The intermediate in the four-index transform also exhibits the highest reuse factor (see Fig. 2(d)). Most notably, the put/acc reuse factor for this matrix is indistinguishable from other matrices. This is due to a parallelization scheme in which all updates to a block are coordinated to be computed by a single process. However, the consumption of this intermediate is not optimized in the same fashion, leading to high reuse factors. Through the effective scheduling of work that exploits this reuse, there is promise for further improvements in the code. The largest tensor contraction in each iteration also involves matrices that exhibit high reuse factors. During the iterative CCSD calculation, most update operations result in reuse factors greater than or equal to 1. However, a non-trivial fraction of the operations result in less than 1 reuse factor for GA_Get operations.

Approximately 90% of the arrays in the calculation are live only about 5% of the total execution time (see Fig. 2(e)). In addition, almost no data are live for longer than two-thirds of the execution time. This is due to the fact that the CCSD calculation consists of two distinct sub-calculations. The durations of the read-only windows for CCSD are similar to those of HF/DFT.

8   Evaluation of Various Fault Tolerance Schemes

In this section, we evaluate the fault tolerance schemes presented in Section 5 for the two modules characterized earlier. The information collected on the creation and

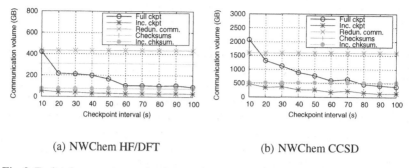

(a) NWChem HF/DFT   (b) NWChem CCSD

Fig. 3. Fault tolerance communication overhead as a function of the checkpoint intervals

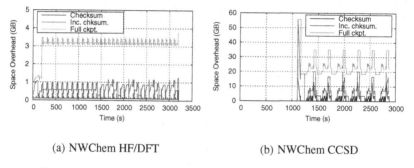

(a) NWChem HF/DFT   (b) NWChem CCSD

Fig. 4. Fault tolerance space overhead as a function of the execution time

destruction of arrays and their associated sizes was used to evaluate the checkpointing schemes. Total volume of communication between each collective synchronization is used to evaluate the redundant data communication approach. Any calls that modify a GA matrix are intercepted to mark the matrix as modified. This has an impact on the cost of the checksum-based approach to tolerating faults.

Fig. 3 depicts the total data volume overhead for the various approaches. Note that the checkpoint interval does not affect the checksum-based and redundant communication approaches. The cost of replication-based approaches is proportional to the total data size and scales linearly with number of processors. This is especially true for strong-scaling calculations of interest in this domain. While checksum computations typically involve $O(\log p)$ steps in tree-based approaches, algorithms that are linear in the message size per process exist for large message sizes [31]. Thus, the total data size is a faithful representation of checkpoint and checksum-based approaches—both for the runs evaluated and to compare expected behavior on larger systems. For both DFT and CCSD modules, the checksum-based approach incurs higher overhead than other schemes, except for full checkpointing with small checkpoint intervals. This is expected from the small size of the read-only windows. When the read-only windows are large, the checksum-based approach performs better because a large number of temporary arrays transition less frequently between read-only and modification phases. The redundant communication approach performs worse due to the reuse factors encountered.

Fig. 3(b) shows the cost associated with the schemes for the CCSD module. Incremental checkpointing expectedly improves upon full checkpointing. The frequent transitions in data and the associated small read-only windows (as shown in Fig. 2(f)) cause the checksum-based approach to be the most expensive. However, unlike HF/DFT, CCSD's data access behavior results in high reuse factors for many arrays. This increases the cost of redundant communication-based fault tolerance. Fig. 3(b) demonstrates this approach is as expensive as the checksum-based approach.

Space overheads of the various schemes are shown in Fig. 4. Full and incremental checkpointing and redundant data communication approaches all incur the same space overhead and are represented as "checkpointing" in the figure. The data movement overhead of the checksum-based approach is compensated for by the low space overhead observed. In particular, there are large reductions in space overheads for the HF/DFT calculation. The peak space overhead is higher for the CCSD calculation due to the larger arrays encountered, which must be duplicated when being modified.

The incremental checksum approach combines the best features of checkpointing- and checksum-based approaches. The data volume overhead is comparable to incremental checkpointing in both calculations. The space overhead, while higher than a pure checksum-based approach, is much lower than other schemes.

9   Conclusions

Designing fault tolerance into existing applications is a non-trivial task. We presented our approach to evaluating various fault tolerance schemes in the context of representative modules in a large computational chemistry application. We developed a methodology to identify the data access characteristics of the application modules. To the best of our knowledge, this is the first-ever characterization of its kind for applications in the computational chemistry domain. The choice of fault tolerance scheme is influenced not only by performance implications but also the ease with which it can be incorporated into the application. We believe such an analysis is essential in understanding the trade-off between the implementation effort and the benefits achieved. In addition, this characterization has spurred efforts to improve the implementation beyond fault tolerance and could be a benchmark for design of other runtime components. As future work, we are investigating similar analysis of applications employing other data abstractions, such as Portable, Extensible Toolkit for Scientific Computation (PETSc) [24].

Acknowledgments. This work was supported by the U.S. Department of Energy via Grant 47590. A portion of the research was performed using the Molecular Science Computing (MSC) capability at EMSL, a national scientific user facility sponsored by the Department of Energy's Office of Biological and Environmental Research and located at Pacific Northwest National Laboratory (PNNL). PNNL is operated by Battelle for the U.S. Department of Energy under contract DE-AC05-76RL01830.

References

1. EMSL Basis Set Exchange, `https://bse.pnl.gov/bse/portal`
2. HPCToolkit, `http://hpctoolkit.org`
3. MPI, `http://www.mpi-forum.org`
4. Ali, N., Krishnamoorthy, S., Govind, N., Palmer, B.: A redundant communication approach to scalable fault tolerance in PGAS programming models. In: 19th Euromicro International Conference on Parallel, Distributed, and Network-Based Computing, pp. 24–31 (February 2011)
5. Ali, N., Krishnamoorthy, S., Halappanavar, M., Daily, J.: Tolerating correlated failures for generalized cartesian distributions via bipartite matching. In: ACM International Conference on Computing Frontiers (May 2011)
6. Bartlett, R.J., Musiał, M.: Coupled-cluster theory in quantum chemistry. Reviews of Modern Physics 79(1), 291–352 (2007)
7. Bosilca, G., Delmas, R., Dongarra, J., Langou, J.: Algorithm-based fault tolerance applied to high performance computing. Journal of Parallel and Distributed Computing 69(4), 410–416 (2009)
8. Cappello, F., Guermouche, A., Snir, M.: On communication determinism in parallel HPC applications. In: 19th International Conference on Computer Communications and Networks, pp. 1–8 (August 2010)
9. Carns, P.H., Latham, R., Ross, R.B., Iskra, K., Lang, S., Riley, K.: 24/7 characterization of petascale I/O workloads. In: Proceedings of the First Workshop on Interfaces and Architectures for Scientific Data Storage, pp. 1–10 (September 2009)
10. Chen, Z., Dongarra, J.: Algorithm-based checkpoint-free fault tolerance for parallel matrix computations on volatile resources. In: Proceedings of the 20th International Parallel & Distributed Processing Symposium (April 2006)
11. Cullen, J.M., Zerner, M.C.: The linked singles and doubles model–an approximate theory of electron correlation based on the coupled-cluster ansatz. The Journal of Chemical Physics 77(8), 4088–4109 (1982)
12. Elnozahy, E.N., Alvisi, L., Wang, Y.M., Johnson, D.B.: A survey of rollback-recovery protocols in message-passing systems. ACM Computing Surveys 34(3), 375–408 (2002)
13. Elnozahy, E.N., Plank, J.S.: Checkpointing for peta-scale systems: A look into the future of practical rollback-recovery. IEEE Transactions on Dependable and Secure Computing 1(2), 97–108 (2004)
14. Graham, S.L., Kessler, P.B., McKusick, M.K.: Gprof: A call graph execution profiler. In: Proceedings of the 1982 SIGPLAN Symposium on Compiler Construction, vol. 17(6), pp. 120–126 (1982)
15. Harrison, R.J., et al.: Toward high-performance computational chemistry: II. a scalable self-consistent field program. Journal of Computational Chemistry 17(1), 124–132 (1996)
16. Helgaker, T., Jorgensen, P., Olsen, J.: Molecular Electronic-Structure Theory. John Wiley & Sons Ltd., Chichester (2004)
17. Jong, W.A., et al.: Utilizing high performance computing for chemistry: parallel computational chemistry. Physical Chemistry Chemical Physics 12(26), 6896–6920 (2010)
18. Kohn, W., Sham, L.J.: Self-consistent equations including exchange and correlation effects. Physical Review 140(4A), A1133–A1138 (1965)
19. Martin, R.M.: Electronic Structure: Basic Theory and Practical Methods. Cambridge University Press, Cambridge (2004)
20. Nieplocha, J., Palmer, B., Tipparaju, V., Krishnan, M., Trease, H., Aprà, E.: Advances, applications and performance of the global arrays shared memory programming toolkit. International Journal of High Performance Computing Applications 20(2), 203–231 (2006)

21. Nieuwejaar, N., Kotz, D., Purakayastha, A., Sclatter Ellis, C., Best, M.: File-access characteristics of parallel scientific workloads. IEEE Transactions on Parallel and Distributed Systems 7(10), 1075–1089 (1996)

22. Parr, R.G., Yang, W.: Density-Functional Theory of Atoms and Molecules. Oxford University Press, Inc., New York (1989)

23. Perdew, J.P., Schmidt, K.: Jacob's ladder of density functional approximations for the exchange-correlation energy. In: AIP Conference Proceedings, vol. 577(1), pp. 1–20 (2001)

24. PETSc, http://www.mcs.anl.gov/petsc/petsc-as/

25. Purvis, G.D., Bartlett, R.J.: A full coupled-cluster singles and doubles model–the inclusion of disconnected triples. The Journal of Chemical Physics 76(4), 1910–1918 (1982)

26. Roth, P.C.: Characterizing the I/O behavior of scientific applications on the Cray XT. In: Proceedings of the International Workshop on Petascale Data Storage, Reno, NV, pp. 50–55 (2007)

27. Schroeder, B., Gibson, G.A.: Understanding failures in petascale computers. Journal of Physics: Conference Series 78(1) (2007)

28. Schulz, M., de Supinski, B.R.: P^NMPI tools: A whole lot greater than the sum of their parts. In: Proceedings of the ACM/IEEE Conference on Supercomputing, pp. 1–10 (2007)

29. Shende, S.S., Malony, A.D.: The TAU parallel performance system. International Journal of High Performance Computing Applications 20(2), 287–311 (2006)

30. Szabo, A., Ostlund, N.S.: Modern Quantum Chemistry. McGraw-Hill Inc., New York (1996)

31. Thakur, R., Rabenseifner, R., Gropp, W.: Optimization of collective communication operations in mpich. International Journal of High Performance Computing Applications 19(1), 49–66 (2005)

32. Valiev, M., et al.: NWChem: A comprehensive and scalable open-source solution for large scale molecular simulations. Computer Physics Communications 181(9), 1477–1489 (2010)

High-Performance Numerical Optimization on Multicore Clusters

Panagiotis E. Hadjidoukas[1], Constantinos Voglis[1], Vassilios V. Dimakopoulos[1], Isaac E. Lagaris[1], and Dimitris G. Papageorgiou[2]

[1] Department of Computer Science
{phadjido,voglis,dimako,lagaris}@cs.uoi.gr
[2] Department of Materials Science and Engineering
dpapageo@cc.uoi.gr
University of Ioannina, Ioannina, Greece, GR-45110

Abstract. This paper presents a software infrastructure for high performance numerical optimization on clusters of multicore systems. At the core, a runtime system implements a programming and execution environment for irregular and adaptive task-based parallelism. Building on this, we extract and exploit the parallelism of a global optimization application at multiple levels, which include Hessian calculations and Newton-based local optimizations. We discuss parallel implementations details and task distribution schemes for managing nested parallelism. Finally, we report experimental performance results for all the components of our software system on a multicore cluster.

Keywords: task parallelism, message passing, numerical differentiation, global optimization.

1 Introduction

Numerical optimization is a useful tool that has been widely used on many scientific problems such as space trajectory calculation and computation of optimal shapes for automobile or aircraft components. Optimization problems, especially global ones, have high computational demands because of the substantial execution time and the possibly multiple local mimima of the objective function to minimize. Exploitation of parallelism at several levels such as function evaluations, numerical computations and the optimization algorithms themselves can drastically reduce the time required to find a solution.

The *Multistart* method is a standard and widely used scheme for dealing with global optimization problems. According to this method, a local optimization procedure is applied to a number of randomly selected points. Local optimization can be based on the Newton method with Hessian modification, a general and powerful method for multidimensional non-linear optimization that makes use of first and second derivatives of the objective function. This, in turn, introduces further computational complexity as derivative estimation via finite differentiation requires a number of function evaluations.

E. Jeannot, R. Namyst, and J. Roman (Eds.): Euro-Par 2011, LNCS 6853, Part II, pp. 353–364, 2011.
© Springer-Verlag Berlin Heidelberg 2011

Task-based parallelism, as expressed by the master-worker programming paradigm, can be an effective approach for a cluster-aware implementation of global optimization methods such as Multistart. Function evaluations are mapped to tasks and assigned to the workers. The dynamic load balancing of the model further enhances its suitability. A naive implementation of the model, however, cannot meet all the requirements that Multistart imposes. First, the large expected number of spawned tasks (typically on the order of 10^6) affect the scalability as the single master becomes a bottleneck. Secondly, the exploitation of nested parallelism requires advanced runtime techniques, able to provide efficient management of processing elements. Additionally, it is important to have a hardware-independent solution that transparently uses multi-threading to fully exploit the physically shared memory of SMP/multi-core systems.

In this paper, we present a software infrastructure that deals with all the above limitation issues that concern the parallelization of the Multistart method. At the core of the system there is TORC, a novel runtime environment for programming and executing irregular and adaptive master-worker applications on multi-core SMPs and clusters of such machines. As such, TORC targets both message passing and shared memory programs by exporting an API that provides ease of programming and transparent load balancing without requiring any interaction with the low-level message passing primitives. Building on TORC, we design a standalone numerical differentiation software package (PNDL) that provides routines for gradient and Hessian computations. We manage to extract parallelism at all possible levels in a straightforward and seamless manner, while we present several task distribution schemes, which are combined with the work stealing mechanisms of TORC. Finally, we present the parallelization of a Newton-based Multistart method using both TORC and PNDL to execute multiple local optimizations and gradient/Hessian calculations. The experimental evaluation on a dedicated multicore cluster demonstrates the efficiency of our system.

The rest of this paper is organized as follows: Section 2 gives a brief introduction to the non-linear global optimization problem. Section 3 discusses the parallelization issues of Multistart. Sections 4 and 5 present the TORC tasking library and PNDL. Experimental evaluation and related work are reported in Sections 6 and 7 respectively. We conclude with a discussion in Section 8.

2   Numerical Optimization

The task of numerical optimization is to locate (approximate) a minimizer of a generally multidimensional objective function. The mathematical formulation is

$$\min_{x \in \mathbb{R}^n} f(x) \tag{1}$$

where $x \in \mathbb{R}^n$ is a real vector and $f : \mathbb{R}^n \to \mathbb{R}$ the objective function. There exist a plethora of applications in physics, chemistry, engineering, and economics that can be formulated as optimization problems. Predicting the tertiary protein structures, defining optimal sea routes, calculating bound states for few body systems, tuning all kinds of machine learning models and identifying the seismic properties of a piece of the earths crust, are all examples of real world applications that can be tackled as optimization problems.

An optimization algorithm is a sequential procedure that, beginning from a starting point $x_0 \in S$, generates a sequence of iterates $\{x_k\}_{k=0}^{\infty}$ that terminates when the solution point is approximated with a prescribed accuracy. In deciding how to move from one iterate x_k to the next the algorithm uses information about the function at x_k (function value, first or second order derivatives). A general class of optimization algorithms use second order derivative information of the objective function and use it to build and minimize a quadratic model around the current iteration. The main representative of this class is the *Newton* method. At each iterate, the Newton method makes use of first and second order derivative information to proceed to the next point. This can be achieved using a *line search* algorithm which searches along a descent direction $p_k \in \mathbb{R}^n$ for an iterate with lower function value. The distance to move along p_k can be found by solving the following one-dimensional minimization problem that is to find a step length α that minimizes $f(x_k + \alpha p_k)$. The main computational cost of a single Newton iteration is determined by the objective function and the derivatives calculation that are used to compute the search direction.

In many cases derivatives cannot be expressed analytically because the underlying functions are represented by large and complicated computer codes. In these cases finite differencing is an approach for calculating the first and second order derivatives of an $n-$dimensional objective function at a point x by examining the objective function behavior on small finite perturbations around x. The number of function evaluations depends on the order of the derivative (first or second) and on the requested accuracy (the larger accuracy the more function evaluations). For the gradient vector at least $n + 1$ function evaluations are required and for the Hessian at least $n(n + 1)/2$. Two of the most popular formulas for approximating gradient and Hessian, using central differences are summarized below:

$$\frac{\partial f(x)}{\partial x_i} \approx \frac{f(x + \epsilon e_i) - f(x - \epsilon e_i)}{2\epsilon} \tag{2}$$

$$\frac{\partial^2 f(x)}{\partial x_i \partial x_j} \approx \frac{f(x + \epsilon e_i + \epsilon e_j) - f(x - \epsilon e_i + \epsilon e_j)}{4\epsilon^2} - \frac{f(x + \epsilon e_i - \epsilon e_j) + f(x - \epsilon e_i - \epsilon e_j)}{4\epsilon^2}$$

where e_i is the $i-$th unit vector and ϵ a small positive scalar. Finite differencing is a perfect candidate for parallel execution. All function evaluations in Eq.(2), $f(x + \epsilon e_i)$ and $f(x + \epsilon e_i + \epsilon e_j)$, can be performed independently and in parallel.

The Newton procedure locates a minimizer efficiently with quadratic convergence speed. However, there is no guarantee that this minimizer will be the one with the lowest function value in all S, as the minimizer may stick at a local minimum. This requirement introduces the problem of *global optimization*, one of the most difficult problems in applied mathematics. Searching for the global minimum is a quite challenging, yet extremely useful task for all applications mentioned in the beginning of this Section. It is proven, in the multidimensional case, that it is impossible to guarantee the globally optimal value will be found in finite time. All that can be assured is that the probability of locating the global minimizer approximates 1. One of the oldest and most popular schemes for

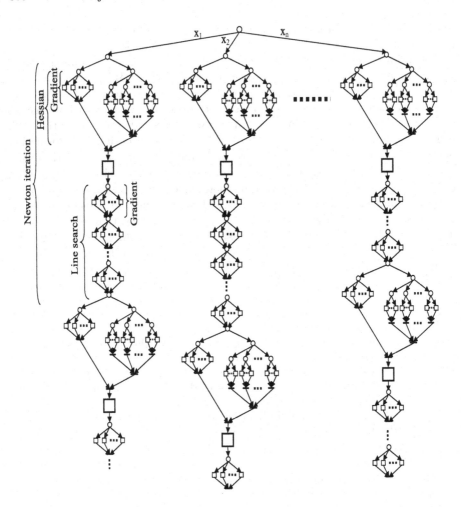

Fig. 1. Execution task graph for Multistart using finite difference derivatives

dealing with global optimization problems is the *Multistart* method. According to this method, a local search procedure $\mathcal{L}$ is executed for each point in a sample generated from a uniform distribution over the search space S. Albeit simple in principle, *Multistart* is the heart of more sophisticated global optimization algorithms such as *clustering methods*[1,2].

3   Multistart Parallelism Issues

In the framework of global optimization based on numerical differentiation, there exist several levels of parallelism that can be exploited in order to accelerate the method. Fig. 1 illustrates the execution task graph of the Multistart method. Each circle corresponds to code that spawns parallelism, which can be expressed and instantiated with lower-level tasks. Tasks at the innermost level are represented with squares and correspond to serial code and specifically to either single function evaluations or sequential direct linear algebra operations. Therefore, the paths of the graph represent operations that can be performed in parallel while their meeting points represent the completion of all tasks in a team with the satisfaction of all data and control dependencies.

Initially, the application runs the Multistart method and spawns first-level ($L1$) tasks. These perform the Newton *local search* method to multiple independent initial points (x_i) and execute iterations until the convergence criterion is met. In each iteration, the tasks first proceed with the derivative calculation, spawning two second-level ($L2$) tasks that compute the gradient and Hessian respectively. The gradient computation includes a number of function evaluation ($L3$) tasks. The Hessian computation, however, exploits an additional level of parallelism by assigning the numerical calculation of each partial derivative to a ($L3$) task that can spawn two to nine function evaluation ($L4$) tasks, depending on the desired accuracy and the bounds. Local search continues with a sequential task that performs the required matrix modification and the solution of the linear system. The iterative line search method follows, exploiting each time a single level of parallelism for the gradient computation. For a large number of initial points, a gradual execution of Multistart can be performed by applying the Newton method to bunches of points. In such case, the execution task graph is repeated until the desired number of points has been processed.

Multistart is a highly irregular parallel application: first, the local search method is applied concurrently to multiple points, the number of which may not be exactly divided by the number of available processors. Secondly, the execution time of local search exhibits significant variation as the number of iterations required for convergence depends on the randomly selected initial point. Similarly, the line search method is performed for an unknown before number of iterations. Irregularity is found even at the innermost level of parallelism (Hessian calculation), as the number of function evaluations for the derivative computation at a specific point also depends on the imposed bounds on the variables. According to the above, the execution times for finding a minimum for each initial point are neither balanced nor known beforehand. Derivative estimation via finite differencing is computationally expensive for several applications where the time for a single function call is substantial. Therefore, the highly irregular nested parallelism of Multistart must be exploited at all possible levels, without making any assumption about the number of available processors.

4   TORC Runtime Library

TORC [3] implements a task-based programming and runtime environment that makes the development of master-worker applications almost trivial. Although TORC supports several features, due to lack of space we briefly present only those related to the parallelization of the Multistart method.

TORC assumes that a single application consists of multiple MPI processes with private memory when running on the cluster. Furthermore, it uses multi-threading to exploit the multiprocessor/multicore cluster nodes. A *task* represents a work unit that is independent of its execution vehicle, i.e. the MPI process or thread. A spawned task can be submitted for execution to *any* MPI process; the programmer may specify the target process in the task creation routine. When a parent task blocks, its underlying vehicle can proceed to the execution of other ready-to-run tasks. This means that a TORC application can run successfully even if only a single-threaded process is used. Furthermore, each process can have multiple worker threads. Therefore, the same application code can run on any combination of MPI processes and threads, exploiting at runtime the presence of physically shared memory, if available.

Due to the decoupling of tasks and execution vehicles, multiple levels of task parallelism are inherently supported and any child task can become a master and spawn new tasks. Therefore, TORC enables the programmer to express hierarchical and recursive task parallelism naturally, which would be otherwise quite difficult to implement. In the task creation routine (`torc_task()`), the user specifies the task function, the number of arguments this function receives and an argument list. For each argument, its size and data type is required. In addition, an intent attribute must be also supplied, similarly to the IN, OUT and INOUT intent attributes of Fortran 90. Any data movement is performed transparently to the user. After task creation, a master task calls the `torc_waitall()` routine to suspend itself until all child tasks have finished and their results have arrived. Several master-worker applications may have global data that is initialized by the master and then broadcast to the workers. The `torc_bcast()` routine allows any task to broadcast global data to all MPI processes, thus avoiding unnecessary data transfers.

As task stealing is inherently supported by TORC, the programmer has only to decide about the task distribution scheme, by querying the execution environment and then specifying the node or worker where each task will be initially submitted for execution. The scheduling loop of a worker thread is activated when its current task finishes or blocks. A worker extracts and executes the task that is at the front of its local ready queue. If this is empty, the worker tries to steal a task from the rest of the intra-node ready queues. If inter-node task stealing is enabled, it issues requests for work to remote nodes in sequential order. Task stealing is always performed from the back of the ready queues. The stealing of a task from a remote queue includes the corresponding data movement, unless the task returns to its parent node. Inter-node task stealing is optional and must be explicitly enabled based on the load imbalance of the parallel application. On the other hand, intra-node task stealing is always active. A more detailed description of the TORC library can be found in [3].

5   PNDL and Parallel Multistart Implementation

The parallel implementation of the numerical differentiation library for multi-core clusters has been based on the tasking model that TORC provides. For each function evaluation, a task is created, with main input argument a vector x and

```
! first level
subroutine pndlhf(f, x, n, iord, hes)
external f, driver
integer n, iord
double precision x(n), hes(n,n), xx(n)
common /data/ xx
...
<set xx(I) = x(I)> ! create copy of x
call torc_bcast(xx, n, MPI_DOUBLE_PRECISION)
iworker = torc_worker_id()
nworkers = torc_num_workers()
<for each derivative>
   call torc_task(iworker, driver, ..)
   istride1 = <# function values required>
   iworker = mod(iworker+istride1,nworkers)
call torc_waitall()
end
```

```
! second level
subroutine driver(f, n, ...)
double precision xx(n)
common /data/ xx(n)
...
iworker = torc_worker_id()
nworkers = torc_num_workers()
istride2 = 1
<for each required function value>
   call torc_task(iworker, f, ..)
   iworker = mod(iworker+istride2,nworkers)
call torc_waitall()
<compute partial derivative h(i,j)>
end
```

Fig. 2. Outline of a PNDL Hessian calculation with exploitation of two levels of parallelism using the STRIDE distribution scheme

result the computed function value $f(x)$. The core routine that PNDL implements for Hessian computations is: pndlhf(f,x,n,iord,hes), where f is the function to be differentiated, x the vector containing the point of calculation, n the dimensionality of the function, *iord* the requested order of accuracy, and *hes* the resulting Hessian matrix.

The parallel routines that PNDL exports to MPI programs have been redesigned for a master-worker execution mode. The calling process initializes the input parameters and receives the computed derivatives. When a PNDL routine is invoked, the primary task initially broadcasts the input vector x, through the use of a common block. If the routine has a single level of parallelism, function evaluation tasks are spawned and distributed cyclically to the workers. After task completion, the primary task uses the gathered function values to compute the derivatives. Although a reduction operation can be used, the adopted scheme preserves the sequential order of calculations and, thus, avoids rounding errors.

The above scheme, however, may increase significantly the memory requirements of PNDL for the estimation of second order derivatives of functions with a large number of variables, which can be of the order of thousands for specific problems. To handle this issue, we exploit nested parallelism; each element of the Hessian is calculated by a first-level task, which issues function calls through second-level tasks. The number of first-level tasks is equal to $(n(n+1)/2)$ and each of them spawns 2 to 9 second-level tasks, according to user parameters. Memory usage is drastically reduced because the number of *active* first-level tasks, which reserve stack space for the results, never exceeds the number of available workers. This is achieved because second-level tasks are inserted in the front of the ready queues and thus have higher execution priority than first-level tasks, which are inserted at the end.

The runtime architecture of TORC allows for several task distribution schemes: Fig. 2 outlines the hierarchical parallel implementation of the **pndlhf** routine using the STRIDE scheme, which divides equally the number of function evaluations among the available workers. The first argument of the task creation routine denotes the identifier of the worker thread where the task will be submitted to. The parent task distributes the first-level tasks using a variable stride (`istride1`) that is determined by the (known beforehand) number of second-level tasks that correspond to each task. Next, each first-level task distributes the inner tasks to consecutive workers (`istride2=1`), starting from the worker where that task runs on. The STRIDE scheme is, however, suitable only for dedicated homogeneous clusters and may result in an excessive number of messages for high-dimensional functions. To overcome these issues we have introduced a dynamic task distribution scheme, called GLTS, which distributes the first-level tasks cyclically across the processors (`istride1=1`) and submits the second-level tasks locally (`istride2=0`) with task stealing enabled. GL is another task distribution scheme that differs from GLTS in that task stealing is used only at the intra-node level, i.e. between workers that belong to the same process. Finally, LLTS is a variant of the GLTS scheme that submits even the first-level tasks locally and specifically in the ready queue of the worker that issued the PNDL routine (both strides are equal to zero).

The parallelization of the Newton method relies on two PNDL routines that compute the required gradient and Hessian matrices. These routines can be executed concurrently, as an additional level of parallelism, through the spawning of two TORC tasks. This, however, requires appropriate modifications in PNDL, due to the usage of the common block for broadcasting the input vector. Therefore, we use an array of input vectors in the common block and each PNDL function call is dynamically assigned a unique identifier that specifies an available entry of this array. Parallel Multistart takes advantage of the reentrancy of PNDL functions to issue multiple local searches concurrently starting from randomly chosen initial points. As the number of points increases, the small serial fraction of the Newton method becomes negligible and the effective utilization of parallel hardware is further improved. For Multistart, we have followed MLTS, a modified LLTS distribution scheme: tasks at the first-level of parallelism, i.e. Newton optimizations, are distributed cyclically across the workers. Parallelism at all inner levels is submitted locally, with inter-node task stealing enabled. Ideally, each local search will be performed exclusively by a single worker. Idle workers will try to steal and execute tasks that belong to the first-level of parallelism and will participate in the execution of remotely issued PNDL routines only when the number of remaining optimizations is less than the number of workers.

6   Performance Experiments

In this section we present experimental results from application executions on a dedicated 16-node Sun Fire x4100 cluster with Gigabit Ethernet, each node with 2 dual-core AMD Opteron 275 CPUs. The software setup includes Linux 2.6, GCC 4.3 and MPICH2. In all experiments we use the multithreaded configuration, running a single process with multiple workers on each cluster node.

Our system targets mostly medium to coarse-grained tasks for remote execution. As an indication, for the specific platform used for our experimental evaluation, the task execution overhead is measured approximately 0.1ms for a zero-argument task; this overhead however decreases with the number of tasks due to the overlap of task creation, data movement and task execution. Thus, the overhead for the single task case depends on the latency of the interconnection network, while the overall minimum overhead depends on the maximum bandwidth. In addition, the minimum overhead for a given number of tasks is a linear function of the argument size. In contrast, within a multi-core node we support very fine-grained tasks efficiently.

Parallel Hessian. We present two sets of synthetic experiments that calculate the Hessian with $O(h^4)$ precision without imposing bounds on the variables. The first set of experiments (E1, E2) uses a test function with 20 variables and leads to a total of 820 objective function calls. We have arranged for function evaluation time to be 100ms and 1000ms via appropriate artificial delays. The second set (E3, E4) uses a 100-dimensional test function with artificial delays of 10ms and 100ms. The number of function evaluations for this set is 20200. Both experiments are designed to cover a wide range of practical situations and correspond to medium and large problem sizes. They are representative of applications with many dimensions and/or substantial function execution time.

Figs. 3-4 and 5-6 present the results from the two sets of experiments with the 4 task distribution schemes (STRIDE, GLTS, GL, and LLTS). For the first set, we observe that the speedup increases with the computational cost of the test function, due to the higher computation-to-communication ratio. The slight decrease in performance is attributed to several factors: a small serial fraction of code in the PNDL function, the overhead for broadcasting the point and the load imbalance when the number of function evaluations is not exactly divided by the number of workers. Although all distribution schemes exhibit comparable performance up to 32 processors, GLTS achieves the highest speedup on 64 processors, with LLTS and GL to follow. The lowest speedup corresponds to the STRIDE scheme because of its large number of explicit messages. For the 100-dimensional function (Figs. 5 and 6), the obtained speedup of GLTS almost coincides with the ideal for both cases. In this set of experiments, the lowest speedup values are observed for LLTS, due to the bottleneck at the single queue where the 5050 first-level tasks are submitted for execution.

Parallel Multistart. In order to evaluate parallel Multistart we use a test function with 10 variables, artificial delays that range from 1ms to 1s, and the modified LLTS task distribution scheme. Fig. 7 depicts the speedup for a single starting point, which represents a worst-case but unlikely to occur scenario in global optimization problems. We observe that the Newton method fails to scale as the number of workers increases, regardless of the function evaluation time. This is mostly attributed to the small sequential task ($\simeq 2\%$) of the Newton method. The speedup can be further affected by the communication overheads, especially when the computational cost of the objective function is low. For function evaluation time equal to 1s, however, the measured speedup is very close to the maximum theoretical speedup as defined by Amdahl's law. Figs. 8 to 10 show the speedup of Multistart for 16, 64 and 1024 optimizations. The attained

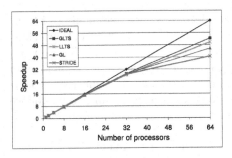

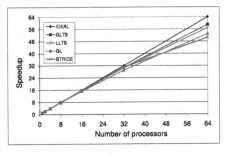

Fig. 3. Speedup for experiment E1 (variables=20, delay=100ms)

Fig. 4. Speedup for experiment E2 (variables=20, delay=1000ms)

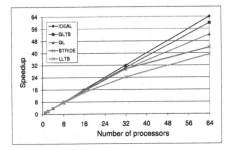

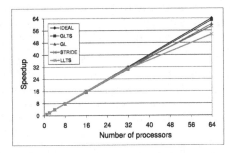

Fig. 5. Speedup for experiment E3 (variables=100, delay=10ms)

Fig. 6. Speedup for experiment E4 (variables=100, delay=100ms)

speedup increases with the number of optimizations, especially if this exceeds the number of available processing cores. For 1024 optimizations, the speedup almost coincides with the ideal for both 10ms and 100ms evaluation time. The performance results are in accordance with those obtained for a real application case that deals with the protein folding problem [4].

7   Related Work

Although many parallel local and global optimization algorithms were proposed in the last decades (e.g. [5,6], only a handful of actual systems exist. One of the most widely used scientific software programs, MATLAB, presented its first parallel optimization solution in 2009 [7]. In the pioneer work of [8] an interval global optimization method is implemented using dynamic load balancing. PGO [9] is a general parallel computing based on the Genetic Algorithm. In PGO, the parallel (and heterogeneous) computing framework is organized as a global master-slave system using a central database management system for storing all the data during optimization progress. Oriented in interoperability, the MHGrid platform [10] exploits meta-heuristics based search methods and Grid computing to enable the transparent sharing of heterogeneous and dynamic resources offering a versatile Global optimization framework. MANGO [11] is a middleware that involves the development of an extensible and flexible multiagent platform, in which

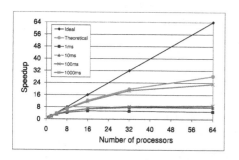

Fig. 7. Speedup for 1 local search

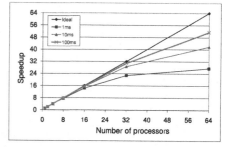

Fig. 8. Speedup for 16 local searches

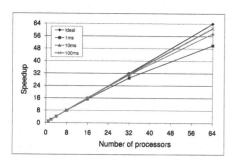

Fig. 9. Speedup for 64 local searches

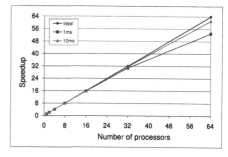

Fig. 10. Speedup for 1024 local searches

autonomous agents can solve global optimization problems in cooperation. Finally, PaGMO[12] is a open source multi-threaded software that offers a plethora of local and global optimization codes exploiting modern multi-core architectures. In contrast to our infrastructure, none of the above supports hierarchical and multi-level task parallelism. In addition, our system is platform-agnostic supporting transparently both shared and distributed memory architectures.

Despite the availability of several software packages for estimating derivatives numerically (e.g. [13,14]) their implementation is sequential. The only parallel numerical Hessian implementations we are aware of are [15] and [16], mainly used for computational chemistry.

8   Conclusions

We presented a system for efficient exploitation of nested and irregular parallelism in non-linear optimization problems. At the core of our system is TORC, a runtime library that supports adaptive task-based parallelism on clusters of multicores/SMPs. Using TORC, we manage to extract and execute the multiple levels of parallelism inherent in the Multistart optimization method, performing thus Newton-based local searches, gradient and Hessian calculations and function evaluations in parallel.

Our ongoing work includes the integration of additional numerical optimization techniques into our infrastructure. We also work on the efficient parallelization of a real application case, concerning the protein folding problem. Finally, we plan to extend the applicability of our system to computational grids and GPGPU environments.

References

1. Rinnooy Kan, A.H.G., Timmer, G.T.: Stochastic global optimization methods part I: Clustering methods. Math. Progr. 39, 27–56 (1987)
2. Voglis, C., Lagaris, I.E.: Towards ideal multistart, A stochastic approach for locating the minima of a continuous function inside a bounded domain. Applied Math. and Comput. 213, 216–229 (2009)
3. Hadjidoukas, P.E., Dimakopoulos, V.V.: TORC: a tasking library for multicore clusters. Technical Report TR-2011-6, Dept. of Computer Science, University of Ioannina, Greece (2011)
4. Voglis, C., Hadjidoukas, P.E., Dimakopoulos, V.V., Lagaris, I.E., Papageorgiou, D.G.: Task-parallel global optimization with application to protein folding. In: 9th Int'l Conf. on High Perf. Comput. and Simul., Istanbul, Turkey (2011)
5. Schutte, J.F., Reinbolt, J.A., Fregly, B.J., Haftka, R.T., George, A.D.: Parallel global optimization with the particle swarm algorithm. Int'l J. Numerical Methods in Engin. 61(13), 2296–2315 (2004)
6. Byrd, R.H., Eskow, E., van der Hoek, A., Schnabel, R.B., Oldenkamp, K.P.B.: A Parallel global optimization method for solving molecular cluster and polymer conformation problems. In: 7th Siam Conf. on Parallel Processing for Scientific Comput., pp. 72–77. SIAM, Philadelphia (1995)
7. Kozola, S.: Improving optimization performance with parallel computing. MATLAB Digest (2009)
8. Eriksson, J., Lindstrom, P.: A parallel interval method implementation for global optimization using dynamic load balancing. Rel. Comput. 1/2, 77–91 (1995)
9. He, K., Zheng, L., Dong, S., Tang, L., Wu, J., Zheng, C.: PGO: a parallel computing platform for global optimization based on genetic algorithm. Computers & Geosciences 33(3), 357–366 (2006)
10. Wahib, M., Munetomo, M., Munawar, A., Akama, K.: Mhgrid: Towards an ideal optimization environment for global optimization problems using grid computing. In: 8th Int'l Conf. on Parallel and Distr. Comput., Applic. and Technologies, pp. 167–168. IEEE Computer Society, Washington, DC (2007)
11. Günay, A., Öztoprak, F., Birbil, Ş., Yolum, P.: Solving global optimization problems using MANGO. In: 3rd KES Int'l Symp. on Agent and Multi-Agent Systems: Technologies and Applic., Upsalla, Sweden, pp. 783–792 (2009)
12. Biscani, F., Izzo, D., Yam, C.: A global optimisation toolbox for massively parallel engineering optimisation. In: 4th Int'l Conf. on Astrodynamics Tools and Techniques, Madrid, Spain (2010)
13. GSL, GNU Scientific Library (2010), http://www.gnu.org/software/gsl/
14. NAG Fortran Library, D04 Numerical Differentiation, subroutine D04AAF
15. Krishnan, M., Alexeev, Y., Windus, T.L., Nieplocha, J.: Multilevel parallelism in computational chemistry using Common Component Architecture and Global Arrays. In: ACM/IEEE Supercomp. Conf., Seattle, WA, p. 23 (2005)
16. Staveley, M.S., Poirier, R.A., Bungay, S.D.: An evaluation of parallel numerical Hessian calculations. In: High Perf. Comput. Symp., Kingston, ON, Canada, pp. 196–214 (2009)

Parallel Monte-Carlo Tree Search for HPC Systems

Tobias Graf[1,*], Ulf Lorenz[2], Marco Platzner[1], and Lars Schaefers[1,**]

[1] University of Paderborn, Paderborn, Germany
{slars@,platzner@,tobiasg@mail.}uni-paderborn.de
[2] TU Darmstadt, Darmstadt, Germany
lorenz@mathematik.tu-darmstadt.de

Abstract. Monte-Carlo Tree Search (MCTS) is a simulation-based search method that brought about great success to applications such as Computer-Go in the past few years. The power of MCTS strongly depends on the number of simulations computed per time unit and the amount of memory available to store data gathered during simulation. High-performance computing systems such as large compute clusters provide vast computation and memory resources and thus seem to be natural targets for running MCTS. However, so far only few publications deal with parallelizing MCTS for distributed memory machines. In this paper, we present a novel approach for the parallelization of MCTS which allows for an equally distributed spreading of both the work and memory load among all compute nodes within a distributed memory HPC system. We describe our approach termed UCT-Treesplit and evaluate its performance on the example of a state-of-the-art Go engine.

Keywords: UCT, HPC, Monte-Carlo Tree Search, distributed memory.

1 Introduction

Monte-Carlo tree search (MCTS) is a simulation-based search method that brought about great success in the past few years regarding the evaluation of stochastic and deterministic two-player games. MCTS learns a value function for game states by consecutive simulation of complete games of self-play using randomized policies to select moves for either player. Especially in the field of Computer Go, an Asian two-player board game, MCTS highly dominates over traditional methods such as $\alpha\beta$ search [12]. MCTS may be classified as a sequential best-first search algorithm [17], where "sequential" indicates that simulations are not independent of each other, as is often the case with Monte-Carlo algorithms. Instead, statistics about past simulation results are used to guide future simulations along the search space's most promising paths in a best-first manner. This dependency and the need to store and share the statistics among

---

[*] Authors are listed in alphabetical order.
[**] This work is supported by Microsoft Research Ltd. through a Phd-Scholarship.

E. Jeannot, R. Namyst, and J. Roman (Eds.): Euro-Par 2011, LNCS 6853, Part II, pp. 365–376, 2011.

all computation entities makes parallelization of MCTS for distributed memory environments a highly challenging task.

Parallelization of traditional $\alpha\beta$ search is a pretty well solved problem, e.g., see [6][10]. While for $\alpha\beta$ search it is sufficient to map the actual move stack to memory, MCTS requires us to keep a consecutively growing search tree representation in memory. On SMP machines, sharing a single search tree representation in memory is straight-forward and has already been proven to be very effective for MCTS parallelization [3][7]. However, sharing a search tree as the central data structure in a distributed memory environment is rather involved and only few approaches have been investigated so far [2].

In this paper, we present a novel approach for the parallelization of MCTS for distributed high-performance computing (HPC) systems. Our algorithm spreads a single search tree representation among all compute nodes (CNs) and guides simulations across CN boundaries using message passing. We map search tree nodes to randomized hash values, and the hash values to CNs in an equally distributed fashion which makes spreading tree nodes a straight-forward procedure [14][8]. A comparable approach used with traditional $\alpha\beta$ search was termed transposition driven scheduling (TDS) [15]. Computing more simulations in parallel than cores are available allows us to overlap communication times with additional simulations. We evaluate the performance of our parallelization technique on a real-world application, our high-end Go engine Gomorra. Gomorra has proven its strength at the Computer Olympiad 2010 in Kanazawa, Japan. In summary, we make the following contributions:

- Our algorithm makes efficient use of *all memory resources* available in the cluster. This is in strong contrast to formerly investigated parallelization methods that either extensively duplicate data [3][2] or resort to using only a fraction of a cluster's overall memory capacity.
- We provide a *flexible parallelization framework* not only for MCTS but for history-dependent simulation processes in general. Our framework is adjustable to light and heavy-weight simulations as well as for different kinds of networks, targeting an optimal exploitation of available resources.
- Compared to other MCTS parallelizations for distributed memory systems, our approach *reduces the loss of information* that inevitably results from parallelizing simulations.

The reminder of the paper is structured as follows: Section 2 introduces to the basic MCTS algorithm and reviews related work in parallelizing MCTS. Our novel parallelization approach for MCTS is presented in Section 3. Section 4 evaluates our algorithm and details the experimental setup and results achieved. Section 5 concludes the paper and gives an outlook to future work.

2 MCTS: Background and Related Work

In this section we provide some background of MCTS techniques and review related work. First, we give a brief introduction to basic MCTS and then focus on efforts to parallelize MCTS algorithms.

2.1 Basic MCTS

We present the most basic MCTS algorithm used for two-player zero-sum games with complete information. While we concentrate on two-player games and especially the game of Go for reasons of comparability with the work of others, we want to note that our approach is applicable to the wider class of Markov Decision Processes (MDP), e.g., see [13]. The so-called UCT algorithm (short for *Upper Confidence Bounds applied to trees* [13]) is a modern variant of MCTS and yields the experimentally best results for most of the current Go programs. Algorithm UCT shows a pseudo code representation of UCT.

Algorithm UCT: Basic UCT-Algorithm for two-player zero-sum games

Data: $G := (S, A, \Gamma, \delta, r)$, with S being the set of all possible states (i.e. game positions), A the set of actions (i.e. moves) that lead from one state to the next, a function $\Gamma : S \to \mathcal{P}(A)$ determining the subset of available actions at a state, the transition function $\delta : S \times A \to \{S, \emptyset\}$ specifying the follow-up state for a state-action pair where $\delta(s) = \emptyset$ iff $s \notin \Gamma(s)$ and a reward function $r : S_t \to \{0, 1\}$ assigning a binary reward to each terminal state $S_t := \{s \in S | \Gamma(s) = \emptyset\}$. A set $T \subseteq S$ contains all states that have a memory representation. Counters $N_{s,a}$ and $W_{s,a}$ are kept in memory for all states $s \in T$ and their corresponding actions $a \in \Gamma(s)$. We further set $N_s := \sum_{a \in \Gamma(s)} N_{s,a}$.

input : A state $s_0 \in S$ and a time limit
output: An action $a \in \Gamma(s_0)$

$T \leftarrow s_0; t \leftarrow 0;$
while *Time available* **do**
    if $s_t \in T$ **then**
        // in-tree policy:
        $a_{t+1} \leftarrow \text{argmax}_{a \in \Gamma(s)} \texttt{actionValue}(W_{s_t,a}, N_{s_t,a}, N_s);$
        $N_{s_t,a_{t+1}} \leftarrow N_{s_t,a_{t+1}} + 1;$
        $s_{t+1} \leftarrow \delta(s_t, a_{t+1});$
        $t \leftarrow t + 1;$
    else
        $T \leftarrow T \cup s;$ // Expand memory representation of search tree
        reward $\leftarrow$ playout $(s_t);$
        update (reward); // Update all W_{s_i,a_i} for $0 \le i \le t$ appropriately
        $t \leftarrow 0;$ // Start a new simulation
    end
end
return $\text{argmax}_{a \in \Gamma(s_0)}, N_{s_0,a};$
Function $\texttt{actionValue}$ *(w, v, V)*
    return $\frac{w}{v} + 2C\sqrt{\frac{2 \log(V)}{v}};$ // With C being a constant

---

UCT takes a state and a time limit as inputs and returns an action. As long as the time limit is not exceeded, UCT computes search tree simulations and builds up statistics about their outcomes for visited states. Being a so-called anytime algorithm, UCT can be interrupted any time and returns the best action found so far. As memory is generally limited, practical implementations build up simulation statistics for near-root tree nodes only. An effective way proposed in [4] is the generation of a memory tree T by starting with the root node and adding the first node not already covered by T during each simulation. This method leads to an efficient and predictable memory usage, as the memory tree likely grows in the most interesting branches and a maximum of one tree node is added with each additional simulation. Accordingly, simulation guidance based on node statistics is only possible for nodes covered by T. Once a simulation leaves T a randomized heuristic policy π_H is used for action selection until a terminal game position, i.e., a leaf of the real game tree, is reached. We call the randomized heuristic policy *playout policy* and the history-dependent one used for nodes covered by T *in-tree policy*.

Practical MCTS approaches may be divided into the four steps in-tree, expand, playout, and update, that form one simulation and that are repeated in a loop as illustrated in Fig. 1. In Section 3 we will use these steps to form work packages that can be distributed across a cluster.

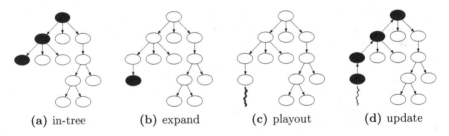

(a) in-tree   (b) expand   (c) playout   (d) update

Fig. 1. Building blocks of MCTS

UCT handles the task of selecting an action as a multi-armed-bandit problem (MAB) and is designed following prior investigations on MAB in [1]. As shown in Algorithm UCT, all data needed by UCT to select an action in the in-tree phase are the simulation statistics of all possible actions available at that states. Thus, for practical implementations it makes sense to store the statistics of all those actions together to optimize memory access patterns. This becomes even more important when considering the distribution of the search tree in a cluster, where all data necessary for an action selection step should be stored on a single CN to minimize communication overhead.

Researchers investigating the Go playout policy π_H follow two principal directions. While some researchers concentrate on handcrafted, computationally light-weight policies [4][9], others advocate more heavy-weight policies learned off-line from records of expert games or games of self-play with many simulations [5][18][11]. In either case, the computation of playouts is completely independent of former simulation results and therefore a perfect place to look for work-packages that may be distributed across a cluster. For Go, the playout step is typically dominating the simulation runtime, especially in the early stages of a game where the depth of the search tree T remains much smaller than the real game tree depth. The runtime dominance of the playout step is even more pronounced when heavy-weight playout policies are being used.

2.2 Parallelization of MCTS

The most common parallelization methods presented so far are termed Tree-Parallelization, Leaf-Parallelization and Slow-Tree-Parallelization [3], [2]. Fig. 2 illustrates these methods together with our algorithm UCT-Treesplit proposed in this paper. Tree-Parallelization is most common on SMP machines as one search tree representation is shared among several compute cores. Each core performs one simulation and updates the shared tree representation using atomic instructions. Leaf-Parallelization and Slow-Tree-Parallelization are suited for distributed memory machines. Leaf-Parallelization handles the in-tree part on only one CN, and computes multiple playouts on remote CNs once a leaf of the search tree representation T is reached. Slow-Tree-Parallelization performs rather independent searches on all CNs but occasionally synchronizes statistics of near-root tree nodes.

Among these methods, Slow-Tree-Parallelization is currently excelling for distributed memory systems. One drawback of this method is that no effort is made at all to exploit the increased amount of memory available within a cluster. Instead, all CNs try to keep a nearly identical copy of the search tree representation. However, simply distributing the search tree across all cluster nodes would result in very costly remote read/write operations, slowing down the simulation dramatically. In the following section we present a novel approach that combines both possibilities stated above resulting in an efficient parallelization of MCTS for distributed memory systems.

3   The UCT-Treesplit Algorithm for Parallel MCTS

We concentrate on homogeneous HPC systems with a fast interconnect (e.g., 10GBit Ethernet or Infiniband) consisting of N compute nodes (CNs), each having $C > 1$ compute cores that share the CN's entire memory, a model that fits most common HPC systems. We use MPI for message passing and devote one core (IO) of each CN for a thread handling message passing and work package distribution while the remaining $C - 1$ cores (workers) are bound to worker threads. Fig. 3 illustrates the setup of a CN.

The IO and worker cores communicate using thread-safe ring-buffers (Transfer Buffer In/Out) that reside in shared memory. An infinite loop running on the IO core reads available messages containing work packages from the network link and stores them in a buffer. Afterward, a work package scheduler distributes the received packages among the workers' ring-buffers, balancing the work load.

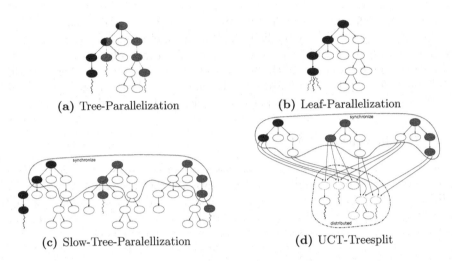

(a) Tree-Parallelization

(b) Leaf-Parallelization

(c) Slow-Tree-Paralellization

(d) UCT-Treesplit

Fig. 2. Overview of MCTS parallelization methods

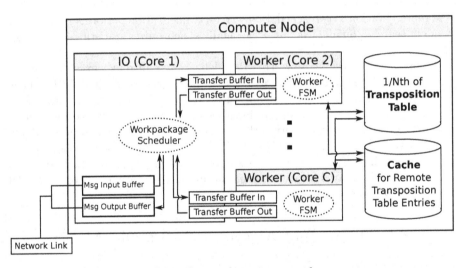

Fig. 3. Setup of a compute node

Workers start computation once they receive a package and, if required, send response messages back to the IO core using the corresponding buffer. In turn, the IO core frequently collects messages from the workers' ring-buffers and forwards them to the network link appropriately.

During their computation, workers require access to state-action statistics. As for some search domains it may be possible that equal states are reached through different paths, we use a lock-free but thread-safe transposition table[7] TR capable of storing size(TR) nodes of the search "tree" representation. We assume the existence of a hash-function hash : $S \rightarrow \mathbb{N}$ assigning a unique and equally distributed hash value to any search tree node. A straight forward method for computing an index $I_{TR}(s)$ into TR for a search tree node s is given by:

$$I_{TR}(s) = \text{hash}(s) \bmod \text{size}(TR)$$

Distributing TR on N CNs of a cluster may be done by storing on each CN i for $1 < i < N$ a transposition table TR_i of size $\text{size}(TR)/N$. An index to this distributed table is a tuple of a CN i and a local index I_{TR_i} to TR_i. Both can be computed as follows:

$$i(s) = \text{hash}(s) \bmod N \qquad (1)$$
$$I_{TR_i}(s) = (\text{hash}(s)/N) \bmod (\text{size}(TR)/N) \qquad (2)$$

Beside the partial transposition table, Fig. 3 also shows a cache for entries of remote transposition tables allowing for faster access to statistics of frequently visited states.

The key technique of our approach is the spreading of a single search tree among all CNs while overlapping communication with the computation of additional simulations. We break simulations into small work packages that can be computed on different cores. Moreover, cores computing work packages of one simulation do not need to be on the same CN. Message passing is used to guide simulations over CN boundaries.

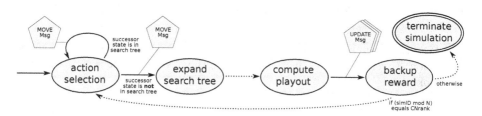

Fig. 4. Finite state machine for our distributed simulation (Worker FSM)

Fig. 4 illustrates our distributed simulation process as a finite state machine (FSM). During the in-tree part of a simulation, several action selection steps take place. Each of those steps can be computed without the need to communicate with other CNs by storing the statistics about all actions available in one state together in memory. However, between two consecutive action selection steps a simulation may move to another CN through a MOVE message. The dotted arrows in Fig. 4 represent state transitions that always happen on a single CN while solid arrows may cross CN boundaries. Those arrows are annotated with the corresponding messages that are sent in case the CN changes. Note that the UPDATE message is likely sent to more than one CN. In fact, it is send to all CNs visited by the simulation as statistics need to be updated on all these CNs. The states of the FSM are the work packages that make up the computational load.

Each of S_{par} simulations running in parallel suffers loss of information represented by the results of the $S_{par} - 1$ other simulations that would be available in a sequential UCT version. Obviously this impairs the search quality [16] and urges us to keep S_{par} as small as possible. Furthermore, we duplicate and occasionally synchronize frequently visited tree nodes on all CNs to reduce the communication overhead. In total the algorithm requires us to determine the values of three parameters:

- N_{dup}: The number of simulations that must have passed a state before it gets duplicated on all CNs.
- N_{sync}: The number of simulations that lead to the synchronization of a shared state. Each time one action of a shared state s has been visited at least N_{sync} times after the last synchronization, all values of actions of s are synchronized.
- O: An overload factor used to compute the number of simulations $S_{par} := (C - 1)NO$ that run in parallel on a system with N compute nodes.

Note that UCT-Treesplit may be configured to behave comparably to Slow-Tree-Parallelization by sharing all nodes immediately and choosing an appropriate threshold N_{sync} for triggering synchronization. On the other extreme, UCT-Treesplit behaves like Tree-Parallelization if tree nodes are never shared.

The search process begins by sharing the root node among all CNs. Then, each CN starts S_{par}/N simulations. A unique identifier simID $\in \{0, S_{par} - 1\}$ is assigned to each simulation. The data structure describing a simulation consists of:

- The **state stack** containing all states visited and actions taken during simulation. Together with each state, we store the CN where the action selection took place.
- The **simulation identifier** simID.

Procedure WorkerMainLoop gives a pseudo code representation of the main loop running on each worker core. Once a worker receives a MOVE message it searches for the current state's statistics in the cache and transposition table, respectively (line 6). If no entry exists, the worker **expands** the search tree by adding a new entry for the state. Immediately afterwards, a **playout** is computed and an UPDATE message is sent to all CNs visited in the course of the simulation (line 9). In case state statistics are found, an **action is selected** as in the sequential algorithm (line 7) leading to a new state s'. If the statistics to s' are already shared or are located in the CN's transposition table, the computation continues on the same CN. Otherwise, a MOVE message is sent to the CN storing the statistics of s'. Upon receipt of an UPDATE message, statistics for all states for which an action was selected on this CN get **updated** (line 12). MPI assigns a rank number CNrank $\in \{0, N\}$ to each CN. If the CN's rank equals (simID mod N), the worker initiates a new simulation.

Procedure WorkerMainLoop

```
   //
 1 while Time available do
 2  │  if an incoming message is available then
 3  │  │   M_out ← ∅;
 4  │  │   msg ← inTransferQueue.dequeue();
 5  │  │   if msg is a move-message then
 6  │  │   │   if msg.stateStack.top() is already in transposition table then
 7  │  │   │   │   M_out ← actionSelection(msg);
 8  │  │   │   else
 9  │  │   │   │   M_out ← expandAndPlayout(msg);
10  │  │   │   end
11  │  │   else// an update-message arrived
12  │  │   │   M_out ← update(msg);
13  │  │   end
14  │  │   outTransferQueue.enqueue(M_out);
15  │  end
16 end
```

---

4 Experiments

In this section, we present the experimental setup and the results achieved with our Go engine Gomorra that incorporates the novel UCT-Treesplit algorithm.

4.1 Setup

Our computer Go engine Gomorra implements several state of the art enhancements over basic UCT and proved its playing strength previously in several games against the currently strongest computer Go programs. In our experiments different instances of Gomorra play against each other on a 19x19 board size, giving each player 10 minutes to make all its moves in a game. We choose the following values for the three UCT-Treesplit parameters: $N_{dup} := 16$, $N_{sync} := 100$ and $O := 3$. Note that the optimal values will depend on parameters of the compute resources such as network latency and bandwidth as well as on the ratio of processor speed to work package size. Although reducing N_{dup} decreases the communication overhead for single simulations because less CN hops take place, the overhead for synchronizing shared nodes increases because more nodes are shared. However, few hops per simulation allow for keeping O small. Furthermore, lower values of N_{sync} lead to increased network traffic as more synchronization messages are sent.

For our experiments we use a cluster consisting of 60 CNs, each one equipped with 2 Intel Xeon X5650 CPUs (12 cores in total) running at 2.67 GHz and 36 GByte of main memory. The CNs are connected by a 4xSDR Infiniband network. We use OpenMPI for message passing between the CNs.

4.2 Results

Since UCT-Treesplit is communication intensive we first measure the average load on the worker cores. A low load indicates that work packages cannot be communicated fast enough and that the network is too slow in relation to the average work package size, number of cores and the core's clock rate. As MOVE and UPDATE messages are rather tiny in our experiments, we identify the network latency as the limiting parameter. With the given settings for N_{dup}, N_{sync} and O, we have to restrict the number of used cores per CN to 6 to achieve reasonable loads for higher numbers of CNs. Fig. 5a shows the measured core loads for different numbers of CNs at different phases of the game. Considering different phases of a game is important since a 19x19 Go game lasts for about 250-350 moves, and playouts at the end of the game require less computation.

Next, we are interested in the achievable simulation rates, measured in simulations per second. The simulation rate directly influences the achievable playing strength. Fig. 5b displays the scalability of the simulation rate over the number of compute nodes. It can be seen that UCT-Treesplit scales well for up to 32 CNs in early game phases while after move 200 the performance decreases considerably with higher numbers of CNs.

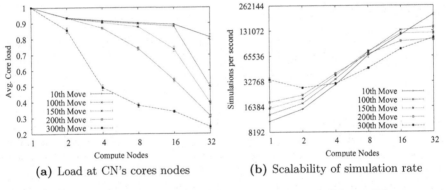

(a) Load at CN's cores nodes   (b) Scalability of simulation rate

Fig. 5. Scalability of simulations with increasing number of CNs at different game phases

The most important metrics, however, is the gain in playing strength achievable with increasing compute resources. Table 1a presents the achieved winning percentages of Gomorra playing against a copy of itself that can rely on a doubled amount of simulations computed per move decision. To conduct these experiments in reasonable time, we use a 4 core SMP machine except for the results marked with * for which we use a 12 core SMP machine and for the results marked with ** for which we use a 24 core SMP machine equipped with 132 GB of main memory to be able to store the vast amount of statistics. Comparable experiments were done in [2](Table 1) for the Go program MoGo. Although the absolute playing strength of MoGo is superior to Gomorra's, our Go engine seems to scale better with higher simulation numbers.

Table 1. Evaluation of Gomorra's playing strength

(a) Running on a single compute node.

N_s: Number of sim/move	Winrate $2N_s$ vs N_s	Games played
1,000	85.8 ±1.6%	500
4,000	86.0 ±1.6%	500
16,000	79.6 ±1.8%	500
128, 400*	82.4 ±1.7%	500
256, 800**	83.9 ±3.0%	149

(b) Using parallel UCT-Treesplit.

N: Number of CNs	Winrate 2N vs N	Games played
1*	53.3 ±2.0%	600
2	73.0 ±1.8%	600
4	61.3 ±2.0%	600
8	53.8 ±2.0%	600
16	46.1 ±2.3%	486

Table 1b shows the achieved winning rates of Gomorra playing against a copy of itself using the UCT-Treesplit algorithm on a varying number of compute nodes. In the 2 vs 1 node experiment marked with *, the single node version has no additional work for building MPI messages, moving simulations between compute cores, etc., explaining the rather small advantage of the double node version.

5   Conclusion and Future Work

In this paper, we investigate a novel approach to parallelize MCTS on distributed memory HPC systems. We present a way to share a single game tree representation efficiently among all compute nodes and evaluate the behavior of the new UCT-Treesplit algorithm in a high-end Go engine. We show that, for the game of Go, UCT-Treesplit scales up to 16 nodes.

An explanation for the diminishing gains UCT-Treesplit achieves with increasing compute nodes as shown in Table 1b could be the high number of simulations that are computed in parallel. With an overload factor of 3, we compute 480 simulations in parallel on 32 CNs. Richard B. Segal measured the scaling of the Go program Fuego with increasing numbers of parallel simulations and different time settings in [16]. His experiments showed a major decrease in playing strength if more than 128 threads are used for short time settings. However, giving Fuego more time per move yielded good results for even 512 parallel simulations. Thus, we may expect better scalability of UCT-Treesplit by just increasing the search time.

Another important observation and possible explanation for the scalability limitations is the diminishing increase in simulation rate at the end of games, i.e., when the remaining search tree and thus the work packages become too small, as shown in Fig. 5b. As part of future work we will try to keep the simulation rate high and reduce the communication overhead occurring during each simulation, for example by smoothly decreasing N_{dup} and increasing N_{sync} when it comes to the end of a game. Furthermore, we will study the influence and optimal settings for the various parameters involved.

References

1. Auer, P., Cesa-Bianchi, N., Fischer, P.: Finite-Time Analysis of the Multiarmed Bandit Problem. In: Machine Learning, vol. 47, pp. 235–256. Kluwer Academic, Dordrecht (2002)
2. Bourki, A., Chaslot, G.M.J.-B., Coulm, M., Danjean, V., Doghmen, H., Hoock, J.-B., Hérault, T., Rimmel, A., Teytaud, F., Teytaud, O., Vayssiére, P., Yu, Z.: Scalability and Parallelization of Monte-Carlo Tree Search. In: International Conference on Computers and Games, pp. 48–58 (2010)
3. Chaslot, G.M.J.-B., Winands, M.H.M., Jaap van den Herik, H.: Parallel Monte-Carlo Tree Search. In: Conference on Computers and Games, pp. 60–71 (2008)
4. Coulom, R.: Efficient Selectivity and Backup Operators in Monte-Carlo Tree Search. In: van den Herik, H.J., Ciancarini, P., Donkers, H.H.L.M(J.) (eds.) CG 2006. LNCS, vol. 4630, pp. 72–83. Springer, Heidelberg (2007)
5. Coulom, R.: Computing Elo Ratings of Move Patterns in the Game of Go. ICGA Journal 30(4), 198–208 (2007)
6. Donninger, C., Kure, A., Lorenz, U.: Parallel Brutus: The First Distributed, FPGA Accelerated Chess Program. In: 18th International Parallel and Distributed Processing Symposium. IEEE Computer Society, Los Alamitos (2004)
7. Enzenberger, M., Müller, M.: A Lock-Free Multithreaded Monte-Carlo Tree Search Algorithm. In: van den Herik, H.J., Spronck, P. (eds.) ACG 2009. LNCS, vol. 6048, pp. 14–20. Springer, Heidelberg (2010)
8. Feldmann, R., Mysliwietz, P., Monien, B.: Distributed game tree search on a massively parallel system. In: Monien, B., Ottmann, T. (eds.) Data Structures and Efficient Algorithms. LNCS, vol. 594, pp. 270–288. Springer, Heidelberg (1992)
9. Gelly, S., Wang, Y., Munos, R., Teytaud, O.: Modifications of UCT with Patterns in Monte-Carlo Go. Technical Report 6062, INRIA (2006)
10. Himstedt, K., Lorenz, U., Möller, D.P.F.: A twofold distributed game-tree search approach using interconnected clusters. In: Luque, E., Margalef, T., Benítez, D. (eds.) Euro-Par 2008. LNCS, vol. 5168, pp. 587–598. Springer, Heidelberg (2008)
11. Huang, S.-C., Coulom, R., Lin, S.-S.: Monte-Carlo Simulation Balancing in Practice. In: Conference on Computers and Games, pp. 81–92 (2010)
12. Donald Knuth, E., Moore, R.W.: An Analysis of Alpha-Beta Pruning. In: Artificial Intelligence, vol. 6, pp. 293–327. North-Holland Publishing Company, Amsterdam (1975)
13. Kocsis, L., Szepesvári, C.: Bandit Based Monte-Carlo Planning. In: Fürnkranz, J., Scheffer, T., Spiliopoulou, M. (eds.) ECML 2006. LNCS (LNAI), vol. 4212, pp. 282–293. Springer, Heidelberg (2006)
14. Lorenz, U.: Parallel controlled conspiracy number search. In: Monien, B., Feldmann, R.L. (eds.) Euro-Par 2002. LNCS, vol. 2400, pp. 420–430. Springer, Heidelberg (2002)
15. Romein, J.W., Plaat, A., Bal, H.E., Schaeffer, J.: Transposition table driven work scheduling in distributed search. In: National Conference on Artificial Intelligence, pp. 725–731 (1999)
16. Segal, R.B.: On the Scalability of Parallel UCT. In: International Conference on Computer and Games, pp. 36–47 (2010)
17. Silver, D.: Reinforcement Learning and Simulation-Based Search in Computer Go. PhD thesis, University of Alberta (2009)
18. Silver, D., Tesauro, G.: Monte-Carlo Simulation Balancing. In: International Conference on Machine Learning, pp. 945–952 (2009)

Petascale Block-Structured AMR Applications without Distributed Meta-data

Brian Van Straalen, Phil Colella, Daniel T. Graves, and Noel Keen

Applied Numerical Algorithms Group,
Lawrence Berkeley National Laboratory,
Berkeley, CA 94720, USA

Abstract. Adaptive mesh refinement (AMR) applications to solve partial differential equations (PDE) are very challenging to scale efficiently to the petascale regime.

We describe optimizations to the Chombo AMR framework that enable it to scale efficiently to petascale on the Cray XT5. We describe an example of a hyperbolic solver (inviscid gas dynamics) and an matrix-free geometric multigrid elliptic solver. Both show good weak scaling to 131K processors without any thread-level or SIMD vector parallelism.

This paper describes the algorithms used to compress the Chombo metadata and the optimizations of the Chombo infrastructure that are necessary for this scaling result. That we are able to achieve petascale performance without distribution of the metadata is a significant advance which allows for much simpler and faster AMR codes.

1 Introduction

PDE solvers using adaptive mesh refinement, AMR, on block structured grids, e.g. [3, 4], are among the most challenging applications to adapt to massively parallel computing environments. Because the grids can be anywhere in the domain, metadata is required to describe where the data lives and what processor is responsible for it. Standard Chombo metadata is not distributed- all processors keep a redundant index of the distributed data layout. Previous results [10] have shown that Chombo AMR with scales well to 10K processors. The size (in memory) of the metadata in Chombo made further scaling impossible without significant metadata redesign. Other AMR infrastructures have distributed their metadata among the processors. PARAMESH [7] and SAMRAI [12] do this for large problems. BoxLib [5], in the CASTRO code [1], does not distribute metadata but extensively threads their code and uses bigger grids and were able to scale up to 200K processors using one grid per processor. In this paper, we use Chombo's flat MPI method and we compress Chombo's metadata. We show good weak scaling to petascale with many more boxes per processor (77 in the hyperbolic case, 53 in the elliptic case). Because we do not distribute our metadata, we avoid substantial increases in code complexity and communication time.

E. Jeannot, R. Namyst, and J. Roman (Eds.): Euro-Par 2011, LNCS 6853, Part II, pp. 377–386, 2011.
© Springer-Verlag Berlin Heidelberg 2011

2   AMR Applications

Block-structured AMR, developed by Berger and Oliger [4,3] for computational
gas dynamics, is a multi-scale algorithm that achieves high spatial and temporal
resolution in localized regions of dynamic multidimensional numerical simulations. A broad range of physical phenomena modeled by PDE exhibit multi-
scale behavior where variations in the solution occur over scales that are much
smaller than the overall problem domain. Examples include flame fronts arising in the burning of hydrocarbon fuels, nuclear burning in supernovae, effects
of localized features in orography or bathymetry on ocean currents, tracking of
tropical cyclones, localized kinetic effects for plasma physics problems, and, in
general, small scale effects due to nonlinear instabilities. In each of these problems, the fundamental mathematical description is given in terms of various combinations of PDE of classical type (elliptic, parabolic, hyperbolic). The Berger
and Oliger AMR algorithm organizes refined regions into rectangular structured
grids of several hundred to several thousand grid points per grid. High-resolution
structured-grid methods (typically expressed as stencils) are used to advance the
solution in time. Furthermore, the overhead of managing the irregular data is
amortized over a relatively large number of floating point operations on the rectangular grids. For time-dependent problems, refinement is performed in time as
well as space. Each level of spatial refinement has its own stable time step, with
the time steps on a level constrained to be integer multiples of the time steps on
all finer levels.

2.1   Chombo AMR Framework

AMR applications require a long-term sustained investment in software infrastructure to create scalable solvers that are capable of utilizing the full capabilities
of the largest available HPC platforms. We have created a framework for implementing scalable parallel AMR calculations called Chombo [6] that provides an
environment for rapidly assembling portable, high-performance AMR applications for a broad variety of scientific disciplines.

Chombo is a fully instrumented C++ library. There are a set of timer macros
that can be used to time functions or sections of code. These timers attempt
to use native instructions on the target architecture in order to minimize the
overhead of collecting detailed performance data.

In the standard Chombo framework, there is metadata associated with each
computational region on a adaptive (called a "Box"). Each Box contains the integer locations of the lower left and upper right corners of the region. A collection
of these regions along with their processor mapping (a "DisjointBoxLayout") is
represented internally by a

```
Vector< pair<Box, int> >
```

This metadata is not distributed and therefore grows with the size of the problem. In previous scaling studies [10], the memory cost of this metadata was

shown to become prohibitive in the 8K-32K processor range. A large part of the current work is to compress this metadata without distributing it. This allows the Chombo framework to scale to 131K processors for both elliptic and hyperbolic benchmarks without without the large memory cost and without the communication cost associated with distributed metadata.

3   Benchmarking Methodology

In many applications that use PDE solvers, the primary motivation for using large numbers of processors is to achieve weak scaling. Even with AMR, many leading scientific problems remain out of reach due to inadequate grid resolution. In those cases, increasing the number of processors is used to increase the spatial resolution of the grids using the minimum number of processors necessary to fit the problem into the available memory. Therefore, we focus on a methodology for constructing weak-scaled AMR benchmarks because this methodology models the dominant use-case for scientific problems that employ this computational method. We use two different examples for our benchmark, an explicit Godunov method for gas dynamics and a multigrid solver to solve Poisson's equation. These two are reasonable proxies for the two components of most complete AMR applications: explicit solvers for hyperbolic equations and implicit solvers for elliptic and parabolic equations.

3.1   Replication Scaling Benchmarks

Following [10] we use two benchmarks based on replication scaling. We take a grid hierarchy and data for a fixed number of processors and scale it to higher concurrencies by making identical copies of the hierarchy and the data, see Figure 1. The full AMR code (processor assignment, problem setup, etc.) is run without any modifications to guarantee it is not directly aware of the replicated grid structure. Replication scaling tests most aspects of weak scalability, is simple to define, and provides results that are easy to interpret. Thus, it is a very useful tool for understanding and correcting impediments to efficient scaling in an AMR context.

3.2   Poisson Benchmark

We first benchmarked an AMR solver for Poisson's equation, based on a cell-centered multilevel discretization of Laplacian in three dimensions [8]. The solver itself used multigrid iteration suitably modified for an AMR grid hierarchy [11,2]. The benchmark applied ten iterations of the AMR-multigrid V-cycle, which is typical of the number of iterations required for the solver to converge, and corresponds to 1700 flops/grid point. This is a very demanding application from the standpoint of parallelism – requiring multiple communication and synchronization steps per multigrid iteration. The algorithmic features of this benchmark are typical of broad range of elliptic solvers arising in applications using AMR.

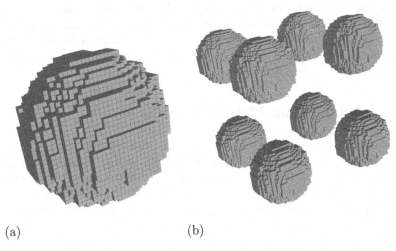

(a)   (b)

Fig. 1. (a) Grids at the finest AMR level used in the hyperbolic gas dynamics bench-mark – these grids cover the shock front of a spherical explosion in 3D. (b) Replicated grids at the finest AMR level used in the weak scaling performance study of the hyperbolic gas dynamics benchmark. There are 14902 boxes per processor at the finest level before replication. Each box is size 16^3.

The grids used as the basis of for the Poisson replication benchmark are shown in Figure 3.3. There are three levels of AMR with a refinement ratio of four between each level. There is one unknown per grid point for a total of 15M grid points in the configuration with no replication.

3.3   Hyperbolic Gas Dynamics Benchmark

We benchmarked an explicit method for unsteady inviscid gas dynamics in three dimensions that is based on an unsplit PPM algorithm [13, 9]. This algorithm requires approximately 6000 flops/grid point. Since it is an explicit method, com-munication between processors is required only once per time step. We used the implementation of this method from the Chombo software distribution without significant modification. The grids used as the basis for the hyperbolic bench-mark are shown in Figure 1.

The benchmark used three levels of AMR with a factor of 4 refinement between levels and with refinement in time proportional to refinement in space. We use fixed-sized 16^3 grids and and five unknowns per grid point, with 10^9 grid point updates performed for the single coarse-level time step. None of the grids at any level were changed during any of the time steps, i.e., there was no grid adaptation in time which is sometimes called "regridding". In the results given here, we are only timing the cost of computing a single coarse-level time step, which includes all intermediate and fine time steps on all AMR levels but excludes the problem setup and initialization times.

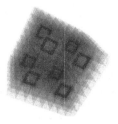

Fig. 2. Grids used in the Poisson benchmark before replication. The red is the level 0 grids, the green is level 1, the blue is level 2. This shows a 2x2 replication. There are 1280 boxes per processor at the finest level before replication. Each box is size 32^3.

4   Optimizing AMR for Scalability

To achieve our performance results for the two 3D Chombo applications discussed, several important changes were made to the standard code. A run-length compression method was used to greatly reduce the memory overhead associated with the metadata for the grids. There were also application-specific optimizations. For example, for our hyperbolic application, we optimized inter-level coarse-fine interpolation objects to take advantage of our new metadata structure. For our elliptic solver, we carefully control the number of communication steps necessary and greatly reduce the number of all-to-all communications in the multigrid algorithm for AMR.

4.1   Memory Performance: Compression

Moore's Law continued unabated for CPU design, but the gap between memory capacity and memory latency has grown every year. In such an environment it makes sense to work with the necessary metadata in an application in a compressed format and utilize the excess of processor cycles to uncompress this information on-the-fly as the processor needs it. This also makes better use of processor-to-memory resources, while decompression can happen in very fast local register storage.

Standard Chombo metadata holds the grid data for each level as explicit vectors of pairs of each box and its associated processor assignment. As the number of boxes becomes large, the memory associated with this representation of the grids grows linearly since this description is not distributed among processors. In [10], it was found that, for a typical Chombo application, the memory usage becomes untenable at between 8K and 16K processors (where the total number of boxes was between 1M and 10M).

We compress the metadata by first stipulating that every patch on a level must be of fixed size. We then create a bitmap of the domain coarsened by the box size and put a 1 where there is to be a box and a 0 where there is none. This bitmap is compressed using run-length compression. The load balancing is done by simply dividing the patches up evenly between processors. If there are N patches per processor, the first N in lexicographic order go to processor 0, the next N to processor 1 and so on. For applications where the load on a patch is more variable, a more flexible load balancing scheme may be necessary.

The results of this change in representation were striking. Figure 3 shows the memory usage for a sample weak scaling run of a gas dynamics solver. The problem has 77 boxes per processor at the finest level (at 196K processors, this amounts to 15.2 million boxes). Figure 5 shows the memory usage for a sample weak scaling run of an elliptic solver with 53 boxes per processor at the finest level (at 98K processors this amounts to 6.55 million boxes). We track the memory that Chombo allocates and measure the memory that the operating system is using for the application. The amount of memory reported by the operating system is substantially higher at high concurrencies and is largely due to MPI memory overhead. In both cases, the metadata compression was necessary to run at the highest concurrency, otherwise the memory of the compute nodes was exhausted.

Figure 4 shows a line labeled "MPI overhead". This is still a speculation on our part, but many tools were used to eliminate and quantify the use of memory in the benchmark applications.

4.2   Run Time Performance

To achieve better scaling of Chombo run time performance at higher concurrencies, optimizations were done for both the hyperbolic and elliptic solvers. For the hyperbolic solver, changes were made to the definition of some inter-level

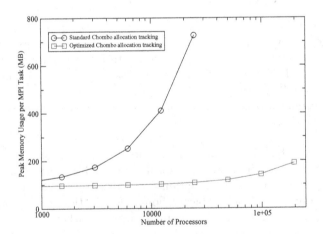

Fig. 3. Memory performance of Chombo 3D inviscid gas dynamics solver before and after metadata compression. In both cases, there are three levels of refinement (factor of 4 between levels) and 77 boxes per processor at the finest level. The standard Chombo solver was not able to run at the highest concurrencies because of memory requirements.

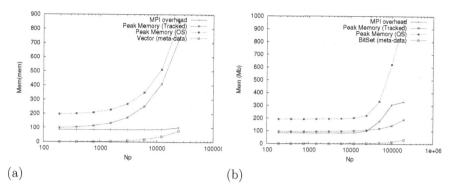

(a)   (b)

Fig. 4. A closer look at memory performance of Chombo 3D inviscid gas dynamics solver (a) before and (b) after metadata compression. In both cases, there are three levels of refinement (factor of 4 between levels) and 77 boxes per processor at the finest level. The original solver (a) was not able to run at the highest concurrencies because it ran out of memory.

objects to account for the new fixed box size representation of the layouts. These changes allowed excellent weak scaling results to 196K processors.

The standard Chombo elliptic solver required several optimizations to facilitate scaling to 98K. We used the fact that the equation is solved in residual-correction form to minimize how often inhomogeneous inter-level interpolation is done in the solve. Standard Chombo also does an extra coarse-fine interpolation before the refluxing step. Previous to our optimizations, there were 8 inhomogeneous coarse-fine interpolations per multigrid v-cycle. We were able to reduce this to only 2 inhomogeneous coarse-fine interpolations per multigrid v-cycle. The standard Chombo Poisson solver does not do box aggregation. Once the input grids are no longer coarsenable, a bottom solver is called. We introduce box aggregation and take multigrid down to a two-cell grid. At the bottom level we simply do two Gauss-Seidel relaxations. This saved a lot of communication time at higher concurrencies because other bottom solvers require substantial all-to-all communication to calculate norms. The AMR Multigrid is this case turns into a true multigrid solve. We reduced the amount of communication at relaxation steps by only doing ghost cell exchanges every other relaxation step. This did not substantially affect the multigrid convergence. The run time comparison between standard Chombo and optimized Chombo for 10 elliptic solves (each with 7 multigrid v-cycles) is given in figure 5. The standard Chombo solver was not able to run at the higher concurrencies because the memory requirements were too large for the machine.

Excellent weak scaling is observed on Jaguar (Cray XT5) to 196K processors for the hyperbolic problem and 98K for the elliptic problem as is shown in Fig. 5. There are 77 boxes per processor for the hyperbolic problem, and 53 boxes per processor for the elliptic problem.

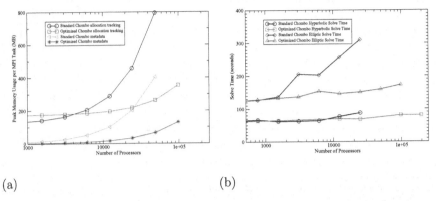

(a)   (b)

Fig. 5. a)Memory performance of Chombo 3D elliptic solver before and after metadata compression. In both cases, there are three levels of refinement (factor of 4 between levels) and 53 boxes per processor at the finest level. The standard Chombo solver was not able to run at the highest concurrencies because of memory requirements. b)Solve time of Chombo 3D hyperbolic and elliptic solvers before and after optimization. In both cases, there are three levels of refinement (factor of 4 between levels). The largest hyperbolic benchmark was run at 196K cores.

As the problem size per processor remains constant for these weak scaling experiments, we can estimate that the number of flops per processor are also constant. For the hyperbolic solver, the number of total flops is estimated to be 7.4e10. At a concurrency 196K, with a solve time of 80.5 seconds, the aggregate flop rate is 181 TFlops. For the multigrid elliptic solver, with the estimated number of flops at 4.4e10, a solve time of 172.7 seconds, and 98K MPI processors, the aggregate flop rate is 25 TFlops.

5   Summary and Conclusions

We present petascale weak scaling results for two key AMR applications. With some modifications of the metadata representation of standard Chombo, we were able to show good weak scaling for both hyperbolic and elliptic problems without having to distribute metadata. Fully local metadata is maintained through the use of compressing the metadata format and decompressing the information as it is utilized in the calculation. The computational cost of working with loss-less compression techniques is not measurable in these applications. This results in efficient memory usage at a slight increase in local processing cycles. This approach involves far less complexity than distributed metadata designs that must balance communication and local caching algorithms, which are application-specific.

In general, we feel that coding theory will become an increasingly important aspect of HPC computing on future platforms where memory and communication costs will increasingly be the science-limiting characteristic of these machines. Trading off excess compute cycles for more efficient memory usage and lower communication costs will be a more common theme in future HPC codes.

While this has been a great success, we recognize that ultimately the flat MPI parallelism model is not extensible to the billion-way concurrency models needed to achieve exascale performance. We consider this work to be but one part of a larger exascale computing strategy where metadata is *not* distributed at the MPI parallelism layer. Within an MPI rank there are several further levels of parallelism to be exploited. The first level is threading the load currently handled sequentially by each MPI rank. The next level is fine-grain parallelism within the dimensional loops within a box. Both areas are being worked on currently. Finally there are instruction-level parallelism models to make use of vector processing within a larger threading model. These are all orthogonal optimization efforts that must all succeed to meet the goal of exaflop simulations.

Acknowledgments. We would like James Hack and the OLCF Resource Utilization Council for access to *Jaguar* via the Applied Partial Differential Equations Center project. The authors were supported by the Office of Advanced Scientific Computing Research in the Department of Energy under Contract DE-AC02-05CH11231.

References

1. Almgren, A., Bell, J., Kasen, D., Lijewski, M., Nonaka, A., Nugent, P., Rendleman, C., Thomas, R., Zingale, M.: Maestro, castro and sedona – petascale codes for astrophysical applications. In: SciDAC 2010. J. of Physics: Conference Series (2010)
2. Almgren, A.S., Buttke, T., Colella, P.: A fast adaptive vortex method in three dimensions. J. Comput. Phys. 113(2), 177–200 (1994)
3. Berger, M., Collela, P.: Local adaptive mesh refinement for shock hydrodynamics. J. Computational Physics 82, 64–84 (1989)
4. Berger, M., Oliger, J.: Adaptive mesh refinement for hyperbolic partial differential equations. Journal of Computational Physics 53, 484–512 (1984)
5. BoxLib Reference Manual, https://seesar.lbl.gov/anag/eb/reference-manual/boxlib.html
6. Colella, P., Graves, D., Ligocki, T., Martin, D., Modiano, D., Serafini, D., Straalen, B.V.: Chombo software package for AMR applications: design document, http://davis.lbl.gov/apdec/designdocuments/chombodesign.pdf
7. MacNeice, P., Olson, K.M., Mobarry, C., deFainchtein, R., Packer, C.: Paramesh: A parallel adaptive mesh refinement community toolkit. Computer Physics Communications 126, 330–354 (2000)
8. Martin, D., Colella, P., Graves, D.T.: A cell-centered adaptive projection method for the incompressible Navier-Stokes equations in three dimensions. Journal of Computational Physics 227, 1863–1886 (2008)
9. Miller, G., Colella, P.: A conservative three-dimensional Eulerian method for coupled solid-fluid shock capturing. Journal of Computational Physics 183, 26–82 (2002)

10. Straalen, B.V., Shalf, J., Ligock, T., Keen, N., Yang, W.-S.: Parallelization of structured, hierarchical adaptive mesh refinement algorithms. In: IPDPS:Interational Conference on Parallel and Distributed Computing Systems (2009)
11. Thompson, M.C., Ferziger, J.H.: An adaptive multigrid technique for the incompressible Navier-Stokes equations. Journal of Computational Physics 82(1), 94–121 (1989)
12. Wissink, A.M., Hornung, R.D., Kohn, S.R., Smith, S.S., Elliot, N.: Large scale parallel structured amr calculations using the samrai framework. In: SC 2001 Conference on High Perfomrance Computing (2001)
13. Woodward, P.R., Colella, P.: The numerical simulation of two-dimensional fluid flow with strong shocks. Journal of Computational Physics 54, 115–173 (1984)

Accelerating Anisotropic Mesh Adaptivity on nVIDIA's CUDA Using Texture Interpolation

Georgios Rokos[1], Gerard Gorman[2], and Paul H.J. Kelly[1]

[1] Software Performance Optimisation Group,
Department of Computing
[2] Applied Modelling and Computation Group,
Department of Earth Science and Engineering,
Imperial College London,
South Kensington Campus, London SW7 2AZ, United Kingdom,
{georgios.rokos09,g.gorman,p.kelly}@imperial.ac.uk

Abstract. Anisotropic mesh smoothing is used to generate optimised meshes for Computational Fluid Dynamics (CFD). Adapting the size and shape of elements in an unstructured mesh to a specification encoded in a metric tensor field is done by relocating mesh vertices. This computationally intensive task can be accelerated by engaging nVIDIA's CUDA-enabled GPUs. This article describes the algorithmic background, the design choices and the implementation details that led to a mesh-smoothing application running in double-precision on a Tesla C2050 board. Engaging CUDA's texturing hardware to manipulate the metric tensor field accelerates execution by up to 6.2 times, leading to a total speedup of up to 148 times over the serial CPU code and up to 15 times over the 12-threaded OpenMP code.

Keywords: anisotropic mesh adaptivity, vertex smoothing, parallel execution, CUDA, metric tensor field, texturing hardware.

1 Introduction

Mesh adaptivity is an important numerical technology in Computational Fluid Dynamics (CFD). CFD problems are solved numerically using unstructured meshes, which essentially represent the discrete form of the problem. In order for this representation to be accurate and efficient, meshes have to be adapted according to some kind of error estimation. Furthermore, this error estimation may also encode information about a possible spatial orientation of the problem under consideration, in which case we say that the underlying dynamics is anisotropic and the error estimation is described using a metric tensor field.

One sophisticated adaptation technique, suitable for anisotropic problems, is Vertex Smoothing. Adapting a mesh to an error estimation involves an enormous amount of floating-point operations which can push even the most powerful processing units to their limits. The CUDA platform offers great computational power at relatively low cost. These properties make it a perfect candidate for

E. Jeannot, R. Namyst, and J. Roman (Eds.): Euro-Par 2011, LNCS 6853, Part II, pp. 387–398, 2011.
© Springer-Verlag Berlin Heidelberg 2011

accelerating mesh adaptation. We wrote a new application framework which implements Pain's smoothing algorithm [7] along with the proposal by Freitag et al. [6] for its parallel execution, enabling mesh adaptation to be accelerated on CUDA GPUs. The main objectives achieved through this project can be summarised as follows:

– This is the first adaptive mesh algorithm implemented on CUDA as far as the authors are aware.
– It resulted in an application running in double-precision on a Fermi-based Tesla board up to 15 times faster than on a 12-core server.
– The key optimisation proved to be the use of texturing hardware to store and interpolate the metric tensor field, which offered speed-ups of up to 6.2 times over the simple CUDA code.

The rest of the article is organised as follows: Section 2 contains a comprehensive description of the algorithmic background and Section 3 describes how the application was designed and implemented. Section 4 presents performance graphs comparing the serial code against OpenMP and CUDA versions. We conclude this paper and discuss ideas for future work in Section 5.

2   Background

2.1   PDEs, Meshes and Mesh Quality

The Finite Element Method (FEM) is a common numerical approach for the solution of PDEs, in which the problem space is discretised into smaller sub-regions, usually of triangular (in 2D) or tetrahedral (in 3D) shape. These sub-regions, referred to as elements, form a mesh. The equation is then discretised and solved inside each element. Common discretisation techniques often result in low quality meshes and this affects both convergence speed and solution accuracy [5]. A posteriori error estimations on the PDE solution help evaluate a quality functional [10] and determine the low-quality elements, which a mesh-improving algorithm tries to "adapt" towards the correct solution. Unstructured meshes, i.e. meshes in which a node can be connected to an arbitrary number of other nodes, offer greater numerical flexibility but their more complex representation is followed by higher computational cost [8].

2.2   Anisotropic PDEs

A problem is said to be "anisotropic" if its solution exhibits directional dependencies. In other words, an anisotropic mesh contains elements which have some (suitable) orientation, i.e. size and shape. The process of anisotropic mesh adaptation begins with a (usually automatically) triangulated mesh as input and results in a new mesh, the elements of which have been adapted according to some error estimation. This estimation is given in the form of a metric tensor field, i.e. a tensor which, for each point in the 2-D (or 3-D) space, represents

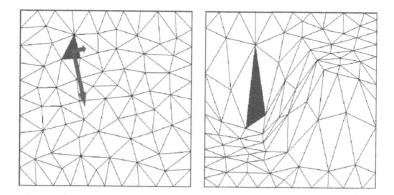

Fig. 1. Example of anisotropic mesh adaptation. The initial red triangle is stretched according to the metric tensor value (green arrow).

the desired length and orientation of an edge containing this point. As was the case with the PDE itself, the metric tensor is also discretised; more precisely, it is discretised node-wise. The value of the error at an in-between point can be taken by interpolating the error from nearby nodes. An example of adapting a mesh to the requirements of an anisotropic problem is shown in Figure 1.

Adapting a mesh so that it distributes the error uniformly over the whole mesh is, in essence, equivalent to constructing a uniform mesh consisting of equilateral triangles with respect to the non-Euclidean metric $M(\mathbf{x})$. This concept can be more easily grasped if we give an analogous example with a distorted space like a piece of rubber that has been stretched (see Figure 2). In this example, our domain is the piece of rubber and we want to solve a PDE in this domain. According to the objective functional we used, all triangles in the distorted (stretched) piece of rubber should be equilateral with edges of unit length. When we release the rubber and let it come back to its original shape, the triangles will look compressed and elongated.

The metric tensor M can be decomposed as

$$M = Q \Lambda Q^T$$

where Λ is the diagonal matrix, the components of which are the eigenvalues of M and Q is an orthonormal matrix consisting of eigenvectors Q^i. Geometrically, Q represents a rotation of the axis system so that the base vectors show the direction to which the element has to be stretched and Λ represents the amount of distortion (stretching). Each eigenvalue λ^i represents the squared ideal length of an edge in the direction Q^i [8].

2.3   Vertex Smoothing and the Algorithm by Pain et al.

Vertex smoothing is an adaptive algorithm which tries to improve mesh quality by relocating existing mesh vertices. Contrary to other techniques, which we

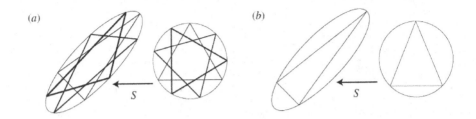

Fig. 2. Example of mapping of triangles between the standard Euclidean space (left shapes) and metric space (right shapes). In case (a), the elements in the physical space are of the desired size and shape, so they appear as equilateral triangles with edges of unit length in the metric space. In case (b), the triangle does not have the desired geometrical properties, so it does not map to an equilateral triangle in the metric space.(Figure from [8])

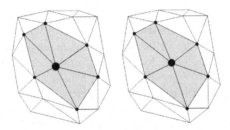

Fig. 3. Vertex smoothing example. The vertex under consideration is the one marked with a big black circle. The local problem area is the light-orange one. Left figure shows the cavity before smoothing. Right figure shows the result of local smoothing. (figure from [9])

discuss in Section 5, it leaves the mesh topology intact, i.e. connectivity between nodes does not change. All elements affected by the relocation of one vertex form an area called a cavity. A cavity is defined by its central, free vertex and all incident vertices. A vertex smoothing algorithm tries to equidistribute the quality among cavity elements by relocating the central vertex to a new position. Optimisation takes into account only elements belonging to the cavity, which means that only one vertex is considered for relocation at a time. An example of optimising a cavity is shown in Figure 3 [9].

The scope of optimisation is the cavity, therefore vertex smoothing is a local optimisation technique. The algorithm moves towards the global optimum through a series of local optimisations. The local nature of vertex smoothing leads to the need for optimising a cavity over and over again. After having smoothed a vertex, smoothing an incident vertex in the scope of its cavity may change the quality of the first cavity. Because of this property, the algorithm has to be applied a number of times in order to bring things to an equilibrium.

Running a smoothing algorithm in parallel can be done as dictated by the framework proposed by Freitag et al. [6]. In a parallel execution, we cannot smooth arbitrarily any vertices simultaneously. When a vertex is smoothed, all adjacent vertices have to be locked at their old positions. This means that no two adjacent vertices can be smoothed at the same time. In order to satisfy this requirement and ensure hazard-free parallel execution, mesh vertices have to be coloured so that no two adjacent vertices share the same colour. All vertices of the same colour form an independent set, which means that they are completely independent from each other and can be smoothed simultaneously. This is a classic graph colouring problem, with the graph being the mesh, graph nodes being mesh vertices and graph edges being mesh edges.

Pain et al. proposed a non-differential method to perform vertex smoothing [7]. A cavity C_i consists of a central vertex V_i and all adjacent vertices V_j. Let L_i be the set of all edges connecting the central vertex to all adjacent vertices. The aim is to equate the lengths of all edges $\in L_i$ (recall that the optimal cavity is the one in which all triangles are equilateral with edges of unit length with respect to some error metric). The length of an edge l in metric space is defined as $r_l = \left(u_l^T M_l u_l\right)$, where M_l is the value of the metric tensor field in the middle of the edge.

Let $\hat{p}^i$ be the initial position of the central vertex and p^i the new one. Then, the length of an edge in the standard Euclidean space is $u_l = p^i - y_l^i$, where y_l^i is the position of a non-central cavity vertex V_j. Also, it is important to use relaxation of p^i for consistency reasons, using $x^i = wp^i + (1 - w)p^i, w \in (0, 1]$. In this project, $w = 0.5$. We define $q^i = \sum_{l \in L_i} M_l y_l^i$ and $A^i = \sum_{l \in L_i} M_l$ and introduce a diagonal matrix D^i to ensure diagonal dominance and insensitivity to round-off error:

$$D_{jk}^i = \begin{cases} max A_{jj}^i, (1 + \sigma) \sum_{m=1, m \neq j} \mid A_{jm}^i \mid, & \text{if } j = k \\ 0, & \text{if } j \neq k \end{cases}$$

In this project, $\sigma = 0.01$. Then, x^i can be found by solving the linear system

$$(D^i + A^i)(x^i - \hat{p}^i) = w(q^i - A^i \hat{p}^i).$$

In the case of boundary vertices, i.e. vertices which are allowed to move only along a line (the mesh boundary), a modification of the above algorithm has to be used. The restriction that the vertex can only move along a line means that the new position x^i can be calculated using the equation

$$x^i = a_C^i u_l^i + \hat{p}^i,$$

where u_l^i is the unit vector tangential to the boundary line and a_C^i is the displacement along this line measured from the initial position $\hat{p}^i$ of the vertex. a_c^i can be calculated from the equation

$$(D^i + \hat{M}^i)a_c^i = wg^i,$$

where $\hat{M}^i = u_l^{i^T} \sum_{l \in L_i} M_l u_l^i$ and $g^i = \sum_{l \in L_i} u_l^{i^T} M_l(x^i - \hat{p}^i)$.

2.4   CUDA's Texturing Hardware

Texturing hardware is an important heritage left by the graphics-processing roots of CUDA. Reading data from texture memory can have a lot of performance benefits, compared to global memory accesses. Texture memory is cached in a separate texture cache (optimised for 2D spatial locality), leaving more room in shared memory/L1 cache. If memory accesses do not follow the patterns required to get good performance (as is the case with unstructured problems), higher bandwidth can be achieved provided there is some locality on texture fetches. Additionally, addressing calculations are executed automatically by dedicated hardware outside processing elements, so that CUDA cores are not occupied by this task and the programmer does not have to care about addressing [3,4].

The most important texturing feature is interpolation. Textures are discretised data from a (theoretically) continuous domain. In graphics processing, a texture value may be needed at a coordinate which falls between discretisation points, in which case some kind of texture data filtering has to be performed. Interpolating values from the four nearest discretisation points is the most common type of texture filtering, called linear filtering. In two dimensions, the result $tex(x, y)$ of linear filtering is:

$$tex(x, y) = (1 - \alpha)(1 - \beta)T[i, j] + \alpha(1 - \beta)T[i + 1, j] +$$
$$+ (1 - \alpha)\beta T[i, j + 1] + \alpha\beta T[i + 1, j + 1],$$

where α is the horizontal distance of point (x, y) from the nearest texture sample (discretisation point) $T[i, j]$ and β is the vertical distance. The key point is that this calculation can be automatically performed by dedicated texturing hardware outside multiprocessors. Interpolation performed by this specialised hardware is done faster than performing it in software. Apart from freeing CUDA's multiprocessors to perform other tasks, it also decreases the size of the computational kernel by occupying fewer registers, which is quite important for the maximum achievable warp occupancy.

3   Design and Implementation

The application we developed targets nVIDIA's Fermi architecture (compute capability 2.0). Double-precision arithmetic was preferred over single-precision in order to make the algorithm more robust to the order in which arithmetic operations take place (a quite common problem in numerical analysis) and reduce round-off errors. The application adapts 2D meshes using the vertex smoothing scheme proposed by Pain et al. [7]. By performing vertex smoothing, node connectivity remains constant and there is no need to re-colour the mesh after every iteration. Graph colouring was implemented using a single-threaded and greedy colouring algorithm, called First Fit Colouring [1], which runs adequately fast and colours the mesh with satisfactorily few colours (7-8 on average).

The mesh is represented using two arrays: an array V of vertices (a vertex is simply a pair of coordinates) and an array C of cavities. A cavity in the i-th

Listing 1.1. Setting up texture memory.

```
#if defined(USE_TEXTURE_MEMORY)
  texture<float4 , 2, cudaReadModeElementType> metricTex;

  cudaChannelFormatDesc channelDesc =
    cudaCreateChannelDesc(32, 32, 32, 32, cudaChannelFormatKindFloat);
  cudaMallocArray(&cudaMetricField , &channelDesc , textDim , textDim);
  cudaMemcpyToArray(cudaMetricField , 0, 0, hostMetricField ,
    textDim * textDim * sizeof(float4), cudaMemcpyHostToDevice);

  metricTex.normalized = true;
  metricTex.filterMode = cudaFilterModeLinear;
  metricTex.addressMode[0] = cudaAddressModeClamp;
  metricTex.addressMode[1] = cudaAddressModeClamp;

  cudaBindTextureToArray(metricTex, cudaMetricField , channelDesc);
#else
  cudaMalloc((void **) &cudaMetricField ,
    metricDim * metricDim * 4 * sizeof(float));
  cudaMemcpy(cudaMetricValues , hostMetricValues ,
    metricDim * metricDim * 4 * sizeof(float), cudaMemcpyHostToDevice);
#endif
```

position of C is the cavity defined by vertex V[i], i.e. the cavity in which V[i] is the central vertex, and (this cavity) is in turn an array containing the indices in array V of all vertices which are adjacent to V[i]. E.g. if vertex V[0] is connected to vertices V[3], V[5], V[10] and V[12], then C[0] is the cavity in which V[0] is the free vertex and C[0] = 3, 5, 10, 12. There is also a simple representation of independent sets, each one being an array containing the indices of all vertices belonging to that set.

As was described in Section 2, the metric tensor field is discretised vertex-wise, so it could be represented by extending the definition of a vertex to include the metric tensor value associated with that vertex, in addition to the vertex's coordinates. However, looking at the smoothing algorithm, it becomes apparent that the middle of an edge (where the metric value is needed) will most probably not coincide with a discretisation point, so the metric value has to be found by interpolating the values from nearby discretisation points. If the field is stored as a texture, interpolation can be done automatically by CUDA's texturing units. Linear filtering is quite suitable and yields good interpolation results, even when the metric tensor field has a lot of discontinuities.

More insight into the role of the metric tensor field reveals that it is just an estimation or indication about the desired orientation of an element and we have observed that it does not have to be as accurate as possible. For this reason, single-precision representation (double-precision is not supported for textures) is more than enough and the data structure used to store it can be a 2D array, organised using the GPU's blocked texture storage layout. In order to convert the unstructured representation to an array, we super-sample the initial field with adequate resolution and store these samples in an array. The initial, auto-generated mesh is anyway quite uniform, i.e. elements tend to be equilateral

Listing 1.2. Accessing a metric tensor field value.

```
#if defined (USE_TEXTURE_MEMORY)
    float4 floatMetric = tex2D(metricTex, iCoord, jCoord);
#else
    double iIndex = jCoord * metricDim, jIndex = iCoord * metricDim;
    int i = floor(((metricDim - 1) / metricDim) * iIndex);
    int j = floor(((metricDim - 1) / metricDim) * jIndex);
    iIndex -= i; jIndex -= j;

    if(i == metricDim - 1)       // top or bottom boundary
      metric = cudaMetricValues[metricDim*(metricDim-1) + j] * (1-jIndex) +
               cudaMetricValues[metricDim*(metricDim-1) + (j+1)] * jIndex;
    else if(j == metricDim - 1)  // left or right boundary
      metric = cudaMetricValues[(i+1)*metricDim - 1] * (1-iIndex) +
               cudaMetricValues[(i+2)*metricDim - 1] * iIndex;
    else
      metric = cudaMetricValues[i*metricDim + j] * (1-iIndex)*(1-jIndex) +
               cudaMetricValues[i*metricDim + (j+1)] * (1-iIndex)*jIndex +
               cudaMetricValues[(i+1)*metricDim + j] * iIndex*(1-jIndex) +
               cudaMetricValues[(i+1)*metricDim + (j+1)] * iIndex*jIndex;
#endif
```

triangles and vertices are equally spaced from each other, a state which is not very different from a 2D-array representation.

In 2D, the metric tensor field is a 2×2 matrix, so it can be represented as a 4-element vector of single-precision floating-point values. Copying data from host to device as textures is demonstrated in Listing 1.1. Retrieving the value of the metric tensor field at any point in the mesh is just a texture fetch, as can be seen in Listing 1.2, which also contrasts the addressing and interpolation overhead we avoid.

Subsequent adaptation attempts will have to use the unstructured representation of the field. After adapting the mesh, we re-solve the PDE and make new error estimations, which lead to a new metric tensor field, discretised at the nodes of an anisotropic, non-uniform mesh. In this case, different resolution will be needed in different areas of the mesh and we have to follow the unstructured approach. This does not reduce the significance of using texturing hardware. The first adaptation attempt is the one which really needs to be sped-up, as it needs $\Theta(number\_of\_vertices)$ iterations to converge, inducing the most extensive changes to the mesh. After that, the mesh will have, more or less, acquired its final "shape", so subsequent attempts will only need a few iterations to improve it.

In devices of CUDA's compute capability 2.0 and above, the on-chip memory is used both as shared memory and L1 cache. The unstructured nature of anisotropic mesh adaptivity has not allowed us to use shared memory explicitly. On the other hand, a hardware-managed L1 cache exploits data locality more conveniently. Configuring the on-chip memory as 16KB of shared memory with 48KB of L1 cache can be done [3] by preceding the kernel invocation with a statement like:

$cudaFuncSetCacheConfig(optimizationKernel, cudaFuncCachePreferL1);$.

Listing 1.3. OpenMP execution

```
for(int indSetNo = 0; indSetNo < numberOfSets; indSetNo++) {
  vertexID iSet[] = indSets[indSetNo];

#pragma omp parallel for private(setIterator)
  for(int setIterator = 0; setIterator < verticesInSet; setIterator++) {
    cavityID cavity = iSet[setIterator];
    if(!meshCavities[cavity].isOnBoundary())
      newCoords = relocateInnerVertex(...);
    else
      newCoords = relocateOuterVertex(...);
  }
}
```

Listing 1.4. CUDA execution

```
for(int indSetNo = 0; indSetNo < numberOfSets; indSetNo++) {
  dim3 numBlocks(ceil((double) verticesInSet / threadsPerBlock));
  kernel<<<numBlocks, threadsPerBlock>>>(indSets[indSetNo]);
  cudaThreadSynchronize();
}

__device__ void kernel(IndependentSet iSet) {
  int vertex = blockIdx.x * blockDim.x + threadIdx.x;
  if(vertex < verticesInSet) {
    cavityID cavity = iSet[vertex];
    if(!meshCavities[cavity].isOnBoundary())
      newCoords = relocateInnerVertex(...);
    else
      newCoords = relocateOuterVertex(...);
  }
}
```

Parallel execution is based on the independent sets. The way cavities are
assigned to OpenMP resp. CUDA threads can be seen in Listing 1.3 resp. Listing 1.4. Recall from the description of the vertex smoothing algorithm that
boundary vertices are smoothed using a variation of the main algorithm. In order
to avoid thread divergence, which is problematic for a CUDA kernel, boundary
vertices are put into dedicated independent sets, so that a dedicated set contains
vertices of the same "kind".

4   Experimental Evaluation

All experiments were run on node CX1 of Imperial College's HPC supercomputer, hosting two Intel "Gulftown" six-core Xeon X5650 CPUs (2.8GHz), 24GB
RAM and a nVIDIA Tesla C2050 graphics board. The operating system was Red
Hat Enterprise Linux Client release 5.5 running Linux kernel 2.6.18. CPU code
was compiled with GCC version 4.1.2 giving the -O3 flag, whereas for GPU code
we used CUDA SDK 3.1 and CUDA compilation tools, release 3.1, V0.2.1221,
with the -O2 flag. Experiments were done using the nVIDIA Forceware driver,
version 260.19.29.

We have compared the running time between a single-threaded execution, a 12-threaded OpenMP execution and CUDA execution with and without engaging texturing hardware. Timing results include the time it takes to copy data between host and device, but no measurement includes the time it takes to read in the unstructured grid from the disk, construct the mesh, colour it or write back the results to the disk, because these tasks are always performed in a single-threaded fashion on the host side. On the other hand, the time to copy data between host and device is trivial because these transfers take place only twice during an execution (copying the initial mesh to the device at the beginning and copying the adapted mesh back to the host at the end) and when there are thousands of iterations this time is amortised.

The optimisation kernel occupies 59 registers in the non-textured version and 51 registers in the textured one. Using the Occupancy Calculator [2] and experimental measurements, it was found that the best CUDA execution configuration is 32 threads per block, which gives an occupancy of 33.3%. In both cases, occupancy is very low, suggesting the future optimisation of breaking down the kernel into smaller parts.

Table 1 presents the amount of time each version of the code needs to perform 1,000 iterations over various meshes. Figure 4 shows the relative speedup between these versions. The 12-threaded OpenMP version is steadily ~ 10 times faster than the serial code. The non-textured CUDA version runs on average 24 times faster than the serial code (peaking at 42 times) and 2.5 times faster than the OpenMP version (peaking at 4.7 times). Enabling texturing support, the CUDA code runs on average twice as fast as its non-textured counterpart (peaking at 6.2 times). Compared to the host side, it runs on average 60 times faster than the serial CPU version (peaking at 148 times) and 6 times faster than the OpenMP version (peaking at 15 times).

The high performance divergence and the unpredictable (to some extent) behaviour of a CUDA implementation come as a consequence of the highly unstructured nature of the problem. We expect substantial differences from one mesh to another in terms of achievable data locality, partitioning of data in global

Table 1. Comparison of the execution time in seconds between the serial, the 12-threaded OpenMP, the non-textured CUDA and the textured CUDA versions, performing 1,000 iterations over meshes of variable size

Number of mesh vertices	CPU 1 Thread	CPU 12 Threads	CUDA (no texturing)	CUDA (texturing)
25,472	20.76	2.156	1.243	0.610
56,878	46.82	4.676	2.281	1.163
157,673	144.3	15.27	5.902	3.172
402,849	510.3	52.36	15.03	8.249
1,002,001	777.6	75.34	28.17	5.664
4,004,001	3,203	318.0	134.2	21.64
5,654,659	8,921	983.9	210.0	114.2

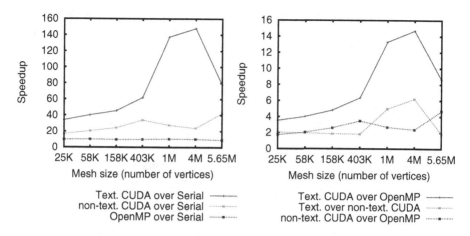

Fig. 4. Relative speedup between the serial, the OpenMP, the non-textured CUDA and the textured CUDA versions

memory and degree of coalescence of memory accesses. This uncertainty could be mitigated by implementing a two-level mesh partitioning scheme: one topological partitioning of the mesh into mini-partitions, so that the whole mini-partition fits in the on-chip memory, and a second logical partitioning (graph colouring) inside each mini-partition for the purpose of correct parallel execution.

5   Conclusions and Future Work

The aim of this project was to determine the extent to which a Fermi-based GPU can still be efficient when it has to deal with unstructured problems. The experimental results show that the capabilities of this architecture extend well beyond the borders of structured applications, which are the norm in evaluating and demonstrating processing hardware. A single Tesla C2050 board outperformed 12 Nehalem cores by many times and there is still room for improvement, as is suggested by the low warp occupancy and the scope for improved data locality.

When it comes to texturing hardware, it was shown that it offers substantial amounts of computational power and can more than double performance in problems with appropriate characteristics, like the metric tensor field of anisotropic mesh adaptivity. The assistance of texturing hardware in the 3D version of the problem (a 3D implementation is planned for future work) is expected to be even more valuable. In a 3D metric tensor field we have to interpolate the values from the 8 nearest points and doing so requires (compared to 2D problems) double the data volume to be fetched from memory and three times more arithmetic.

Apart from the aforementioned optimisations, this project leaves many other topics open to further study. Perhaps, the most interesting direction is the implementation of a much heavier and of higher quality smoothing algorithm, based on differential methods, which is called optimisation-based smoothing [5]. Additionally, vertex smoothing is usually combined with other adaptive methods, such as regular refinement, edge flipping and edge collapsing [6,7]. The development of a mesh improving application which manipulates all these techniques and the assessment of its performance on CUDA (and possibly other high-performance architectures through OpenCL) is in progress.

References

1. Al-Omari, H., Sabri, K.E.: New graph coloring algorithms. Journal of Mathematics and Statistics (2006)
2. nVIDIA Corporation: CUDA GPU Occupancy Calculator,
   http://developer.download.nvidia.com/compute/cuda/
   CUDA_Occupancy_calculator.xls
3. nVIDIA Corporation: nVIDIA CUDA Programming Guide, Version 3.1. Tech. rep. (2010)
4. nVIDIA Corporation: nVIDIA CUDA Reference Manual, Version 3.1. Tech. rep. (2010)
5. Freitag, L., Jones, M., Plassmann, P.: An Efficient Parallel Algorithm for Mesh Smoothing. In: International Meshing Roundtable, pp. 47–58 (1995)
6. Freitag, L.F., Jones, M.T., Plassmann, P.E.: The Scalability Of Mesh Improvement Algorithms. In: Ima Volumes in Mathematics and its Applications, pp. 185–212. Springer, Heidelberg (1998)
7. Pain, C.C., Umpleby, A.P., de Oliveira, C.R.E., Goddard, A.J.H.: Tetrahedral mesh optimisation and adaptivity for steady-state and transient finite element calculations. Computer Methods in Applied Mechanics and Engineering 190(29-30), 3771–3796 (2001),
   http://www.sciencedirect.com/science/article/B6V29-42RMN8G-5/2/
   8c901884db5fd4b67d19b63fb9691284
8. Piggott, M.D., Farrell, P.E., Wilson, C.R., Gorman, G.J., Pain, C.C.: Anisotropic mesh adaptivity for multi-scale ocean modelling. Philosophical Transactions of the Royal Society A: Mathematical, Physical and Engineering Sciences 367(1907), 4591–4611 (2009),
   http://rsta.royalsocietypublishing.org/content/367/1907/4591.abstract
9. Rokos, G.: ISO Thesis: Study of Anisotropic Mesh Adaptivity and its Parallel Execution. Imperial College, London (2010)
10. Vasilevskii, Y., Lipnikov, K.: An adaptive algorithm for quasioptimal mesh generation. Computational Mathematics and Mathematical Physics 39(9), 1468–1486 (1999)

Introduction

Wolfgang Karl, Samuel Thibault, Stanimire Tomov, and Taisuke Boku

Topic chairs

The recent years have seen great research interest in exploiting GPUs and accelerators for computations, as shown by the latest TOP500 editions, whose very top entries are fully based on their use. Their potential computation power and energy consumption efficiency are appealing, but programming them however reveals to be very challenging, as not only task offloading and data transfer issues show up, but programming paradigms themselves appear to need reconsideration. Fully taping into this new kind of computation resource thus stands out as an open issue, particularly when conjointly using regular CPUs and several accelerator simultaneously. This is why we welcome this year the opening of a new "GPU and Accelerators Computing" topic along the collection of Euro-Par topics. The focus of this topic covers all areas related to accelerators: architecture, languages, compilers, libraries, runtime, debugging and profiling tools, algorithms, applications, etc.

The topic attracted numerous submissions, among which 4 papers were selected for publication. They cover various aspects, such as application-side optimizations, multiple GPU data transfer management, as well as low-level performance analysis.

In "Model-Driven Tile Size Selection for DOACROSS Loops on GPUs", Peng Di and Jingling Xue (from the University of New South Wales, Australia) propose a performance model for tiled and skewed SOR-loops on NVIDIA GPUs, and provide an evaluation of the model accuracy. The model is then used to automatically tune the tile size in a very reduced amount of time, compared to performing measurements of actual runs.

Bertil Schmidt, Hans Aribowo and Hoang-Vu Dang (from the Nanyang Technological University, Singapore); propose a new hybrid format for sparse matrices in "Iterative Sparse Matrix-Vector Multiplication for Integer Factorization on GPUs". After presenting various existing sparse formats, they analyze the shape of the matrices derived from Number Field Sieve problems, and consequently define a new format composed of several slices encoded in a few well-known formats. They present a dual gpu implementation of sparse matrix-vector multiplication (SpMV) with overlapped communications.

"Lessons Learned from Exploring the Backtracking Paradigm on the GPU", from John Jenkins, Isha Arkatkar, John D. Owens, Alok Choudhary, and Nagiza Samatova (from the North Carolina State University, the University of California, Davis, and the Northwestern University, USA), describes how very low performance can get on GPUs when implementing backtracking paradigms.

E. Jeannot, R. Namyst, and J. Roman (Eds.): Euro-Par 2011, LNCS 6853, Part II, pp. 399–400, 2011.

It provides a detailed analysis of the reasons for this and how to optimize at best. It thus provides interesting guidelines for such class of applications.

Last but not least, in "Automatic OpenCL Device Characterization: Guiding Optimized Kernel Design", Peter Thoman (from the University of Innsbruck, Austria) presents CLbench, a suite of micro-benchmarks which aims at characterizing the performance of implementations of the OpenCL standard, including classic arithmetic and memory performance, but also branching behavior and runtime overhead. The results being provided for 3 OpenCL implementations over 7 hardware/software configuration, they can be used to guide optimizations.

Model-Driven Tile Size Selection for DOACROSS Loops on GPUs*

Peng Di and Jingling Xue

Programming Languages and Compilers Group, School of Computer Science and
Engineering, University of New South Wales, Sydney, Australia

Abstract. DOALL loops are tiled to exploit DOALL parallelism and
data locality on GPUs. In contrast, due to loop-carried dependences,
DOACROSS loops must be skewed first in order to make tiling legal and
exploit wavefront parallelism across the tiles and within a tile. Thus, tile
size selection, which is performance-critical, becomes more complex for
DOACROSS loops than DOALL loops on GPUs. This paper presents a
model-driven approach to automating this process. Validation using 1D,
2D and 3D SOR solvers shows that our framework can find the tile sizes
for these representative DOACROSS loops to achieve performances close
to the best observed for a range of problem sizes tested.

1 Introduction

GPGPUs have become one of the most powerful and popular platforms to exploit
fine-grain parallelism in high performance computing. Recent research on developing programming and compiler techniques for GPUs focuses on (among others) general programming principles [5,9,14], cost modeling and analysis [1,15,3],
automatic code generation [2,11], and performance tuning and optimization
[4,6,12,19]. However, these research efforts are almost exclusively limited to
DOALL loops. In practice, DOACROSS loops play an important role in many
scientific and engineering applications, including PDE solvers [13], efficient preconditioners [7] and robust smoothers [8]. Presently, Pluto [2] seems to be the
only framework that can map sequential DOACROSS loops to CUDA code automatically for NVIDIA GPUs. This is done by applying loop skewing and tiling
with user-supplied tile sizes (for a user-declared grid of thread blocks of threads).

DOALL loops are tiled to exploit DOALL parallelism and data locality on
GPUs. Unlike DOALL loops, DOACROSS loops must be skewed first to ensure that the subsequent tiling transformation preserves the loop-carried dependences. Furthermore, performing skewing and tiling allows wavefront parallelism
to be exploited both across the tiles and within a tile. Tile size selection, which is
performance-critical on GPUs, are more complex for DOACROSS than DOALL
loops due to parallelism-inhibiting loop-carried dependences and more complex
interactions among the GPU architectural constraints. Thus, it is not practical

---

* This research is supported by an Australian Research Council Grant DP110104628.

E. Jeannot, R. Namyst, and J. Roman (Eds.): Euro-Par 2011, LNCS 6853, Part II, pp. 401–412, 2011.
© Springer-Verlag Berlin Heidelberg 2011

to rely on the user to pick the right tile sizes to optimize code through improving processor utilization and reducing synchronization overhead. Existing tile size techniques proposed for caches in CPU architectures do not apply [10,20].

This paper makes the following contributions:

- We present (for the first time) a model for estimating the execution times of tiled DOACROSS loops running on GPUs (Section 3);
- We introduce a model-driven framework to automate tile size selection for tiled DOACROSS loops running on GPUs (Section 4);
- We evaluate the accuracy of our model using representative 1D, 2D and 3D SOR solvers and show that the tile sizes selected lead to the performances close to the best observed for a range of problem sizes tested (Section 5).

2   Parallelization of DOACROSS Loops on GPUs

We describe a scheme for mapping sequential DOACROSS loops to CUDA code on GPUs. Our illustrating example is a 1D SOR-like solver. This scheme is the same as that supported by Pluto [2] except tiles are mapped to thread blocks in a different way in order to achieve better load balance.

```
for(i1=1;i1<=I1;i1++)
  for(i2=1;i2<=I2;i2++)
    A[i2]=(A[i2-1]+A[i2]+A[i2+1])/3;
```

Fig. 1. Sequential loop nest for the 1D SOR solver

Loop Transformations. In Pluto, parallelizing an n-dimensional DOACROSS loop nest L consists of mapping it into a 2n-dimensional loop nest as follows:

$$\rho : \mathbb{Z}^n \mapsto \mathbb{Z}^{2n}, \rho(i) = \begin{pmatrix} t \\ e \end{pmatrix} = (t_1, \ldots, t_n, e_1, \ldots, e_n)^T = \begin{pmatrix} \mathcal{W} \lfloor \mathcal{TS}(i) \rfloor \\ \mathcal{WS}(i) \end{pmatrix} \quad (1)$$

$$\mathcal{T} = \begin{bmatrix} m_1 & 0 & \ldots & 0 \\ 0 & m_2 & \ldots & 0 \\ \vdots & \vdots & \ddots & \vdots \\ 0 & 0 & \ldots & m_n \end{bmatrix}_{n \times n}^{-1}, \quad \mathcal{W} = \begin{bmatrix} 1 & 1 & 1 & \ldots & 1 & 1 \\ 0 & 1 & 0 & \ldots & 0 & 0 \\ \vdots & \vdots & \vdots & \ddots & \vdots & \vdots \\ 0 & 0 & 0 & \ldots & 0 & 1 \end{bmatrix}_{n \times n} \quad (2)$$

The mapping process for the 1D SOR solver in Figure 1 is illustrated in Figure 2. The mapping ρ [18] is realized by composing a loop skewing $\mathcal{S}$, a loop tiling $\mathcal{T}$ and another loop skewing $\mathcal{W}$. First, the iteration space of L is skewed by a unimodular transformation $\mathcal{S} \in \mathbb{Z}^{n \times n}$. Second, the skewed iteration space is tiled into n-dimensional rectangles of size $m_1 \times \cdots \times m_n$ by $\mathcal{T}$. $\mathcal{S}$ is chosen to guarantee the legality of tiling so that all loop-carried dependences in L are preserved [18]. At this point, a 2n-dimensional loop nest is created such that the first n loops (called *tile loops*) enumerate the tiles and the inner n loops (called *element loops*) enumerate the iterations within a tile. Finally, another skewing

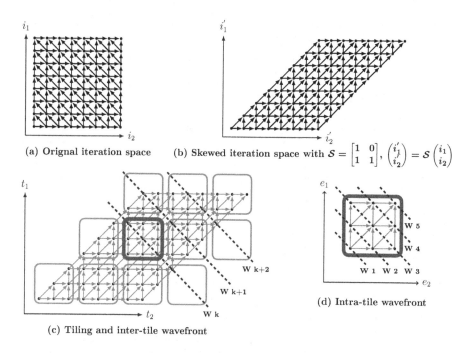

(a) Orignal iteration space   (b) Skewed iteration space with $\mathcal{S} = \begin{bmatrix} 1 & 0 \\ 1 & 1 \end{bmatrix}$, $\begin{pmatrix} i'_1 \\ i'_2 \end{pmatrix} = \mathcal{S} \begin{pmatrix} i_1 \\ i_2 \end{pmatrix}$

(c) Tiling and inter-tile wavefront

(d) Intra-tile wavefront

Fig. 2. Exploiting wavefront parallelism for 1D SOR on GPUs

transformation $\mathcal{W}$ is applied to the iteration spaces of both sets of loop nests to expose wavefront parallelism across the tiles and within a tile. In either loop nest, the first loop is sequential and the remaining $n - 1$ loops are DOALL. We will speak of inter-tile wavefronts and intra-tile wavefronts (as shown in Figure 2).

Mapping to GPUs. A NVIDIA GPU consists of a number of *streaming multiprocessors (SMs)*, each of which contains a number of processor cores called *streaming processors (SPs)*. All SPs in one SM share a local memory and a set of registers. GPU programming is enabled through CUDA. A kernel is executed by a grid of threads organized as *thread blocks* (known as a *thread organization*). A thread block is scheduled to execute on any one of the SMs (as a whole). The threads in a block are partitioned into 32-thread *warps*, which are units of execution (on the SPs of the SM to which the block is mapped). The threads in one block can synchronize through `syncthreads` and communicate through shared memory. The inter-block synchronization is not directly supported.

Figure 3 gives the CUDA code for the 1D SOR solver parallelized as shown in Figure 2. Tiles are mapped to thread blocks and individual loop iterations in a tile are mapped to the threads in a block. All inter-tile wavefronts are executed sequentially to satisfy inter-tile (or inter-block) dependences. Hence, the `syncblocks` macro as introduced in Pluto at ④. In Pluto, the tiles in an $(n - 1)$-dimensional inter-tile wavefront are distributed over a 2D grid of thread blocks of size `gridDim.x` × `gridDim.y` cyclically along two of the $n-1$ dimensions

```
// inter-tile loop nest
for(t1=Lt1;t1<=Ut1;t1++){
   for(t2=Lt2(t1)+blockIdx.x;t2<=Ut2(t2);t2+=gridDim.x){
      // intra-tile loop nest
      Code for shared memory coalesced loads
      __syncthreads(); // ① barrier for the loads       DOALL
      for(e1=Le1(t1,t2);e1<=Ue1(t1,t2);e1++){
         for(e2=Le2(t1,t2,e1)+threadIdx.x;e2<=Ue2(t1,t2,e1);e2+=blockDim.x){
            i2=h(t1,t2,e1,e2);
            A[i2]=(A[i2-1]+A[i2]+A[i2+1])/3;
         }
         __syncthreads(); // ② barrier for each intra-tile wavefront
      }
      Code for shared memory coalesced stores
      __syncthreads(); // ③ barrier for the stores
   }
   __syncblocks(); // ④ barrier for each inter-tile wavefront
}
```

Fig. 3. CUDA code for 1D SOR on GPUs

of the wavefront. This can cause load imbalance for large tiles since a wavefront has irregular boundaries. In this paper, the tile coordinates in such a wavefront are "linearized" much like how the subscripts of a multi-dimensional array are. Then the tiles are mapped to a 1D grid of thread blocks of size gridDim.x cyclically to achieve better load balance (with gridDim.y=1 always). As a result, all thread blocks in an inter-tile wavefront can be potentially executed in parallel but the tiles within a block are always executed sequentially.

The loop iterations in a tile are distributed as in Pluto to a 3D thread block of size blockDim.x × blockDim.y × blockDim.z cyclically. To improve memory coalescing, all data read by a tile are first loaded from device memory to shared memory at ① and all those written in a tile are stored back to device memory at ③. Like inter-tile wavefronts, all intra-tile wavefronts are executed sequentially. Hence, the syncthreads instruction at ②. The iterations in an intra-tile wavefront that are assigned to different threads can execute in parallel.

3   Execution Time Modeling

We parameterise an execution time model for a tiled DOACROSS loop nest in order to automate tile size selection. Initially, we assume that all tiles are full. We consider first intra-tile execution (Section 3.1) and then inter-tile execution (Section 3.2). In Section 3.3, we estimate the parameters used. In Section 3.4, we discuss briefly how to mitigate the effects of border tiles on performance.

3.1   Intra-tile Execution

This section focuses on estimating the execution time, $T_{\mathcal{TILE}}$, for a single (full) tile, denoted $\mathcal{TILE}$. As shown in (1) and Figures 2 and 3, the loop iterations in a tile are indexed by $(e_1, \ldots, e_n)$, where e_1 enumerates all intra-wavefronts within the tile. As illustrated in Figure 2, $T_{\mathcal{TILE}}$, which can be broken down into the time on loading the input data at ①, the time on executing the tile, and the time on storing the results back ③, is approximated by:

$$T_{\mathcal{TILE}} = \sum_{e1=L_{e1}}^{U_{e1}} T_{e1} + T_{mem} + 2\sigma_{thd} \tag{3}$$

The first term $\sum_{e1=L_{e1}}^{U_{e1}} T_{e1}$ is the computation cost of $\mathcal{TILE}$ estimated as a sum of the execution times T_{e1} of all its intra-tile wavefronts with $e1$ ranging over these wavefronts starting from the smallest given by the lower bound L_{e1} of loop $e1$ to the largest given by the upper bound U_{e1} of loop $e1$ along dimension $e1$. The second term T_{mem} denotes the memory latency consumed by the memory accesses at the code before ① and the code before ③. The last term $2\sigma_{thd}$ denotes the overhead of the two syncthreads at ① and ③, where σ_{thd} is dependent on the number of threads used, i.e., blockDim.x × blockDim.y × blockDim.z.

The execution time T_{e1} of the intra-tile wavefront indexed by $e1$ is given by:

$$T_{e1} = \alpha \times G_{e1} + \beta \times H_{e1} + \sigma_{thd} \tag{4}$$

GPUs execute instructions with warps as units of execution and hide memory latency through interleaving of thread blocks. In the scheme shown in Figure 3, the warps are never idle when executing a wavefront as all memory accesses happen before and after the execution of $\mathcal{TILE}$. Thus, the first term represents the workload for computing the wavefront indexed by e_1, which is estimated to be proportional to G_{e1}, the number of 32-thread warps executed at the wavefront. In addition, the first term implicitly considers the effects of bank conflicts on the execution time of $\mathcal{TILE}$. However, the same G_{e1} may result when the number of loop iterations, H_{e1}, in the wavefront indexed by $e1$ varies (due to division by 32). To accommodate its impact on performance, T_{e1} is fine-tuned by including the second term $\beta \times H_{e1}$, which attempts to differentiate the effects of varying H_{e1} values on performance. Note that G_{e1} and H_{e1} may vary from wavefront to wavefront as shown in Figure 2(d). Given an intra-tile wavefront, both can be precisely calculated. The last term σ_{thd} is the overhead of syncthreads at ②.

By substituting T_{e1}^i in (4) into (3), we obtain:

$$T_{\mathcal{TILE}} = \sum_{e1=L_{e1}}^{U_{e1}} (\alpha \times G_{e1} + \beta \times H_{e1} + \sigma_{thd}) + T_{mem} + 2\sigma_{thd} \tag{5}$$

which is illustrated graphically in Figure 4. As highlighted, G_{e1} varies across the wavefronts with less work being done at the beginning and end of the computation process for $\mathcal{TILE}$ when its wavefronts are executed.

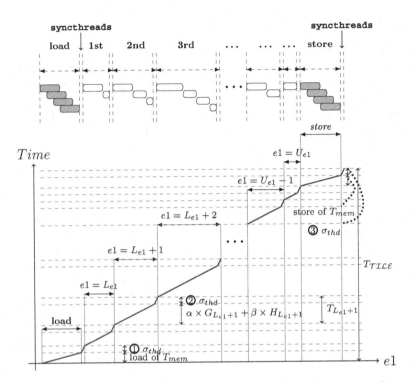

Fig. 4. Execution of the N_{e1} wavefronts of $\mathcal{TILE}$ along dimension e_1 according to (5)

The memory access latency T_{mem} can be estimated by:

$$T_{mem} = \gamma \times N_{mem} \qquad (6)$$

where N_{mem} denotes the number of loads and stores made in $\mathcal{TILE}$. Note that N_{mem} is not related to memory coalescing since that would make its estimation dependent on the actual data layout at run time. This simple approximation seems to be adequate as T_{mem} is not a dominant term in (3) for the following reasons combined. First, DOACROSS loops are usually not bandwidth-bound. Second, optimal tile sizes found are large in order to exploit two levels of wavefront parallelism. Finally, the memory accesses performed by warps can overlap. We will return to this issue briefly at the end of Section 5.

By substituting T_{mem} given in (6) into (5), simplifying and letting

$$L_{\mathcal{TILE}} = m_1 \times \cdots \times m_n = \sum_{e1=L_{e1}}^{U_{e1}} H_{e1} \qquad (7)$$

we obtain the following estimated execution time of $\mathcal{TILE}$:

$$T_{\mathcal{TILE}} = \alpha \times \sum_{e1=L_{e1}}^{U_{e1}} G_{e1} + (U_{e1} - L_{e1} + 3) \times \sigma_{thd} + \beta \times L_{\mathcal{TILE}} + \gamma \times N_{mem} \quad (8)$$

with $U_{e1} - L_{e1} + 1$ syncthreads instructions executed at ② inside the wavefront, one at ① and one at ③, as shown in Figure 3.

3.2   Inter-tile Execution

A DOACROSS loop nest is parallelized into a CUDA kernel. The execution time, T_{total}, for the kernel, i.e., for its inter-tile wavefronts is estimated by:

$$T_{total} = \sum_{t1=L_{t1}}^{U_{t1}} T_{t1} + \sigma_{ker} \tag{9}$$

The first term is the computation cost of all tiles in the kernel estimated as a sum of the execution times T_{t1} of all its inter-tile wavefronts with $t1$ starting from the lower bound L_{t1} of loop $t1$ to the upper bound U_{t1} of loop $t1$ along dimension $t1$. The second term σ_{ker} is the kernel startup cost.

Thus, $U_{t1} - L_{t1} + 1$ is the number of tiles contained in the inter-tile wavefront indexed by $t1$. If all P SMs execute simultaneously up to B thread blocks each, then the number of tiles, denoted I_{t1}, contained in a thread block is:

$$I_{t1} = \lceil \frac{U_{t1} - L_{t1} + 1}{B \times P} \rceil = \lceil \frac{U_{t1} - L_{t1} + 1}{\text{gridDim.x}} \rceil \tag{10}$$

where B is decided by the GPU architectural constraints and kernel code according to the CUDA programming guide as demonstrated in Table 1.

The execution time T_{t1} is determined by the slowest among the P SMs with the other SMs idle waiting at the syncblocks macro at ④. As a result, we have:

$$T_{t1} = \sum_{i=1}^{I_{t1}} T_{t1}^i + \sigma_{blk} \tag{11}$$

$$T_{t1}^i = \begin{cases} B \times T_{TILE} & 1 \leq i \leq I_{t1} - 1 \\ \lceil \frac{(U_{t1} - L_{t1} + 1)\%(B \times P)}{P} \rceil \times T_{TILE} & i = I_{t1} \end{cases} \tag{12}$$

where T_{t1}^i is the execution time of the i-th tiles in all B thread blocks by the slowest SM and σ_{blk} is the overhead of syncblocks at ④ (to be measured below).

3.3   Parameter Estimation

We determine the six parameters used in T_{total} given in (9) for a tiled loop nest L: σ_{ker}, σ_{thd}, σ_{blk}, α, β and γ. We do so for a given thread organization (determined by gridDim and blockDim) so that its tile size selection can be automated (Section 4). For NVIDIA GPUs, there are at most $16B_{max}$ different thread organizations because (1) there are B_{max} different 1D grid layouts with gridDim.x $= B \times P$, where $B \leq B_{max} \leq B_{SM} = 8$ as shown in Table 1, and (2) the number of threads per block, i.e., blockDim.x $\times$ blockDim.y $\times$ blockDim.z is one of the 16 possibilities contained in $\{32, 64, \ldots, 512\}$.

Table 1. Determining B for an NVIDIA Tesla C1060 GPU. An item in Column 1 that depends on hardware, kernel code or both is indicated with an H, S or B appropriately.

Description	Name
Warp Size (H)	$W = 32$
Max Number of Active Warps per SM (H)	$W_{SM} = 32$
Max Number of Active Threads per SM (H)	$T_{SM} = 1024$
Max Number of Active Blocks per SM (H)	$B_{SM} = 8$
Shared Memory per SM (H)	$S_{SM} = 16KB$
Number of 32-bit Registers per SM (H)	$R_{SM} = 16K$
Threads per Thread Block (S)	K
Register Usage per Thread Block (S)	R_{TB}
Shared Memory Usage per Thread Block (S)	S_{TB}
Warps per Thread Block (B)	$W_{TB} = \lceil \frac{K}{W} \rceil$
Thread Blocks Limited by Warps (B)	$B_W = \min(B_{SM}, \lfloor \frac{W_{SM}}{W_{TB}} \rfloor)$
Thread Blocks Limited by Registers (B)	$B_R = \lfloor \frac{R_{SM}}{R_{TB}} \rfloor$
Thread Blocks Limited by Shared Memory (B)	$B_S = \lfloor \frac{S_{SM}}{S_{TB}} \rfloor$
Thread Blocks (B)	$B = \min(B_W, B_R, B_S)$

Architectural Parameters: σ_{ker}, σ_{thd} and σ_{blk}. These overheads are small (relative to the execution time of a loop nest L) and are measured for a GPU architecture as follows. First of all, σ_{ker} is the startup overhead of the kernel for L, which can be obtained through running an empty version of the kernel (with the computations in L removed) for a given thread organization. In fact, as $\sigma_{ker} \ll T_{total}$, treating it as a small constant for all thread organizations does not affect in practical terms how the relative performances of L are ranked for all combinations of thread organizations and tile sizes used. As for syncthreads executed at ①, ② and ③, it is lightweight on NVIDIA GPUs. Its overhead σ_{thd} depends on the number of threads per block, i.e., blockDim.x × blockDim.y × blockDim.z and is measured as in [16]. There are only 16 cases to consider as blockDim.x × blockDim.y × blockDim.z is a multiple of 32 ranging from 32 to 512. Finally, the syncblocks macro is invoked at the end of each inter-tile wavefront at ④. Its overhead σ_{blk}, which is higher than σ_{thd}, depends mainly on the number of thread blocks contained in an inter-tile wavefront, i.e., gridDim.x $= B \times P$. The effects of different blockDim.x × blockDim.y × blockDim.z values on σ_{blk} are negligible. As $B \leq B_{max} \leq B_{SM} = 8$, syncblocks is measured as in [17] for a few, i.e., up to B_{max} different gridDim.x values.

Program-Dependent Parameters: α, β and γ. Once the values of σ_{ker}, σ_{thd} and σ_{blk} are determined, the given loop nest L is simplified to possess one inter-tile wavefront with exactly $B \times P$ thread blocks consisting of only full tiles. This ensures that all P SMs have exactly the same workload so that these three program-dependent parameters can be accurately measured.

The three parameters are found for each of up to $16B_{max}$ different thread organizations as mentioned earlier (where $B_{max} \leq 8$). In each case, the simplified loop nest L is executed for a total of n times, each with a different tile size. Let T_i be the execution time corresponding to the tile size S_i used. Given a tile size S_i, all parameters in T_{total} except α, β and γ are now known. We can find the values of α, β and γ by performing a linear curve fitting using the least-square method for T_{total} with respect to the n execution times, $T_1, \ldots, T_n$, obtained.

1	Compute the register usage per thread, R_T, using any tile size and thread organization.
2	**for** each tile size $m = (m_1, \ldots, m_n)$ that satisfies the tile size constraint
3	Let S_{TB} (shared memory usage per block) be set as the shared memory usage per tile
4	**for** each $t = (\texttt{blockDim.x}, \texttt{blockDim.y}, \texttt{blockDim.z})$ that satisfies the blockDim constraint
5	Let $R_{TB} = R_T \times \texttt{blockDim.x} \times \texttt{blockDim.y} \times \texttt{blockDim.z}$
6	Let $B = \min(B_W, B_R, B_S)$, where B_W, B_R and B_S are computed in Table 1.
7	Evaluate T_{total} given in (9) for the current tile size m and the current thread organization specified by $\texttt{gridDim} = B \times P$ and $\texttt{blockDim} = t$
8	**if** $T_{total} < T_{best}$ // T_{best} is initialized to ∞
9	$T_{best} = T_{total}$
10	Record m as the best tile size so far (and set $\texttt{gridDim.x} = B \times P$ and $\texttt{blockdDim} = t$)

Fig. 5. An algorithm for automating tile size selection

3.4 Border Tiles

A border tile may execute faster than a full tile. If the i-th inter-tile wavefront that induces T_{t1}^i in (11) contains non-full border tiles, then T_{t1}^i may overapproximate the actual execution time of the wavefront. We can improve this inaccuracy with an estimate of $0.5 \times T_{\mathcal{TILE}}$ as the execution time of a border tile $\mathcal{TILE}$ by assuming that the average size of border tiles is half of a full tile.

4 Model-Driven Tile Size Selection

Given the estimated execution time of T_{total} in (9) for a tiled loop nest L as input, we employ an "educated" search to find automatically and efficiently an optimal tile size $m = (m_1, \ldots, m_n)$ for L and an associated thread organization, determined by $\texttt{gridDim}$ and $\texttt{blockDim}$, used for realizing the optimal tiling. In this paper, a *tile layout* is determined by a tile size and a thread organization.

4.1 The Algorithm

We use two kinds of constraints to prune the search space:

Tile Size Constraint. The tile size, i.e., $L_{\mathcal{TILE}} = m_1 \times \cdots \times m_n$ is bounded from below by a data reuse rate $D = \frac{L_{\mathcal{TILE}}}{N_{mem}}$ (where N_{mem} is introduced in (6)) and from above by the size of shared memory. For DOACROSS loops, large tile sizes lead to higher data reuse rates. Thus, D must be larger than an empirical minimum threshold to ensure better intra-tile data locality.

blockDim Constraint. In NVIDIA GPUs, $\texttt{blockDim.x} \times \texttt{blockDim.y} \times \texttt{blockDim.z}$ represents the number of threads per block. According to [16], the SP performance usually suffers with too many or too few threads. Furthermore, $\texttt{blockDim.x} \times \texttt{blockDim.y} \times \texttt{blockDim.z}$ must be no smaller than the number of iterations contained in the largest intra-tile wavefront to ensure that every thread has some work to do. Thus, some small and large values of $\texttt{blockDim.x} \times \texttt{blockDim.y} \times \texttt{blockDim.z}$ can be ignored.

Our algorithm for automating tile size selection for L is outlined in Figure 5. Recall that as shown in Figure 3, all tiles in a thread block are executed sequentially. Thus, for every type of resource listed in Table 1, the amount consumed by a block is calculated on a per-tile basis. The basic idea is to perform an "educated" search when going through all tile layouts to find the one with the smallest execution time T_{total}. In line 1, the register usage per thread, denoted R_T, is measured independently of tile layouts used. This is because in each case the same code as shown in Figure 3 is compiled for each thread by NVIDIA's nvcc compiler. Finding R_T this ways speeds up the process for calculating R_{TB} in line 5. Similarly, in line 3, S_{TB} does not depend on blockDim. Once R_{TB} and S_{TB} are known, B_W, B_R and B_S are computed in line 6 as per Table 1.

4.2 The Framework

We have implemented our tile size selection technique using a combination of the Clan polyhedral representation extractor, Pluto's polyhedral parallel tiling infrastructure and CLooG code generator, as shown in Figure 6.

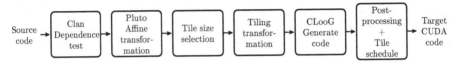

Fig. 6. A model-driven tile size selection framework

Our tile size selection module is invoked in the third step in the sequence.

5   Experiments

We use three representative DOACROSS kernels, 1D (3-point), 2D (5-point) and 3D (7-point) SOR solvers, to demonstrate the accuracy and efficiency of our tile size selection framework on an NVIDIA Tesla C1060 GPU (c.f. Table 1). Four problem sizes are discussed for each kernel, representing 12 different optimization problems for which best tile layouts (tile size/blockDim combinations) are solved.

Accuracy. It is impractical to measure the accuracy of our tile size selection framework for a kernel by comparing the actual execution time of the best tile layout found with the execution times of all tile layouts.

We have decided to evaluate this work empirically as is often done in automated performance tuning. For each of the 12 optimization problems discussed here, we have randomly sampled 1000 different tile layouts. The largest relative error (between the estimated execution time T_{total} and actual execution time) is observed to be within 6.05%. To see this graphically, the relative errors of 100 sampled tile layouts for each optimization problem are plotted in Figures 7 − 9. Let us look at the actual performance gap between the tile layout found by us and the best-performing one in each case. Let us consider a generic optimization problem O. Let T_{total}^m and R_{total}^m be the estimated and actual execution times of any tile layout m for O with the relative error being e_m. In particular, let T_{total}^{opt} and R_{total}^{opt} be the estimated and actual execution times of the the the best tile layout opt predicted for O with the relative error being e_{opt}. The (worst)

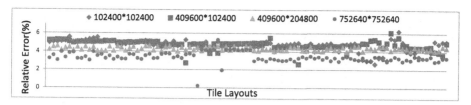

Fig. 7. 1D SOR: relative errors for 100 tile layouts in each of the four problem sizes

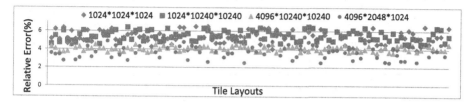

Fig. 8. 2D SOR: relative errors for 100 tile layouts in each of the four problem sizes

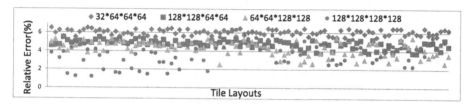

Fig. 9. 3D SOR: relative errors for 100 tile layouts in each of the four problem sizes

performance gap between *opt* and the best-performing one, m, is bounded by $(\frac{1+e_m}{1+e_{opt}} - 1) \times 100\%$, when $R_{opt} - R_m = T^{opt}_{total}/(1 + e_{opt}) - T^m_{total}/(1 + e_m)$, i.e., e_m is the largest, where $R_{opt} > R_m$. Based on our sampled tile layouts, the performance gaps are found to be 5.29%, 0.51%, 2.10% and 4.92% for the four problem sizes of 1D SOR (displayed from left to right in Figure 7), 4.33%, 1.31%, 5.14% and 2.01% for the four problem sizes of 2D SOR (Figure 8), and 0.66%, 2.28%, 6.05% and 1.30% for the four problem sizes of 3D SOR (Figure 9).

Search Time. Our algorithm is efficient in finding the best tile layout for a loop nest (on an Intel Xeon 2.0 GHz CPU). When tiling an n-dimensional loop nest that represents an $(n-1)$-D SOR solver with a tile size $m = (m_1, \ldots, m_n)$, m_1 represents the time dimension and $m_2, \ldots, m_n$ represent the $n-1$ spatial dimensions for the underlying mesh. Due to loop skewing, the worksets of different time slices of a tile are also skewed [10]. Thus, the data reuse rates of a tile for the 1D, 2D and 3D SOR solvers are expressed as a function of m and are bounded from above by m_2, $\frac{m_2 m_3}{m_2 + m_3}$ and $\frac{m_2 m_3 m_4}{m_2 m_3 + m_2 m_4 + m_3 m_4}$, respectively, when $m_1 \to \infty$. For the 1D SOR solver, the data reuse rate induces a tile size constraint: $\frac{L_{TILE}}{N_{mem}} \geq 300$, where the threshold 300 is empirically set (Section 4.1). The search time is 238 secs over a search space of 3×10^6 tile layouts. For the 2D SOR solver, the tile size constraint is $\frac{L_{TILE}}{N_{mem}} \geq 6$. The search time is 369 secs over a search space of 3.2×10^6 tile layouts. For 3D SOR, the data reuse rate imposes $\frac{L_{TILE}}{N_{mem}} \geq 1$. The search time is 503 secs over a search space of 4.7×10^6 tile layouts.

References

1. Baghsorkhi, S.S., Delahaye, M., Patel, S.J., Gropp, W.D., Hwu, W.-m.W.: An adaptive performance modeling tool for GPU architectures. In: PPoPP 2010, pp. 105–114. ACM Press, New York (2010)
2. Baskaran, M.M., Ramanujam, J., Sadayappan, P.: Automatic C-to-CUDA Code Generation for Affine Programs. In: CC 2010, pp. 244–263 (2010)
3. Choi, J.W., Singh, A., Vuduc, R.W.: Model-driven autotuning of sparse matrix-vector multiply on GPUs. In: PPoPP 2010, pp. 115–126 (2010)
4. Cui, H., Wang, L., Xue, J., Feng, X., Yang, Y.: Automatic library generation for blas3 on gpus. In: IPDPS 2011 (2011)
5. Cui, H., Xue, J., Wang, L., Yang, Y., Feng, X., Fan, D.: Extendable pattern-oriented optimization directives. In: CGO 2011, pp. 107–118 (2011)
6. Di, P., Wan, Q., Zhang, X., Wu, H., Xue, J.: Toward harnessing doacross parallelism for multi-gpgpus. In: ICPP 2010 (2010)
7. Fischer, S.: A parallel SSOR preconditioner for lattice QCD. Computer Physics Communications 98(1-2), 20–34 (1996)
8. Hackbusch, W.: Iterative solution of Large Sparse Systems of Equations. Applied Mathematical Sciences. Springer, Heidelberg (1993)
9. Hong, S., Kim, H.: An analytical model for a GPU architecture with memory-level and thread-level parallelism awareness. In: ISCA 2009, p. 152 (June 2009)
10. Huang, Q., Xue, J., Vera, X.: Code tiling for improving the cache performance of PDE solvers. In: ICPP 2003, pp. 615–625 (2003)
11. Lee, S., Min, S.-J., Eigenmann, R.: OpenMP to GPGPU: a compiler framework for automatic translation and optimization. In: PPoPP 2009, pp. 101–110 (2009)
12. Liu, Y., Zhang, E.Z., Shen, X.: A Cross-Input Adaptive Framework for GPU Programs Optimization. In: IPDPS 2009, pp. 16–19 (2009)
13. Quarteroni, A., Valli, A.: Numerical Approximation of Partial Differential Equations. Springer, Heidelberg (1994)
14. Ryoo, S., Rodrigues, C.I., Baghsorkhi, S.S., Stone, S.S., Kirk, D.B., Hwu, W.-m.W.: Optimization principles and application performance evaluation of a multithreaded GPU using CUDA. In: PPoPP 2008, pp. 73–82 (2008)
15. Volkov, V., Demmel, J.W.: Benchmarking GPUs to tune dense linear algebra. In: SC 2008, pp. 1–11 (2008)
16. Wong, H., Papadopoulou, M.M., Sadooghi-Alvandi, M., Moshovos, A.: Demystifying GPU microarchitecture through microbenchmarking. In: ISPASS 2010, pp. 235–246 (2010)
17. Xiao, S., Feng, W.-C.: Inter-block GPU communication via fast barrier synchronization. In: IPDPS 2010, pp. 1–12 (2010)
18. Xue, J.: Loop Tiling for Parallelism. Kluwer Academic Publishers, Dordrecht (2000)
19. Yang, Y., Xiang, P., Kong, J., Zhou, H.: A GPGPU compiler for memory optimization and parallelism management. In: PLDI 2010, p. 86 (May 2010)
20. Yuki, T., Renganarayanan, L., Rajopadhye, S., Anderson, C., Eichenberger, A.E., O'Brien, K.: Automatic creation of tile size selection models. In: CGO 2010, p. 190 (2010)

Iterative Sparse Matrix-Vector Multiplication for Integer Factorization on GPUs

Bertil Schmidt[1,2], Hans Aribowo[1], and Hoang-Vu Dang[1]

[1] School of Computer Engineering, Nanyang Technological University, Singapore
{asbschmidt,hans.aribowo,hvdang}@ntu.edu.sg
[2] Institut für Informatik, Johannes Gutenberg University Mainz, Germany
bertil.schmidt@uni-mainz.de

Abstract. The Block Wiedemann (BW) and the Block Lanczos (BL) algorithms are frequently used to solve sparse linear systems over GF(2). Iterative sparse matrix-vector multiplication is the most time consuming operation of these approaches. The necessity to accelerate this step is motivated by the application of these algorithms to very large matrices used in the linear algebra step of the Number Field Sieve (NFS) for integer factorization. In this paper we derive an efficient CUDA implementation of this operation using a newly designed hybrid sparse matrix format. This leads to speedups between 4 and 8 on a single GPU for a number of tested NFS matrices compared to an optimized multi-core implementation.

Keywords: SpMV, CUDA, Block Wiedemann, RSA, Number Field Sieve, Factorization.

1 Introduction

The Number Field Sieve (NFS) is the current state-of-the-art integer factorization method. It requires the solution of a large sparse linear system over GF(2) (called the linear algebra step). Presently there are two efficient algorithms to solve such a large sparse linear system, namely Block Wiedemann (BW) [8] and Block Lanczos (BL) [15]. Both algorithms have a common time consuming operation: iterative sparse matrix vector multiplication (SpMV).

Recent integer factorization efforts have been using CPU clusters to solve the large sparse linear system [1,13]. The RSA-768 factorization [13], for example, reported a runtime of 3 months for the linear algebra step on a cluster with 48 AMD dual hex-core CPUs. Previous works on parallelizing the linear algebra step focused on using CPU clusters and grids [2,10,11,12]. In this paper, we investigate how a Fermi GPU [17] and the CUDA programming model [16] can be used to accelerate the costly iterative SpMV for matrices derived from NFS.

The memory access pattern in the SpMV operation generally consists of regular access patterns over the matrix and irregular access patterns over the vector. The irregular access pattern over the vector is a challenge that is pronounced more on the GPU than on the CPU, because of the smaller cache and the restrictive memory access pattern requirement to achieve maximum performance.

E. Jeannot, R. Namyst, and J. Roman (Eds.): Euro-Par 2011, LNCS 6853, Part II, pp. 413–424, 2011.

However, a high-end GPU has an order-of-magnitude higher bandwidth than a high-end CPU; e.g. a GeForce GTX 580 has 192.4 GB/s memory bandwidth, while an Intel Core-i7 has a maximum of 25.6 GB/s memory bandwidth.

SpMV on the GPU has been explored previously in several papers [3,6,7,14] for matrices derived from scientific computing applications. However, sparse matrices derived from NFS have generally different properties, i.e. they are larger, have a few dense rows and have many extremely sparse rows. The large size of the matrix causes the BL and BW algorithms to require a large number of SpMV iterations. This means that the time spent for matrix preprocessing and matrix data transfer to the GPU memory are negligible compared to the total runtime. Thus, approaches to the SpMV on GPUs for NFS matrices may be different from previously published GPU SpMV approaches.

This paper is organized as follows. Section 2 describes several published sparse matrix formats and their GPU performance when used with NFS matrices. Section 3 presents our new formats specifically designed for NFS matrices and their CUDA implementation for the Fermi GPU architecture. We compare our result to an Intel Nehalem CPU with the publicly available CADO-NFS [9] software in Section 4. Finally, Section 5 concludes the paper.

2   SpMV on GF(2) for NFS Matrices Using Existing Formats on GPUs

In this section, we review a few relevant previously published sparse matrix formats on GPUs and study their performance when applied to sparse matrices over $GF(2)$ derived from integer factorization with NFS.

We consider a sparse binary matrix A of size $N \times N$ and a dense vector X of size $N \times n$ bit, where n is called the *blocking factor*. Typical blocking factors are of the form of $64 \cdot k$, $k \in \mathbb{N}$. Note that doubling the blocking factor roughly halves the number of SpMV iterations required but doubles the input vector size.

For all $0 \leq i \leq N - 1$ let $c\_index[i]$ be a column index of $A[i]$ which contains the indices of the non-zero entries of row i. Then, the following pseudocode shows a single SpMV iteration of A with input vector X and result vector Y.

```
SPMV( Input: c_index,X; Output:Y )
    for (i=0 ; i<N ; i++)
       Y[i] = 0
       for (j=0; j<c_index[i].size(); j++)
          ind = c_index[i,j]
          Y[i] = Y[i] XOR X[ind]
    end
```

The costly operations in the SpMV pseudocode are the memory accesses for loading $c\_index[i,j]$, $X[ind]$, $Y[i]$ and storing $Y[i]$. To speed up those operations on any architecture a common approach is to design a cache-friendly order of accessing the memory. The order is especially important for the vectors X and Y, since their memory locations might be accessed multiple times.

CUDA implementations generally store both matrix and vectors in high latency global memory. Memory accesses to A and Y can usually be coalesced. Memory accesses to X are random and non-coalesced, but texture memory can be used to take advantage of texture cache. We now briefly review the CUDA implementation of a number of SpMV formats published in previous papers [3,14].

Coordinate list (COO). For each non-zero, both its column and row indices are explicitly stored. The Cusp implementation [4] stores elements in sorted order of row indices ensuring that entries with the same row index are stored contiguously. This format is well suited with respect to storage space for very sparse matrices with many empty rows, since the storage size is strictly proportional to the number of non-zero elements. Implementing SpMV on CUDA with this storage format requires doing atomic updates to the Y vector from parallel threads, which leads to a low performance. The Cusp implementation attempts to solve this problem by using parallel segmented reduction on shared memory within a warp and block before writing to Y. However, because shared memories are only visible to threads within the same block, results from different blocks still need to be combined in the global memory.

Compressed Sparse Row (CSR). Non-zeros are sorted by the row index, and only their column indices are explicitly stored in a column array. Additionally, the vector row_start stores indices of the first non-zero element of each row in the column array. The CSR Cusp implementation assigns one warp to each matrix row. Each thread in a warp computes the result of one non-zero at a time and then the warp moves to the next 32 elements. A parallel reduction operation is performed within a warp to get the final result of the row.

ELLPACK (ELL). Let K be the maximum number of non-zero elements in any row of the matrix. Then, for each row, ELL stores exactly K elements (extra padding is required for rows that contain less than K non-zero elements). As in the CSR format, elements are sorted by row index and only column indices are explicitly stored. In this format, the column array is stored in transposed manner allowing coalesced memory access. The storage size of the ELL format is proportional to $K \times N$. The Cusp ELL implementation assigns one thread per row. Each thread iterates K times accumulating the sum of the respective elements into a register. This format outperforms CSR if the number of non-zeros per row is relatively even. When the number of non-zero elements per row is uneven, overhead from extra padding elements increases the memory usage and decreases performance.

Sliced ELLPACK (SLE). This format partitions the matrix into horizontal slices of S adjacent rows [14]. Each slice is stored in ELLPACK format. The

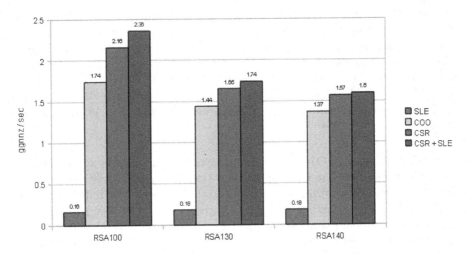

Fig. 1. SpMV performance comparison between sparse matrix formats for various NFS matrices in terms of giga non-zeros per second on a Tesla C2070 GPU with 64 bit blocking factor

Table 1. Properties of utilized NFS matrices resulting from factorizing integers with 100, 130, 140 and 170 digits, respectively

(a) Small Matrices

	RSA-100	RSA-130
Size	$284,836 \times 284,996$	$1,698,881 \times 1,699,041$
Non-zeros	26,274,784	192,416,939
Max row weight	118,252	731,247
Min row weight	11	11
Average row weight	92.24	113.2

(b) Large Matrices

	RSA-140	RSA-170
Size	$3,576,848 \times 3,577,008$	$10,463,019 \times 10,463,197$
Non-zeros	347,915,287	994,785,014
Max row weight	1,327,624	5,582,861
Min row weight	11	3
Average row weight	97.26	95.08

maximum number of non-zeros may be different for each slice. An additional array *slice_start* is used to index the first element in each slice. The matrix rows are usually sorted by the number of non-zeros per row in order to move rows with similar number of non-zeros together. Since there is to our knowledge no open-source SLE CUDA implementation, we have developed our own code. Our SLE CUDA implementation is similar to ELLPACK i.e. 1 thread per row. A requirement is that the height of each slice has to be divisible by the warp size

(32). This format adapts well to many sparse matrix types, and improves the memory usage compared to ELLPACK. However, there is still some overhead due to padding. The Variable-Height SLE [14] format can be used to reduce the overhead further.

We have modified the NVIDIA Cusp library to adapt with GF(2) operation and performed the comparison between the above formats on matrices from RSA-100, RSA-130 and RSA-140 factorization. A summary of these matrices is shown in Table 1. The result of our comparison in Figure 1 shows that CSR outperforms other formats. SLE is slower because of the uneven number of non-zeros per row in the dense part of the matrix. However, SLE performs better than CSR on the sparse part of the matrix. Thus, using CSR for the dense part and SLE for the sparse part improves the performance.

3   New Formats for SpMV on GPUs for NFS Matrices

As a preprocessing step, we reorder the rows of the matrix by their *row weight*, in non-increasing order. The row weight of row j of A is defined as the total number of non-zero elements in row j. We then partition the sorted matrix rows into at most four consecutive parts. Each part uses a different format. The different formats are optimized for the sparseness properties of each partition as shown in Figure 2. For the densest part, we use a dense format. When the matrix gets less dense, we switch to another format which we call Sliced COO. Sliced COO has three variants, small, medium, and large. Our formats are now described in more detail.

3.1   Dense Format

The dense format is used for the dense part of the matrix. This format uses 1 bit per matrix entry. Within a column, 32 matrix entries are stored as a 32 bit integer. Thus, 32 rows are stored as N consecutive integers.

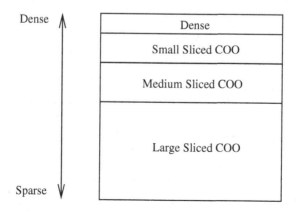

Fig. 2. Partitioning of a row-sorted NFS matrix into four formats

Each CUDA thread works on a column. Each thread fetches one element from the input vector in coalesced fashion. Then, each thread checks the 32 matrix entries one by one. When the matrix entry is a non-zero, the thread performs a XOR operation between the element from the input vector and the partial result for the row. This means each thread only accesses the input vector once to do work on up to 32 non-zeros. The partial result from each thread needs to be stored and combined to get the final result for the 32 rows. These operations are performed in CUDA shared memory.

The 32 threads in a warp share 32 shared memory entries to store the partial results from the 32 rows. Since all threads in a warp execute a common instruction at a time, access to these 32 entries can be made exclusive. The result from each warp in a thread block is combined using reduction on shared memory. The result from each thread block is combined using an atomic XOR operation on global memory.

When the blocking factor is larger than 64, access to the shared memory needs to be reorganized to avoid bank conflicts. Each thread can read/write up to 64 bit data at a time to the shared memory. If a thread is accessing 128 bit data for example, two read/write operations need to be performed. Thus, there will be bank conflicts if we store 128 bit data on contiguous addresses. We can avoid bank conflicts by having the threads in a warp first access consecutive 64 bit elements representing the first halves of the 128 bit elements. Then, the threads again access consecutive 64 bit elements representing the second halves of the 128 bit elements. The same modification can be applied to other formats as well.

The dense format is used for 32 to 64 dense rows of the RSA-170 matrix, which comprises about 15-24% of the total non-zeros in the matrix. This translates to a memory usage of at most 0.36 bytes per non-zero.

3.2   Sliced COO

The Sliced COO format is adapted from the CADO-NFS software for CPUs [9]. The aim is to reduce irregular accesses to the input vector and increase the texture cache hit rate. Sliced COO stores the column index and the row index of each non-zero. A number of consecutive rows form a slice. Non-zeros within a slice are sorted by their column index.

For each non-zero, two bytes are used to store the column index. However, two bytes are not enough for large RSA matrices. Thus, we further divide a slice into groups. Group i contains non-zeros with column index between $i \times 2^{16}$ and $(i + 1) \times 2^{16} - 1$. An additional array stores the starting position of each group in the slice.

One thread block works on a slice, one group at a time. Neighboring threads work on neighboring non-zeros in the group. Each thread works on more than one row. Thus, each thread needs some storage to store the partial result and combine them with the result from the other threads. Since neighboring non-zeros may or may not come from the same row, we cannot share the entries in shared memory among the threads in a warp with exclusive access. Thus, shared memory is either partitioned among threads or shared using atomic XOR operations. Based on the way we allocate the shared memory, we further divide the sliced COO format into three different subformats: small, medium, and large.

Small Sliced (SS) COO. In this subformat, each thread has one exclusive entry in shared memory to store the partial result for each row. The assignment of the shared memory is organized such that each thread in a warp accesses only one bank and there is no bank accessed by more than one thread. Thus, there is no bank conflict. A reduction operation on shared memory is required to combine partial results from each thread.

The maximum number of rows per slice is calculated as *size of shared memory per SM in bits / (number of threads per block * blocking factor)*. In Fermi, the size of shared memory per SM is 48 KB. We use 512 threads per block for 64 bit blocking factor which gives 12 rows, and 256 threads per block for 128 and 256 bit blocking factor, which gives 12 and 6 rows, respectively. Hence, one byte per row index is sufficient for this subformat.

Medium Sliced (MS) COO. In this subformat, each thread in a warp gets an entry in the shared memory to store the partial result for each row. However, this entry is shared with the threads in other warps. Access to the shared memory uses an atomic XOR operation. Each thread in a warp accesses only one bank, avoiding bank conflicts. A reduction operation on shared memory is required to combine the 32 partial results.

The maximum number of rows per slice is calculated as *size of shared memory per SM in bits / (32 * blocking factor)* where 32 is the number of threads in a warp. This translates to 192, 96, and 48 rows per slice for blocking factor of 64, 128, and 256 bit, respectively. Hence, one byte per row index is sufficient for this format.

Large Sliced (LS) COO. In this subformat, the result for each row gets one entry in shared memory, which is shared among all threads in the thread block. Access to shared memory uses an atomic XOR operation. Thus, there will be bank conflicts. However, this drawback can be compensated by a higher texture cache hit rate.

The maximum number of rows per slice is calculated as *size of shared memory per SM in bits / blocking factor*. This translates to 6144, 3072, and 1536 rows per slice for blocking factor of 64, 128, and 256 bit, respectively. We use two bytes for the row index.

3.3   Determining the Cut-Off Point of Each Format

To determine which format to use, we compare the performance of two consecutive formats in terms of giga non-zeros (*gnnz*) per second, starting with the dense format and the SS-COO format. The two formats start from the same row (starting from the first row) and work on the minimum number of rows possible. For the dense format, the minimum number of rows is 32. For the SS-COO format (and its variants), the minimum number of rows is the number of rows in a slice times the number of multiprocessors in the GPU, since one thread block works on one slice and one thread block is assigned to one multiprocessor.

The next comparison depends on the result of the current comparison. If the dense format performs better, we decide to use it for rows 1 to 32, and we continue comparing the dense format and the SS-COO format starting from row 33. However, if the SS-COO format performs better, we compare its performance with the next format, MS-COO, starting from the same row, and so on. The idea is to stop considering the denser format once the sparser format outperforms it. Once we get to the comparison between MS-COO and LS-COO, and LS-COO performs better, we don't need to do any further comparisons. LS-COO should be used for the rest of the matrix.

For Sliced COO format, it needs to be noted that when one slice is assigned to each multiprocessor, the load for one multiprocessor may be much higher than the other multiprocessors. This is because the matrix rows have been reordered by their weight in a non-increasing order, so the first slice contains more non-zero entries than the rest. Thus, we need to further reorder the rows such that each multiprocessor gets the same level of load.

3.4 Dual-GPU Implementation

Two GPUs are connected to PCIe slots and communicate to each other directly using the NVIDIA GPUDirect^{TM} v2.0. To balance the workload between the two GPUs, we partition the matrix rows into two smaller matrices so that each has a similar number of non-zeros. Each smaller matrix is assigned to one GPU. Each GPU computes the multiplication with the complete input vector. The result from each GPU is half of the result vector.

As we need to perform multiple SpMV iterations, we have to combine the result from each GPU before moving on to the next multiplication. Our goal is to reduce the overhead of communication by overlapping computation and communication. We also take advantage of the fact that bi-directional PCIe data transfer has a higher bandwidth than uni-directional. Each half-matrix is divided further into several sub-matrices so that the computation and communication can be interleaved. This is illustrated in Figure 3. Note that the number of rows in each sub-matrix doesn't have to be equal to each other. The dense sub-matrix has fewer rows than the sparse sub-matrices.

There are two events where non-overlapping computation and communication occur. The non-overlapping computation occurs in the multiplication of the first sub-matrix. The non-overlapping communication occurs when the GPU sends the result of the last sub-matrix. To reduce the time spent on non-overlapping computation and communication, we do the multiplication in the order of the sparseness of the sub-matrix, sparsest first. The sparse sub-matrix multiplication is fast to compute because it has few non-zeros, so non-overlapping computation is minimized. The dense sub-matrix takes longer to compute, but has fewer rows than the sparse sub-matrices. Thus, the transfer size for the result is smaller, and non-overlapping communication is minimized.

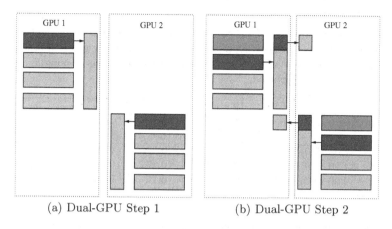

(a) Dual-GPU Step 1   (b) Dual-GPU Step 2

Fig. 3. (a) GPU1 and GPU2 perform the SpMV iteration on the first sub-matrix and store the result in their own device memory. (b) After computing the result of the first sub-matrix, the result is transferred to the other GPU using GPUDirect^{TM} v2.0. At the same time, GPU1 and GPU2 compute the next sub-matrix multiplication. In this step, two operations are executed at the same time. Bi-directional data transfer is utilized on each GPU.

In Figure 3, the boxes on the left and on the right side represent the two GPUs, the horizontal bars at each side represent small sub-matrices of each half-matrix and the vertical bars represent the result vector. The input vector is not shown in the diagram.

The process shown in Figure 3 continues until each GPU finishes every sub-matrix and the last result is transferred to the other GPU. At this point, both GPUs have the same result vector, and the next SpMV iteration can be started.

If GPUDirect^{TM} v2.0 is not available, data transfer between GPUs is performed via the CPU. In this case, we can allocate a CPU vector to receive the result from each GPU before transferring it to the other GPU. The bi-directional data transfer can still be utilized, but the data transfer is less efficient because of the additional communication with the CPU.

4   Results

We have evaluated our implementation on an NVIDIA Tesla C2070 with 6 GB RAM and an NVIDIA GTX580 with 1.5 GB RAM. We have compared the GPU performance with the open-source CADO-NFS [9] program running on Intel Core i7-920 CPU with 12 GB DDR3-1066 memory. The RSA-140 and RSA-170 matrix (see Table 1) are used for performance evaluation. These matrices have been created by CADO-NFS [9] and Msieve [5,18], respectively. The performance is measured in terms of gnnz/s. CADO-NFS contains several CPU optimized SpMV implementations using multi-threading and SSE instructions. In Table 2 and 3, we have included the performance for the basic format (based on CSR)

Table 2. Performance in terms of gnnz per second for a single SpMV iteration with the RSA-140 matrix on one and two GPUs. The speedups compared to the multi-threaded CADO-NFS bucket implementation on an Intel Core i7-920 are given in brackets. The GPU memory required to store the sparse matrix and the corresponding bytes per non-zero (nnz) are also reported.

Blocking factor	GTX580 (speedup)	C2070 (speedup)	2 x C2070 (speedup)	Core i7-920		GPU memory (bytes/nnz)
				basic 8 threads	bucket 4 threads	
64	7.65 (7.0)	5.20 (4.8)	10.38 (9.5)	0.32	1.09	1177 MB (3.55)
128	4.20 (6.6)	2.83 (4.4)	5.64 (8.8)	0.27	0.64	1205 MB (3.63)
256	2.80 (8.0)	1.85 (5.3)	3.60 (10.3)	0.17	0.35	1207 MB (3.64)

Table 3. Performance in terms of gnnz per second for a single SpMV iteration with the RSA-170 matrix on one and two GPUs. The speedups compared to the multi-threaded CADO-NFS bucket implementation on an Intel Core i7-920 are given in brackets. The GPU memory required to store the sparse matrix and the corresponding bytes per non-zero (nnz) are also reported.

Blocking factor	C2070 (speedup)	2 x C2070 (speedup)	Core i7-920		GPU memory (bytes/nnz)
			basic, 8 threads	bucket, 4 threads	
64	4.28 (4.9)	8.38 (9.5)	0.31	0.88	2748 MB (2.90)
128	2.78 (5.6)	5.37 (10.7)	0.26	0.5	2967 MB (3.13)
256	1.88 (7.0)	3.68 (13.6)	0.16	0.27	3000 MB (3.16)

and the CPU-cache optimized bucket format of CADO-NFS with the number of threads that gives the best performance. Speedups of our GPU implementation are given compared to the faster bucket format.

Table 2 shows the result for the RSA-140 matrix. GTX580 achieves the best performance, with speedups between 6.6 to 8.0 over the Core i7-920 depending on the blocking factor used. C2070 achieves speedups between 4.4 to 5.3. Note that C2070 has the ECC (error correcting codes) memory disabled, since ECC reduces the GPU performance. As described in [11], a checkpointing approach could be used to replace the necessity of ECC memory. Comparing to the performance of existing sparse matrix formats with 64 bit blocking factor in Figure 1, our NFS optimized format is 3.25 times faster than the best performing format at a similar memory consumption on the same hardware. The dual-GPU implementation achieves speedups between 1.94 to 1.99 compared to the single-GPU performance.

Table 3 shows the result for the RSA-170 matrix. C2070 achieves similar performance to the RSA-140 matrix. Since storing the RSA-170 matrix requires almost 3 GB, the GTX580 cannot be used in this case due to lack of RAM. The dual-GPU implementation achieves speedups between 1.93 to 1.96 compared to the single-GPU performance.

Table 4 and 5 show that the performance in terms of gnnz per second decreases when the matrix gets sparser. The dense format is not used in RSA-140 matrix,

Table 4. Performance in terms of gnnz per second for each of the three sub-format partitions of the RSA-140 matrix on a C2070. The percentage of non-zeros (nnz) per partition is given in bracket.

Blocking factor	SS-COO	MS-COO	LS-COO
64	13.73 (20%)	8.30 (26%)	3.72 (54%)
128	7.16 (31%)	3.46 (6%)	2.16 (63%)
256	3.95 (29%)	2.25 (7%)	1.47 (64 %)

Table 5. Performance in terms of gnnz per second for each of the four sub-format partitions of the RSA-170 matrix on a C2070. The percentage of non-zeros (nnz) per partition is given in bracket.

Blocking factor	Dense	SS-COO	MS-COO	LS-COO
64	13.66 (24%)	9.66 (11%)	8.13 (13%)	2.77 (52%)
128	9.66 (15%)	7.53 (24%)	3.23 (6%)	1.86 (55%)
256	7.00 (15%)	4.21 (21%)	2.34 (6%)	1.33 (58 %)

since the dense rows are not dense enough for the dense format to outperform the SS-COO format. The MS-COO and LS-COO performance degrade when the blocking factor is increased from 64 to 128 and 256 bit. This is caused by the increased number of bank conflicts and serialization of atomic XOR operations on larger blocking factors. Thus, the SS-COO format gets a higher percentage of non-zeros with 128 and 256 bit blocking factor.

5   Conclusion and Future Work

We have presented our implementation of iterative SpMV for NFS matrices on GPUs with the CUDA programming language. Our single and dual-GPU implementation have been described in detail and both show promising improvements over an optimized CPU implementation. Our GPU implementation takes advantage of the variety of sparseness in NFS matrices to produce suitable formats for different parts.

Our current dual-GPU implementation on two C2070s is able to deal with NFS matrices containing up to 3 billion non-zeros. To deal with even larger matrices such as the RSA-768 [13] matrix ($\approx$ 28 billion non-zeros) and the SNFS-1024 [1] matrix ($\approx$ 10 billion non-zeros), we are currently working on extending our approach to a GPU-cluster.

Acknowledgements. We would like to thank Martin Krone for making the RSA-170 matrix available to us. We are also grateful to Emmanuel Thomé for his advises that enabled us to obtain performance numbers for CADO-NFS.

References

1. Aoki, K., Franke, J., Kleinjung, T., Lenstra, A.K., Osvik, D.A.: A Kilobit Special Number Field Sieve Factorization.. In: ASIACRYPT (2007)
2. Aoki, K., Shimoyama, T., Ueda, H.: Experiments on the linear algebra step in the number field sieve. In: Miyaji, A., Kikuchi, H., Rannenberg, K. (eds.) IWSEC 2007. LNCS, vol. 4752, pp. 58–73. Springer, Heidelberg (2007)
3. Bell, N., Garland, M.: Efficient Sparse Matrix-Vector Multiplication on CUDA. NVIDIA Technical Report NVR-2008-004, NVIDIA Corporation (December 2008)
4. Bell, N., Garland, M.: Cusp: Generic Parallel Algorithms for Sparse Matrix and Graph Computations, version 0.1.0 (2010),
   `http://cusp-library.googlecode.com`
5. Bonenberger, D., Krone, M.: Factorization of rsa-170. Tech. rep., Ostfalia University of Applied Sciences (2010),
   `http://public.rz.fh-wolfenbuettel.de/~kronema/pdf/rsa170.pdf`
6. Boyer, B., Dumas, J.G., Giorgi, P.: Exact Sparse Matrix-Vector Multiplication on GPU's and Multicore Architectures. CoRR abs/1004.3719 (2010)
7. Choi, J.W., Singh, A., Vuduc, R.W.: Model-driven autotuning of sparse matrix-vector multiply on GPUs. SIGPLAN Not. 45, 115–126 (2010)
8. Coppersmith, D.: Solving Homogeneous Linear Equations Over GF(2) via Block Wiedemann Algorithm. Mathematics of Computation 62 (1994)
9. Gaudry, P., et al.: CADO-NFS (2010), `http://cado-nfs.gforge.inria.fr/`
10. Hwang, W., Kim, D.: Load Balanced Block Lanczos Algorithm over GF(2) for Factorization of Large Keys. In: HiPC, pp. 375–386 (2006)
11. Kleinjung, T., Nussbaum, L., Thomé, E.: Using a grid platform for solving large sparse linear systems over GF(2). In: 11th ACM/IEEE International Conference on Grid Computing (Grid 2010), Brussels Belgique (October 2010)
12. Kleinjung, T., et al.: A Heterogeneous Computing Environment to Solve the 768-bit RSA Challenge. Cluster Computing (2010)
13. Kleinjung, T., et al.: Factorization of a 768-Bit RSA Modulus. In: International Crytology Conference, pp. 333–350 (2010)
14. Monakov, A., Lokhmotov, A., Avetisyan, A.: Automatically Tuning Sparse Matrix-Vector Multiplication for GPU Architectures. In: HiPEAC, pp. 111–125 (2010)
15. Montgomery, P.L.: A Block Lanczos Algorithm for Finding Dependencies Over GF(2). In: Theory and Application of Cryptographic Techniques, pp. 106–120 (1995)
16. Nickolls, J., Buck, I., Garland, M., Skadron, K.: Scalable Parallel Programming with CUDA. Queue 6, 40–53 (2008)
17. Nickolls, J., Dally, W.J.: The GPU Computing Era. IEEE Micro. 30, 56–69 (2010)
18. Papadopoulos, J.: Msieve (2010), `http://sourceforge.net/projects/msieve/`

Lessons Learned from Exploring the Backtracking Paradigm on the GPU

John Jenkins[1,2], Isha Arkatkar[1,2], John D. Owens[3],
Alok Choudhary[4], and Nagiza F. Samatova[1,2,*]

[1] North Carolina State University, Raleigh, NC 27695, USA
samatova@csc.ncsu.edu
[2] Oak Ridge National Laboratory, P.O. Box 2008, Oak Ridge, TN 37830, USA
[3] University of California, Davis, Davis, CA 95616, USA
[4] Northwestern University, Evanston, IL 60208, USA

Abstract. We explore the backtracking paradigm with properties seen as sub-optimal for GPU architectures, using as a case study the maximal clique enumeration problem, and find that the presence of these properties limit GPU performance to approximately 1.4–2.25 times a single CPU core. The GPU performance "lessons" we find critical to providing this performance include a *coarse*-and-*fine*-grain parallelization of the search space, a low-overhead *load-balanced* distribution of work, global memory latency hiding through *coalescence, saturation,* and *shared memory utilization,* and the use of GPU *output buffering* as a solution to irregular workloads and a large solution domain. We also find a strong reliance on an efficient global problem structure representation that bounds any efficiencies gained from these lessons, and discuss the meanings of these results to backtracking problems in general.

1 Introduction

The backtracking paradigm, a depth-first search method that finds solutions in a memory efficient manner, is ubiquitous in computing. A few examples include constraint satisfaction in AI [11], frequent itemset mining in data mining [6], maximal clique enumeration in graph mining [16], k-d tree traversal for ray tracing in graphics [9], and logic programming languages such as Prolog. Backtracking typically constructs optimal solutions from candidate solutions, thus forming a search tree that the backtracking traverses. Backtracking is oftentimes at the core of the problems that are combinatorial by nature and, therefore, compute-and-memory-intensive. For many such problems, performing a breadth-first search of the search tree is infeasible due to memory requirements. For instance, frequent itemset mining, as exhibited by the *Apriori* algorithm and its variants [1], becomes infeasible for large input domains.

---

* Corresponding author.

E. Jeannot, R. Namyst, and J. Roman (Eds.): Euro-Par 2011, LNCS 6853, Part II, pp. 425–437, 2011.
© Springer-Verlag Berlin Heidelberg 2011

To reduce backtracking's computational requirements, several strategies have been explored. Pruning the search tree by eliminating non-candidate subtrees, such as in α-β game-tree pruning, avoids unnecessary computation. Likewise, an efficient data model for problem representation (e.g., bitmaps) enables efficient use of intermediate data structures. Finally, parallel implementation of backtracking search on HPC multi-node, multi-core architectures offers scalability for large problem domains (e.g., parallel maximal clique enumeration (MCE) in graphs [16]).

Recent advancements in parallel computing architectures have opened up possibilites for more computationally- and energy-efficient algorithms. In particular, graphics processing units (GPUs) have been maturing not only for graphics applications, but also for general-purpose computations[1] [14]. Some computational motifs perform effectively on a GPU, while the effectiveness of others is still an open issue. For instance, Lee et al. note an average speedup of 2.5× of various algorithms on the GPU vs. optimized Nehalem implementations, and both Lee et al. and Vuduc et al. highlight memory-bound algorithms on the GPU that perform at the same level or worse than the corresponding CPU implementation [12,18].

Despite some of the successes of recent computational dwarfs on GPUs, the mapping of the backtracking paradigm onto the GPU architecture has been recognized as a notoriously difficult problem for a number of reasons. Table 1 names a number of difficulties that a mapping of a backtracking problem to the GPU could encounter, leading to a vastly inefficient use of the GPU memory hierarchy and SIMD-optimized GPU multi-processors.

There have, however, been algorithms successfully mapped onto the GPU, though with major departures from the general case of backtracking. The most visible example is in ray tracing, where k-d tree acceleration structures are used to compute ray intersections by traversing the tree in a depth-first fashion [5,10]. The tree is computed in full before traversals, and the stack-based representation is eliminated by explicitly computing transitions through the tree. However, these properties cannot be assumed in many backtracking problems, much less in the general case.

Our goal, therefore, is to investigate the parallelization of the backtracking paradigm on the GPU. To do this, we analyze the components of *difficult* backtracking problems and propose *tree-level* and *node-level* parallelizations of search space traversal, as well as *buffer-based* output. At best, given the performance of other computational motifs and the nature of the backtracking problem, we cannot expect an order of magnitude increase in performance. Rather, a more realistic performance goal is to perform at one to two times the CPU performance, which opens up the possibility of building future backtracking algorithms on heterogeneous hardware (such as CPU-GPU clusters) and performing workload-based optimizations.

---

[1] All further discussion will be based on Nvidia's CUDA architecture.

2   Motivation

As mentioned, backtracking is a depth-first exploration of a problem space, where states represent partial solutions. At each step, either the partial solution is expanded to another possible solution, or it is determined that a solution cannot possibly lie on this path, and the search *backtracks* to a previous state. Some backtracking problems are harder than others, and it is the characteristics of the harder ones that are of the most interest. Table 1 summarizes these characteristics compared to optimal GPU conditions.

Table 1. Opposing algorithm and hardware characteristics

	Backtracking	GPU optimal
Problem Instance	Irregular access	Regular access with locality
Work Unit	Memory, computation variable	Constant size, perfectly SIMD
Output	Exponential size (if enumerative), hard to estimate	Polynomial size, apriori
Search Space	Tree-based, unbalanced	Fixed, apriori (if applicable)

The problem instance can lend itself to irregular access patterns, making it difficult for GPU algorithms to coalesce memory accesses. One example of this is an adjacency list representation of graphs, where vertices may link to arbitrary other vertices. This problem has been recognized by attempts to perform graph algorithms such as breadth-first search on the GPU [8]. In many cases, graphs are too large to use an adjacency matrix representation.

In many problems, the search node, or work unit, is variable in both memory and computational requirements, making load balancing, enforcing thread convergence, and efficiently utilizing processors and storage mechanisms difficult. One example is an instance of constraint satisfaction, where solutions are subsets of a very large set.

The output size of enumerative problems can be exponential with respect to the problem size. For instance, finding all maximal cliques in a graph has a worst-case exponential output size [13]. This can limit acceleration of a GPU-based method due to overhead in CPU-GPU memory transfers.

Finally, the search tree in many backtracking problems is unpredictable, making it difficult to divide the work evenly. For example, in the context of MCE, current parallel methods rely on communication between compute nodes to load balance and distribute [16], whereas on GPUs thread blocks are optimized to perform independently of each other.

3   Backtracking Case Study: Bron-Kerbosch MCE

3.1   Algorithm Overview

A *clique* of a graph is a subset of the vertex set in which there is an edge connecting each pair of vertices in the set, and a *maximal clique* is a clique that is not contained in any other, larger clique. Maximal clique enumeration (MCE) is ubiquitous in real world problems. Examples of the uses of MCE include identification of common secondary structure elements of proteins [7], detection of protein-protein interaction complexes [19], clustering of similar mass spectrometry spectra [17], and detection of social heirarchy from email communications [15]. Thus, efficient MCE algorithms are of high value.

The MCE algorithm by Bron and Kerbosch (BK) employs a backtracking strategy that embodies the properties in Table 1, constructively building *maximal cliques* of an input graph [3]. Each subtree being traversed has a *compsub* list, or a list representing the current clique, and each search node consists of two data structures, collectively known as a *candidate path*:

1. *candidate*—the vertices connected to all vertices in *compsub*: these may be added to *compsub* to create a new clique; and
2. *not*—the vertices connected to all vertices in *compsub* that would create a redundant clique if added.

Procedure 1: enum(`cp_stack`, `compsub`): traverse subtree(s) in `cp_stack`, using global `compsub`. Both CPU and GPU use multiple stacks and split among compute elements to achieve *coarse-grain* parallelism.

```
1  // process-per-stack on CPU, warp-per-stack on GPU
2  while not empty(cp_stack) do
3      cp ← pop(cp_stack)
4      update compsub
5      if empty(not(cp)) and empty(cand(cp)) then
6          output compsub
7      else
8          spawn(cp_stack, cp)
9  // CPU -- steal work from other stacks
10 // GPU -- assign stack on CPU side to split work with
11 load_balance(cp_stack)
12 if not empty(cp_stack) then
13     goto 2
```

Procedures 1 and 2 show the enumeration routine. The variable `cp_stack` is a stack data structure, pushing and popping *candidate path* structures in depth-first fashion, in lieu of a recursive representation of backtracking. The stack(s) are initially populated with size-one cliques, that is, a vertex and its neighbors, where the *not* and *candidate* lists are determined lexicographically by vertex label. Until the stack is empty, a process gets the current *candidate path*, and either outputs its *compsub* in the case of a maximal clique, or iteratively creates new *candidate paths* by choosing a vertex to expand (which is added to *compsub* when the new *candidate path* is visited) and computing new *candidate* and *not* lists based on adjacency to the selected vertex. Search tree pruning is performed by the addition of `fixv`, reducing by a large degree the number of subtrees leading to redundant cliques.

Procedure 2: spawn(cp_stack, cp): expand candidate path cp onto stack cp_stack, the GPU splits the procedures in lines 2, 9, 10, 14 to achieve *fine-grain* parallelism.

```
 1 // finding fixv is warp-level parallel on GPU
 2 fixv ← minimum disconnected vertex to vertices in cand(cp)
 3 if fixv in not(cp) then
 4     cv ← first vertex in cand(cp) not adjacent to fixv, or nil
 5 else
 6     cv ← fixv
 7 while cv ≠ nil do
 8     // filter(cond_fn, list) is warp-level parallel on GPU
 9     not(newcp) ← filter(adjacent_to_fixv, not(cp))
10     cand(newcp) ← filter(adjacent_to_fixv, cand(cp))
11     push(cp_stack,newcp)
12     move cv to not(cp)
13     // finding next cv is warp-level parallel on GPU
14     cv ← next vertex not adjacent to fixv, or nil
15 // CPU -- service load balance requests from other processes
```

3.2   Algorithm Parallelization

On both the CPU and GPU variants of the algorithm, *coarse-grain* parallelization is achieved by performing Procedure 1 for many stacks, partitioned among processes. For the GPU variant of the algorithm, *fine-grain* parallelization is performed on the warp level by performing lines 2, 9, 10, and 14 of Procedure 2 in parallel. To find the minimum disconnected vertex fixv, each thread in the warp takes vertices in the *not* and *candidate* list in strides, recording the minimum non-connectivity counts, then the warp performs a prefix-sum-like operation to retrieve the global minimum. For instance, for a *not* and *candidate* list of total size n, thread zero computes the local minimum of vertices at offsets 0, 32, etc. thread one computes the local minimum at offsets 1, 33, etc. until all n vertices have been processed. To perform the filter operation, the warp steps through the *not* and *candidate* lists in strides, testing connectivity to fixv, and uses a prefix-sum to compute the correct offsets to output connected vertices to fixv. To determine the next cv, each thread in the warp takes a vertex of the remaining *candidate* in strides, testing connectivity, and performs a prefix-sum-like operation to return the correct vertex. Shared memory is used to store warp-wide variables such as *candidate path* information as well as buffers for performing the prefix-sum operations. All shared memory accesses utilize the broadcasting mechanism, where each thread accesses the same memory bank, and avoid bank conflicts for operations such as the prefix-sum. *candidate paths* are also loaded into shared memory in two ways: partially and in full. The partial load method is used when finding fixv, loading the *candidate* and *not* lists in warp-level chunks and iteratively testing connectivity between those vertices and the thread-local vertex. Loading the *candidate path* in full allows performance of all operations on it in shared memory, at the cost of lower occupancy from increased storage requirements (the *candidate path* is size-bound by the maximum vertex degree).

Unlike CPUs, GPUs do not have the capability of outputting directly to disk, so a more complex method of handling output data must be considered. Furthermore, in enumerative problems such as MCE, it is infeasible to store all output solutions in GPU memory at once, so there must be some intermediate CPU-GPU transfers. Naïvely, each stack's *compsub* could be transferred after each expansion iteration to the CPU, where the valid solutions are extracted and output. However, such a method would suffer from low density of usable output. The more efficient way is to use atomic operations to reserve space from a pre-allocated output buffer and allow blocks to continue expanding states until the buffer is full, decoupling the strict expand-then-output algorithm structure. This allows warps to run more independently of each other, expanding multiple states until a stopping condition is reached. If the output buffer is not large enough, then the method reduces to the first solution, or worse. Also, the need to atomically access and update a single variable across many warps (the buffer "lock") can incur a performance penalty, one that is offset by reducing the amount of data sent to the CPU. For the GPU version of the BK algorithm, the size of the output buffer is the number of concurrent subtrees times a heuristic maximum clique size, determined using vertex degrees.

Load-balancing on the CPU is performed by adding *work-stealing*, requesting work from other processes at the end of Procedure 1 and re-entering the loop if work is recieved, and servicing work requests at the end of Procedure 2 if there is work to give. On the GPU, a very simple method of load balancing is performed. Each warp keeps a count of the number of nodes on its stack, stored contiguously to the output buffer. At the end of each iteration (full buffer), this list is transferred to the CPU with the buffer, sorted, and then pairs of blocks with empty stacks and blocks with large stacks *share* work, moving the bottommost half to the block with a previously empty stack. Since the number of processes is not large (typically in the hundreds), the cost of sorting and transferring the load-balance pairs is very small compared to the algorithm (about a single percent), so benefit gained from performing the sort on the GPU would be minimal. For completeness, we expect to move this routine to the GPU in the near future.

4   Benchmarking

4.1   Input Graphs

To benchmark the parallelized BK algorithm, a few graphs of varying characteristics have been chosen (see Table 2), including a functional gene-gene association network (ava80), climate network with Sea Level Pressure profiles between spatial grid points (slp) over the last 60 years, and a few synthetic

Table 2. Input graphs

Graph	Origin	# Vertices	# Edges	# Maximal Cliques
ava80	Biological	193,568	2,260,872	395,306
slp	Climate	10,512	679,056	365,605
rmat1	Synthetic	8,192	723,849	5,823,741
rmat2	Synthetic	32,768	3,809,695	21,903,896

graphs (`rmat-series`) generated using *GTgraph* [2], under the Recursive Matrix Graph Model (R-MAT) [4], a scale-free random graph generator.

4.2 GPU vs. Multi-core CPU Timing

Two differing GPU implementations are shown in Fig. 1, representing partially and fully loading a node into shared memory, compared to single-core and quad-core CPU implementations. While the GPU methods outperform the single-core CPU method in all cases, relative performance is varied against the non-load-balanced CPU version and worse than the load-balanced method. The GPU method with partial node loading performs between 1.4× and 2.25× the single-core CPU method, but up to 3× worse than the load-balanced quad-core CPU method (the speedup of ava80 is disregarded due to the very short run-time). In terms of distribution of time, the GPU transferral of cliques and load balancing accounted for between one and two percent of enumeration time, except in ava80, which was closer to ten percent. The time taken to transfer the cliques is about one percent of total time, so the buffering methodology is quite efficient compared to the enumeration process.

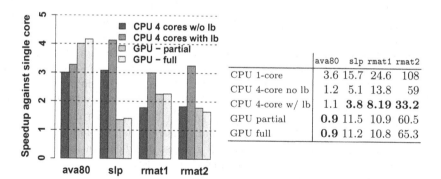

	ava80	slp	rmat1	rmat2
CPU 1-core	3.6	15.7	24.6	108
CPU 4-core no lb	1.2	5.1	13.8	59
CPU 4-core w/ lb	1.1	**3.8**	**8.19**	**33.2**
GPU partial	**0.9**	11.5	10.9	60.5
GPU full	**0.9**	11.2	10.8	65.3

Fig. 1. Comparison of BK algorithm between CPU and GPU. NVIDIA Tesla C2050 for GPU and Intel Xeon X5355 Quad Core 2.66GHz for CPU. Left: speedups relative to single-core performance. Right: actual time (in seconds).

5   Lessons Learned

5.1   Coarse vs. Fine-Grain Parallelization

CPU parallelized backtracking methods utilize *coarse-grain* parallelization, where multiple subtrees are explored in parallel, rather than parallelizing the work-unit itself. CPU threads/processes are *heavy-weight* in comparison to GPU threads, as they fully utilize the CPU when running and have higher context switch overhead. They also have no direct hardware dependency on other threads, as opposed to CUDA's warp-based architecture.

Coarse-grain parallelization of backtracking algorithms on the GPU is, in a naïve sense, simple to port. Call each unit executing a subtree a *process* and partition the global memory of the GPU, one "stack" for each process. To saturate the GPU hardware for a one-thread-per-process representation, a huge number of processes would be needed, leaving little to no space per process. Since the search tree and search nodes of many problems is non-uniform and cross-subtree data is non-contiguous, there would be no coalescing and high divergence, both bottlenecks to GPU performance.

For GPUs, *fine-grain* parallelization of the search nodes, or the performance of lines 2, 9, 10, and 14, is essential. In fact, the *fine-grain* implementation of the BK algorithm performs over $100\times$ faster than the naïve *coarse-grain* method on the GPU, due to the aforementioned divergence rate and lack of coalescing. The *fine-grain* parallelization helps to prevent divergence by computing on similar work-units and enables read/write coalescence on *candidate paths*.

In terms of warp-divergence, the algorithm is reasonably efficient, occuring when *candidate paths* are small and when control code is run (such as thread zero updating a warp-level variable in shared memory). When an adjacency matrix data structure is used, the number of unique code paths for the algorithm is at most three, but for other representations (see Sec. 5.3), the diverging paths can be up to warp size. However, for the hash-table representation, this happens rarely. As a raw percentage of total branches, diverging branches occur 15–20% of the time.

A parallelization is useless if there is a poor work distribution strategy. Figure 2 shows the effect of the load-balancing strategy used in the GPU algorithm on `rmat2`. The last iterations suffer from work of too small granularity to be effectively load-balanced. Also, the `slp` graph fails to be effectively load-balanced, due to much larger cliques with relatively smaller branch factor than the other graphs (that is, much of the tree consists of linear chains), explaining the poor speedup compared to the CPU. Across all graphs, the effects of load-balancing on the GPU are not nearly as beneficial as on the CPU.

5.2   Global Memory Latency Hiding

Global memory latency on the GPU poses a challenge for performing memory-bound algorithms such as BK. Each CPU thread has a relatively large cache space to work with, helping to hide memory latency. GPUs do not have this

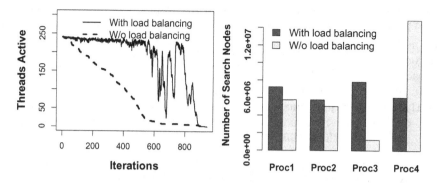

Fig. 2. The effect of load balancing over the algorithm, performed on `rmat2`. Left: GPU load-balancing. Right: CPU load balancing by process. Bars represent number of nodes expanded by process.

luxury, as a large number of lightweight threads leave little thread-level caching capability, so backtracking methods on the GPU have to rely on other strategies.

Latency on GPU memory operations can be hidden through a combination of coalescing, a large number of processes, and effective utilization of shared memory. Having a large number of processes, and thus a high multiprocessor occupancy, allows for some to work on the same multiprocessor while others wait for memory requests, pipelining memory operations. However, fully pipelining requires a very large number of processes, which may not be feasible. Figure 3 shows the effect of adding more concurrent subtrees. While time is decreased, the amount by which it is decreased is sub-linear, due to underutilization of hardware for small numbers of processes, non-even distribution of work, inefficiencies in load-balancing and cache contention; cache misses increased roughly proportional to the number of processes.

The coalescence rate of the algorithm is approximately 20% on average, a very small number compared to optimal. One reason has to do with the global graph data structure, see Sec. 5.3. Another is the nature of the algorithm. The

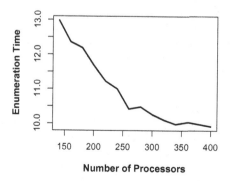

Fig. 3. The effect of adding more processes to enumeration time on the GPU

average size of *candidate path* structures for the graphs tested were no more than six. This is easy to understand, as every *candidate path* representing maximal and near-maximal cliques will be small in size. In other words, opportunities for coalescence are small. This also explains the small change in performance of loading full *candidate paths* into shared memory; only a few of the search nodes actually recieve the benefit. Finally, the low coalescence rate contributes greatly to the poor performance of the algorithm relative to the CPU. Since the GPU relies on coalescing to optimize memory usage, a coalescence rate that low cannot compete with the caching capabilities of the CPU.

5.3   A Reliance on Problem Instance Representation

Backtracking algorithms have control over itermediate structures and how they are used, such as the *candidate path*, but unfortunately, there is little that can be done in a problem-independent manner to optimize global problem instances with irregular access patterns, such as graphs. This reliance on a sub-optimal problem structure is a major impediment to GPU-based algorithms, where the penalty of accessing memory without locality is much higher than on the CPU.

To minimize the penalty for accessing such structures, optimizations such as variable-packing and wide reads (such as 16 byte vs. 4 byte) can help reduce the number of these memory operations by packing data for use in register memory, such as loading multiple vertices from a graph's adjacency list. Also, utilizing the texture and constant memory of the GPU, both of which are cached, can lead to performance improvements, though the amount of each type of memory is limited and thus cannot be used for large problem instances.

For the BK algorithm, three graph representations are used, depending on memory requirements, to minimize the number of memory operations. An adjacency list and adjacency matrix are used for large and small graphs, respectively. For graphs of sizes too large for an adjacency matrix, a hash-table of neighboring vertices is used, using a simple bitwise operation between the vertex label and the table size. With the hash table, often a single index into the hash table is required to determine connectivity of two vertices, being both memory and size efficient and reducing divergence. Vertices with small degree (< 10) use a list rather than a hash table to reduce the memory footprint. To increase the chances of coalescence, parallel connectivity queries by a warp always have one vertex in common, so a similar area of the data is being accessed.

In the BK algorithm, the number of accesses to the graph are directly proportional to the number of memory operations performed on search nodes, so even with a perfect, coalesced, non-diverging algorithm on the search nodes, about 50% of the memory operations are still uncoalesced and can cause divergence, which is a large bottleneck to GPU performance. A small experiment run testing random graph connectivity queries reported a 12% coalescence rate, smaller than that in Section 5.2. It is expected that the memory inefficiency of the graph data structures is a primary cause for poor performance relative to the CPU's higher tolerance of differing access patterns.

5.4   Generality of Backtracking Properties with Respect to GPU-Based Algorithms

Given a backtracking problem, the properties listed and the other lessons learned from the study of the BK algorithm can bring about meaningful insights on the feasibility of parallelizing the problem on the GPU. Having properties such as a problem instance supporting locality of access, a more regular work unit, or a more regular search tree would enable methods that would otherwise be infeasible to perform. Of course, these properties are specific to the algorithm and it is difficult to say whether a particular work unit can be parallelized in a fine grain manner or not, but given a "baseline," these algorithms can be effectively analyzed with respect to their ability to be performed on the GPU.

For example, k-d tree construction is an application in the same class as ours; Zhou et al.'s GPU implementation has important differences from our problem that allows it to compete successfully with state-of-the-art CPU implementations [20]. First, their problem domain has sufficient space to perform the tree construction in breadth-first order, eliminating the need for a stack-based representation and allowing more parallel computations. They also stress the effect of *fine-grain* parallelism on aspects of their algorithm, which we also find important. Finally, their algorithm has more computational requirements that can help hide GPU memory latency with a large enough number of threads, unlike our algorithm which is highly memory-bound. Given these properties, their algorithm competes quite well with CPU-based implementations.

6   Conclusions / Future Work

An attempt at parallelizing the backtracking paradigm, presuming the worst-case attributes against GPU performance, was presented. This problem inspires a number of future directions, despite the inability to provide good performance of MCE against a CPU. Like the k-d tree traversal algorithm on the GPU, parallelizing depth-first algorithms that do not follow the worst-case characteristics highlighted is a promising research question, one that can, under the right representation, hope to compete with or even beat their CPU-based implementations. Furthermore, other computational motifs have yet to be examined for a massively parallel machine such as a GPU. Also, as demand for general-purpose computing support on current and next-generation GPU architectures continues to grow, some of the bottlenecks (such as memory latency) may be sufficiently dealt with, leading to algorithms that could not otherwise be effectively performed on the GPU. Of course, CPU architectures continue to grow to support more throughput and parallelism, while pushing cache sizes. In either case, evaluating new and well-worn computational paradigms on state-of-the-art hardware architectures is a constant need for those who rely on them.

Acknowledgements. We would like to thank Dr. W. Hendrix for useful discussions. Experiments were conducted in part on the ARC cluster support in part by NSF-CRI 0958311 and NVIDIA donations. This work was supported in part by the U.S. Department of Energy, Office of Science (SciDAC SDM Center and SciDAC Institute for Ultrascale Visualization), DOE DE-SC0005340, DOE DE-FG02-08ER25848, DE-FC02-10ER26002/DE-SC0004935, NSF CCF-1029166, CCF-1017399, IIS-0905205, and CCF-0938000. Oak Ridge National Laboratory is managed by UT-Battelle for the LLC U.S. D.O.E. under contract no. DEAC05-00OR22725.

References

1. Agrawal, R., Srikant, R.: Fast algorithms for mining association rules. In: Proc. of the 20th VLDB Conference, pp. 487–499 (1994)
2. Bader, D.A., Madduri, K.: GTgraph: A suite of synthetic random graph generators, https://sdm.lbl.gov/~kamesh/software/GTgraph/
3. Bron, C., Kerbosch, J.: Algorithm 457: Finding all cliques of an undirected graph. Communications of the ACM 16(9), 575–577 (1973)
4. Chakrabarti, D., Zhan, Y., Faloutsos, C.: R-MAT: A recursive model for graph mining. In: SIAM International Conference on Data Mining, pp. 442–446. SIAM, Philadelphia (2004)
5. Foley, T., Sugerman, J.: KD-Tree acceleration structures for a GPU raytracer. In: Graphics Hardware 2005, pp. 15–22 (July 2005)
6. Gouda, K., Zaki, M.J.: Efficiently mining maximal frequent itemsets. In: Proc. of the 2001 IEEE International Conference on Data Mining, pp. 163–170 (2001)
7. Grindley, H.M., Artymiuk, P.J., Rice, D.W., Willett, P.: Identification of tertiary structure resemblance in proteins using a maximal common subgraph isomorphism algorithm. Journal of Molecular Biology 229(3), 707–721 (1993)
8. Harish, P., Narayanan, P.J.: Accelerating large graph algorithms on the GPU using CUDA. In: Aluru, S., Parashar, M., Badrinath, R., Prasanna, V.K. (eds.) HiPC 2007. LNCS, vol. 4873, pp. 197–208. Springer, Heidelberg (2007)
9. Havran, V.: Heuristic Ray Shooting Algorithms. PhD thesis, Czech Technical University in Prague (2001)
10. Horn, D., Sugerman, J., Houston, M., Hanrahan, P.: Interactive k-d tree GPU raytracing. In: Proc. of the 2007 Symposium on Interactive 3D Graphics and Games, pp. 167–174 (2007)
11. Kumar, V.: Algorithms for constraint-satisfaction problems: A survey. AI Magazine 13(1), 32–44 (1992)
12. Lee, V.W., Kim, C., et al.: Debunking the 100X GPU vs. CPU myth: An evaluation of throughput computing on CPU and GPU. In: Int'l Symposium on Computer Architecture, pp. 451–460 (2010)
13. Moon, J., Moser, W.: On cliques in graphs. Israel J. of Math. 3, 23–28 (1965)
14. Owens, J.D., Houston, M., Luebke, D., Green, S., Stone, J.E., Phillips, J.C.: GPU computing. Proceedings of the IEEE 96(5), 879–899 (2008)
15. Rowe, R., Creamer, G., Hershkop, S., Stolfo, S.J.: Automated social hierarchy detection through email network analysis. In: 9th WebKDD and 1st SNA-KDD 2007 Workshop on Web Mining and Social Network Analysis (2007)
16. Schmidt, M.C., Samatova, N.F., Thomas, K., Park, B.-H.: A scalable, parallel algorithm for maximal clique enumeration. JPDC 69(4), 417–428 (2009)

17. Tabb, D.L., Thompson, M.R., Khalsa-Moyers, G., VerBerkmoes, N.C., McDonald, W.H.: Ms2grouper: group assessment and synthetic replacement of duplicate proteomic tandem mass spectra. Journal of the American Society for Mass Spectrometry 16(8), 1250–1261 (2005)
18. Vuduc, R., Chandramowlishwaran, A., Choi, J., Guney, M., Shringarpure, A.: On the limits of GPU acceleration. Hot Topics in Paralellism 35(5) (2010)
19. Zhang, B., Park, B.-H., Karpinets, T., Samatova, N.F.: From pull-down data to protein interaction networks and complexes with biological relevance. Bioinformatics 24(7), 979–986 (2008)
20. Zhou, K., Hou, Q., Wang, R., Guo, B.: Real-time KD-tree construction on graphics hardware. ACM Transactions on Graphics 27(5), 1–126 (2008)

Automatic OpenCL Device Characterization: Guiding Optimized Kernel Design

Peter Thoman, Klaus Kofler, Heiko Studt,
John Thomson, and Thomas Fahringer

University of Innsbruck

Abstract. The OpenCL standard allows targeting a large variety of CPU, GPU and accelerator architectures using a single unified programming interface and language. While the standard guarantees portability of functionality for complying applications and platforms, performance portability on such a diverse set of hardware is limited. Devices may vary significantly in memory architecture as well as type, number and complexity of computational units. To characterize and compare the OpenCL performance of existing and future devices we propose a suite of microbenchmarks, uCLbench.

We present measurements for eight hardware architectures – four GPUs, three CPUs and one accelerator – and illustrate how the results accurately reflect unique characteristics of the respective platform. In addition to measuring quantities traditionally benchmarked on CPUs like arithmetic throughput or the bandwidth and latency of various address spaces, the suite also includes code designed to determine parameters unique to OpenCL like the dynamic branching penalties prevalent on GPUs. We demonstrate how our results can be used to guide algorithm design and optimization for any given platform on an example kernel that represents the key computation of a linear multigrid solver. Guided manual optimization of this kernel results in an average improvement of 61% across the eight platforms tested.

1 Introduction

The search for higher sustained performance and efficiency has, over recent years, led to increasing use of highly parallel architectures. This movement includes GPU computing, accelerator architectures like the Cell Broadband Engine, but also the increased thread- and core-level parallelism in classical CPUs [9]. In order to provide a unified programming environment capable of effectively targeting this variety of devices, the Khronos group proposed the OpenCL standard. It includes a runtime API to facilitate communication with devices and a C99-based language specification for writing device code. Currently, many hardware vendors provide implementations of the standard, including AMD, NVIDIA and IBM.

The platform model for OpenCL comprises a *host* – the main computer – and several *devices* featuring individual *global memory*. Computation is performed

E. Jeannot, R. Namyst, and J. Roman (Eds.): Euro-Par 2011, LNCS 6853, Part II, pp. 438–452, 2011.
© Springer-Verlag Berlin Heidelberg 2011

by invoking data-parallel *kernels* on an N-dimensional grid of *work items*. Each point in the grid is mapped to a *processing element*, and elements are grouped in *compute units* sharing *local memory*. Broad acceptance of the standard leads to the interesting situation where vastly different hardware architectures can be targeted with essentially unchanged code. However, implementations suited well to one platform may – because of seemingly small architectural differences – fail to perform acceptably on other platforms. The large and increasing number of hardware and software targets and complex relationships between code and performance changes make it hard to gain an understanding of how some algorithm will perform across the full range of platforms.

In order to enable automated in-depth characterization and comparison of OpenCL hardware and software platforms we have created a suite of microbenchmarks – uCLbench. It provides programs measuring the following data points:

Arithmetic Throughput. Parallel and sequential throughput for all basic mathematical operations, and many built-in functions defined by the OpenCL standard. When available, native implementations (with reduced accuracy) are also measured.

Memory Subsystem. Host to device, device to device and device to host copying bandwidth. Streaming bandwidth for on-device address spaces. Latency for memory accesses to global, local and constant address spaces. Also determines existence and size of caches.

Branching Penalty. Impact of divergent dynamic branching on device performance, particularly pronounced on GPUs.

Runtime Overheads. Kernel compilation time and queuing delays incurred when invoking kernels of various code volume.

2   Benchmark Design and Methodology

Before examining the individual benchmarks composing the uCLbench suite the basic goals that shaped our design decisions need to be established. The primary purpose of the suite is to characterize and compare the low-level performance of OpenCL devices and implementations. As such, we did not employ device-specific workarounds to ameliorate problems affecting performance on some particular device, since the same behavior would be encountered by actual programs. Another concern is providing implementers with useful information that can support them in achieving good performance over a broad range of devices. Particularly the latency and branching penalty benchmarks are designed with this goal in mind.

There are three main implementation challenges for uCLbench:

1. **Ensure accuracy.** The benchmarks need to actually measure the intended quantity on all devices tested, and it must be possible to verify the computations performed.
2. **Minimize overheads.** Overheads are always a concern in microbenchmarks, but with the variety of devices available to OpenCL they can be

hard to avoid. E.g. a simple loop that is negligible in its performance impact on a general purpose CPU can easily dominate completion time on a GPU.

3. **Prevent compiler optimization.** Since kernel code is compiled at runtime using the compiler provided by the OpenCL implementation, we have no control over the generated code. Thus it is imperative to design the benchmarks in a way that does not allow the compiler to perform unintended optimizations. Such optimizations could result in the removal of operations that should be measured.

There is an obvious area of conflict between these three goals. It is particularly challenging to prevent compiler optimization while not creating significant overheads that could compromise accuracy – even more so when the same code base is used on greatly differing hardware and compiled by different closed-source optimizing compilers.

2.1   Arithmetic Throughput

As a central part of the suite, this benchmark measures the arithmetic capabilities of a device. It includes primitive operations as well as many of the complex functions defined in the OpenCL standard. Two distinct quantities are determined: the device-wide throughput that can be achieved by independent parallel execution as well as the performance achieved for sequentially dependent code. All measurements are taken for scalar and vector types, and, if available, both native (less accurate) and default versions of complex functions are considered.

To enable result checking and prevent compiler optimization, input and output are performed by means of global memory pointers, and the result of each operation is used in subsequent ones. The loop is manually unrolled to minimize loop overheads on all devices. Automatic unrolling can not be relied upon to achieve repeatable results for all platforms and data/operation types.

The kernel is invoked with a local and global range of one work item to determine the sequential time required for completion of the operation, and with a local range of $loc = $ CL_DEVICE_MAX_WORK_GROUP_SIZE and a global range of CL_DEVICE_MAX_COMPUTE_UNITS*loc items to calculate device-wide throughput.

2.2   Memory Subsystem

Current GPUs and accelerator devices have a memory design that differs from the deep cache hierarchies common in CPUs.

Bandwidth. While global GPU memory bandwidth per-chip is high, due to the degree of hardware parallelism the memory bandwidth available per processing element can be insufficient [13]. Another bottleneck for current GPUs is the PCIe slot intermediating host memory and device.

Many GPUs and accelerators attempt to ameliorate these issues by providing a manually controlled, multi-layered memory subsystem. In OpenCL, this concept is represented by separate address spaces: private, local, constant, global and host memory.

For this reason, the benchmark is divided in two major parts: one for on-device memory layers and one for memory traffic between host memory and device. To test bandwidth for on-device memory the benchmark invokes kernels streaming data from one layer back into the same layer. We also discern differences between scalar and various vectorized types, as the latter might be optimized.

Host $\leftrightarrow$ device bandwidth measurement does not require any kernel, instead it uses the runtime API for copying data from/to the device's global memory or inside device global memory. For device/host communication, two options are considered: the first generates a buffer and commands the OpenCL runtime to transfer it (`clEnqueueWriteBuffer`), the second *maps* a device buffer into the host memory and works directly on the returned pointer.

For the streaming kernel, overheads were a major concern. This was addressed by using fast add operations to forestall optimization, and by maximizing the ratio of read/write memory accesses.

Latency. In addition to bandwidth, knowledge about access latency is essential to effectively utilize the available OpenCL memory spaces. Depending on the device used, only some or none of the accesses may be cached, and latency can vary by two orders of magnitude, from a few cycles up to several hundreds.

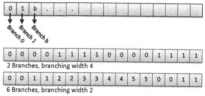

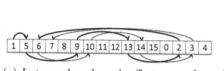

(a) Latency benchmark offset array for a cache line size of 4 elements

(b) Branch penalty measurement layout options

Fig. 1. Patterns used for latency and branch penalty benchmark, respectively

The latency benchmark uses a specifically designed index array to perform a large number of indirect memory accesses. The index array contains address offsets chosen to cause jumps larger than cache line size, and end on a zero entry after traversing the entire array, as illustrated in Fig. 1(a).

Some input-dependent computation and output has to be performed to prevent optimization, which is achieved by accumulating offsets. Manual loop unrolling is used to minimize overheads. When measuring local memory latency a large number of repeated traversals is required.

2.3 Branching Penalty

On some OpenCL devices divergent dynamic branching on work items leads to some or all work being serialized. The impact can differ with the amount and topological layout of diverging branches on the work range. Since the effect on algorithm performance of this penalty can be severe [6] we designed a microbenchmark to determine how devices react to various branch counts and layouts.

The benchmark kernel is provided with an array of floating point numbers equal in length to the amount of work items. Each item then takes a branch depending on the number stored in its assigned location. Fig. 1(b) illustrates how brancharray configurations can be used to test a varying number of branches and different branch layouts.

2.4   Runtime Overheads

Compared to traditional program execution, the OpenCL model introduces two potential sources of overhead. Firstly, it is possible to compile kernels at runtime, and secondly there is an amount of time spent between queuing a kernel invocation and the start of computation. These overheads are measured in uCLBench using the OpenCL profiling event mechanism – we define the invocation overhead as the elapsed time between the CL_PROFILING_COMMAND_QUEUED and CL_PROFILING_COMMAND_START events, and the compilation time as the time spent in the clBuildProgram call. The actual kernel execution time is disregarded for this benchmark, and the accuracy of the profiling events is implementation defined (see Table 1).

3   Device Characterization – Results

To represent the broad spectrum of OpenCL-capable hardware we selected eight devices, comprising four GPUs, three CPUs and one accelerator. Their device characteristics as reported by OpenCL are summarized in Table 1.

NVIDIA TESLA 2050. The GF100 Fermi chip in this GPGPU device contains 14 compute units with a load/store unit, a cluster of four special function units as well as two clusters of 16 scalar processors each. The scalar processor

Table 1. OpenCL devices benchmarked

Device	Tesla2050	Radeon5870	GTX460	GTX275	2x X5570	2x Opt.2435	2xCellPPE	2xCellSPE
Implementation	NVIDIA	AMD	NVIDIA	NVIDIA	AMD	AMD	IBM	IBM
Operating System	CentOS5.3	CentOS5.4	CentOS5.4	Win 7	CentOS5.4	CentOS5.4	YDL 6.2	YDL 6.2
Host Connection	PCIe 2.0	PCIe 2.0	PCIe 2.0	PCIe 2.0	-	-	-	On-chip
Type	GPU	GPU	GPU	GPU	CPU	CPU	CPU	ACCEL
Compute Units	14	20	7	30	16	12	4	16
Max Workgroup	1024	256	1024	512	1024	1024	256	256
Vect.Width Float	1	1	4	1	4	4	4	4
Clock (MHz)	1147	1400	850	1404	2933	2600	3200	3200
Max.Alloc. (MB)	671	256	512	220	1024	1024	757	757
Images	Yes	Yes	Yes	Yes	No	No	No	No
Kernel Args	4352	4352	1024	4352	4096	4096	256	256
Alignment	64	128	128	16	128	128	1	1
Cache	R/W	None	R/W	None	R/W	R/W	R/W	None
Cache Line	128	-	128	-	64	64	128	-
Cache Size (KB)	224	-	112	-	64	64	32	-
Global Mem (MB)	3072	1024	2048	877	3072	3072	3072	3072
Constant (KB)	64	64	64	64	64	64	64	64
Local Type	Scratch	Scratch	Scratch	Scratch	Global	Global	Global	Scratch
Local (KB)	48	32	48	16	32	32	512	243
Timer Res. (ns)	1000	1000	1	1000	1	1	37	37

clusters can work on different data using different threads that issue the same instruction, a method referred to as Single Instruction Multiple Thread (SIMT).

AMD Radeon HD5870. The Cypress GPU on this card has 20 compute units containing 16 Very Long Instruction Word (VLIW) [5] processors with an instruction word length of five. To benefit from the VLIW architecture in OpenCL the programmer should use a vector data type such as `float4`.

NVIDIA GeForce GTX460. The GTX460 contains a GF110 Fermi GPU which comprises 7 compute units. These compute units are similar to the ones on the TESLA 2050, with one important difference. Each compute unit consists of 3 SIMT clusters fed by 2 superscalar scheduling units.

NVIDIA GeForce GTX275. This graphics card is based on the GT200 GPU which has 30 compute units containing 8 scalar processors which work in SIMT manner.

Intel Xeon X5570. The Intel Xeon X5570 features 4 physical CPU cores with simultaneous multithreading (SMT) leading to a total of 8 logical cores. The Xeons used in our benchmarks are mounted on an IBM HS22 Blade featuring two CPUs with shared main memory, resulting in a single OpenCL device with a 16 compute units.

AMD Opteron 2435. The Opteron 2435 CPUs used in this paper are mounted on a two-socket IBM LS22 Blade. Each of them contains 6 cores leading to a total of 12 compute units.

IBM PowerXCell 8i. The accelerator device in our benchmarks consists of two PowerXCell 8i processors mounted on an IBM QS22 Blade. In OpenCL a Cell processor comprises two devices: A CPU (the PPE of the Cell) and an accelerator (all SPEs of the Cell). The two Cell PPEs, each featuring SMT, contain four compute units, the eight SPE cores of the two Cell chips add up to 16 compute units.

3.1   Arithmetic Throughput

We have gathered well over 3000 throughput measurements using the uCLBench arithmetic benchmark. A small subset that provides an overview of the devices and contains some of the more significant and interesting results will be presented in this section.

Fig. 2(a) shows the number of floating point multiplications per second measured on each device and the theoretical maximum calculated from the hardware specifications. The first thing to note is the large advantage of GPUs in this metric, which necessitates the use of separate scales to portrait all devices meaningfully.

Looking at the effective utilization of hardware capabilities, the GPUs also do well. The Fermi cards reach over 99% utilization. The other GPUs still go

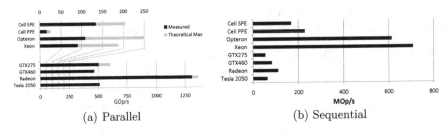

Fig. 2. Floating point multiplication throughput

over 80% while the two x86 CPUs fail to reach the 50% mark. IBM's OpenCL performs a bit better, achieving slightly over 65% of the theoretical maximum throughput on both PPEs and SPEs.

While throughput of vectorized independent instructions is important for scientific computing and many multimedia workloads, some problems are hard to parallelize. The performance in such cases depends on the speed at which sequentially dependent calculations can be performed, which is summarized in Fig. 2(b). The CPUs clearly outperform GPUs and accelerators here, providing a solid argument for the use of heterogeneous systems.

Vectorization. Figures 3(a) and 3(b) show the relative performance impact of manual vectorization using the `floatN` OpenCL datatypes. With a single work item all devices benefit from vectorization to some extent. Since all three CPUs deliver the same relative performance, they are consolidated.

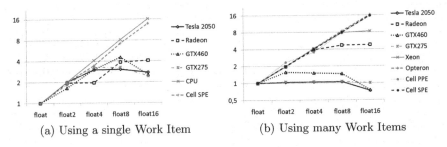

Fig. 3. Vectorization Impact

When the full amount of work items is used there are two clearly visible categories. The NVIDIA GPUs effectively gather individual work items into SIMD groups and thus show no additional benefit from manual vectorization, vectors with 16 elements even slow down execution. The GTX460 result is counterintuitive, but can be explained by scheduling constraints introduced by the superscalar architecture.

3.2   Memory Subsystem

The memory subsystems of the benchmarked OpenCL devices diverge in two areas – availability of dedicated global device memory and structure of the on-chip memory. The GPU devices feature dedicated global memory while for all other devices the global device address space resides in host memory. Furthermore, the local memory on GPUs and Cell SPUs is a manually managed scratchpad while on CPUs it is placed inside the cache hierarchy.

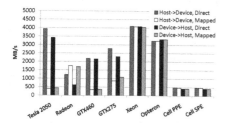
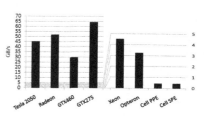

(a) Bandwidth transferring data from/to the device

(b) Global device memory bandwidth

Fig. 4. Bandwidth measurements

Bandwidth. The bandwidth measured between host and devices is shown in Fig. 4(a). For CPUs data is simply copied within the main memory, while for GPUs it has to be transferred over the PCIe bus. Therefore the bandwidth measured for CPUs is higher in this benchmark,and the results of the two CPUs using the AMD implementation correspond to their main memory bandwidth. All NVIDIA GPUs perform similarly, whereas the Radeon is far behind them using direct memory while it is faster when using mapped memory. The Cell processor achieves very low bandwidth although it is equipped with fast memory, a result that we attribute to an immature implementation of the IBM OpenCL runtime.

A second property we measured is the bandwidth of the devices' global memory. As shown in Fig. 4(b) the GPUs lead the benchmark due to their wide memory interface. The GTX275 outperforms the Radeon as well as the newer NVIDIA GPUs although the theoretical memory bandwidth of the latter ones is slightly higher. All CPUs achieve the same bandwidth as in the host ↔ device benchmark since host and device memory are physically identical.

Looking further into the memory hierarchy we measure the bandwidth of a single compute unit to its local memory. Since all compute units on a device can access their local memory concurrently, the numbers provided need to be multiplied by the compute unit count to calculate the local memory bandwidth of the whole device. We measured the bandwidth in four ways: in the first case only one work item accesses the memory, in the second the maximum launchable number is used. These two variants were used on local memory that has been statically declared inside a kernel function as well as to local memory passed as

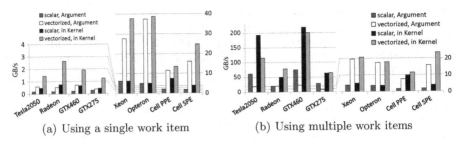

Fig. 5. Bandwidth of one compute unit to its local memory

an argument to the kernel function. Furthermore, all benchmarks were performed using scalars and vector data types. Fig. 5(a) shows the result of the benchmarks using only one work item while Fig. 5(b) displays the values for the full amount. GPU scratchpad memories are clearly designed to be accessed by multiple work items, and with parallel access performance increases by up to two orders of magnitude.

In contrast to the GPUs the Cell SPE scratchpad memory can be used efficiently in the sequential benchmark, parallelizing the access has only a minor impact on the speed. On the CPU side, all systems exhibit unexpected slowdowns with multiple work items. We believe that this is caused by superfluous cache coherency operations due to false sharing [11]. All CPUs benefit from vector types, and all devices can utilize higher bandwidth to the local memory when it is statically declared inside the kernel function.

Latency. One purpose of the multiple address spaces in OpenCL is allowing access to lower latency memory pools. This is particularly important on GPUs and accelerators, where global memory is often uncached. As shown in Fig. 6(b) absolute access latency to global memory is almost an order of magnitude larger on GPUs and accelerators than on CPUs. Additionally CPUs can rely on their highly sophisticated cache hierarchies to reduce the access times even further. The impact of caching is shown in Fig. 6(a) which shows the relative time to access a data item of a certain size in comparison to the previously measured latency to the global memory. This depiction clearly identifies the number of caches featured by a device, as well as their usable size in OpenCL. Non-Fermi GPUs as well as the Cell SPE do not feature any automated caching of data in global memory resulting in equal access time for all tested sizes.

Local and Constant memory latency is significantly smaller on all devices. On the CPUs it corresponds to L1 cache latency as expected. All four GPUs show very similar results to access the local memory, while the Fermi based chips outperform the Radeon and GTX275 in accessing the constant memory by approximately six and three times, respectively. The accelerator's behavior more closely resembles a CPU than a GPU regarding local latency, resulting in the largest difference between global and local timings. The SPEs are the only device to achieve significantly lower latency for `const` than `local` accesses.

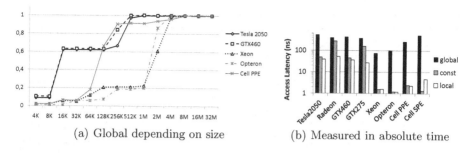

(a) Global depending on size   (b) Measured in absolute time

Fig. 6. Memory access latency

3.3   Branching Penalty

We measured the time taken to process the branch penalty testing kernel with one to 128 branches relative to the time required to complete a single branch. All CPUs remain at the same performance level regardless of the number of divergent branches. This is expected, as CPUs do not feature the SIMT execution model that results in a branching penalty. The Cell SPE accelerator also does not exhibit any penalty. The situation is more interesting for the GPUs, which show a linear increase in runtime with the number of branches until a cutoff point. In case of all NVIDIA GPUs, this point is reached at 32 divergent branches, and it takes 64 branches on the Radeon. This measurement coincides perfectly with the *warp size* reported for each GPU, which is the number of SIMT threads that are grouped for SIMD execution.

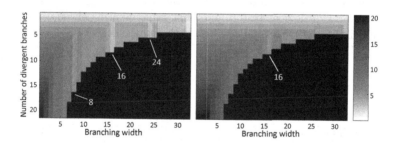

Fig. 7. Branching penalty with varying branch width. NVIDIA GPUs / Radeon

Fig. 7 summarizes the results obtained varying both branch count and topo-logical layout of branches in the local range. A darker color indicates longer kernel runtime, and the lower right part is black since it contains infeasible com-binations of branching width and branch count. Generally, grouping branches to-gether improves performance. In fact, the hardware behaves in a very predictable way: if the condition $branchingWidth * branchCount \geq warpSize$ is fulfilled, further increases in the branch count will not cause performance degradation. On NVIDIA GPUs, multiples of 8 for the branch width are particularly advan-tageous, and the same is true for multiples of 16 on the Radeon. For GTX275 and Radeon this value is equal to the reported SIMD width of the architecture. This is not the case for the Fermi-based NVIDIA GPUs, where a SIMD width of 16 is generally assumed, yet their behavior remains unchanged.

3.4 Runtime Overheads

Invocation overheads remain below 10 microseconds on the tested x86 CPUs as well as the Fermi GPUs. The two IBM Parts and the GTX275 take around 30 and 50 microseconds, respectively. The Radeon HD5870 requires approximately 450 microseconds from enqueueing to kernel startup.

We measured compilation times below 1 second for all mature platforms, scaling linearly with code size. The IBM platform has larger compilation times, particularly for the SPEs, reaching 30 seconds and more for kernels beyond 200 lines of code.

4   Guiding Kernel Design

In this section we evaluate the usefulness and accuracy of the device characterization provided by the uCLbench suite on a real-world kernel. Performance portability is a main concern with OpenCL kernels, with different devices reacting very differently to optimization attempts. While some of these optimizations – such as local work group dimensions – can be auto-tuned relatively easily others require significant manual implementation effort. These latter optimizations are the focus of this chapter. We will demonstrate that the automatic characterization provided by uCLbench reliably identifies promising optimizations for each of the diverse set of devices tested, and that implementing these optimizations consistently improves performance beyond the capabilities of a state-of-the-art optimizing GPGPU compiler.

4.1 The Model Problem

As our test case we selected a simple elliptic partial differential equation, the discrete two-dimensional Poisson equation with Dirichlet boundary conditions in the unit square $\Omega = (0,1)^2$. This problem is given by

$$-\Delta_h u_h(x,y) = f_h^{\Omega}(x,y)$$
$$u_h(x,y) = f_h^{\Gamma}(x,y) \quad \text{for} \quad ((x,y) \in \Gamma_h = \partial\Omega_h)$$

with boundary conditions $f_h^{\Gamma}(x,y)$ and discretization width $h = 1/n$, $n \in \mathbb{N}$ being the number of grid points in each direction.

The central component in a multigrid algorithm for this problem is the relaxation step, an iterative solver for the discretized version of the given equation. Due to its good smoothing effects and parallelization properties (see [12]), we chose an ω-Jacobi kernel.

4.2   Optimizations

To evaluate the predictive power of our benchmark suite, we implemented six kinds of manual optimizations belonging to one of three categories:

Vectorization. The kernel was vectorized for vector widths of 4, 8 and 16. (designated vec4, vec8, vec16)

Branching Elimination. A straightforward implementation of the boundary conditions $f_h^\Gamma(x, y)$ introduces dynamic branching. This optimization eliminates the branching by using oversized buffers and performing a separate step to fix the boundary after each iterative step. (designated BE)

Manual Caching. These kernels manually load their working set into local memory. There are two slightly different versions, *dynamic* and *static*. The former passes the local memory buffer to the kernel as a parameter while the latter statically allocates it inside the kernel. (designated mcDyn and mcStat)

Combinations of these optimizations result in a total of 16 implementations, plus one baseline version with no manual optimization. This does not include the variation introduced by easily tunable parameters like local work group dimensions – for these, we selected the optimal values for each device by exhaustively iterating over all possibilities. Clearly, manually implementing 16 or more versions of each kernel is not generally viable. The following results demonstrate that automatic characterization can be used to assess the impact of an optimization on a given device before implementing it, thus guiding the implementation effort.

4.3   Results

Table 2 lists, for each device, the best performing version of the ω-Jacobi kernel and how much that version improves upon the baseline (Improvement = $(T_{baseline}/T_{best} - 1) * 100$). If the best version combines more than one optimization, the *Primary* column contains the single optimization that has the largest impact on the result, and the *Contribution* column shows the impact of that option on its own. Finally, we present the speedup achieved by a state of the art GPGPU optimizing compiler targeted primarily at the GTX275 architecture [15]. Even when a speedup is achieved, the automatically optimized kernel still does not reach the performance of the version arrived at by guided manual optimization.

The results correspond to several device characteristics identified by our benchmarks. All the devices that do not show any caching behavior in the memory latency tests – Radeon, GTX275 and Cell SPE – benefit from manual caching, and as determined by the local memory bandwidth results static allocation is generally as fast or faster than the alternative (Section 3.2). Branch elimination is most beneficial on those devices that show a high variance in the branch penalty benchmark, while not having any impact on CPUs with their flat branching profile (Section 3.3). In this regard, the minor Cell PPE speedup due to BE seems

Table 2. Most Effective Optimizations per Device and their Speedup

Device	Best Version	Improvement	Primary	Contrib.	GPGPU Compiler
Tesla 2050	vec4	14%	-	-	slowdown
Radeon	vec8_BE_mcStat	63%	vec8	29%	crashed
GTX460	vec4_BE	10%	vec4	9%	slowdown
GTX275	BE_mcStat	31%	mcStat	19%	22%
Xeon	vec8	42%	-	-	slowdown
Opteron	vec16	79%	-	-	slowdown
Cell PPE	vec16_BE	39%	vec16	25%	slowdown
Cell SPE	vec16_BE_mcStat	192%	mcStat	78%	90%

counter-intuitive, but its arithmetic throughput results indicate that it suffers penalties in branching code due to its long pipeline and in-order execution.

The impact of vectorization is well predicted by the characterization results (Section 3.1). While the NVIDIA compiler and devices do a good job at automatically vectorizing scalar code, on all other platforms the impact of manual vectorization is large. Our vectorization benchmark results correctly indicate the most effective vector length for each device.

5   Related Work

Microbenchmarks have a long history in the characterization of parallel architectures. The Intel MPI Benchmarks (IMB) [7] are often used to determine the performance of basic MPI operations on clusters. For OpenMP, the EPCC suite [2] measures the overheads incurred by synchronization, loop scheduling and array operations. Bandwidth is widely measured using STREAM [8], and our memory bandwidth benchmark implementation is based on its principles.

A major benefit of using OpenCL is the ability to target GPU devices. Historically these were mostly used for graphics rendering, and benchmarked accordingly, particularly for use in games. A popular tool for this purpose is 3DMark [10]. When GPU computing first became widespread Stanford University's GPUbench suite [1] provided valuable low-level information. However, it predates the introduction of specific GPU computing languages and platforms, and therefore only measures performance using the restrictive graphics programming interface. In depth performance analysis of one particular GPU architecture has been performed by Wong et al. [14].

Recently the SHOC suite of benchmarks for OpenCL was introduced [4]. While it contains some microbenchmarks, it is primarily targeted at measuring mid- to high-level performance. It does not try to identify the individual characteristics of mathematical operations or measure the latency of access to OpenCL address spaces. Conversely, our suite is aimed at determining useful low-level characteristics of devices and includes exhaustive latency and arithmetic performance measurements as well as a benchmark investigating dynamic branching penalties. We also present results for a broader range of hardware, including an accelerator device.

The Rodinia Heterogeneous Benchmark Suite [3] predates wide availability of OpenCL, therefore separately covering CUDA, OpenMP and other languages with distinct benchmark codes. Also, unlike uCLbench, Rodinia focuses on determining the performance of high-level patterns of parallelism.

6   Conclusion

The uCLbench suite provides tools to accurately measure important low-level device properties including: arithmetic throughput for parallel and sequential code, memory bandwidth and latency to several OpenCL address spaces, compilation time, kernel invocation overheads and divergent dynamic branching penalties. We obtained results on eight compute devices which reflect important hardware characteristics of the platforms and, in some cases, show potential for improvement in the OpenCL implementations.

The automatic device characterization provided by uCLbench is useful in quickly gaining an in-depth understanding of new hardware and software OpenCL platforms, exposing undisclosed microarchitectural details such as dynamic branching penalties. We have shown that the measured characterisics can be used to guide manual optimization by identifying the most promising optimizations for a given device. Applying these transformations to an ω-Jacobi multigrid relaxation kernel results in an average improvement of 61% across the devices tested.

Acknowledgments. This research is partially funded by the Austrian Research Promotion Agency (FFG), project number: 824925. We thank Dr.techn. Peter Zinterhof from the University of Salzburg for giving us access to a Tesla cluster.

References

[1] Buck, I., Fatahalian, K., Hanrahan, P.: GPUBench (2004)

[2] Bull, J.M.: Measuring synchronisation and scheduling overheads in openmp. In: Proc. of 1st Europ. Workshop on OpenMP, pp. 99–105 (1999)

[3] Che, S., Boyer, M., Meng, J., Tarjan, D., Sheaffer, J.W., Lee, S., Skadron, K.: Rodinia: A benchmark suite for heterogeneous computing. In: IEEE Workload Characterization Symposium, pp. 44–54 (2009)

[4] Danalis, A., Marin, G., McCurdy, C., Meredith, J.S., Roth, P.C., Spafford, K., Tipparaju, V., Vetter, J.S.: The scalable heterogeneous computing (shoc) benchmark suite. In: GPGPU 2010: Proc., pp. 63–74. ACM, New York (2010)

[5] Fisher, J.A.: Very long instruction word architectures and the ELI-512. In: Proceedings of the 10th Annual International Symposium on Computer Architecture, pp. 140–150. ACM, New York (1983)

[6] Fung, W.L., Sham, I., Yuan, G., Aamodt, T.M.: Dynamic warp formation and scheduling for efficient gpu control flow. In: MICRO 40, pp. 407–420. IEEE Computer Society, Washington, DC, USA (2007)

[7] MPI Intel. Benchmarks: Users Guide and Methodology Description. Intel GmbH, Germany (2004)

[8] McCalpin, J.D.: Memory bandwidth and machine balance in current high performance computers. IEEE Comp. Soc. Tech. Comm. on Computer Architecture (TCCA) Newsletter, pp. 19–25 (December 1995)

[9] Olukotun, K., Hammond, L.: The future of microprocessors. Queue 3(7), 26–29 (2005)

[10] Sibai, F.N.: Performance analysis and workload characterization of the 3dmark05 benchmark on modern parallel computer platforms. ACM SIGARCH Computer Architecture News 35(3), 44–52 (2007)

[11] Torrellas, J., Lam, M.S., Hennessy, J.L.: False sharing and spatial locality in multiprocessor caches. IEEE Transactions on Computers 43(6), 651–663 (1994)

[12] Trottenberg, U., Oosterlee, C.W., Schueller, A.: Multigrid. Academic Press, London (2001)

[13] Volkov, V., Demmel, J.W.: Benchmarking gpus to tune dense linear algebra. In: SC 2008, pp. 1–11. IEEE Press, Piscataway (2008)

[14] Wong, H., Papadopoulou, M., Sadooghi-Alvandi, M., Moshovos, A.: Demystifying gpu microarchitecture through microbenchmarking. In: ISPASS, pp. 235–246 (2010)

[15] Yang, Y., Xiang, P., Kong, J., Zhou, H.: A gpgpu compiler for memory optimization and parallelism management. In: Proceedings of the 2010 ACM SIGPLAN conference on Programming language design and implementation, PLDI 2010, pp. 86–97. ACM, New York (2010)

Author Index